Criminal and Environmental Soil Forensics

Karl Ritz • Lorna Dawson • David Miller
Editors

Criminal and Environmental Soil Forensics

Editors
Karl Ritz
National Soil Resources Institute
Natural Resources Department
School of Applied Sciences
Cranfield University
Cranfield
Bedfordshire MK43 0AL
UK

Lorna Dawson
The Macaulay Institute
Craigiebuckler
Aberdeen AB 15 8QH
UK

David Miller
The Macaulay Institute
Craigiebuckler
Aberdeen AB15 8QH
UK

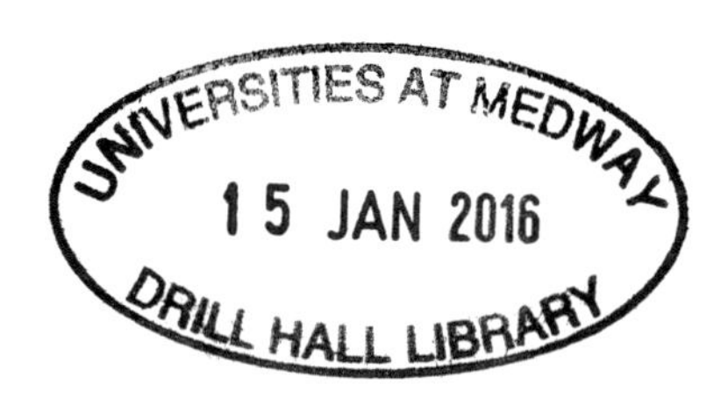

ISBN 978-1-4020-9203-9 e-ISBN 978-1-4020-9204-6

Library of Congress Control Number: 2008937475

Cover images (top left to bottom right): Lidar image of a search area; sampling soil from shoe impression at crime scene; optical laser scan of footprint impression in sand; polished section of concrete; trace evidence from urban soil including glass fibres embedded in aggregates; trace evidence from soil including pollen grain. Images all derived from material in this volume.

Printed on acid-free paper

springer.com

Foreword

Having crossed Hadrian's Wall at Carlisle for the short journey from England into Scotland needed to present as guest speaker before the 2007 Soil Forensics Conference in Edinburgh, so capably organised by the Macaulay Institute, it was kind of Professor Karl Ritz from Cranfield University to invite me to follow up observations made there with a foreword to the resulting book; this fabulous compendium of ground-breaking, international science emerging from your important, biennial conference.

As the articles following show, it is a fast-moving science but one taking many different forms and routes, in many different places around the world; places where climate, topography, ecosystems and even cultures vary hugely. However, amongst all that diversity of physical context or intellectual effort, the influence of the soil is the one common scientific denominator, whilst its application to criminal cases of homicide and unlawful killing, a recurrent reminder of man's inhumanity to man, is naturally pre-eminent.

The first, encouraging counterpoint to those grim atrocities against which forensic science is so often deployed is found in magnificent human ingenuity, as it builds remorselessly upon one scientific progression or invention after another to construct reliable mosaic pictures of the past out of whatever fragmentary remains are left us in the present, fit to persuade a court. The second is that universal thirst for justice which drives everyone engaged in the investigative and court process (police officers, forensic scientists, lawyers and ordinary citizens alike) to excavate answers out of those who would bury wickedness.

What my own offering to the conference was about ('March of the Gladiators – Scientists entering the Arena of Lawyers') were gentle words of caution from a former criminal litigator, linked to the dangers from false positives. Those fields of forensics applied in police work are of course far wider than just soil-based (my Police Authority providing a typical example, as recent signatory to a novel 14-force/authority consortium across England & Wales, buying-in a whole range of forensic science 'packages' to support criminal investigation). There are valuable cautionary tales to be had here from the unfortunate experiences of others, presenting forensic evidence before the stern scrutiny of courts, whether from the fallibility of fingerprints or what dangers lie in DNA.

The bottom line was simple – do not expect the courts to receive your findings with uncritical gratitude or the mild controversy of academia. (After all, the life, liberty and reputations of real people hang on your words). But, if your scientific and personal integrity take utter objectivity and self-critical rigour as their companion guardian angels, then you may yet be able to withstand the formalised assault which surely will – and perfectly properly – be launched in the courtroom against your findings and opinions. (Don't forget, a challenge potentially to be maintained – on and off – for several years thereafter).

Your vital role as experts for the administration of justice is invaluable, much appreciated and growing, but do not expect the honour of this arduous responsibility to be afforded you more easily than any other participant, or indeed (citing the tragic, historical example of Sir Bernard Spilsbury) to be provided you as convenient vehicle for winning personal and professional glory.

Neither are our scientific colleagues the sole audience for the deep science found within this book. If the Oxford dictionary defines '*symbiotic*' as the adjective applying to "*a close association of two interdependent animal or plant species, persons or groups*", then those distinguished crime writers also present at the conference were evidence of just such an arrangement; keen as they obviously are to keep up with every last development and weave it into their latest plot no sooner than scientists have announced it.

Not only was it great pleasure to share a table at Conference with such charming and learned scientific folk as Professor Karl Ritz, Professor David Miller or Dr Lorna Dawson, it was also a privilege to be able to debate there in person with celebrated writers like Ian Rankin, whose '*Inspector Rebus*' books are so firmly Scottish but win an international following. Why, having spent most of the conference apologising widely for being a lawyer (one who has – in his time and as an advocate – commissioned, championed, challenged, tested or actually found the forensic evidence relied upon in criminal cases), it gives genuine relief now to claim my attendance there was really made as some sort of writer instead.

Happily, since the publication of *Mosaic (The Pavement that Walked) – the novel by Clive Ashman*[1] – I can. This fictionalised reconstruction of the biggest unsolved crime-cum-art-theft in British archaeology (the overnight theft from post-war Yorkshire, in 1948, of a Roman mosaic floor first found in 1941 then left carefully wrapped in alluvial soil 'till archaeologists could return and prepare it for lifting seven years later) makes it no longer impertinent to claim membership of that happy band. Its plot also offers me points in common to debate with all those field archaeologists which so many forensic soil scientists turn out to be when you scratch them, all those plant biologists who had confined themselves to reconstructing the landscape of ancient Britain from a drilled-out earth core, until that fateful day when they were called out by the police to a scene of crime for the very first time.

To every reader, however you got here and whatever your motivation for taking up and opening "Criminal & Environmental Soil Forensics", let me commend to

[1] Voreda Books, 2008. ISBN 9780955639807. voredabooks@hotmail.co.uk

you a book that captures the very latest research from all over the world. A learning to be applied diligently and built upon carefully, to bring to justice the most intolerable forms of human wrong-doing imaginable. To be applied in the honest belief that – wherever in the world they are – the greater the deterrence that exists for the potential wrong-doer; the greater the grinding certainty for those that really do go on to offend of facing successful detection, prosecution, conviction and detention; then the greater the likelihood for the rest of us of delivering a time of reckoning which can respect the deceased, salve the bereaved and reassure all our communities of their future safety.

Clive Alcock LLB Solicitor
Chief Executive, Cumbria Police Authority
June 2008
clive.alcock@cumbria.police.uk

Preface

Soils are present on the outermost layer of the Earth's terrestrial landmass and as such cover a large (but declining) proportion of the planetary surface, playing a pivotal role in the functioning of the contemporary Earth system. Human civilisations are irrevocably bound to them, as they serve as a platform for habitation, and are literally fundamental to food, fibre and fuel production, provide a source of raw materials and act as an archaeological repository. Soils also provide a wider range of ecosystem goods and services, supporting all terrestrial habitats, cycling carbon and nutrient elements, storing and purifying water, acting as a biodiversity reservoir, and regulating atmospheric gases.

Soils are amongst the most complex of known systems. Surface soils comprise a diverse mixture of inorganic and organic materials which are physically structured in a heterogeneous but characteristic manner across some twelve orders of magnitude, from micrometres to megametres. The biomass that they support belowground, which is predominantly microbial, significantly exceeds that aboveground. Subsoils, and the interface with the bedrock (the regolith), are less complex but also have characteristic properties and geographic distribution, as does the fundamental geology.

Soil science has advanced a great deal in the past two decades, and we know increasingly more about the distribution and properties of soils, how they function, and the significance of their fundamental importance. Ironically, the increasing urbanisation of current civilisation, and reduced connections with farming and food production, is resulting in a progressive decline in the appreciation of the importance of soil by the majority of the populace. Yet, humans interact with soils wittingly for sound reasons, and sometimes unwittingly when operating nefariously.

The variety in the constitution, distribution and function of soils provides an intriguing basis, and great potential, for research and application in a forensic context. Their analysis and interpretation can provide intelligence, insight and evidence in the forensic arena at a wide range of scales. This volume, based upon contributions to the Second International Conference on Environmental and Criminal Soil Forensics, held in Edinburgh in 2007, explores the conceptual and practical interplay of soils across scientific disciplines, and investigative and legal spectra. The 32 chapters that follow show that the increasing convergence of a wide range of knowledge and application is leading to a thriving collaboration across disciplines of criminal and environmental soil forensics, with common perspectives but complementary approaches. The chapters have been grouped broadly into five themes: concepts, evidence, geoforensics, taphonomy and technology. However, the interdisciplinary nature of much of the material means that such apparently discrete structuring should only be used as a guideline. This challenge when aiming to organise the material in a simple manner implies to us as editors that soil forensics is indeed a discipline that is starting to mature.

July 2008 Karl Ritz, Lorna Dawson, and David Miller

All material in the chapters is the responsibility of the respective authors, and any views expressed therein do not necessarily represent those of the editors, their organisations or the publisher.

Acknowledgements

A volume that contains such diverse material as this one requires input, effort and support from many people over and above the 97 authors. We sincerely thank the cadre of independent experts who graciously gave their time and provided insightful and thoroughly professional reviews of the chapters. We gratefully acknowledge the superb professional support from Maryse Walsh and Melanie van Overbeek at Springer. We particularly thank Jane Morrice for her excellent assistance in editing and proof-reading the book contents.

Thanks are also given to the Macaulay Institute and Cranfield University for their institutional support in the running of the Second International Conference on Environmental and Criminal Soil Forensics in 2007, and in the preparation of this book.

Contents

Contributors

Morgan Adams
Centre for Research in Energy and Environment, The Robert Gordon University, Schoolhill, Aberdeen AB10 1FR, UK

Colin Aitken
School of Mathematics, The University of Edinburgh, James Clerk Maxwell Building, The King's Buildings, Mayfield Road, Edinburgh EH9 3JZ, UK

Derek Auchie
Law Department, The Robert Gordon University, Garthdee Road, Aberdeen AB10 7QE, UK

Natasha Banning
The Centre for Land Rehabilitation, University of Western Australia, Perth, Australia

David Barclay
Caorainn, Laide IV22 2NP, UK

Lorraine Barry
School of Geography, Archaeology and Palaeoecology, Queen's University, Belfast, Belfast BT7 1NN, UK

Matthew Bennett
School of Conservation Sciences, Bournemouth University, Talbot Campus, Fern Barrow, Poole BH12 5BB, UK

Laura Benninger
Faculty of Science, University of Ontario Institute of Technology, 2000 Simcoe St N, Oshawa ON, L1H 7K4, Canada

Jochen Berger
Institute for Soil Science and Land Evaluation, University of Hohenheim, Emil-Wolff-Strasse 27, 70599 Stuttgart, Germany

Mark Brewer
Biomathematics and Statistics Scotland, The Macaulay Institute,
Craigiebuckler, Aberdeen AB15 8QH, UK

Gemma Broadbridge
Centre for Forensic Sciences, Centre for Archaeology,
Anthropology and Heritage, School of Conservation Sciences, Bournemouth
University, Bournemouth BH12 5BB, UK

Maarten Broekmans
Geological Survey of Norway, Department of Industrial Minerals and Ores,
N-7491 Trondheim, Norway

Peter Bull
Oxford University Centre for the Environment, University of Oxford,
South Parks Road, Oxford OX1 3QY, UK

Holly Campbell
Helford Geoscience LLP, Menallack Farm, Treverva, Penryn,
Cornwall TR10 9BP, UK

Gerardo Carpio - Diaz
Centro Mallqui, Ilo, Peru

David Carter
Department of Entomology, College of Agricultural Sciences and Natural
Resoruces, University of Nebraska – Lincoln,
202 Plant Industry Building, Lincoln, Nebraska, USA

John Cassella
Department of Forensic Science, Faculty of Sciences, Staffordshire University,
Stoke-on-Trent, Staffordshire ST4 2DE, UK

Konstantinos Christidis
Centre for Research in Energy and Environment, The Robert Gordon University,
Schoolhill, Aberdeen AB10 1FR, UK

Georgina Cox
Centre for Forensic Sciences, Centre for Archaeology, Anthropology
and Heritage, School of Conservation Sciences, Bournemouth University,
Bournemouth BH12 5BB, UK

Lorna Dawson
The Macaulay Institute, Craigiebuckler, Aberdeen AB15 8QH, UK

Peter De Forest
John Jay College of Criminal Justice/CUNY, 445 West 59th Street, New York,
USA

Kerith-Rae Dias
The Centre for Forensic Science, University of Western Australia, Perth, Australia

Jessica Djohari
Centre for Forensic Sciences, Centre for Archaeology, Anthropology and Heritage, School of Conservation Sciences, Bournemouth University, Bournemouth BH12 5BB, UK

Laurance Donnelly
Halcrow Group Ltd., Deanway Technology Centre, Wilmslow Road, Handforth, Cheshire SK9 3FB, UK

Sarah Dunkerley
Oxford University Centre for the Environment, University of Oxford, South Parks Road, Oxford, OX1 3QY, UK

Isabel Fernandes
Department and Centre of Geology, Faculty of Science, University of Porto, Rua do Campo Alegre 687, 4169-007 PORTO, Portugal

Sabine Fiedler
Institute for Soil Science and Land Evaluation, University of Hohenheim, Emil-Wolff-Strasse 27, 70599 Stuttgart, Germany

Rob Fitzpatrick
Centre for Australian Forensic Soil Science, CSIRO Land and Water, Private Bag No 2, Glen Osmond, South Australia

Shari Forbes
Faculty of Science, University of Ontario Institute of Technology, 2000 Simcoe St N, Oshawa ON, L1H 7K4, Canada

Sean Forrester
Centre for Australian Forensic Soil Science, CSIRO Land and Water, Private Bag No 2, Glen Osmond, South Australia

Jeanne Freudiger-Bonzon
Faculty of Geosciences and the Environment, University of Lausanne, Switzerland

Patricia Furphy
Centre for Forensic Sciences, Centre for Archaeology, Anthropology and Heritage, School of Conservation Sciences, Bournemouth University, Bournemouth BH12 5BB, UK

Silvia Gonzalez
School of Biological and Earth Sciences, Liverpool John Moores University, Byrom Street, Liverpool, L3 3AF, UK

Kenneth Gow
Centre for Research in Energy and Environment, The Robert Gordon University, Schoolhill, Aberdeen AB10 1FR, UK

Olga Gradusova
Russian Federal Centre of Forensic Research, Building 2, Khokhlovsky Lane 13, 109028 Moscow, Russian Federation

Conor Graham
School of Geography, Archaeology and Palaeoecology, Queen's University Belfast, Belfast BT7 1NN, UK

Matthias Graw
Institute of Legal Medicine, University of Munich, Nußbaumstrasse 26, 80336 Munich, Germany

Paul Greenwood
The Centre for Land Rehabilitation, University of Western Australia, Perth, Australia

Sonia Guillen
Centro Mallqui, Ilo, Peru

Ian Hanson
Centre for Forensic Sciences, Centre for Archaeology, Anthropology and Heritage, School of Conservation Sciences, Bournemouth University, Bournemouth BH12 5BB, UK

Mark Harrison
National Policing Improvement Agency, Wyboston Lakes, Great North Road Wyboston, Bedfordshire MK44 3AL, UK

Claire Hodgson
Centre for Forensic Sciences, Centre for Archaeology, Anthropology and Heritage, School of Conservation Sciences, Bournemouth University, Bournemouth BH12 5BB, UK

Jacqui Horswell
The Institute of Environmental Science and Research Limited (ESR), Kenepuru Science Centre, Porirua, New Zealand

David Huddart
School of Biological and Earth Sciences, Liverpool John Moores University, Byrom Street, Liverpool, L3 3AF, UK

Robert Janaway
Archaeological Sciences, School of Life Sciences, University of Bradford, Bradford, West Yorkshire BD7 1DP, UK

Thomas Jellis
Oxford University Centre for the Environment, University of Oxford, South Parks Road, Oxford OX1 3QY, UK

John Jervis
Applied and Environmental Geophysics Group, School of Physical Sciences and Geography, Keele University, Staffordshire ST5 5BG, UK

Robert Jones
National Soil Resources Institute, Natural Resources Department, School of Applied Sciences, Cranfield University, Cranfield, Bedfordshire MK43 0AL, UK

Antoinette Keaney
School of Geography, Archaeology and Palaeoecology, Queen's University Belfast, Belfast BT7 1NN, UK

Helen Kemp
Stable Isotope Laboratory, Scottish Crop Research Institute, Invergowrie, Dundee DD2 5DA, UK

Lynne Macdonald
The Macaulay Institute, Craigiebuckler, Aberdeen AB15 8QH, UK

Kristine Martens
Laboratory of Applied Geology and Hydrogeology, Gent University, Krijgslaan 281 – S8, B-9000 Gent, Belgium

Robert Mayes
The Macaulay Institute, Craigiebuckler, Aberdeen AB15 8QH, UK

Suzzanne McColl
School of Biomolecular Sciences, Liverpool John Moores University, Byrom Street, Liverpool L3 3AF, UK

Jennifer McKinley
School of Geography, Archaeology and Palaeoecology, Queen's University Belfast, Belfast BT7 1NN, UK

Wolfram Meier-Augenstein
Stable Isotope Laboratory, Scottish Crop Research Institute, Invergowrie, Dundee DD2 5DA, UK

David Miller
The Macaulay Institute, Craigiebuckler, Aberdeen AB15 8QH, UK

Ruth Morgan
UCL Jill Dando Institute of Crime Science, 2nd Floor Brook House, Torrington Place, London WC1E 7HN, UK

Andrew Morrisson
School of Life Sciences, The Robert Gordon University, St Andrews Street, Aberdeen AB25 1HG, UK

Stephen Mudge
School of Ocean Sciences, Bangor University, Menai Bridge LL59 5AB, UK

Ekaterina Nesterina
Russian Federal Centre of Forensic Research, Building 2, Khokhlovsky Lane 13, 109028 Moscow, Russian Federation

Julia Newberry
Department of Natural and Social Sciences, University of Gloucestershire, Swindon Road, Cheltenham GL50 4AZ, UK

Katharine Nichols
Oxford University Centre for the Environment, University of Oxford, South Parks Road, Oxford OX1 3QY, UK

Fernando Noronha
Department and Centre of Geology, Faculty of Science, University of Porto, Rua do Campo Alegre 687, 4169-007 PORTO, Portugal

Simon Officer
Centre for Research in Energy and Environment, The Robert Gordon University, Schoolhill, Aberdeen AB10 1FR, UK

Jennifer Orr
Centre for Forensic Sciences, Centre for Archaeology, Anthropology and Heritage, School of Conservation Sciences, Bournemouth University, Bournemouth BH12 5BB, UK

Rachel Parkinson
The Institute of Environmental Science and Research Limited (ESR), Kenepuru Science Centre, Porirua, New Zealand

Duncan Pirrie
Helford Geoscience LLP, Menallack Farm, Treverva, Penryn, Cornwall TR10 9BP, UK

Pat Pollard
Centre for Research in Energy and Environment, The Robert Gordon University, Schoolhill, Aberdeen AB10 1FR, UK

Simon Pollard
Sustainable Systems Department, School of Applied Sciences, Cranfield University, Bedfordshire MK43 0AL, UK

Matthew Power
Intellection UK Ltd, North Wales Business Park, Abergele LL22 8LJ, UK

Gopala Prabhu
Centre for Research in Energy and Environment, The Robert Gordon University, Schoolhill, Aberdeen AB10 1FR, UK

Jamie Pringle
Applied and Environmental Geophysics Group, School of Physical Sciences and Geography, Keele University, Staffordshire ST5 5BG, UK

Mark Raven
Centre for Australian Forensic Soil Science, CSIRO Land and Water
Private Bag No 2, Glen Osmond, South Australia

John Reffner
John Jay College of Criminal Justice/CUNY, 445 West 59th Street, New York, USA

Karl Ritz
National Soil Resources Institute, Natural Resources Department,
School of Applied Sciences, Cranfield University, Cranfield,
Bedfordshire MK43 0AL, UK

Peter Robertson
Centre for Research in Energy and Environment, The Robert Gordon University, Schoolhill, Aberdeen AB10 1FR, UK

James Robertson
Australian Federal Police, GPO Box 401, ACT 2601, Australia

Gavyn Rollinson
Camborne School of Mines, University of Exeter, Cornwall Campus, Penryn, Cornwall TR10 9EZ, UK

Jasmine Ross
The Macaulay Institute, Craigiebuckler, Aberdeen AB15 8QH, UK

Alastair Ruffell
School of Geography, Archaeology and Palaeoecology, Queen's University Belfast, Belfast BT7 1NN, UK

George Sensabaugh
School of Public Health, 50 University Hall, MC# 7360, University of California, Berkeley, California, USA

Jay Siegel
Forensic and Investigative Sciences, Indiana and Purdue Universities, 402 N Blackford Street, LD 326, Indianapolis IN, USA

Alvin Smucker
Soil Biophysics, Department of Crop and Soil Sciences,
Michigan State University, East Lansing, Michigan MI, USA

Karl Stahr
Institute for Soil Science and Land Evaluation, University of Hohenheim,
Emil-Wolff-Strasse 27, 70599 Stuttgart, Germany

Kathryn Stokes
Centre for Forensic Science, University of Western Australia,
35 Stirling Highway, Crawley 6009, Australia

Tracey Temple
Department of Applied Science, Security and Resilience,
Defence Academy of the UK, Cranfield University, Shrivenham SN6 8LA, UK

Mark Tibbett
Centre for Land Rehabilitation, School of Earth and Geographical Sciences,
The University of Western Australia,
35 Stirling Highway, Crawley WA 6009, Australia

George Tuckwell
Stats Limited, Porterswood House, St Albans, Hertfordshire AL3 6PO, UK

Arpad Vass
Oak Ridge National Laboratory, PO Box 2008, Oak Ridge, TN 37831, USA

Stewart Walker
Forensic and Analytical Chemistry, School of Chemistry, Physics and Earth Sciences, Flinders University, GPO Box 2100, Adelaide, South Australia 5001, Australia

Kristine Walraevens
Laboratory of Applied Geology and Hydrogeology, Gent University,
Krijgslaan 281 – S8, B-9000 Gent, Belgium

Brooke Weinger
John Jay College of Criminal Justice/CUNY, 445 West 59th Street, New York, USA

Anna Williams
Department of Applied Science, Security and Resilience,
Defence Academy of the UK, Cranfield University, Shrivenham SN6 8LA, UK

Andrew Wilson
Archaeological Sciences, School of Life Sciences, University of Bradford,
Bradford, West Yorkshire BD7 1DP, UK

Patricia Wiltshire
Department of Geography and Environment, University of Aberdeen,
Elphinstone Road, Aberdeen AB24 3UF, UK

Przemyslaw Zelazowski
Oxford University Centre for the Environment, University of Oxford,
South Parks Road, Oxford OX1 3QY, UK

Colour Plates

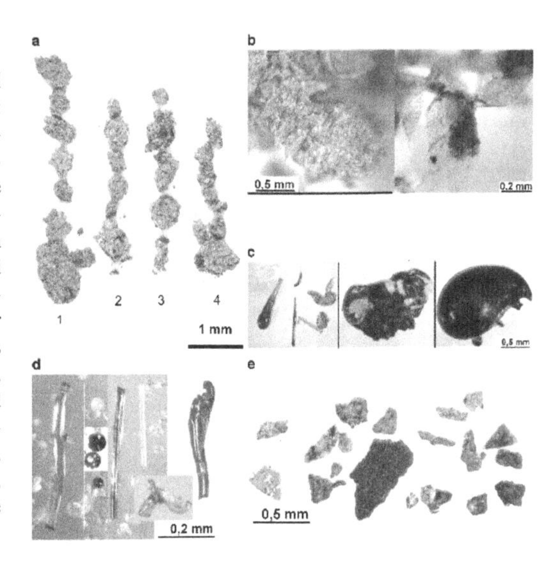

Chap. 5 Fig. 1 Case study I. Soil micro-particles procured from site: (a) Overview of material: 1 – comparative sample; 2 – window sill; 3 – sweater of the victim (owner of the flat), 4 – suspect's boots. (b) Fragment of cormophyte moss inside a soil micro-particle (*left*) and extracted from it (*right*). (c) Glass fibre fragments extracted from comparator samples. (d) Glass fibre fragments extracted from washed soil traces, taken from the sill, the sweater and the boots respectively. (e) Micro-particles similar to paint fragments, taken from the comparative samples, the sill, the suspect's sweater and the suspect's boots

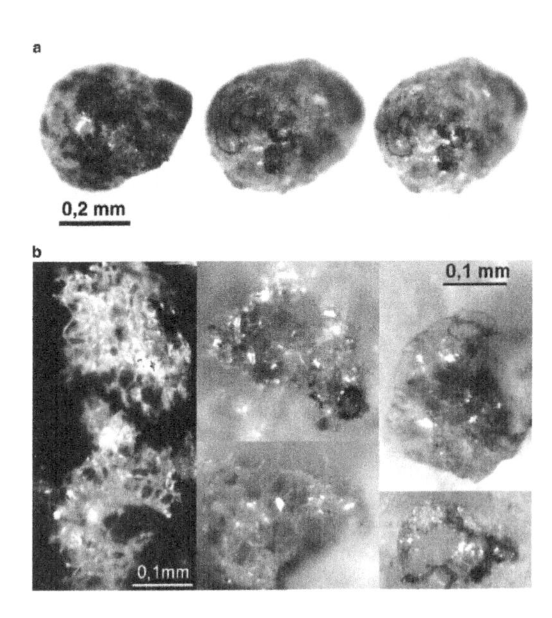

Chap. 5 Fig. 2 Case study II. (a) Trace evidence comprising globules of bituminous perlite after several stages of washing in chloroform. (b) Inner structure of one such perlite particle

Chap. 5 Fig. 3 Case study III. (a) Soil micro-aggregates with embedded glass fibres. (b) Glass fibres with Fe (III) hydroxide. (c) Calx, the source of iron in vivianite formation. (d) Fragments of bones, the source of phosphorus in vivianite formation. (e) Vivianite globule in a soil aggregate (*left*) and extracted from it (*right*)

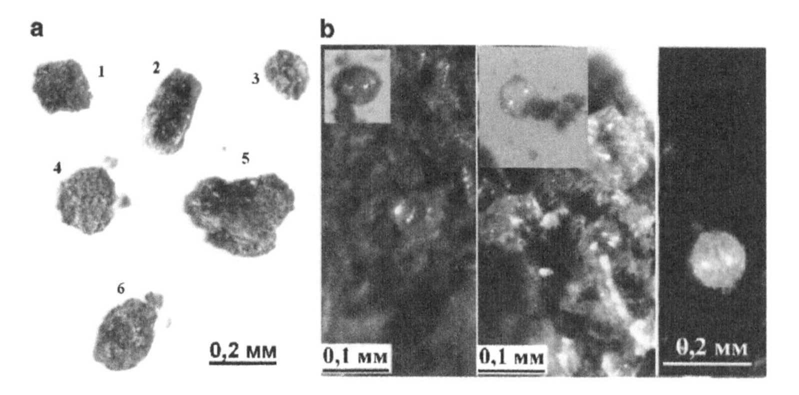

Chap. 5 Fig. 4 Case study IV. Micro-particles procured from the corpse in this case. (a) Vivianite particles separated from soil traces obtained from the corpse. (b) Glass microspheres of fly ash inside (*left* two panels) and separated from (*right*) soil traces obtained from the corpse

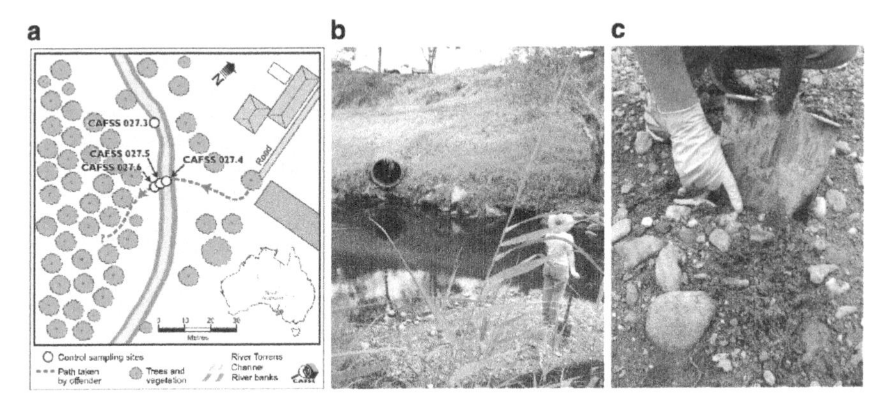

Chap. 8 Fig. 1 The Hit-and-Run case study. (**a**) Schematic map showing locations, aspect, path taken by offender and sampling points with associated reference numbers; (**b**) overview of river bank, person standing at point where Sample CAFSS_027.5 taken; (**c**) close-up view of the soil surface near where a shoe impression matching the sole tread of the shoe worn by the offender

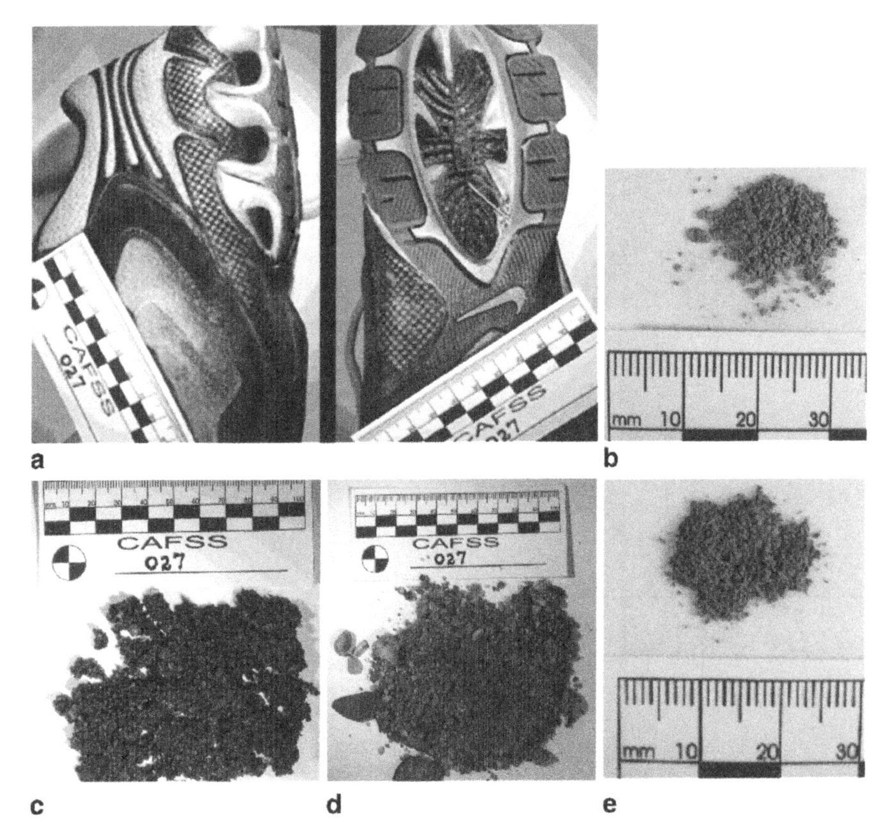

Chap. 8 Fig. 2 Soil samples associated with the hit-and-run case study. (**a**) Shoes from suspect; (**b**) sample scraped from the shoe; (**c**) control soil specimen from the river channel; (**d**) control soil specimen from the bank of river and (**e**) the <50 μm fraction separated from this sample

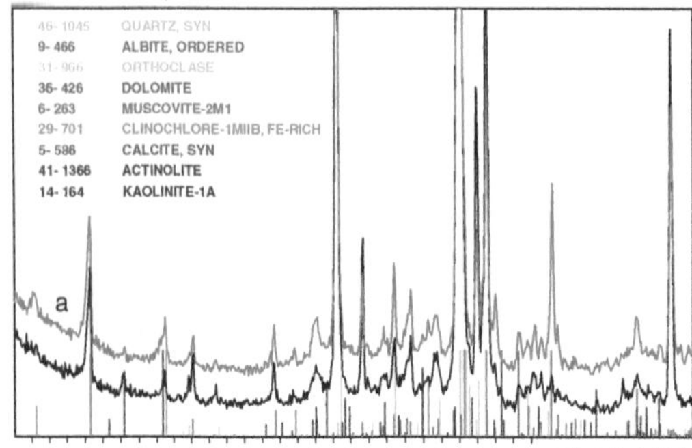

Chap. 8 Fig. 4 Comparisons between X-ray diffraction (XRD) patterns of soil samples from the shoe (**b**) and river bank (<50 μm fraction) (**a**) shown in Figures 1 and 2. The <50 μm fraction was separated from the stony river bank soil by sieving through a 50 μm sieve. Shoe and river bank samples were both ground using an agate mortar and pestle before being lightly pressed into aluminium sample holders for XRD analysis. XRD patterns were recorded with a Philips PW1800 microprocessor-controlled diffractometer using Co Kα radiation, variable divergence slit and graphite monochromator (from Fitzpatrick et al. 2007)

Chap. 8 Fig. 6 Abandoned quarry near Stirling, South Australia. Control soil was sampled in the surface layer (0 to 2 cm) of an anthropogenic soil (i.e. so-called man made soil that has recently formed in the floor of the quarry) between the tire tracks and metal construction (sieve) on the left hand side and at the base of the quarry

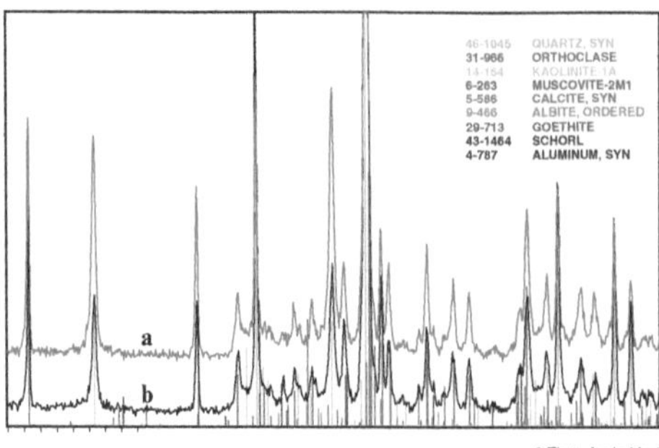

Chap. 8 Fig. 7 Comparisons between representative X-ray diffraction (XRD) patterns of soil samples from the High Street quarry, Stirling (a, AGU1) and material recovered from the boot of the suspect's vehicle (b, AGU2). There are considerable similarities between the XRD patterns indicating that they are most likely to have been derived from the same location

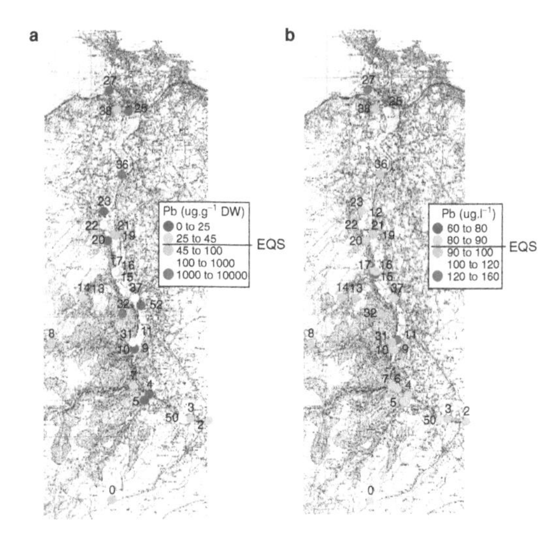

Chap. 10 Fig. 2 Spatial distribution of Pb in **(a)** the sediments/soils and **(b)** waters of the Conwy Catchment. EQS = Environmental Quality Standard. These data do not show a simple gradient away from a single source and potentially imply multiple sources of Pb in this region. Background map images Crown Copyright/ Database Right 2008. An Ordnance Survey/(Datacentre) supplied service

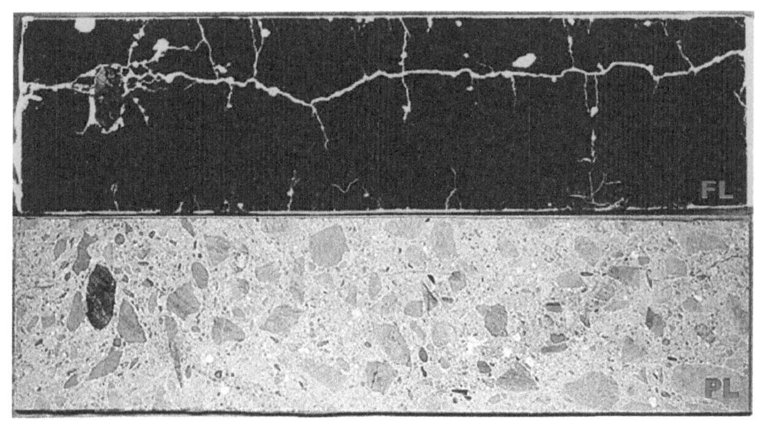

Chap. 11 Fig. 1 Impregnated full core in fluorescence (FL), compared with the same core in plain light (PL). At >1 mm width, the crack is difficult to discern in plain light, with off-shoots and thinner cracks along the side essentially hidden. Core dimensions: 100 × 400 mm

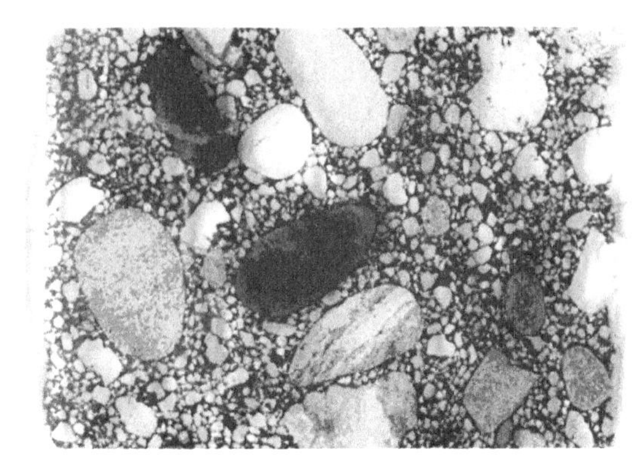

Chap. 11 Fig. 5 Alizarin-R stained thin section, revealing a number of stained coarse aggregate particles obviously containing carbonate. Actual size: 26 × 35 mm

Chap. 11 Fig. 7 Image of impregnated plane section in fluorescence (*left*), and thin section of bituminous concrete in plain light (*right*). Note the surface roughness and the pockets with debris. Core dimensions: 75 × 45 mm, thin section dimensions: 30 × 45 mm

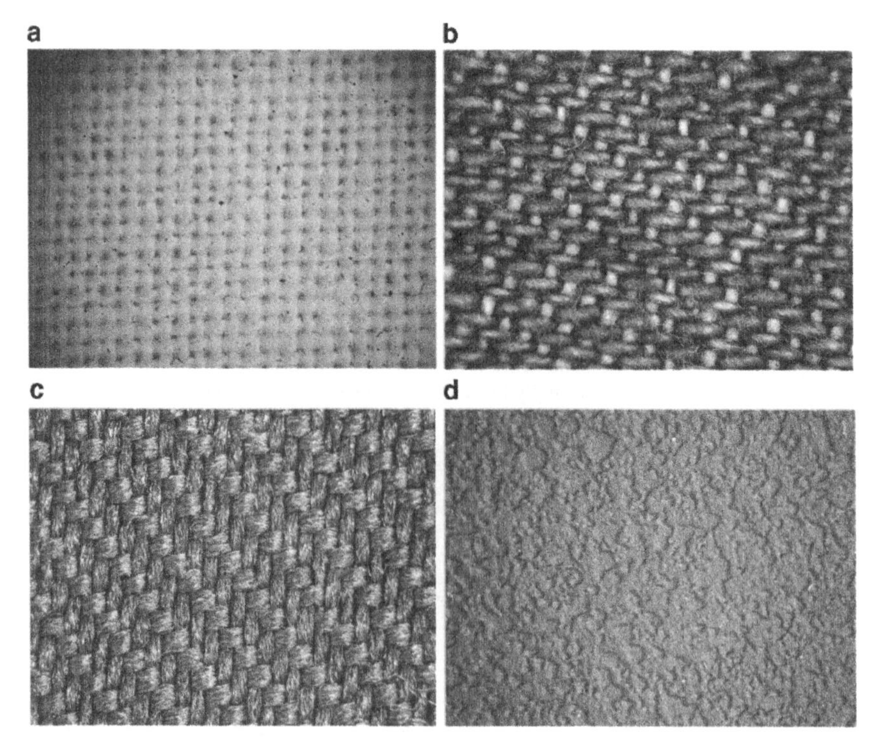

Chap. 14 Fig. 1 Micrographs (1 cm wide) showing **(a)** cotton; **(b)** denim; **(c)** polyester; **(d)** rubber car mat. Note contrasting textures of each of the materials

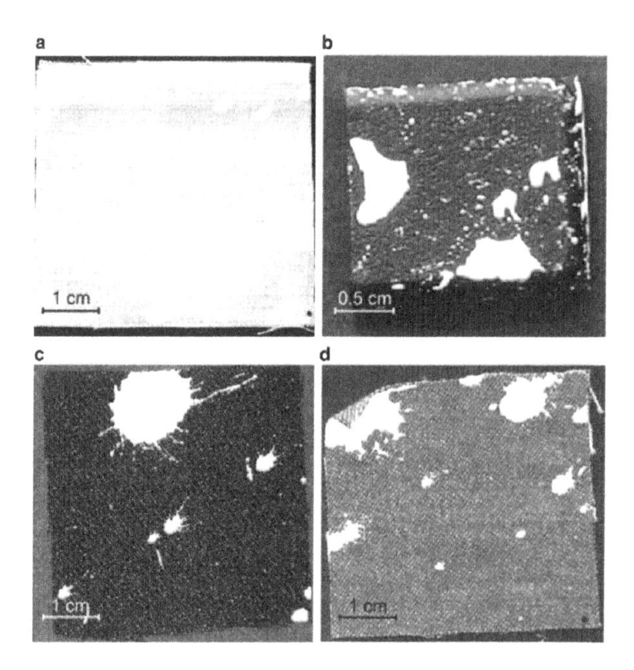

Chap. 14 Fig. 3 Photographs of the cut portions of 5 × 5 cm materials with splashes of kaolinite to be irradiated by XRD. (a) Cotton; (b) rubber; (c) polyester; (d) denim

Chap. 15 Fig. 1 Earthworm burrows through archaeological deposits. Section showing effect of vertical and horizontal burrows through agricultural topsoil into the numerous layers within an Aceramic Neolithic (10,000 BP) plaster mixing pit. Such earthworm action can disrupt attempts to define contexts, accurately carbon date layers and fills, differentiate and date deposits by artefact type and define palaeobotanical assemblages by context. Scale bar = 50 cm (source: Hanson)

Chap. 15 Fig. 2 Contemporary observations at Down House. Test trench in plan revealing the cinders deposited on a horizon at 14–18 cm. A brick fragment can also been seen. The flint layer on which the cinders come to rest can be seen at 18 cm (source: Hanson)

Chap. 15 Fig. 3 Lost from view: Filtering of cinders (originally deposited by Darwin, now white washed and re-used) into the turf root mat at Down House 9 months after deposition. Placed in early August, they were lost from view in 3 months due to grass growth. Earthworm casts are also starting to cover them (source: Hanson)

Chap. 15 Fig. 4 Maggot mass purging from rear of a roe deer 17 days after death: Note the transport of hair away from the body into the leaf litter. Scale bar = 20 cm (source: Hodgson)

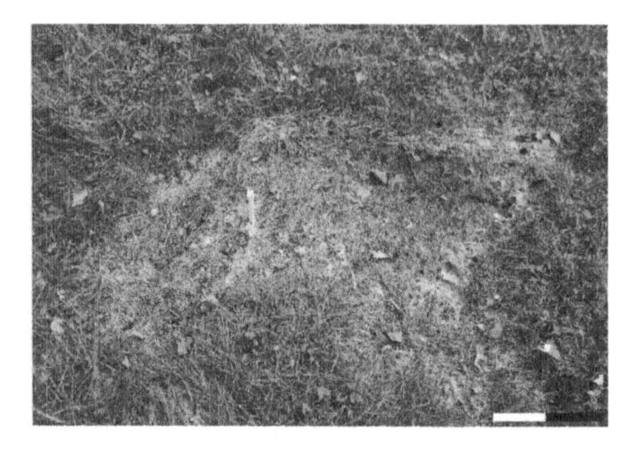

Chap. 15 Fig. 5 Vegetation die-off caused by decomposition of a sika deer: the original position of the cadaver can clearly be seen, giving the body outline. Visible for 8 months after carcass is dispersed by scavenger activity, until obscured by new seasonal vegetation growth. Scale bar = 20 cm (source: Broadbridge)

Chap. 15 Fig. 6 Half section excavation underneath deer deposition site: This located small bones and hair brought down by maggot activity to the horizon between the leaf litter/humic layer and the more compact subsoil. The stain to the soil can clearly be seen (source: Orr and Furphy)

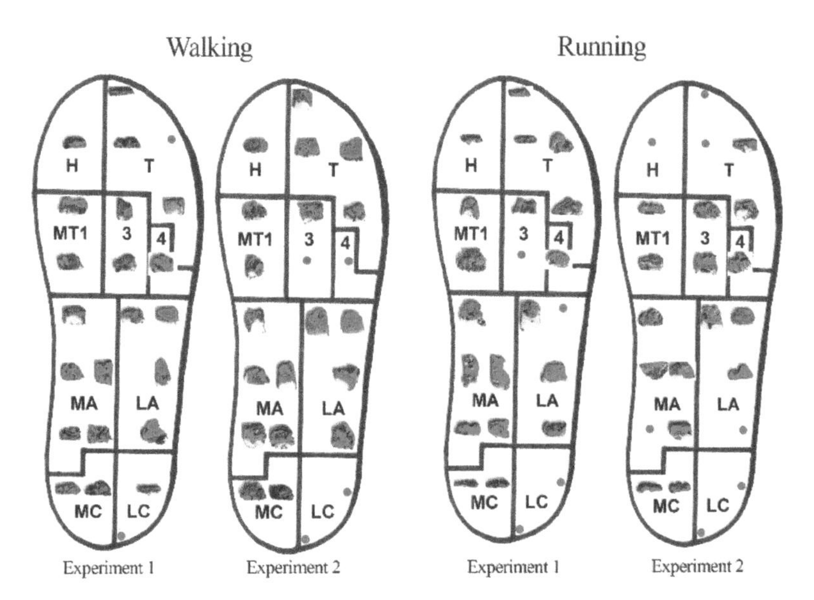

Chap. 16 Fig. 1 Plasticine plugs recovered from each sample point on the right shoe sole for both walking and running experiments (MC = medial calcaneus, LC = lateral calcaneus, MA = medial arch, LA = lateral arch, MT1 = first metatarse, 3 = second and third metatarse, 4 = fourth and fifth metatarse, H = hallux and T = toes)

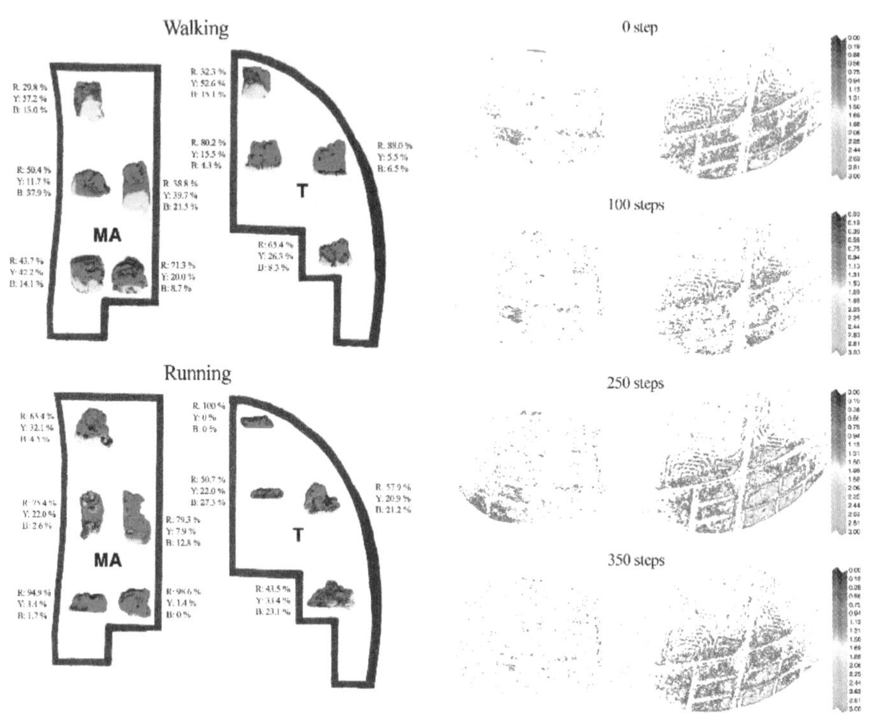

Chap. 16 Fig. 2 The percentage of each layer comprising the Plasticine plugs recovered from the medial arch area and the toes area of the right shoe soles for both walking and running experiments

Chap. 16 Fig. 5 Pixelated image to show the silt-sized material retained on the shoe soles after walking different distances (mean brightness indicates the amount of silt-sized material remaining on the sole)

Chap. 16 Fig. 8 Muddy footwell mat and the distinct footprint in case study

Chap. 18 Fig. 1 Composite terrestrial Lidar image of the search site. Image created by a composite of laser scanned images of the area

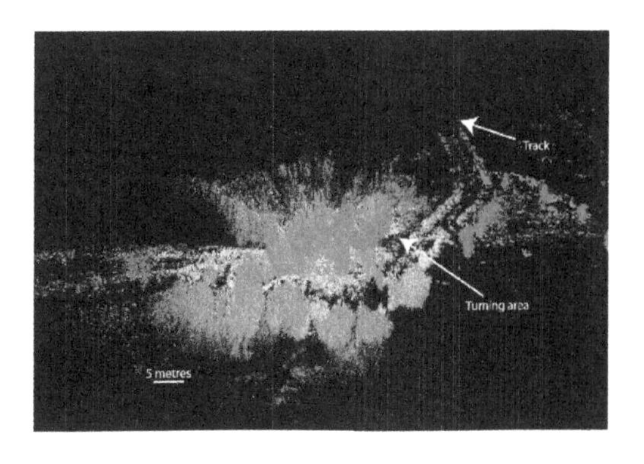

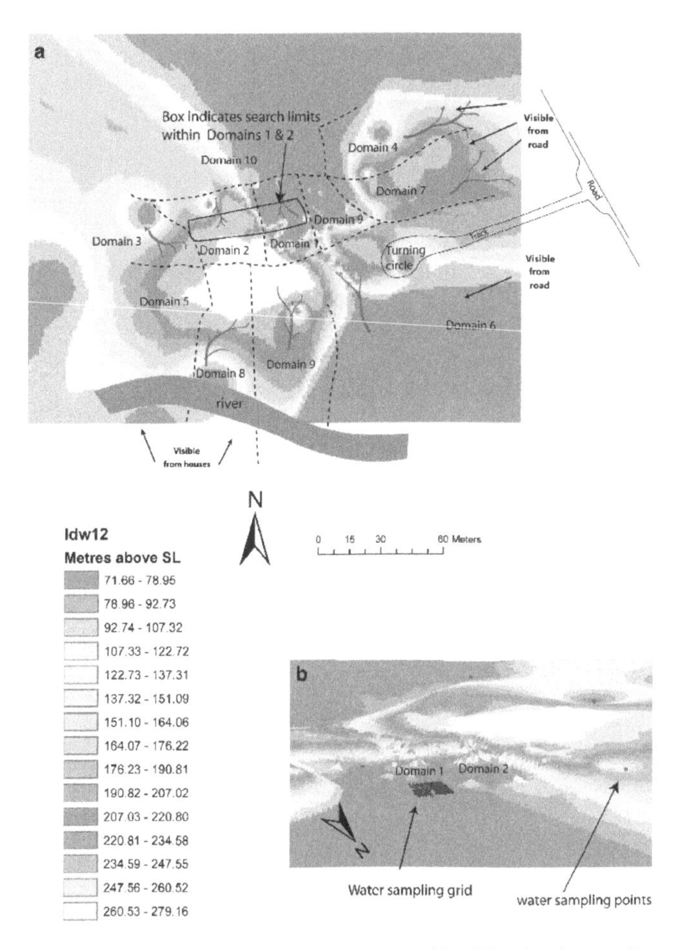

Chap. 18 Fig. 4 Digital elevation models (DEM) generated in GIS using inverse distance weighting (IDW) interpolation of DGPS elevation data (IDW 12 indicates the number (12) of neighbouring DGPS data used in the interpolation procedure). (a) Two-dimensional; (b) three-dimensional. 3D DEM has been overlain with the domain-based search areas shown in Figure 3 described in the text. Highlighted area refers to target search area with Domains 1 and 2. Key locations for water sampling throughout the area along with a larger area in Domain 1 are shown on the 3D DEM. The viewing perspective has been changed to aid landscape interpretation

Chap. 18 Fig. 6 (a) Photograph of vegetation in the search area and use of GPR. Figure is approximately 1.8 m. Field of view is 10 to 15 m; (b) Terrestrial Lidar image of the vegetation generated using a Leica HDS3000 laser scanner. Field of view is 15 to 20 m

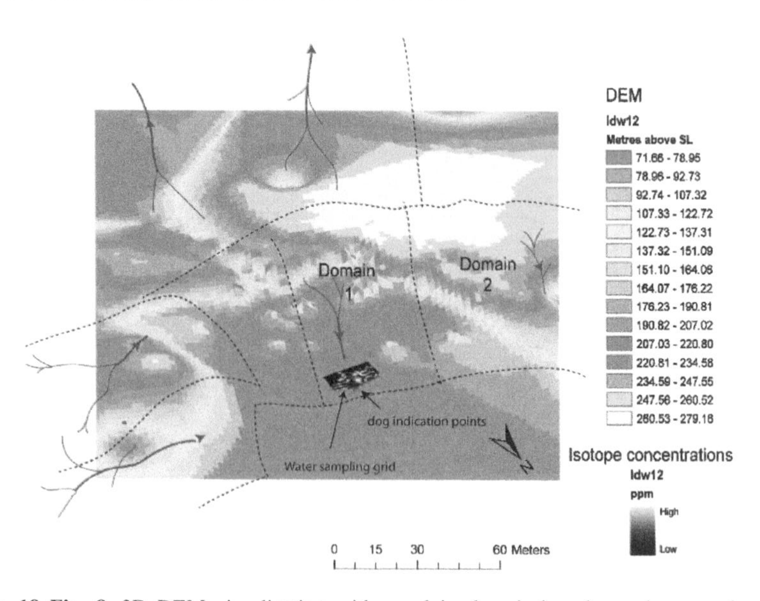

Chap. 18 Fig. 8 3D DEM visualisation with overlain domain-based search areas shown in Figures 3 and 6 described in the text. Target search areas Domain 1 and 2 and the interpolated TOC distribution of the water sampling grid. The DEM and isotope distribution are generated in GIS using inverse distance weighting (IDW) interpolation (IDW 12 indicates the number (12) of neighbouring data used in the interpolation procedure)

Chap. 22 Fig. 1 Overview of site. (a) General view of experimental site with pig graves open while environmental monitoring is being set up; the nearest is Grave 1, the middle is Grave 2 (tomb), the furthest is Grave 3. (b) Soil section from control test pit (0.5 m scale/10 cm divisions). The profile consists of an upper layer of white/grey sand with rock fragments, overlying a red/brown sand horizon with rock fragments and a grey/white sand horizon with dense rock inclusions is at the base

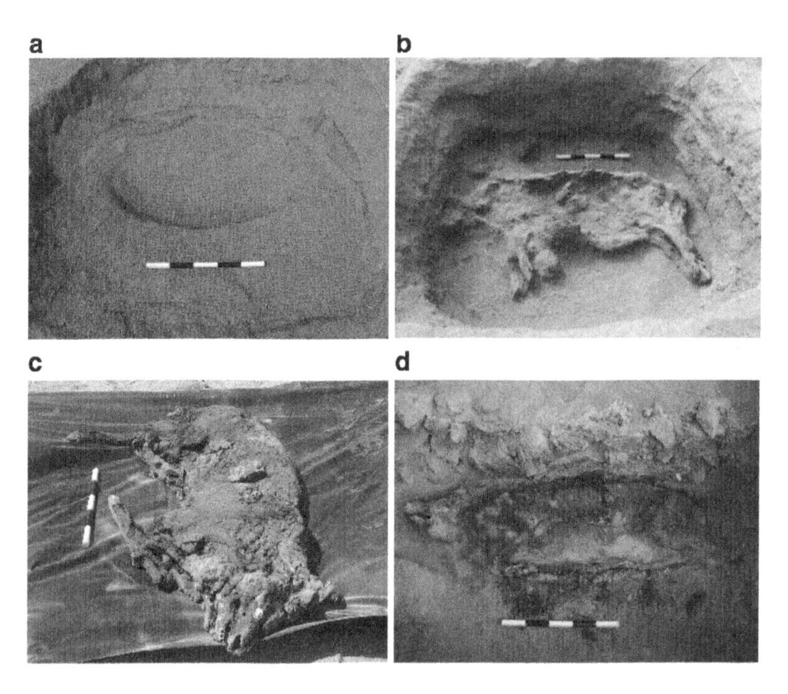

Chap. 22 Fig. 2 Overview of carcass decomposition. (a) Netting buried subsurface above Grave 1 to deter scavengers, which shows soil movement associated with the collapse of the body cavity not apparent at the surface. (b) Carcass in grave after 760 days showing collapse of the body cavity and concretions associated with moisture derived from the decomposing carcass, especially evident adjacent to the cranial trauma. (c) Underside of Pig 3 after excavation, showing trauma and more advanced skeletalisation to the head. (d) Pig in tomb following removal of capping and part of the tomb construction. Dark pupal masses are visible adjacent to belly. All scales are 0.5 m with 10 cm divisions

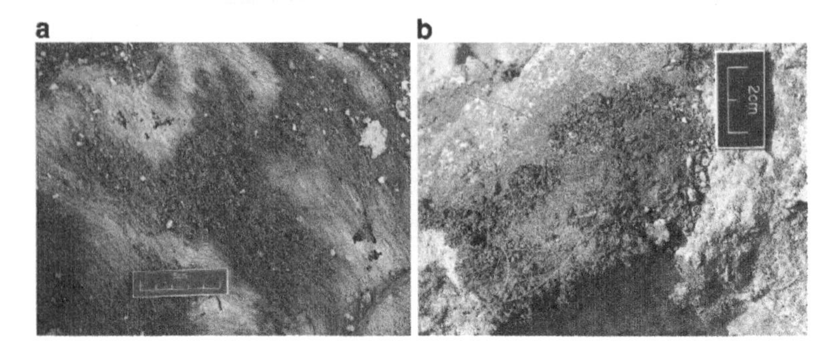

Chap. 22 Fig. 3 Details of pig surfaces after decomposition. (a) Upper surface of Pig 2 flank showing pupal cases, loss of bristle and holes in the skin characteristic of insect damage. (b) Pig decomposition products deposited on the under-surface of the capping within the tomb, including adhering bristle and empty pupal cases

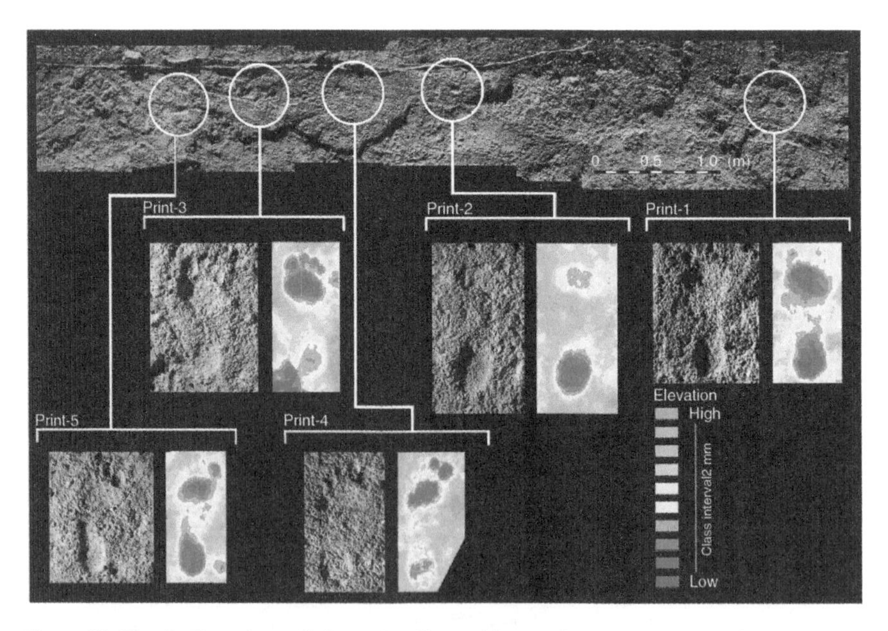

Chap. 28 Fig. 2 Footprint trail from the Cuatrociénegas Basin in Coahuila State, Northern Mexico. Five footprints are shown along a single discontinuous trail, illustrated via photomontage. This trail is exposed in the floor of a travertine quarry. Close-up photographs and isopleth maps created from scanned images are shown for each of the five prints

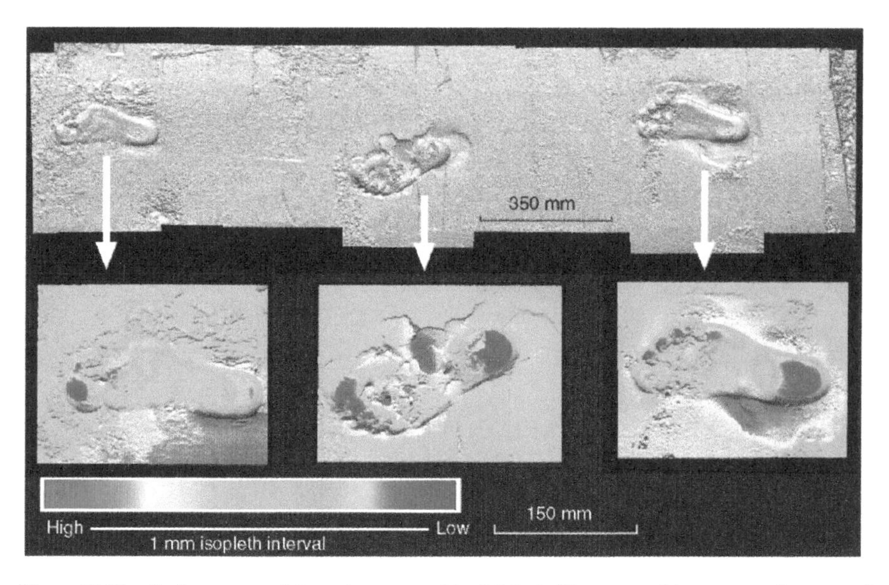

Chap. 28 Fig. 3 Sequence of footprints created by Male 1. Upper panel is a composite scan of the sand tray while the lower panels are close-up scans of each footprint rendered with 1 mm isopleth to enhance visualisation

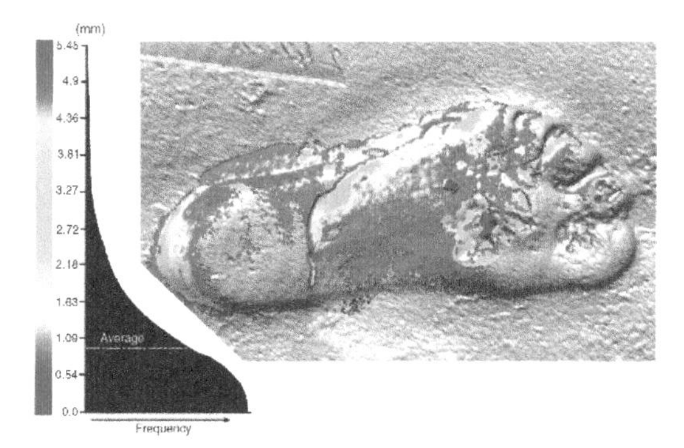

Chap. 28 Fig. 5 This interference image of two scans one taken from the original print and one taken from the cast of that print. Coloured areas are those with a measurable difference between the two scans following image registration in rapidform. The interference image clearly demonstrates that the two scans are not identical. The difference can be accounted for by the casting rather than the scanning process

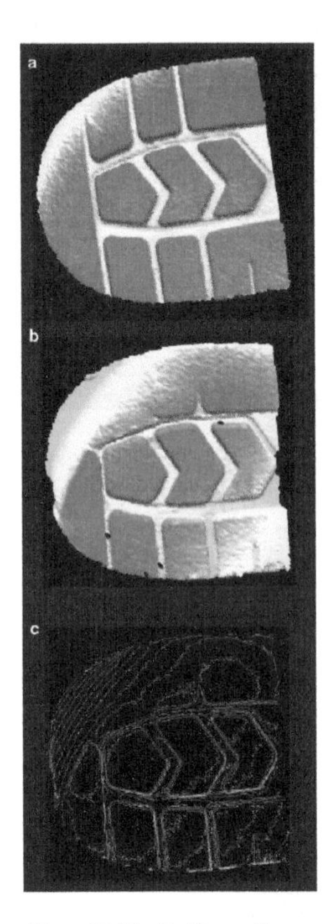

Chap. 28 Fig. 7 Three-dimensional images of two similar right heels showing different levels of wear depicted by the 1 mm colour isopleths. (a) Right heel showing limited wear. (b) Right heel showing pronounced wear on the outer heel edge. (c) Contour map, 1 mm interval, of Heel B allowing direct quantification of the level of wear

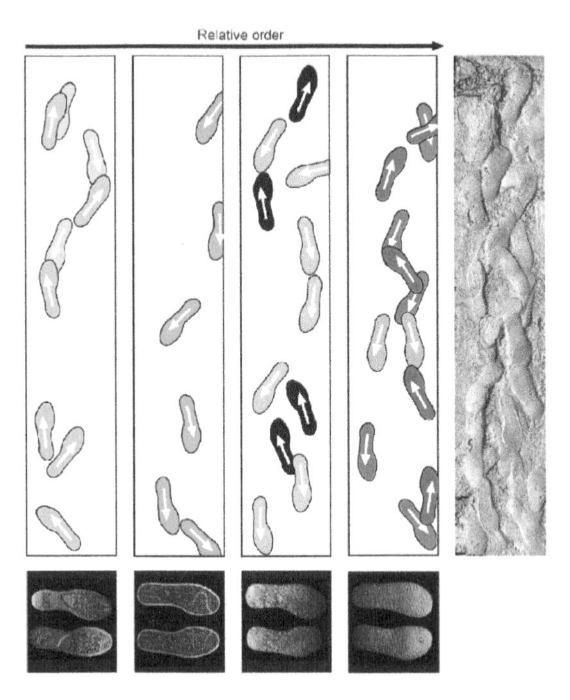

Chap. 28 Fig. 10 Footprint sequence in which the palimpsest on right-hand panel was created

Part I
Concepts

Chapter 1
"Soils Ain't Soils": Context and Issues Facing Soil Scientists in a Forensic World

James Robertson

Abstract The last decade has seen a welcome increased interest on the part of soil scientists, and related professionals, to apply their knowledge and technologies to addressing questions of forensic interest. This chapter discusses some of the challenges which face these professionals as they apply their science in the forensic space, principally arguing that soil science has the most to contribute in answering the 'what happened' question. Given the ubiquitous 'CSI hype' surrounding forensic science it is vital that soil forensic evidence is realistic and based on the sound application of science. The role of the expert witness is also discussed. Finally, the importance of underpinning standards and the need for research and importance of a partnership approach are emphasised.

Introduction

This chapter is based upon a public lecture delivered at the 2nd International Soil Forensics Conference, held in Edinburgh, Scotland in October 2007. In transforming a verbal lecture to a written piece I have omitted most of the visual material used to illustrate my presentation, and material originally included to entertain. However, I have attempted to retain, and further develop, the issues which I believe are important for soil scientists using their science for forensic applications.

Much is sometimes made of the term 'applied science' with an underlying implication that this is in someway 'inferior science' when compared to 'fundamental science'. Interestingly it was Louis Pasteur who stated that "there are no such things as applied sciences, only application of science". Figure 1.1 shows what is known as Pasteur's Quadrant and shows that the keys are the quest for fundamental understanding and consideration of uses. The challenge then for soil scientists is to apply their techniques and quest for fundamental understanding of soils in the forensic domain!

J. Robertson
National Manager Forensic and Data Centres, Australian Federal Police, GPO Box 401, ACT 2601, Australia
e-mail: jim.robertson@afp.gov.au

K. Ritz et al. (eds.), *Criminal and Environmental Soil Forensics*

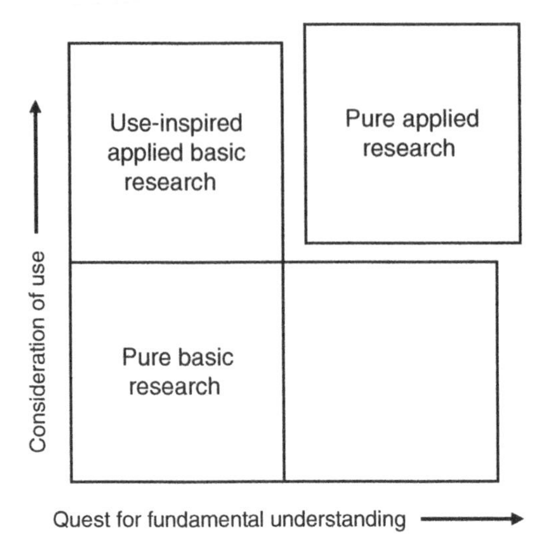

"There are no such things as applied sciences, only application of science". Louis Pasteur.

Fig. 1.1 Pasteur's Quadrant. Adapted from Stokes (1997)

My own interest in soils owes its origins to my background as a graduate of the 1970s in agricultural sciences. Soils formed a significant component of second year agricultural chemistry. I also well recall field visits to study soil horizons and agricultural botany; of course, soils and plants are really inseparable. My post-doctoral work involved looking at the effect of soil compaction on the roots of cereal crops and the effects on crop yields. Hence, it was perhaps no surprise that, when I entered the world of forensic science in the mid 1970s, I should also develop an interest in the examination of soils in a forensic context. The work of my colleagues, students and myself at this time focused on developing a sequential protocol for handling soil samples as they presented in case work. Of course, critical to the *context* for forensic applications is that, more often than not, the forensic scientist has only a very limited evidence sample with which to work. Forensic work would rarely encounter the world of soil horizons and, whilst geological soil maps can be important, forensic samples are usually more akin to mixed surface debris. Included in this early work to establish suitable protocols was specific work looking at how best to treat 'soil' samples for particle analysis (Robertson et al. 1984; Wanogho et al. 1987a,b), which then led to early work on evaluating different instrumental approaches to particle analysis (Wanogho et al. 1989), studies on variability of soils in a control field setting and work on the statistical assessment of these analytical results to assign soil samples correctly to source (Wanogho et al. 1985).

The last grant I obtained before leaving Scotland to move to Australia in 1985 was to look at the presence of residues of plant growth retardants in soils in remedial sites in Glasgow. These plant growth retardants were being used to reduce grass growth in cleared building sites as Glasgow went through urban renewal. The idea was to reduce the number of times the grass would need to be mowed! Interestingly, and perhaps understandably for the time, we did this work in splendid isolation from the world of soil scientists. Indeed it is a relatively recent phenomenon for 'fundamental scientists' to show an interest in applying their science to forensic problems. Some of this undoubtedly has resulted from the need to identify new sources of funding for research, some has resulted from an increased focus on security issues and, increasingly, on environmental impact. Perhaps a little unkindly I might comment on the impact of the TV series such as 'Crime Scene Investigation' (CSI) and the like, as everyone these days wishes to be just like CSI. Whatever the sum total of the reasons for the interest shown by soil scientists in the forensic world, it is indeed welcome. Forensic scientists are by nature generalists and cannot be expected to be specialists in geology, soils, pollen or the wide range of analytical approaches used by the various branches of soil science and geology. Hence, what we need to develop is an *epiphytic* relationship, one where soil science (broadly defined) is the host body and forensic science the epiphyte – in this relationship both derive benefit. The epiphyte cannot survive on its own. I think conceptually this is a very useful model to use to describe the ideal relationship between soil scientists and forensic scientists.

The engagement of soil scientists with the forensic world, and some notable recent major contributions to solving serious crime (hence, raising its profile) is indeed welcome (Porter 2006). Recent evidence of this resurgence of interest and activity can be seen in the literature. Every third year Interpol host an international forensic symposium at which allocated member countries report on a review of the scientific literature for an evidence category. This review covers the 3 years between symposia. One of the evidence categories is forensic geology and soils. As a member of the organising committee for several of these symposia, a recurring discussion through the 1990s was, on the basis of published work, whether or not we could justify retaining soils and forensic geology as a separate evidence category. I am pleased to say that, in the most recent Interpol symposium, Sugita Suzuki from Japan reported on nearly 200 references which included the outcomes from the first International Soil Forensics conference held in Australia in 2006 (Suzuki and Katsumata 2007). Soils and forensic geology are truly alive well and even, dare I say it, flourishing.

I also want to acknowledge the work of Rob Fitzpatrick from Australia's CSIRO Land and Water Division. The founder of the Australian Forensic Soil Centre, through his enthusiasm and energy, he has been the key individual in re-establishing interest in forensic soil examination in Australia and has had an international impact.

Finally, even I have been surprised by the breadth and scope of the content of this symposium. The term 'Soil Forensics' embraces many disciplines and is an excellent example of an application which requires the ability to bring together a diversity of knowledge which cut through interdisciplinary boundaries. More than simply cutting through such boundaries, solutions demand truly interdisciplinary thinking. The unifying aspect is the application to solving problems in the law enforcement and the broader justice system.

The Changing World of Forensic Science

The title of this chapter refers to the context and issues facing soil scientists in a forensic world. Some practitioners coming from a non-traditional forensic background but, in government laboratories, will have gained considerable experience and will have a mature and developed understanding of the environment in which they now choose to operate. For others, and newcomers, it may be helpful to look at how the world of forensic science has changed in the last 20 or so years and how this has altered the forensic operating environment.

Perhaps the greatest single change in the last 30 years has been the emergence of analytical procedures which enable the testing of DNA from biological material. Of course the main focus on 'forensic DNA' is on human DNA but there are forensic applications for animal and plant DNA. In particular, profiling of microbial DNA is finding application in the forensic soil world. It is the 20th anniversary of the first application of DNA analysis in a forensic context. The case of Colin Pitchfork from the UK and the early pioneering work of Sir Alec Jeffries and scientists from the then UK Home Office Forensic Laboratories is well documented. Over the past 20 years technologies have evolved with a couple of drivers and trends. One of these has been the ability to extract and analyse DNA from ever smaller samples. In the late 1980s a bloodstain the size of a large coin or a strong semen sample with sperm were required. Today it is possible to obtain a DNA profile by sampling areas where there is no biological material visible to the naked eye! A parallel, but different driver, has been the ability to also analyse smaller fragments of DNA such that today's technology (short tandem repeats or STRs) can analyse fragments as short as 100 base pairs or slightly smaller. The purpose of this paper is not to review DNA and the reader is referred to Fourney (2007) for a comprehensive review of recent literature dealing with the broader area of forensic biology.

The key message which emerges from the evolution of forensic DNA analysis is that today a DNA profile can be produced from very low levels of biological material. This has raised its own issues for forensic science. In the past, with appropriate information and interpretation, the biological material, bloodstain or semen, could be related to the circumstances of an alleged crime. Prior to DNA the levels of individualisation were quite low (in the 1-in-10s to 1-in-100s). By contrast, today the levels of individualisation are very high (in the 1-in-millions to 1-in-billions!) but, at least for trace biological samples, it is more difficult to be certain that the biological material is relevant to the alleged crime. There is a lack of relevant research to assist in the interpretation of questions such as how and when the biological material was deposited or how long one might reasonably expect such biological material to remain or persist under a variety of environmental circumstances. A recent report on the use of low copy DNA techniques in forensic work has emphasised the difficulty in interpreting not only the technical profile but also when and how the DNA was deposited and, hence, its relevance to an incident under investigation (Home Office 2008).

These sorts of questions, which go to the heart of 'what happened' as compared to 'who dunnit', are the very essence of a criminalistics approach to the forensic science. Put another way, forensic science has to be information in context.

It is relatively easy to extrapolate these issues into the world of soil forensics. It is always a temptation to be drawn to the issue of identification, or with soils at least individualisation before properly, and fully, exploring issues of interpretation. To interpret what significance may be attached to finding two soil samples that may have a common source, it is critical to ensure that samples collected from relevant scenes are adequate and representative. It is also important to consider when soil may have been deposited and whether or not this supports the timeline for an alleged offence or crime.

There is an enormous pressure to attach numbers to all forms of evidence driven by the issue of identity. For soils, most cases will revolve around the 'what happened' and not the 'who dunnit' which will come from biometric identifiers, most commonly DNA and fingerprints. Of course, it is both desirable and useful to be able to show that testing regimes are objective and numbers play a part in this.

The other major advance resulting from DNA testing is that law enforcement has a second major database, after fingerprints, which can help identify the source of a recovered biological material. It is important to remind oneself that these databases are intelligence databases. The 'expert' is still required to interpret the evidence material.

A final comment about DNA and soils, which they have in common, is that there are often mixtures. Sometimes these are easily separated and sometimes they are not.

The second major factor which has influenced the popular image of forensic science is, of course, the emergence in the last 10 years or so of a plethora of forensic programmes on the TV, led by the almost ubiquitous CSI and clones! No doubt these programmes have helped raise the public profile of the forensic sciences but arguably they have left the public *and* even professionals with an artificial view of what forensic science can and cannot do. Putting aside silly and annoying things such as the fact the CSI investigator likes to work in the dark, shine small light sources on everything, pick up evidence before recording it and, of course, all whilst wearing designer clothes, there is a more serious downside. CSI presents an image of instant results with analysis taking seconds to minutes at most. It also portrays forensic scientists as working on single cases whilst this is almost never reality. Although it is still widely argued what influence CSI has on jurors or others, it seems clear that these programmes have created a popular image of what forensic science can do which is at odds with the reality. Those applying soil science techniques to forensic problems should ensure their 'clients' have realistic expectations or they will end up with failed expectations!

Soil scientists also need to learn about the fundamentals of how to operate at a crime scene as part of a multi disciplinary professional team. This works best when individuals understand their role and respect the role of others.

Another aspect of the world today which has changed how we think in the forensic sciences is terrorism and security concerns. In the past, and to a continuing extent, scientists have been more focused on getting complete answers rather than on how long it takes to achieve this outcome. This is not to say scientists are oblivious to the need to complete casework as quickly as possible but, as a profession, there is no doubting there is an image problem around timeliness and backlogs. The argument is that there is a need to have complete answers, and correct answers, to assist the court, with an emphasis on evidence. This is of course true but it does not recognise the competing need to provide information at the earliest possible time to assist in the investigative stage. At this time, quick information may be vital in determining the direction of an investigation or in eliminating people or scenes. The more information forensic science can offer in these early stages, the more value it will be seen to have. There is a dual responsibility to support the investigation phase, which may be with intelligence, and the broader justice system with information which meets the full weight of evidence.

The final major change that has occurred in the last 20 years or so in forensic science is the emergence of laboratory accreditation and credentialing of individuals. Accreditation for laboratories is now routine under a relevant national body such as the Australian National Association of Testing Authorities (NATA), the American Society of Crime Laboratory Director – Laboratory Accreditation Board (ASCLAD-LAB) or the UK's NAMAS. In Australia most of our crime scene examiner groups, as well as our scientific laboratories, hold formal accreditation under the NATA scheme. In the UK there is also a Certified Registered Practitioner scheme which will in the future be closely aligned with a newly established Office of the Forensic Regulator. Laboratory accreditation generally complies with ISO 17025 standards but, in Australia, there are quite onerous supplementary standards around forensic practice. These have now been developed into a forensic module which organisations holding NATA authorisation under a non forensic testing scheme, can gain forensic accreditation as an add-on. One recent example in Australia was where the NSW – Environmental Protection Agency (NSW-EPA) gained forensic accreditation under this new scheme. Such an agency may well have an increasing future role in forensic work through environmental contamination cases, including contamination of soils. At the recent Interpol forensic meeting, the US-EPA presented on environmental contamination and made the point that in the USA such contamination is increasingly coming under the criminal jurisdiction. Hence, there will be an increasing obligation to meet evidence handling and processing standards built around evidence continuity and appropriate occupational health and safety concerns.

The challenge in the future for many organisations and individuals wishing to carry out the various aspects of forensic soil examination, will be to demonstrate they comply with the appropriate standards. A key element of this is taking part in appropriate training and in ongoing proficiency testing. For many aspects of forensic soil examination such tests are either not developed or available or are not necessarily fit for purpose. Nonetheless it will not be acceptable for much longer, if indeed it is currently acceptable, to argue that forensic soil work should not be covered under formal accreditation. For individuals it may suffice to be covered under individual credentialing but individuals can expect rigorous examination relative to forensic expectations. This includes methods meeting forensic standards of method validation.

In summary: forensic science is no longer 'owned' by a very small number of players; in particular 'physical evidence' is being provided by academia and by other relevant government and private laboratories; and there is an increased focus on standards at individual and organisations level.

The Forensic Challenge

The word forensic simply means pertaining to a court of law. In many ways it is this aspect of the forensic arena which often causes scientists the most problems. It is outwith the scope of this paper to examine in depth the dynamics of the working space shared by the police, forensic scientist and the law. There are formal rules which in part define these interactions but there are informal rules which make it quite difficult for the 'outsider' entering the forensic world (Robertson 1995).

As an example, in Australia, there are Codes of Practice for expert witnesses. Standard features of these include:

- An expert witness has an overriding duty to assist the court impartially on matters relevant to the witness's area of expertise and is not an advocate for a party.
- An expert witness must work cooperatively with other witnesses.
- A report by an expert witness must set out all the facts and assumptions on which the expert's opinions are based and must note any matters that qualify those opinions.
- If an expert witness changes opinion, the witness must prepare a supplementary report.

Whilst this may appear quite straight forward, in practice it does not always work as described. Experts can be placed in a difficult position where they are not in possession of sufficient information to place their findings in context. The 'context' from a defence viewpoint may not even be revealed until cross examination in the witness box. Something as apparently simple as writing a report is far from simple. Remember this report should be scientifically accurate but understandable to the triers of fact; not the lawyers, the jurors! Suffice to say in this very short treatment of a complex subject, expert witnesses should beware and should understand the formal and informal rules of engagement.

A scientist will only be accepted by the court as an expert to give an opinion on subject matter which is beyond the normal understanding of a lay juror. Scientists like to speak in a nomenclature and language of their own and it is easy to forget that a lay juror may take an entirely different meaning from a word such as 'consistent' or 'similar' or 'match' to that meant by the scientist.

It could be argued that the expert should use *none* of the above three words. The purpose of any scientific examination and comparison should be to use appropriate and discriminating tests aimed at finding *differences*. Of course, as no two samples are ever identical in all respects, the onus on the expert is to interpret what is a meaningful difference. This can only be done with a good knowledge of inter- and intra-sample variation for the parameter or parameters in questions. 'Similar' is a misleading term because it gives no sense of whether of not differences are meaningful. Similar really means small (but meaningful?) differences exist. Neither is the use of the word 'match' to be encouraged. Apart from being a thin wooden stick, there are many other definitions. The Collins English Dictionary uses words such as 'resembles', 'corresponds to', 'equal', none of which gives any sense of evidentiary value.

The obligation which lies on an expert witness is to give an opinion on the weight and substance which can be placed on their scientific analysis! There will be instances where a highly definitive answer can be given – perhaps where a highly individualistic soil is recovered and a location can be determined. However, in many cases, if not most forensic examples, the expert will be required to interpret what the analytical results may or may not mean in terms of likely common origin. Often this will be made more difficult because of limitations imposed by poor sampling either at the scene or from a questioned item or simply a lack of relevant background or reference information. The importance of the crime scene cannot be overstressed. It is highly desirable, arguably essential, that the soil scientist is able to attend the crime scene and take, or direct the taking of appropriate, known samples for future comparison with recovered, questioned or 'evidential' samples. Regrettably all too often the soil scientist will not collect initial known samples and these will not be fit for purpose. Of course, sometimes the role of the soil scientist is to point the investigation towards a location. Here it may not be possible to collect contemporaneous samples.

Whilst many countries have excellent geological and soil maps, what relevance will this have to a housing estate where the topsoil has been removed and replaced or in a typical suburban garden?

Some have proposed the use of a Bayesian statistical approach to weighting forensic evidence. Here a likelihood ratio (LR) is calculated based on assessing the apparent strength of evidence. This LR can then be changed into a common English phrase which equates to the strength of evidence (Table 1.1).

This sounds like a good idea and the application of a Bayesian approach has many supporters. There is certainly a quite widely accepted view that it provides a useful way to think about those factors which would contribute to the LR, as a way of thinking about evidence. However, this approach has not yet found wide acceptance in Courts where the judiciary seem certain to take a conservative approach to its adoption. If this approach is to gain acceptance in the future, there will need to be

Table 1.1 Some proposed verbal equivalents for likelihood ratios

Verbal equivalent	Likelihood ratio
Very strong evidence to support …	>10,000
Strong evidence to support …	1,000 to 10,000
Moderately strong evidence to support …	100 to 1,000
Moderate evidence to support …	10 to 100
Limited evidence to support …	1 to 10
Limited evidence against …	1 to 0.1
Moderate evidence against …	0.1 to 0.01
Moderately strong evidence against …	0.01 to 0.001
Strong evidence against …	0.001 to 0.0001
Very strong evidence against …	<0.0001

Adapted from Rose (2002).

a large body of research to base an informed estimate of the probabilities which make up the LR. Expert witnesses who use this approach cannot expect to be allowed to offer statements such as 'strongly supports' without having to 'peel back the layers' to explain how a LR is calculated and the underlying assumptions, as well as explaining basic probability theory to a lay audience! Hence, experts are left to attempt to convey weight and substance in lay terms.

Conclusion

There is a requirement for a concerted research effort across all aspects of forensic soil examination. Whilst there is much to be optimistic and enthusiastic about the current renaissance in soil forensics, it would be simplistic and naïve to not suggest much remains to be done. For the new players in the forensic arena the challenges include:

- Understanding a 'criminalistics' approach.
- Understanding how to apply analytical methodologies to non standard samples.
- Understanding the role and obligations of the scientist as an expert witness.
- Understanding the dynamics of working with forensic scientists, lawyers and police.
- Conducting research to assist in the interpretation of 'soil forensics'.

I have briefly touched on some of these but this is not meant to be an in depth treatment of what is a complex subject. My main purpose in raising these challenges is to ensure that, however well meaning individuals may be in wishing to contribute to solving problems in the forensic arena, there is an understanding of the pitfalls which lie in wait for the unwary in the forensic world. The solution requires a partnership between scientists in fundamental areas of science, academia and government forensic laboratories. There is a need to develop protocols and a decision making tree to manage the diverse way in which soils (and related) materials present in forensic applications. There needs to be further research to access the vast array of possible analytical techniques to work out which provide robust and informative answers. Recent debate concerning the use of quartz sand grains and grain surface textures as 'sediment fingerprints' shows that much still remains to be done before there is agreement on appropriate protocols (Bull and Morgan 2007). The latter can only be achieved if the necessary research is carried out to look at the criminalistic elements of variation, transfer and persistence.

Professional practice needs to be underpinned by the development of standard protocols, agreed methods, proficiency testing and appropriate training. In order to achieve a momentum towards these goals it may be useful to consider following the example of other criminalistics groups and establish an international scientific working group (SWG), perhaps SWG-SOIL. This needs to include soil science specialists and forensic science practitioners to achieve the best balance in the application of fundamental sciences in forensic context.

References

Bull PA and Morgan RM (2007). Sediment fingerprints: a forensic technique using quartz sand grains – a response. Science and Justice 47: 141–144.

Fourney R (2007). Recent progress processing biological evidence and forensic DNA profiling – a review 2004–2007. Interpol Web site, www.interpol.int.

Home Office (2008). http://police.homeoffice.gov.uk/news-and-publications/publication/operational-policing/Review_of_Low_Template_DNA_1.pdf. Accessed 20 June 2008.

Porter L (2006). Written on the skin. Chapter 7, Reading the crime scene "The Matthew Holding Case". Pan Macmillan, Australia. pp 284–289.

Robertson J (1995). Forensic science in the adversarial system: current situation and possibilities for change. In: Proceedings of the National Forensic Summit (Eds. A Ross, J Robertson, P Williams and A McFerran). pp 113–135, National Institute of Forensic Sciences, Australia.

Robertson J, Thomas CJ, Caddy B and Lewis JM (1984). Particle size analysis of soils a comparison of dry and wet sieving techniques. Forensic Science International 24: 209–217.

Rose P (2002). Forensic speaker identification. Taylor and Francis, London and New York.

Stokes DE (1997). Pasteur's Quadrant: basic science and technology innovation. Stokes Brookings Institution, Washington DC.

Suzuki S and Katsumata Y (2007). Forensic Geology 2004–2006. Interpol Web site, www.interpol.int.

Wanogho S, Gettinby G, Caddy B and Robertson J (1985). A statistical method for assessing soil comparisons. Journal of Forensic Sciences 30: 864–872.

Wanogho S, Gettinby G, Caddy B and Robertson J (1987a). Some factors affecting soil sieve analysis in forensic science. 1. Dry sieving. Forensic Science International 33: 129–137.

Wanogho S, Gettinby G, Caddy B and Robertson J (1987b). Some factors affecting soil sieve analysis in forensic science. 2. Wet sieving. Forensic Science International 33: 139–147.

Wanogho S, Gettinby G, Caddy B and Robertson J (1989). Determination of particle size distribution of soils in forensic science using classical and modern instrumental methods. Journal of Forensic Sciences 34: 823–835.

Chapter 2
Expert Scientific Evidence in Court: The Legal Considerations

Derek P. Auchie

Abstract In this chapter, the approach to expert evidence in some of the main common law jurisdictions is considered. Evidence given by experts is a special form of evidence and it is treated as exceptional in that there are certain hurdles that must be overcome before it will be allowed to be led. As litigation in some jurisdictions is becoming more and more adversarial in nature, lawyers and judges today are far more sceptical than in the past about the stature of expert evidence. This is particularly so with scientific expert evidence given its status, impact and objectivity in any case where eyewitness evidence is either sparse or unreliable. Consideration is given to the legal criteria applied by courts in civil and criminal cases to the question of whether opinion evidence given by an individual qualifies as expert evidence. The critical concept of admissibility of evidence is examined, in order to discover the conditions under which expert evidence will be permitted in the courtroom. Key cases from the United States, the UK, Australia and Canada are dealt with. Novel scientific evidence is treated as a special case in some jurisdictions, and the conditions for the inclusion of such evidence are discussed. The way in which evidence is presented is also touched upon, since presentation can be as important as substance, particularly in jury cases. This chapter is designed to highlight, in particular for non-legally qualified readers, the legal perspective on the subject of scientific expert evidence. This is an important perspective given that lawyers and judges (as well as jurors) do not approach the justification and presentation of such evidence in the same way as the scientist does. The law has constructed some artificial barriers and checks and balances into this question, in order to prevent the "mystic infallibility" of expert evidence having a disproportionate effect on the decision maker. Knowing what these barriers are, and how to prepare to cross them, is a key part in the preparation of techniques new and old and in the presentation of evidence about them. All of this is useful in considering the presentation of soil forensics expert evidence, since courts adopt a universal approach to all expert evidence, irrespective of the subject matter. As the courts are introduced

D.P. Auchie
Law Department, The Robert Gordon University, Garthdee Road, Aberdeen AB10 7QE, UK
e-mail: d.p.auchie@rgu.ac.uk

K. Ritz et al. (eds.), *Criminal and Environmental Soil Forensics*

to expert evidence from more and more areas of expertise, so they have become increasingly wary of evidence of techniques with which they are unfamiliar. On the other hand, well researched, objectively justifiable and clearly presented expert evidence will pass muster every time. Knowing what the law regards as essential in advance is key to the successful gathering, analysing and presentation of such evidence, and, in turn, to success in the outcome sought.

Introduction and Context

It is clear that giving expert evidence can give rise to a diverse range of experiences, (many less than pleasant) as suggested by the American expert Kenneth S. Cohen:

> I've been battered, I've been flattered, I've been bruised, I've been embarrassed, I've been bloodied, I've been thanked, I've been cajoled, I've been attacked, I've been befriended, I've been accused of acting, I've been called a charlatan, I've been called sneaky, I've been called a whore, but I've continued to survive as a paid expert witness. (Cohen 2007; page ix)

Despite this, the law in all developed legal systems the world over regards such evidence as crucial in many cases it has to consider. It is regarded as special not just because of its impact, but also its nature. Witnesses are not normally permitted to offer their opinion on any matter while in the witness box; they are expected instead to describe what they have seen, heard or experienced. Expert witnesses are the main exceptions to this rule. They will not have directly witnessed any of the events surrounding the case. Instead, they are expected to provide an opinion within their area of expertise, based in most cases on the facts as witnessed by other, non-expert witnesses (except where the expert has had to examine a patient or item for themselves). This then puts them in a peculiar position. For this reason, there are strict rules in all legal jurisdictions regulating when such opinions might be tendered, and in which manner.

One of the most challenging issues in legal scholarship generally is that the law is jurisdiction-dependent. Each legal system has its own rules on every legal subject, for instance criminal law, contract law, divorce law, land law, company law or any other branch of the law. The law of evidence is no exception. This means that unless a comprehensive review of the 150 or so legal systems that exist is carried out (a monumental task) a description of the law in any subject area has to be focused on a particular type of legal system, so that some generalities that apply across the systems within that type can be examined; thankfully, there exists such a classification of legal systems. The two main examples are: common law legal systems and civilian legal systems.

For the purposes of this chapter, the issues that arise out of the presentation of expert evidence in the common law legal systems will be examined. The rules that exist for civilian legal systems are quite different. Suffice to say that in civilian systems, there is no (or little) adversarial atmosphere, and the courts play a much heavier role in the preparation for and the running of the court case. This often translates, for the purposes of expert witnesses, to a list system, where experts are appointed from a list held by the court; expert witnesses may not be hired by the parties. This means that in most civilian jurisdictions there is little or no scope for conflicting expert evidence, one of the main features of a typical common law system. In terms of style, the common law systems (which are adversarial in nature) place the onus on the parties to extract the facts and to argue their case; the court simply decides the dispute on the basis of the information presented to it. In civilian systems (which are inquisitorial in nature) the court plays a significant role in the evidence presentation process, and even in some cases the evidence gathering process also, and so relies less on the work of the parties.

Finding a 'pure' common law system is no easy task, as most carry some features of a typical civilian jurisdiction, such as strong judge case management.

The main systems that come under the common law umbrella are: the United Kingdom (consisting of Scotland on one hand and England and Wales on the other), the United States (with the exception of Louisiana, which is a civilian system), Canada, Australia and some African nations, such as Nigeria and Ghana (for a general discussion of the legal systems of the world, see De Cruz 2007).

The legal rules regulating the use of expert evidence in court are broadly similar across these jurisdictions, and in this paper examples from some of them will be considered.

A number of issues arise, *viz.*

1. Why do experts need to know about the law at all?
2. Who may be an expert witness?
3. What is the purpose of expert evidence?
4. Is all expert evidence permitted in the courtroom?
5. How should expert evidence be presented?

The key focus will be on Question 4 since it is the one that is the most important question in relation to an emerging science such as soil forensics.

In considering how these issues arise, a wide range of cases covering a huge breadth of subject expertise are commented on. Since soil forensics is in its relative infancy, analogies have to be drawn from already decided cases dealing with other scientific fields. There have been, of course, some cases in which soil forensic evidence has been used, but there has been no reported case, known to the author, in which such expertise has been challenged from a reliability perspective in a published judgement. It is clear that the approach of appeal court judges in common law jurisdictions revolves around seeking some kind of objective justification for hearing expert evidence in a particular field. Normally, such justification is to be found in the existence of, for instance, a mathematical formula, or from the fruits of empirical research, such as a database. Given this more or less universal generic approach, it is possible (and in fact essential) to draw analogies from already accepted areas of expertise where that expertise, however well established in the scientific community, comes to being offered for the first time in a court.

Although most of the applications of forensic soil science arise in connection with criminal prosecutions, there are significant areas of application in civil cases, not least in cases involving environmental issues. It is clear that the general approach in most common law jurisdictions on the admissibility and treatment of expert evidence is similar in civil and criminal cases, and some examples of each are woven into this chapter.

Finally, it should be noted that this piece is not intended to be a comprehensive survey of the duties that are incumbent on an expert witness when appearing in court; these duties arise not only from the law, but also from codes of practice and ethics, some created by professional or public bodies around the world (e.g. Freckleton and Selby 2002[†]). While these codes are of importance in practice, they will often not carry determinative weight in court, as they are not formal sources of law, and hence they are not considered here.

The Importance of Knowledge of the Law

It could be argued that expert witnesses do not need to know anything about the legal requirements for giving evidence, as they are not lawyers, and court cases are presented by such professionals. However, there are three key advantages to the expert having some knowledge of the legal process.

Firstly, the giving of evidence can be a daunting experience (as noted above) and familiarity with some of the process might be helpful on the day, if nothing else to settle the nerves of the expert. In other words, it is relevant for background purposes.

Secondly, an understanding of some of the legal issues will allow the expert to prepare for and present his evidence in such a way that it is more likely than not to be accepted and understood. In particular in, in criminal cases, where a witness is giving evidence for the prosecution, the prosecutor will not always have the time to go through the evidence with the witness in advance of the trial in such a way that his testimony is bound to come across smoothly. In addition, the prosecutor will not necessarily have considered all lines of attack that might be employed by the defence lawyer. This can mean that an expert witness who is not versed in the likely lines of enquiry that a defence lawyer will employ, may not be equipped to respond in a coherent and convincing fashion. Where there is an expert employed by the defence, some advance notice of the line to be taken during the questioning of a prosecution expert exists, but this will not prevent a defence lawyer from asking unexpected questions that seek to undermine the evidence being given.

Thirdly, those preparing scientific techniques for the discovery and analysis of material will need (with a view to presenting those techniques in court) to know the kinds of questions the courts will want the answers to, come the day the technique is paraded in front of the judge or jury. This is a particularly important consideration where the field of endeavour involved is novel or relatively underdeveloped or unknown; the courts (and lawyers) are much more likely to question such techniques more closely when compared to well established ones. But the importance of knowing the way a judge's (or juror's) mind works does not stop there; often even in a well

[†] c.f. Chapter 23 for a survey of some of these across a number of common law jurisdictions.

established scientific field, there can be rapid advances in the development of the underlying science, and any development is unlikely to meet with easy approval from a sceptical court, particularly where such scepticism is nurtured by a combative, awkward lawyer. In addition, no established scientific subject is beyond the clutches of the law; examples of some previously well settled areas, such as DNA evidence and fingerprint evidence have recently come under close, and sometimes negative, scrutiny. Freckleton and Selby (2002) examine cases from the US, Canada, Australia and the UK on DNA evidence, and conclude by identifying 23 methodology problems highlighted by decisions of courts (c.f. pp. 541–542). Fingerprint evidence also poses problems, again recognised by Freckleton and Selby (2002; Chapter 15) but also in a real case of mistaken identification of a fingerprint, the Scottish case of Shirley McKie, leading to a report examining the whole fingerprint identification process in Scotland (Scottish Parliament Justice 1 Committee 2007).

For each of these reasons, the expert who is considering the presentation of scientific evidence should have an appreciation of the legal impediments and practical considerations inherent in the traditional adversarial courtroom.

Who May Be an Expert Witness?

A witness who has some knowledge that is outwith the knowledge of the 'trier of fact' (the judge in judge only cases or the jury) and which arises from one or more of the following: qualifications, training, skill, education and experience, is commonly known as an 'expert witness'. It is clear that in general in common law jurisdictions no formal qualifications or training is required. Some examples are: *R v Silverlock* (1894) – England; *White v HMA* (1986) – Scotland; *Weal v Bottom* (1966) – Australia; *R v Wald* (1989) – Canada; *Bales v Shelton* (1990) – USA. This will not normally be an issue in soil forensics cases, since the witness will have formal qualifications and training, but it should be borne in mind.

The terminology can be deceptive, and the proper way to describe such evidence is to refer to it as 'opinion evidence'. In fact, this term marks out a very important starting point in the treatment of expert evidence in adversarial systems.

As alluded to earlier, no one is permitted, as of right, to state an opinion in court. There are certain limited exceptions to this, but on the whole, a witness's opinion is irrelevant, and in fact the court will exclude it as inadmissible (impermissible) evidence. This is because witnesses are supposed to speak to facts, not their opinions. Facts generally comprise: what the witness heard, saw, felt or otherwise experienced. There is one major exception to this rule. Someone with the requisite qualifications, training, skill, education or experience in an area relevant to the issues in the case (an expert witness) is able to state his or her opinion to the court. The reason for labouring this point is that it is important to realise that the evidence of an expert witness is exceptional. For this reason, an expert witness has, technically, no automatic status in a courtroom and may have to persuade the court to allow him to give his opinion. This sets the tone for the treatment of expert testimony.

Of course, in real life, outside the courtroom, there are plenty of individuals who are glad to provide an 'armchair view' in a particular area of expertise. In a court, such a phenomenon is not tolerated. The reason is obvious: the opinion of a lay person holds no intrinsic value. There are some exceptions to this rule, but on the whole, lay witnesses can only speak to what they saw, heard, felt, experienced, not what they think is the case. The expert witness, on the other hand, usually comes to the case cold. In other words, the expert will not have witnessed the crime or event in question and will have had nothing to do with the occurrence of the incident itself; an expert witness is called in to examine the product of the crime/incident. This might involve a physical examination of a victim in a murder investigation or of the accused in an insanity case, or an assessment of the safety of machinery in a health and safety prosecution, or an analysis of a physical sample in the case of forensic evidence. Whatever the field of expertise, the expert is offering his opinion, hence the term 'opinion evidence'. It is pertinent, of course, to note that the expert must base his opinion squarely on the nature and content, as well as the context, of the factual evidence in the case.

The subject matter of the expertise underlying the evidence of an expert can vary widely, and the courts are generally receptive to expert evidence from any discipline; in theory there is no subject matter limitation at all. Some of the more recent additions to the expert armoury are mentioned later.

What Is the Purpose of Expert Evidence?

The purpose of expert evidence is to fill a gap in the knowledge of the trier of fact, where the trier of fact does not have the expertise to form its own opinion. However, in the context of the entirety of the evidence, the expert is just one witness. This role is explained by Lord President Cooper in the Scottish case of *Davie v Magistrates of Edinburgh* (1953):

> Expert witnesses, however skilled or eminent, can give no more than evidence. They cannot usurp the functions of the jury or Judge sitting as a jury…Their duty is to furnish the Judge or jury with the necessary scientific criteria for testing the accuracy of their conclusions, so as to enable the Judge or jury to form their own independent judgment by the application of these criteria to the facts proved in evidence. The scientific opinion evidence, if intelligible, convincing and tested, becomes a factor (and often an important factor) for consideration along with the whole other evidence in the case, but the decision is for the Judge or jury.

This theme is also adopted in England by Lord Justice Lawton in *R v Turner* (1974):

> The fact that an expert witness has impressive scientific qualifications does not by that fact alone make his opinion on matters of human nature and behaviour within the limits of normality any more helpful than that of the jurors themselves; but there is a danger that they may think it does.

So, the expert's role is to give his opinion within his area of expertise, but only where that expertise is required. At the end of the day, however, the evidence of the expert must be weighed with all other evidence in the case; legally, it does not carry any special status, but in practice, of course, it can be critical.

Although the passages above come from UK cases, they reflect the general approach across the main common law jurisdictions.

Is All Expert Evidence Permitted in the Courtroom?

As already mentioned, some expert evidence will be regarded as inadmissible (impermissible). In such cases, the evidence will not even reach the ears and eyes of the trier of fact. So, there may even be expert evidence which purports to directly implicate a suspect in the crime in hand. Where it is declared inadmissible, the jury will never know that it exists. In which circumstances can such an exclusion occur? Again, the main adversarial systems broadly agree on the parameters of the court's discretion, although there are some subtle differences in emphasis from one jurisdiction to another.

Before coming to look at the main admissibility challenges to expert evidence, it is pertinent to consider the procedure generally followed in such cases. The party seeking to have the expert evidence excluded may argue that there should be a '*voir dire*', (in some jurisdictions, for example in Scotland, this procedure is known as a 'trial within a trial') meaning that there should be a hearing before the judge only, and only to consider whether or not the expert evidence should be admitted. Of course, in all cases, the judge making that decision will require to know at least the general tenor of the evidence to be led, in order to decide if it will be admissible. In some cases, he will want to hear the expert's testimony (or a part of it) to fully appreciate the nature and content of what is being proposed. In a judge only case, this creates an odd situation where the judge hears all of the expert's evidence and then decides it is inadmissible and so excludes it; the judge has already heard it, but must pretend that he did not, and must leave that evidence out of account when reaching a decision in the case.

There are a number of arguments that could be made in order to seek to exclude expert evidence from the trier of fact. As indicated above, the evidence must be outwith the knowledge of the trier of fact. In addition, it might be argued that the witness does not have the necessary qualifications, training knowledge or experience in order to be able to state a reliable view on the issue in hand; in other words that the witness is not an expert witness at all (and so is not permitted to offer an opinion). Such arguments are rare and many of the arguments seeking to exclude expert evidence revolve around the reliability of the science underlying such evidence. This is particularly so with new or emerging scientific areas, such as soil forensics.

Evidence Must Have a Reliable Scientific/Technical Underpinning

The expert evidence must be reliable, and if not, the judge can disallow it, if necessary after hearing the evidence itself during a *voir dire*. Different common law jurisdictions have different tests, but they all have a similar thrust to them.

USA

In the USA, perhaps the most influential case internationally in this area is *Daubert v Merrell Dow Pharmaceuticals* (1993). This case was decided by the Supreme Court, the highest and most authoritative court in the US. The case is a decision on the application of the rules on admission of expert evidence in the Federal courts in the USA, namely the Federal Rules of Evidence, Rule 702 states:

> Rule 702. Testimony by Experts:
> If scientific, technical, or other specialized knowledge will assist the trier of fact to understand the evidence or to determine a fact in issue, a witness qualified as an expert by knowledge, skill, experience, training, or education, may testify thereto in the form of an opinion or otherwise, if (1) the testimony is based upon sufficient facts or data, (2) the testimony is the product of reliable principles and methods, and (3) the witness has applied the principles and methods reliably to the facts of the case.

In the case of *Daubert*, the US Supreme Court expanded on this test:

> Faced with a proffer of expert scientific testimony, then, the trial judge must determine at the outset....whether the expert is proposing to testify to (1) scientific knowledge that (2) will assist the trier of fact to understand or determine a fact in issue. This entails a preliminary assessment of whether the reasoning or methodology underlying the testimony is scientifically valid, and of whether that reasoning or methodology properly can be applied to the facts in issue. We are confident that federal judges possess the capacity to undertake this review. Many factors will bear on the inquiry, and we do not presume to set out a definitive checklist or test.

The court then goes onto discuss the following non-exclusive factors to be taken into account (as summarised in the 1993 case of *US v Martinez*):

(1) "Whether the scientific technique can be (and has been) tested.
(2) Whether the technique or theory has been subjected to peer review and publication. While not a *sine qua non* of admissibility, "[t]he fact of publication (or lack thereof) in a peer-reviewed journal thus will be a relevant, though not dispositive, consideration in assessing the scientific validity of a particular technique or methodology on which an opinion is premised."
(3) The known rate of error of the technique and the existence and maintenance of standards controlling the technique's operation.
(4) Whether the technique is generally accepted. "A reliability assessment does not require, although it does permit, explicit identification of a relevant scientific community and an express determination of a particular degree of acceptance within that community." Widespread acceptance can be an important factor in ruling particular evidence admissible, and 'a known technique that has been able to attract only minimal support within the community,' may properly be viewed with scepticism."

The *Daubert* formula has produced a whole field of litigation in the US, with over 40,000 cases citing the formula since 1993 (Westlaw legal database). Of course, US courts will routinely cite the case, but other jurisdictions have referred to it too, including Canada, Australia, New Zealand, the UK and Hong Kong. There are numerous examples of scientific techniques being examined under the *Daubert* microscope in the US, including estimates of the rate of production of methamphetamine (*US v Cavely* (1993)); illicit drug identification evidence (*US v Diaz* (2006)); ink dating techniques (*EEOC v Ethan Allen* (2003)); fingerprint evidence (*US v Llera Plaza* (2002)); toxicology in the case of chemical exposure (*Cavallo v Star Enterprise* (1996)); groundwater contamination (*Carroll v Litton Systems* (1995)); footprint comparison (*US v Ferri* (1985)). These are just a few examples. Cwik and North (2003) provide a comprehensive analysis of the application of *Daubert* across federal and state courts in the US, reviewing hundreds of cases of all types, shapes and sizes. It is clear that mathematical and statistical models attract heightened scrutiny under the *Daubert* formula, at least in part due to their propensity to over-impress (Kaye et al. 2004, page 250).

Australia

In Australia, the test is set out in s.76–80 of the Evidence Act 1995:
"76 The opinion rule:

(1) Evidence of an opinion is not admissible to prove the existence of a fact about the existence of which the opinion was expressed ...

79 Exception: opinions based on specialised knowledge:

> If a person has specialised knowledge based on the person's training, study or experience, the opinion rule does not apply to evidence of an opinion of that person that is wholly or substantially based on that knowledge."

This structure emphasises the role of expert evidence as being exceptional in nature, as noted above. In a decision that pre-dates the Evidence Act provision, but which is used as an elaboration of the duties of the trial judge, the following was said in the case of *R v Bonython* (1984) by Chief Justice King:

> Before admitting the opinion of a witness into evidence as expert testimony, the judge must consider and decide two questions. The first is whether the subject matter of the opinion falls within the class of subjects upon which expert testimony is permissible. This first question may be divided into two parts: (a) whether the subject matter of the opinion is such that a person without instruction or experience in the area of knowledge or human experience would be able to form a sound judgment on the matter without the assistance of witnesses possessing special knowledge or experience in the area, and (b) whether the subject matter of the opinion forms part of a body of knowledge or experience which is sufficiently organized or recognized to be accepted as a reliable body of knowledge or experience, a special acquaintance with which by the witness would render his opinion of assistance to the court. The second question is whether the witness has acquired by study or experience sufficient knowledge of the subject to render his opinion of value in resolving the issues before the court.

This formula was applied in *R v Murdoch* (No.2) (2005), a case involving DNA evidence. The *Daubert* test has been referred to in some Australian cases too, so clearly it holds some influence. Freckleton and Selby (2002), in examining the Australian authorities on expert evidence, make frequent reference to *Daubert* and its principles.

Canada

The four part test for admissibility has been set out by the Supreme Court for Canada in *R. v Mohan* (1993):

(a) Relevance
(b) Necessity in assisting the trier of fact
(c) The absence of any exclusionary rule
(d) A properly qualified expert

On relevance, the court said this:

> Evidence that is otherwise logically relevant may be excluded on this basis, if its probative value is overborne by its prejudicial effect, if it involves an inordinate amount of time which is not commensurate with its value or if it is misleading in the sense that its effect on the trier of fact, particularly a jury, is out of proportion to its reliability.

The court went on:

> There is a danger that expert evidence will be misused and will distort the fact-finding process. Dressed up in scientific language which the jury does not easily understand and submitted through a witness of impressive antecedents, this evidence is apt to be accepted by the jury as being virtually infallible and as having more weight than it deserves.

Again, the *Daubert* test features in some Canadian judgements. These cases, including *R v Mohan* (1993), cited above, and *R v J (J-L)* (1999), are examined by Glancy (2007) who concludes that:

> Canadian law has evolved in the area of expert evidence toward a much more stringent and analytical approach.

UK

In the UK, the *Daubert* formula is not widely referred to, although there are some signs of it creeping in, for example in *R v Dallagher* (2003). However, there is no indication of the *Daubert* formula being accepted wholesale into UK law, and in Scotland, there is no reference to it at all. The UK courts prefer a more general approach, and the exclusion of expert evidence as inadmissible has recently been described as 'exceptional' (see *Keane* (2006) at 563–564, citing a passage from *Cross and Tapper* (1999) at 523).

Synthesis

Although the wording differs from one jurisdiction to another, the overall thrust of admissibility is similar. If we were to draw together some of the specific headings under which the courts may allow a scientific technique to be examined from an admissibility point of view, the list might look something like this:

(i) Necessity of evidence (can the jury reach conclusions on the issue without it?)
(ii) The qualifications of the expert
(iii) The reliability of the science underlying the conclusions
(iv) Testing of the scientific technique
(v) Peer review history
(vi) Error rate
(vii) Acceptance within the scientific community
(viii) Is there a mathematical formula, probability statistic, database or some other objective touchstone

In the USA in particular, the courts have demonstrated a willingness to examine in detail each of these factors, and others, during a *voir dire*, in order to come to a final conclusion on admissibility. This can lead to extensive and detailed scrutiny, perhaps lasting several days, or even weeks. While there is a variance across jurisdictions in the rigour with which the courts approach admissibility examinations of expert scientific evidence (the USA at the most rigorous end and the UK at the other, with Canada and Australia somewhere in between), it would be best practice to assume a fully rigorous examination of the factors above when considering how to defend any proposed expert scientific evidence.

Admissibility of Novel Expert Evidence

The courts have shown a willingness to open the doors to evidence of all kinds, from any background of knowledge, not just the traditional scientific fields. There has been an explosion lately in the kinds of evidence that are presented by experts in courtrooms. Of course, expert scientific evidence has found a place in courts for a long time, and some forms of expert evidence are commonplace and well established in court, including fingerprint evidence, DNA evidence, ballistics evidence and some forms of physical trace evidence. However, where the evidence involves a new technique, even where it is a variation in the area of an already established field, the courts will apply the same basic tests, but in some jurisdictions, they will proceed with extra caution. Some of the new areas of expertise are surprising, including footwear comparison, ear print identification, CCTV footage facial mapping, hypnosis, voice comparison analysis, hair analysis and psychological autopsy evidence (to consider whether someone is likely to have committed suicide or not). Given that soil forensics in many cases uses existing and accepted scientific techniques, but applying them to the analysis of soil, there would seem little room for an argument that soil forensics evidence in a general sense should not be presented as expert evidence. Of course, whether such evidence should be produced in a particular case is another question.

In the USA, given the rigour of the *Daubert* formula as it applies to any expert evidence, there is no sign of a more cautious approach in the case of novel scientific evidence. According to *Cross and Tapper* (Tapper 2004) and the Court of Appeal in *R v Dallagher* (2002), the same appears to be the case in the much less rigorous jurisdiction of England and Wales.

In Australia, again, the *Bonython* case deals with this:

> If the witness has made use of new or unfamiliar techniques or technology, the court may require to be satisfied that such techniques or technology have a sufficient scientific basis to render results arrived at by that means part of a field of knowledge which is a proper subject of expert evidence.

This does not look controversial, but it does signal a more cautious approach in cases where novel evidence is proposed. This was applied again in *R v Murdoch* (No.2) (2005), a case where certain DNA evidence was held to have passed the admissibility test. The *Bonython* court cites earlier examples of new evidence, such as *R v McHardie and Danielson* (1983), a case involving a new mathematical formula used in voice identification.

In Canada, again, in the landmark case of *R. v Mohan* (1994) the Supreme Court offers the following guidance:

> ... expert evidence which advances a novel scientific theory or technique is subjected to special scrutiny to determine whether it meets a basic threshold of reliability and whether it is essential in the sense that the trier of fact will be unable to come to a satisfactory conclusion without the assistance of the expert.

The court in that case approved the approach in an earlier decision (*R v Melaragni* (1992)) in which additional criteria where a new scientific area is the subject matter of the evidence were set out:

(1) Is the evidence likely to assist the jury in its fact-finding mission, or is it likely to confuse and confound the jury?
(2) Is the jury likely to be overwhelmed by the 'mystic infallibility' of the evidence, or will the jury be able to keep an open mind and objectively assess the worth of the evidence?
(3) Will the evidence, if accepted, conclusively prove an essential element of the crime which the defence is contesting, or is it simply a piece of evidence to be incorporated into a larger puzzle?
(4) What degree of reliability has the proposed scientific technique or body of knowledge achieved?
(5) Are there a sufficient number of experts available so that the defence can retain its own expert if desired?
(6) Is the scientific technique or body of knowledge such that it can be independently tested by the defence?
(7) Has the scientific technique destroyed the evidence upon which the conclusions have been based, or has the evidence been preserved for defence analysis if requested?

(8) Are there clear policy or legal grounds which would render the evidence inadmissible despite its probative value?
(9) Will the evidence cause undue delay or result in the needless presentation of cumulative evidence?

This list is not necessarily exhaustive; furthermore, the importance of any one or more of these factors will vary depending upon the particular circumstances of the case.

An excellent example of the treatment of a technique that was relatively recently regarded as 'novel' is that afforded to DNA evidence. One good reason for citing the DNA example is that DNA evidence is, of course, used extensively, especially in the criminal courts. This means that the courts in a number of common law jurisdictions have had to consider admissibility and reliability issues surrounding DNA evidence. This makes DNA evidence a good example of how the courts approach the development of a novel area of expertise. One lesson that does come out of the treatment of DNA evidence is that the courts are sceptical of allowing the introduction of complex scientific principles, particularly in jury cases. It is fair to say that judges have not sought to attack the science behind DNA evidence, rather that certain strictures have been put in place on how it is presented in court. The courts in England have, in particular, addressed these issues over the last 10 years or so. In a string of cases, the English Court of Appeal has considered very closely how such evidence is and should be presented. The main cases in this area are: *R v Adams* (1996), *R v Adams* (No.2) (1998) and *R v Doheny and Adams* (1997). Three important points have arisen from these cases. Firstly, DNA evidence must not be presented in such a way as to fall foul of the by now infamous 'prosecutor's fallacy'. In brief terms, this fallacy arises where evidence is given that, (for instance) only one person in a million within a particular population is estimated to have the same DNA profile as that of the accused (and as that taken from the sample in question) and where this is presented as the equivalent of concluding that the probability of the sample coming from someone other than the accused is one in one million. Taking the UK as an example, with roughly 26 million males living there (assuming the sample in question comes from a male who was in the UK at the time when the sample was deposited) the Court of Appeal has held that the chance of the sample originating from someone other than the accused is, in fact, one in 26 (see *R v Doheny and Adams* (1997), Lord Justice Phillips). The second of the three main issues tackled by the Court of Appeal on DNA evidence is the use of Bayes Theorem as a means of weighing up scientific and non-scientific evidence, in an attempt to mathematically assess the evidence in the case, with a view to reaching a probability figure in relation to the likelihood of the accused's guilt. The Court of Appeal had this to say:

> ... whatever the merits or demerits of the Bayes Theorem in mathematical or statistical assessments of probability, it seems to us that it is not appropriate for use in jury trials, or as a means to assist the jury in their task. In the first place the theorem's methodology requires...that items of evidence be assessed separately according to their bearing on the accused's guilt, before being combined in the overall formula...More fundamentally,

> however, the attempt to determine guilt or innocence on the basis of a mathematical formula, applied to each separate piece of evidence, is simply inappropriate to the jury's task. Jurors evaluate evidence and reach a conclusion, not by means of a formula, mathematical or otherwise, but by the joint application of their common sense and the knowledge of the world to the evidence before them....to introduce Bayes Theorem, or any similar method, into a criminal trial plunges the jury into inappropriate and unnecessary realms of theory and complexity deflecting them from their proper task. (Lord Justice Rose in R v Adams (1996), a view explicitly endorsed by Chief Justice Bingham in R v Adams (No.2) (1998) and by Lord Justice Phillips in R v Doheny and Adams (1997)).

What is interesting about this treatment of Bayes Theorem by the Court of Appeal is not that the theorem is, in itself, flawed, but that as a method of explaining the evidence before the court, it is inappropriate. This demonstrates the influence that non-confusing methods of presentation of evidence has on the attitude of the courts: no matter how sound the theory, it has to be easily presentable to a lay person, otherwise it has no place in a court. The third aspect of the case law in England on DNA evidence comes from the way in which the Court of Appeal has prescribed what can be said about DNA evidence by experts. In the leading case of *R v Doheny and Adams* (1997), the court set out no less than thirteen points of guidance for the prosecution, the expert witness and the judge, on how, in future such evidence should be presented/explained (and how it should not be). The court even sets out a specimen jury direction for future use. In general terms, this indicates that the courts are willing, even in relation to widely used and critical forms of scientific evidence such as DNA evidence, to prescribe how it is to be presented in order to reduce the risk of misrepresentation by the jury. There would seem no reason for the courts to take a less hands on approach in relation to any other 'novel' scientific evidence.

In another English case, *R v Gray* (2003), in the absence of evidence of a sufficient facial characteristics database, evidence derived from facial mapping was held to be unreliable; the court offered the following comments:

> No evidence was led of the number of occasions on which any of the six facial characteristics identified by [the expert] as "the more unusual and thus individual" were present in the general population, nor as to the frequency of the occurrence in the general population, of combinations of these or any other facial characteristics. [the expert] did not suggest that there was any national database of facial characteristics or any accepted mathematical formula, as in the case of fingerprint comparison, from which conclusions as to the probability of occurrence of particular facial characteristics or combinations of facial characteristics could safely be drawn. This court is not aware of the existence of any such database or agreed formula. In their absence any estimate of probabilities and any expression of the degree of support provided by particular facial characteristics or combinations of facial characteristics must be only the subjective opinion of the facial imaging or mapping witness. There is no means of determining objectively whether or not such an opinion is justified. Consequently, unless and until a national database or agreed formula or some other such objective measure is established, this court doubts whether such opinions should ever be expressed by facial imaging or mapping witnesses.

These comments could apply to any evidence, including forensic evidence, which relies on sampling, such as soil forensics. The court identified in *Gray* the absence of a mathematical formula or a database; this suggests that such objective underpinning of new areas of expertise must exist and be paraded in court. For a case where evidence of this nature was accepted, but where there was not the same attempt at a comparison exercise, see *R v Mitchell* (2005). Similar cases in England include *R v Gilfolyle* (2001) (absence of a database in a case involving 'psychological autopsies') and *R v Dallagher* (2003) (absence of proper statistical evidence in an ear print identification case; see also *R v Kempster* (2008) where such evidence was considered, but where there was not a sufficiently precise match).

Given that soil forensics is such a new and developing area, it seems prudent to consider that additional precautions in the gathering and presentation of such evidence is wise, since even in cases where the courts do not formally apply a higher standard for the admission of such evidence, they will always be more susceptible to arguments of unreliability in relation to new forms of evidence which they may never or barely have heard of.

Beyond Admissibility

If evidence that is challenged in a *voir dire* survives that challenge, even where this survival happens in the context of a full hearing where the expert evidence is led and cross-examined, the evidence may still be challenged once it is led in the main trial. Indeed, some or all of the challenges can be repeated. Why should this be so? The answer is easiest explained in the context of a jury trial. In such a case, the jury has not heard any of the evidence that was led during the *voir dire*, in which the judge has rejected the challenges to the evidence. The jury may be more sympathetic to such challenges. They will know nothing of the reasons of the judge for rejecting those challenges, and so cannot be influenced by them. It is also important to note that the jury does have a role in assessing the weight to be applied by them to expert evidence in the context of the evidence in the case as a whole. For example, significant weight may be placed on such evidence, or very little, or even none at all. Many of the factors listed above that might affect admissibility will also be relevant during the evidence weighing stage. This distinction between the admissibility and weight of evidence was highlighted in the case of *Carmichael v Kumho Tire Company* (1998), where the US Supreme Court indicated that evidence that passes the admissibility test might still be described as 'shaky'. Matson et al. (2004), in citing Mandell (1999) on this case explains that:

> … vigorous cross-examination, presentation of contrary evidence and careful instruction on the burden of proof are the traditional and appropriate means of attacking shaky but admissible evidence.

When one considers a judge only case, it is more difficult to see the split between admissibility and weight, but it still exists. So, a judge may rule that evidence is admissible, but that it carries little or no weight when he comes to assessing the evidence.

This seems like a double blow to the expert, allowing the party challenging the evidence to have two bites of the cherry. However, it is clear and logical that admissibility and weight are separate concepts.

How Should Expert Evidence Be Presented?

There is insufficient space here to discuss the presentation of such evidence in full, and the interested reader is referred to more comprehensive sources, such as: Matson et al (2004), which includes chapters on preparing for trial, the courtroom drama and on the business and art of expert witnessing; Cohen (2007), including chapters engagingly entitled "To do the courtroom dance, first learn the steps", "Skeletons in your closet" and "Impeachment is not just for presidents"; Redmayne (2001), including a chapter on presenting probabilities in court; and Freckleton and Selby (2002), which includes chapters on experts reports, examination in chief, and cross-examination of experts. Nonetheless, a few key generic points can be made.

The jury consists of laymen. Even in a judge only case, the judge is a layman as regards any area of expertise covered by expert testimony. With juries in particular, they will not be accustomed (as a judge is) to sitting through long, detailed and technical explanations in a language they do not understand. Concepts need to be explained in an interesting and brief, yet accurate way. This is a challenge. However, it is a critical one, as if the jury does not listen and absorb the evidence, the testimony will have been for nothing.

Essentially, a jury will not be particularly interested in the science behind the conclusions reached by the scientist; they are interested in what those conclusions are and how they apply to the case they are determining. Of course, the science is important, but unless it is under strong challenge on its reliability (during a *voir dire* or beyond the admissibility stage) it should probably be played down. One tactic which might enable that to be done effectively is to use everyday analogies to demonstrate the science.

The use of visual aids such as graphs, charts, DVD/video presentations, picture cards, computer simulations (on such simulations generally see Freckleton and Selby (2002), pp. 472–473) or even live demonstrations, might liven up the evidence without dumbing it down[‡]. However, any unusual method of presentation (anything other than straight oral testimony) should be brought to the attention of the other party to the case well in advance, so that approval can be sought from that party and, if necessary, from the court. There may also be logistical considerations involved in setting up technology or equipment in the court. Another possible technique involves the use of evidence arising from 'tailor made' scientific experiments that could be performed especially for the case in question; these may involve effectively reconstructing the incident in question, as happened in the Scottish case of *Campbell and others v HMA* (2004).

The presentation of evidence to a non-expert in a mock session is a good way to test a new technique.

In addition, the jury will expect an expert witness to present his evidence in an authoritative manner. An expert may have a very good knowledge of his subject area, but it can be just as important to present that knowledge confidently.

[‡] See also Freckleton and Selby (2002, pp. 728–733) on 'diagrams, slides and other forms of demonstrative exhibit'.

This all comes back to one key point: preparation. The better a witness is prepared to deliver his evidence and to answer questions (even the most basic of questions, indeed especially such questions) in a confident and interesting manner, the more chance there is that the jury (or judge) will give that evidence due consideration and weight in reaching a verdict.

Conclusion

The courts in common law jurisdictions are ever more willing to look behind expert scientific evidence. In the context of soil forensic evidence, this means that as it develops, scientists must work back from the admissibility hearing criteria discussed above, and justify their findings in the light of those criteria. In fact, as some scares in relation to what were thought to be well established scientific methodologies show, this need to work back from the likely questions in court applies even when using such methodologies.

In a nutshell: *be prepared for anything*!

Cases

Australia

R v Bonython (1984) 38 SASR 45; 15 A Crim. R. 364
R v McHardie and Danielson [1983] 2 NSWLR 733
R v Murdoch (No.2) [2005] NTSC 76
Weal v Bottom (1966) 40 ALJR 436

Canada

R v J (J-L) [1999] 130 CCC (3d.) 541
R v Melaragni (1992) 73 CCC (3d) 348; (1992) 76 CCC (3d) 78
R v Wald (1989) 47 CCC (3rd) 319
R. v Mohan [1994] 2 SCR 9; (1994) 89 CCC (3d.) 402

United Kingdom

Campbell and others v HMA 2004 SLT 397, 2004 SCCR 220
Davie v Magistrates of Edinburgh 1953 SC 34

R v Adams (No.2) [1998] 1 Cr. App. R. 377
R v Adams [1996] 2 Cr. App. R. 467
R v Dallagher [2003] 1 Cr. App. Rep. 195
R v Doheny and Adams [1997] 1 Cr. App. Rep. 369
R v Gilfolyle [2001] 2 Cr. App. Rep. 57
R v Gray [2003] EWCA Crim 1001
R v Kempster [2008] EWCA Crim 975
R v Mitchell [2005] EWCA Crim 731
R v Silverlock [1894] 2 QB 766
R v Turner [1975] 1 QB 834
White v HMA (1986) SCCR 224

United States

Bales v Shelton 399 SE 2d 78 (1990)
Carmichael v Kumho Tire Company 119 S. Ct. 1167 (1999)
Carroll v Litton Systems 47 F.3d 1164 (4th Cir. 1995)
Cavallo v Star Enterprise 100 F.3d 1150 (4th Cir. 1996)
Daubert v Merrell Dow Pharmaceuticals 509 US 579 (1993)
EEOC v Ethan Allen 259 F.Supp.2d 625; 61 Fed. R. Evid. Ser. 589 (2003)
US v Cavely 318 F.3d 987, 60 Fed. R. Evid. Serv. 1052 (1993)
US v Ferri 778 F 2d 985 (3rd Cir 1985)
US v Diaz (2006) WL 3642181
US v Llera Plaza 57 Fed. R. Evid. Ser. 983 (2002)
US v Martinez 3 F.3d 1191 (8th Cir. 1993).

References

Cohen KS (2007). Expert Witnessing and Scientific Testimony: Surviving in the Courtroom. CRC, Boca Raton, FL.
Cwik CH and North JL (2003). Scientific Evidence Review: Admissibility and Use of Expert Evidence in the Courtroom. American Bar Association, Chicago.
De Cruz P (2007). Comparative Law in a Changing World, 3rd edition. Routledge Cavendish, New York.
Freckleton I and Selby H (2002). Expert Evidence: Law, Practice, Procedure and Advocacy, 2nd edition. Lawbook Company, Pyrmont, New South Wales.
Glancy GD (2007). The admissibility of expert evidence in Canada. Journal of American Academy of Psychiatry and Law 35:350–356.
Kaye DH, Bernstein DB and Mnookin JL (2004). The New Wigmore: A Treatise on Evidence: Expert Evidence (updated 2008). Aspen, New York.
Keane A (2006). The Modern Law of Evidence, 6th edition. Oxford University Press, Oxford.
Mandell MS (1999). Kumho: Some Clarity – but not the Last Word – on experts. The Testifying Expert, June 1999 Vol. 3.

Matson JV, Daou SF and Soper JG (2004). Effective Expert Witnessing: Practices for the 21st Century, 4th edition. CRC, New York.
Redmayne M (2001). Expert Evidence and Criminal Justice. Oxford University Press, Oxford.
Scottish Parliament Justice 1 Committee (2007). 3rd Report Session 2 Inquiry into the Scottish Criminal Record Office and the Scottish Fingerprint Service, 15th February 2007.
Tapper C (2004). Cross and Tapper on Evidence, 10th edition. LexisNexis, London.

Chapter 3
Some Thoughts on the Role of Probabilistic Reasoning in the Evaluation of Evidence

Colin G.G. Aitken

Abstract Various aspects of the role of probabilistic reasoning in the evaluation of evidence are described. These include relative frequencies, discriminating power, significance tests and likelihood ratios, and comments on new developments to aid evidence evaluation. The relevance of all these concepts for soil evaluation is considered as appropriate. It is shown that a procedure based on the likelihood ratio emphasises that information from answers to two opposing and relevant questions needs to be considered. It is shown that the likelihood ratio is the factor which converts a prior odds in favour of a prosecution proposition into a posterior odds. The importance of considering evidence at various levels, source, activity and crime, of propositions are discussed. At present, it is not possible to develop models for likelihood ratios in soil analyses in a way that is available for the elemental analyses of glass fragments or the chemical analyses of drug samples. It is shown that the methodology models variability in characteristics in such a way as to account for variation both between source and within source so that the effect on the odds in favour of the ultimate issue can be measured on a continuous scale. Work that is necessary to be done in order to develop likelihood ratios is highlighted together with the difficulties that are particular to soil analyses.

Introduction

Evidence evaluation, in its theoretical development, requires consideration of evidence from two sources, one associated with the crime scene and one associated with a suspect. This consideration is made with reference to a background population from which estimates of measures of variation from items from the same source (within-source variation) and from items from the different source (between-source variation) may be derived.

C.G.G. Aitken
School of Mathematics, The University of Edinburgh, James Clerk Maxwell Building, The King's Buildings, Mayfield Road, Edinburgh EH9 3JZ, UK
e-mail: c.g.g.aitken@ed.ac.uk

K. Ritz et al. (eds.), *Criminal and Environmental Soil Forensics*

There are three situations of interest. First, the evidence at the crime scene originates at the crime scene and its origin is known. This would be the case for fragments of glass from a window broken in a case of burglary or for soil taken by a scene of crime officer from a crime scene. The evidence from the suspect is of unknown origin. For glass fragments found on his clothing, for example, they may or may not have come from the crime scene. For soil particles, they may or may not have come from the crime scene. The second situation is that evidence at the crime scene does not originate from the crime scene. For soil, foreign particles found at the crime scene may have been brought by the criminal. Evidence associated with a suspect has a known origin, perhaps as testified by the suspect, say from his garden. There is a third situation in which both the sample at the crime scene and from the suspect are of unknown origin. The situation in which both the soil sample found at the crime scene and the sample found on the suspect have origins which are truly known is not of interest statistically. For the first two situations, the evidence whose source is known is denoted as *control* evidence and the evidence of unknown source is denoted *recovered* evidence, though the term *questioned* evidence (Pye, 2007) is also in use. In this third situation, both samples would be recovered evidence. The existence of the soil and its quantity are evidence and can be part of the evaluation. However, comments in this paper relate to measurements of characteristics of the soil such as particle size or elemental composition. Other sources of evidence to be considered include *comparison* samples and *reference* samples (Pye, 2007). For example, in the first case above, soil particles on items of clothing of the suspect, other than those for the recovered evidence, would be comparison samples. Also, the recovered sample may be compared with examples of types held in archive collections; these are reference samples.

Two likelihoods are developed in evidence evaluation: one the likelihood of the evidence from the two sources, if the sources were the same; and the other the likelihood of the evidence if the two sources were different. The ratio of these likelihoods is derived. The value of the evidence is then summarised as being so many times more likely if the evidence came from the same source than if it came from different sources (if the likelihood ratio is greater than 1) with an analogous statement if the likelihood ratio is less than 1.

An investigation into the possibilities of prediction for the provenance of a soil sample for forensic purposes with reference to a spatial database is described in Lark and Rawlins (2008). They develop a statistical approach based on a spatial likelihood function to aid the identification of a provenance which they show has 'considerable potential'. However, they admit that it is still a problem, common to any forensic inference from soil, to relate the variability of reference material collected in an experimental context to forensic specimens whose support is unknown and uncontrolled. The results of microscopic analyses of three samples from different locations by experts is described in Rawlins et al. (2006) who report that the 'collective interpretation was very effective in the assessment of provenance for two of the three sites where the mineralogy and plant communities were distinctive'. For the third site, 'Carboniferous spores from domestic coal were initially interpreted as deriving directly from bedrock'.

Work described in Lark and Rawlins (2008) and in Rawlins et al. (2006) considers the source of one sample, the recovered sample, and compares it to a database to try and identify its source. For evidence evaluation, the comparison is of the two samples with reference to the database, which supports interpretation of variability between, and within, sources. These variabilities are not readily available in soil analyses, and the development towards practical implementation of statistical methodology for evidence evaluation, in this context, requires the development of studies by which they can be obtained.

General principles for evidence evaluation are described, with comments on their relevance for the evaluation of soil evidence and for particular difficulties in relation to soils which need to be considered in such an evaluation.

Terminology and Notation

Certain words, phrases and notation are given here to aid understanding of the underlying ideas in the use of probabilistic reasoning in forensic science. A basic idea is that of the probability of an event. In forensic science, consideration is given to propositions and to evidence, both of which are considered as events to which probabilities may be attached.

Propositions may be put forward by prosecutors and by defenders. Examples include the defendant is guilty (innocent), the defendant was (was not) at the scene of the crime and soil found on clothing of a suspect came (did not come) from the crime scene. The prosecution's proposition is denoted H_p and the defence proposition H_d. These propositions need not be complementary.

Evidence is denoted E. Measurements on the control evidence are denoted E_c. Measurements on the recovered evidence are denoted E_r. All probabilities are conditional. The probability of an event is conditioned on background information, I, about the rest of the world as known by the person who is determining the probability. The separation of an event for which the probability is wanted and the events on which it is conditioned is denoted with a vertical bar |, the former coming to the left of, or before, the bar and the latter coming to the right of, or after, the bar. Examples include:

- $Pr(H_d \mid I)$, the probability of the truth of the defence proposition, given background information
- $Pr(E \mid I)$, the probability of evidence, given I
- $Pr(H_d \mid E; I)$, the probability of the truth of the defence proposition given E and I
- $Pr(E \mid H_d; I)$, the probability of E given the truth of the defence proposition and I. Often, I is omitted for ease of notation but it should not be forgotten

Fallacies

A small value for the probability of the evidence if the person on whom it is found is innocent does not imply a large value for the probability that a person is guilty if the evidence is found on them. The belief that these two probability statements are equivalent is known as the prosecutor's fallacy (Thompson and Schumann, 1987) or the fallacy of the transposed conditional.

A related fallacy is the defence fallacy (Thompson and Schumann, 1987). Given a probability, p, for $Pr(E \mid H_d)$, the defence fallacy is to multiply p by the size of a relevant population, N say, to obtain the expected number of people, Np, with E in that population. It is then argued that a large value of Np renders the evidence meaningless as the defendant is only one amongst approximately Np people. However, before the evidence was presented, the defendant was one of N people. The evidence has reduced the pool of potential criminals from N to approximately Np and is therefore of some worth. An example to illustrate these fallacies has been provided by Darroch (1987).

Consider a town in which a rape has been committed. There are 10,000 men of suitable age in the town of whom 200 work underground at a mine. Evidence is found at the crime scene from which it is determined that the criminal is one of the 200 mineworkers. Such evidence may be traces of minerals which could only have come from the mine. A suspect is identified and traces of minerals, similar to those found at the crime scene are found on some of his clothing. The event that 'mineral traces have been found on clothing of the suspect which is similar to mineral traces found at the crime scene' is the evidence E. In this context H_p is used to denote the proposition that the suspect is guilty and H_d the proposition that he is innocent. These figures are illustrated in Table 3.1.

It is assumed that all people working underground at the mine will have mineral traces similar to those found at the crime scene on some of their clothing. This assumption is open to question but the point about conditional probabilities will still be valid. The probability of finding the evidence on an innocent person may then be determined as follows. There are 9,999 innocent men in the town of whom 199 work underground at the mine. These 199 men will, as a result of their work, have this evidence on their clothing, under the above assumption.

Thus, $Pr(E \mid H_d) = 199/9{,}999 \approx 200/10{,}000 = 0.02$, a small number. The ratio is determined from the column labelled 'H_d' in Table 3.1. However, there are 200 men in the town with the evidence (E) on them of whom 199 are innocent (H_d). Thus $Pr(H_d \mid E) = 199 = 200 = 0{:}995$. The ratio is determined from the row labelled 'Present (E)' in Table 3.1. This illustrates that a small probability of finding the evidence on an innocent person, $Pr(E|H_d)$ in this context, is not the same as a small probability that a person on whom the evidence is found is innocent, $Pr(H_d|E)$.

Table 3.1 Numbers of men of suitable age in the town in which a rape has been committed in each of four categories for presence or absence of the evidence of mineral traces and the propositions of guilt (H_p) or innocence (H_d)

	Propositions		
Evidence	H_p	H_d	Total
Present (E)	1	199	200
Absent	0	9,800	9,800
Total	1	9,999	10,000

However, before the evidence was found the suspect was one of 10,000 people. After the evidence has been found, the suspect is one of 200 people. Thus, the evidence is of value in that it has reduced the size of the population to which the criminal belongs from 10,000 to 200 people.

Probabilities Used for Evaluation of Evidence

Relative Frequencies

The *relative frequency* of a characteristic in a sample is the number of members of the sample in which the characteristic is present divided by the total number of members of the sample. It is used as an estimate of the probability of the characteristic being present in a member of the population of which the sample is representative. Occasionally, a subjective estimate of this probability, rather than a figure derived from observations on a sample, has been used in evidence. The smaller the relative frequency, the stronger the support provided by the evidence for a prosecution proposition that associates a suspect with a crime.

The relative frequency has been used for evidence evaluation since at least the mid-nineteenth century (Meier and Zabell, 1980). Later examples in which the relative frequency of a characteristic in some relevant population was used as evidence include the Dreyfus case, in which the importance of the relative frequency of certain letters of the French alphabet in documents written by Dreyfus was an issue (Darboux et al., 1908) and the *Collins* case (Fairley and Mosteller, 1977) in which subjective estimates of the frequency of occurrence of certain characteristics of a couple that committed a mugging were part of the evidence against a couple who matched those characteristics. In the present day, cases involving DNA profiling also use a relative frequency.

Discriminating Power

Discriminating power provides a measure of the general performance of an evidential type. It does not provide a measure for the evidential value in a particular case. An example of its use for soil discrimination is given in Dudley and Smalldon (1978). The discriminating power of a particular analytical test is defined as the probability that samples from different sources, taken at random from a relevant population, would be discriminated by the test (i.e. recognized as coming from different sources). A high value of discriminating power, that is one that is close to 1, is indicative of a method that is effective at discriminating between samples from different sources.

A general approach for the assessment of the performance of any method of evidence evaluation is given by Dudley and Smalldon (1978). The approach determines the number of false positives (a decision that recovered and control evidence have the same source when they do not) and the number of false negatives (a decision that recovered and control evidence have different sources when they do not). Small values for these indicate a good method of discrimination.

Dudley and Smalldon (1978) determine two statistics for a comparison of two soil samples. The first, the *index of variability, IV*, is the ratio, expressed as a percentage, between the difference and the sum of the two total volumes, per unit weight, of the soil samples. Once these are calculated, the distributions of particle diameters are normalised to enclose unit area by dividing by total volume and the shapes of these distributions compared. Cumulative distributions, plotted as the fraction of the sample over a particular particle diameter, were compared. For samples from the same source it was noted that the *maximum difference, MD* (i.e. the greatest difference found between the fractions of the samples greater than a particular particle diameter for a given size) was in general small compared with the *MD* for comparisons involving two samples from different sources. The two statistics, *IV* and *MD*, were then used together for comparisons.

A study was carried out with 18 pairs of soil samples, 18 different sources and two samples from each source. There were thus 18 within-source comparisons and 153 ($18 \times 17/2$) between-source comparisons. A bivariate normal distribution was fitted to the data of 18 (*IV;MD*) same-source pairs and then 95% and 99% confidence contours were determined for this distribution. The 153 (*IV; MD*) different-source pairs were then plotted on the same axes. For a 5% risk of misclassifying true same-source pairs as coming from different sources, 5 of the 153 (or 3%) different-source pairs lay within the 95% contour. For a 1% risk of misclassifying true same-source pairs, 9 of the 153 (or 6%) different-source pairs lay within the 99% contour. These correspond to discriminating powers of 97% and 94% respectively.

Dudley and Smalldon (1978) attach many caveats to their study. There have been many scientific advances in the last 30 years which make other forms of analyses more appropriate now, especially if the interest lies in the value of evidence in a particular case. However, their study provides a very good illustration of the determination of evidential value when there is no reference database.

Significance Probability

A significance probability (P or P-value) provides a measure of compatibility of observations with a proposition. As an example in forensic science consider continuous data, such as measurements (E_c and E_r) of particle sizes in control and recovered samples, respectively, of soil. The proposition is that E_c and E_r come from the same source. The compatibility of the observations with the proposition is measured by considering the statistic, known as a t-statistic, formed from the difference in the mean measurements on E_c and E_r, relative to the estimated standard error of the difference, a procedure known as a t-test. The larger the value of the statistic, the more incompatible the data are with the proposition that E_c and E_r come from the same source. The significance probability is the probability of obtaining a value of the statistic as large as, or larger than, that observed, if the proposition is true. The proposition of a common source is then equivalent to a proposition that the means of the populations from which the control and recovered evidence come from are equal.

Often the population variances for the control and recovered measurements are assumed to be equal, even when they come from different sources. With the advent of the widespread use of statistical computer packages, it is now relatively straightforward to determine significance probabilities even though the population variances are not assumed equal. A small *P*-value suggests that the observed difference is incompatible or inconsistent with a zero value for the difference in means of the distributions from which E_c and E_r have been assumed to come. Certain conventional values of the significance probability are taken to suggest that the proposition of zero difference is false. These conventional values include 0.10, 0.05. 0.01. Thus, an observed difference with a *P*-value that lies between 0.05 and 0.01 is said to be *significant at the 5% level*. It is not significant at the 1% level. A difference with *P* less than 1% is significant at both the 5% and 1% levels. Occasionally, a result is simply said to be *significant*, with no mention of the level.

It is a matter of subjective judgement as to when to decide to act as if the original proposition is false. It may be decided to do so if the significance probability is less than 0.05. If it is decided to act as if the original proposition is false, then that proposition is said to be rejected. If the decision is to do so if *P* is less than 0.05 then the proposition is said to be rejected at the 5% level. Note that rejection at the 5% level is to reject a proposition because the statistic is such that the probability of a value as large as, or larger than, the observed value is less than 0.05, i.e. that value will occur less than one time in 20 that the experiment (of comparing two groups) is conducted, if the proposition is true. It is expected that about once in every 20 independent experiments with a 5% level the proposition will be rejected when in fact it is true and thus rejected in error.

Such a rejection is known as a type 1 error. The alternative, type 2, error is to fail to reject a proposition when in fact it is false. This error would be to decide to act as if E_c and E_r came from the same source when, in fact, they did not. A significance probability is not the probability that the proposition is true, given the value of the associated statistic. A small value for the significance probability does not mean a small value for the probability of the proposition of common source.

The significance probability is the output of the first (comparison) stage of a two-stage process (Parker and Holford, 1968) which models some evaluative processes used by forensic scientists. If the outcome of the comparison is significant (in the statistical sense) then E_c and E_r are deemed to have come from different sources. If the outcome is not significant then E_c and E_r are deemed similar and perhaps to have come from the same source. It is only 'perhaps' since there may be a type 2 error. This leads to an effect known as the 'fall-off-the-cliff' effect (credited in Evett, 1991, to Smalldon). A comparison which is significant at the 4.9% level leads to an exclusion of the evidence, one which is significant at the 5.1% level leads to the continuation of the evidence in the investigative process.

The second stage is one of assessment of the rarity of the similarity. If the similarity is that of a rare characteristic, then the evidence is stronger than if it is of a common

Table 3.2 Summary statistics for the particle size measurements (in μm) for means of four groups of samples taken from a spade and six groups of samples taken from a car (Derived from Pye 2007)

Origin of groups	Number of groups	Mean particle size	Standard deviation
Spade	4	35.7	7.6
Car	6	28.7	5.0

characteristic. However, it is not obvious how such strength of evidence may be assessed numerically and combined with other evidence.

An example of the use of a significance probability in the context of soil evidence is given in Pye (2007). Table 3.2 gives means and standard deviations of the sizes (in μm) of particles in samples taken from a spade and from a car associated with a murder suspect. The significance probability of a *t*-test (not assuming equal variances) is 0.180. The proposition being tested is that the soils on the spade and from the car came from distributions with the same mean size. This proposition is not rejected at the 10% level. The conclusion is that there is insufficient evidence (at 10% level) of a difference in the population means of the sources (spade and car) of the two samples. This result provides a measure of the similarity of the data from the two groups of samples.

The proposition of equal means acts as a surrogate for a proposition of same source or origin and is not rejected. However, a measure of the rarity of the data from the spade and from the car (the second stage of evidence assessment) is not provided. Reference data that assess the variability in the measurements over some reference population of soil samples are needed in order to determine the rarity. Such data are not always easy to find as it can be problematic to define a reference population from which samples may be taken. The identification of a reference population is a matter of considerable debate in the context of evidence of soils. If such data are available, then a statistic known as the likelihood ratio may be calculated. This is a statistic which takes account of both similarity and rarity in one output and provides a very satisfactory measure of the value of evidence.

Likelihood Ratios

In a slight abuse of terminology, the following discussion refers to 'odds but H_p and H_d need not be complementary propositions. The approach to the evaluation of evidence using a likelihood ratio has, as its basis, the assumption that only two probabilities are of importance, the probability of E if H_p is true and the probability of E if H_d is true. With this assumption, it can be shown that the only way that evidence may be evaluated is as a function of the ratio of these two probabilities. Sometimes these probabilities are known as likelihoods, hence the name 'likelihood ratio'. The use of the word 'likelihood' is particularly appropriate when E is a set of measurements and the probabilities are replaced by probability density functions. The so-called odds form of Bayes Theorem states that

$$\frac{\Pr(H_p|E)}{\Pr(H_d|E)} = \frac{\Pr(E|H_p)}{\Pr(E|H_d)} \times \frac{\Pr(H_p)}{\Pr(H_d)}$$

There are three ratios in this expression. First, consider

$$\frac{\Pr(H_p)}{\Pr(H_d)}$$

This is the prior odds in favour of H_p. In the situation in which it is the odds in favour of a proposition, H_p, that the defendant is guilty before any evidence is presented, the prior odds is the value associated with the dictum *innocent until proven guilty*. The next ratio

$$\frac{\Pr(E|H_p)}{\Pr(E|H_d)}$$

is the ratio of the probability of the evidence if the prosecution's proposition is true to the probability of the evidence if the defence proposition is true, the ratio known as the likelihood ratio and denoted V by Aitken and Taroni (2004).

The third ratio

$$\frac{\Pr(H_p|E)}{\Pr(H_d|E)}$$

is the posterior odds in favour of the prosecution's proposition after E has been presented. These odds are associated with the burden of proof with which a court has to find in favour of the prosecution (in a criminal case) or a plaintiff (in a civil case). Note that the above discussion uses the word 'odds' to describe the right-most and left-most ratios. However, the argument still holds if H_p and H_d are not complementary events.

The likelihood ratio is the factor which converts the prior odds into the posterior odds. It takes values from 0 to infinity (∞). This is in contrast to probabilities which take values between 0 and 1. A value V between 1 and ∞ means that the posterior odds are greater than the prior odds. The evidence is said to support the prosecution's proposition. A value between 0 and 1 means that the posterior odds are less than the prior odds. The evidence is said to support the defence proposition. A value of V close to 1 suggests the evidence is of little value (and hence of little relevance in the context of the US Federal Rule of Evidence 401, for example).

Measurements

Much evidence is of a form in which measurements, such as the elemental compositions of soil, may be taken and for which the data are continuous. Evidence $E = (E_c, E_r)$ is then denoted by (x,y) where x is the control evidence E_c and y is the recovered evidence E_r. The two competing propositions are denoted by H_p and H_d. The value V can be shown formally to be

$$V = \frac{f(x, y|H_p)}{f(x, y|H_d)} = \frac{f(x, y|H_p)}{f(x|H_d)f(y|H_d)} \tag{1}$$

where probabilities Pr are replaced by probability density functions f. The formulae for the likelihood ratio may be developed for the situation in which there are two levels of variation. The first is for variation within a source and the second is for variation between sources. For example, with soil, there is variability between measurements of elemental composition from fragments of soil from the same source and there is variability between measurements of elemental composition from fragments of soil from different sources. The likelihood ratio provides a statistic which assesses both within- and between-source variability.

Considerations of Evidential Source

There are various places in which soil may be found in association with a crime. First, there may be soil at the site at which the body of a murder victim is found. This soil may have its origin at that site or it may have been brought there from elsewhere, either in the commission of the murder, or in a perfectly innocent manner. This deposition site may not be the only location relevant to the murder. In a case involving the transportation and deposition of the victim, soil may have come from the place at which the murder was committed which is then also part of the same crime.

Second there may be soil from some aspect of the environment of a suspect. This soil too may have its origin at that environment. Alternatively, it may have come from somewhere else, such as one or more possible crime scenes as in the example in the previous paragraph, or from some perfectly innocent source.

The data may be measurements of various aspects of the particles. They may be the results of elemental analysis where there can be many elements, in which situation a multivariate statistical problem with high dimensions is obtained.

Evaluation of the evidence is determined by calculating the ratio of the likelihoods of the data under each of two propositions, that of the proposition and that of the defence. As suggested above, there can be transfer of soil in several directions: from a crime scene to the criminal, from the criminal to a crime scene, to or from an innocent source from or to the crime scene or the suspect (or any combination of all of these).

These issues of transfer, persistence and recovery need to be considered. Given proposed circumstances of the crime and the elapsed time since its commission, the likelihoods of soil being transferred from some source, persisting on the object to which it has been transferred and being recovered are of interest. There are uncertainties associated with all of these issues. The transfer of soil and its persistence on the object (e.g. body, clothing) to which it has been transferred will depend on many factors of which soil scientists will be aware. The probabilities of the transfer and of the persistence are measures associated with these uncertainties. These are known as *subjective* probabilities in that they are developed by, and are personal to, the experts involved in the analyses of the evidence. Similarly, the probability that a soil sample is recovered is a subjective probability. The theory which has been developed concerning evidence evaluation explains which factors should be considered in the study of the evidence and how the associated probabilities may be included formally in the evaluation.

Levels of Proposition

There are three levels of proposition, source, activity and crime (Cook et al., 1998a, b, 1999). The characteristic (denoted Γ) of the soil used as evidence in this discussion is assumed to be categorical with known relative frequencies for the categories in some relevant population; the relative frequency of the particular characteristic observed is denoted γ. Likelihood ratios are developed for each of the three levels of proposition. A suspect has been identified for a crime in which it is expected there may have been transfer from the scene to the criminal. There are particles of soil on clothing of the suspect. Measurements of characteristics, say elemental concentrations, of the soil on the clothing of the suspect show similarities in some sense to soil found generally at the crime scene.

Source Level

This level is concerned with the source of the soil on the clothing of the suspect. It is not concerned with how the soil came to be there or whether a crime was committed or not. The simplest form of the likelihood ratio in this context is the reciprocal of the relative frequency of the soil characteristics, if this can be evaluated. The propositions, known as *source level* propositions, to be compared are $H_p(H_d)$, the soil on the suspect's clothing came (did not come) from the crime scene; Evidence E, may be divided into two parts, E_c for the control (c) sample of soil and E_r for the soil on the suspect's clothing. This is the recovered (r) sample. It is assumed that all soil samples are classified without error into well-defined categories. For H_d, the two pieces of evidence, E_c and E_r, are independent. The probability that both are of the observed category for Γ is γ^2. If H_p is true, the control and recovered samples are from the same source; the probability of this category for Γ is γ. The likelihood ratio V is then

$$V = \frac{\gamma}{\gamma^2} = \frac{1}{\gamma} \tag{2}$$

The probability γ is sometimes known as the *random match probability* or *random occurrence ratio* and it, or its reciprocal, is the figure often quoted in court.

Activity Level

This level is concerned with how the soil reached the suspect's clothing. It is not concerned with whether a crime was committed or not. The propositions are $H_p(H_d)$, the suspect was (was not) in the vicinity of the crime scene. These are *activity level* propositions. The soil on the suspect either came from some background source other than the crime scene (with a probability denoted t_0) or the soil was transferred during the commission of the crime (with probability t_1). For ease of exposition, the possibilities of secondary transfer or of mixed samples are not considered here. The probabilities that a person from the relevant population will have no soil or one sample of soil on his clothing from background sources are denoted b_0 and b_1, respectively.

In the numerator of the likelihood ratio, there are two mutually exclusive possibilities to consider, the soil on the suspect's clothing came from some background source and it has, coincidentally, similar characteristics of the particles of soil at the crime scene, or the soil was transferred while at the crime scene. The First of these possibilities has probability $t_0\gamma b_1$. The second has probability $t_1 b_0$. There is no γ term in the second possibility because it is assumed that the characteristics of the soil from the crime scene are those of the soil on the suspect's clothing (H_p true). The probability of the evidence is then:

$$t_0\gamma b_1 + t_1 b_0$$

In the denominator, the suspect is assumed not to have been in contact with the crime scene. The soil on the suspect's clothing came from some background source (with probability 1) and, coincidentally, has characteristics with frequency γ. Thus the probability is:

$$\gamma b_1$$

and

$$V = t_0 + \frac{t_1 b_0}{\gamma b_1} \tag{3}$$

(Evett, 1984). In forensic contexts involving body fluids and DNA profiling, t_0 (which is a probability and hence no greater than 1) is small in relation to $t_1 b_0 = \gamma b_1$ because of the magnitude of γ and may be considered negligible. The probability t_0 may not be negligible in the context of soil analysis if γ is not small. An example in Aitken and Taroni (2004), concerning the transfer of stains of body fluid, shows circumstances in which the value for the activity level can be very different from $1/\gamma$ as given at the source level.

Crime Level

The propositions to be considered are $H_p(H_d)$, the suspect committed (did not commit) the crime. The concepts which have to be included at the crime level are those of innocent acquisition, p, and relevance, r. Innocent acquisition refers to transfer to or from a crime scene from or to a victim in an innocent manner. Relevance concerns the choice of soil samples from a selection of soil samples on the suspect's clothing. Some, all or none of the samples chosen for analyses may have been transferred there from places other than the crime scene and therefore are not of relevance.

The value, V, may be shown (Aitken and Taroni, 2004) to be equal to:

$$V = \frac{r + \gamma(1-r)}{\left[\gamma r + \{p + (1-p)\gamma\}(1-r)\right]} \tag{4}$$

Notice that when $r = 1$ and $p = 1$, $V = 1/\gamma$. The result at the source level is a special case of the result at both the activity level, where transfer and background possibilities are considered, and the crime level where innocent acquisition and relevance are considered.

Requirements for Evidence Evaluation of Soil

There are various factors relating to the evaluation of the evidence of soil which require to be considered before a method of evaluation based on likelihood ratios may be implemented. Some of these have been introduced above.

- Reference data are needed to enable the estimation of between-source variability. Appropriate datasets already exist for elemental concentrations of glass or for chemical concentrations of drugs.
- If no reference data are available, there should be an attempt to estimate subjectively the probability distribution and its parameters for the variability in the characteristics of the soil.
- It is not obvious what is meant by source in a continuum. The determination of between- and within-source probability distributions is difficult. Variability between sources through the choice of a reference data set in relation to a crime scene may need to be done anew for each case.
- If possible, more than one sample of soil from a crime scene and from a suspect should be taken to estimate within-source variability.

There are many difficulties to be overcome before procedures using likelihood ratios will be developed for soil analyses in a forensic context. It needs to be remembered, however, that this methodology does not pretend to provide the 'true' probabilities. Even in its current state of development it is an effective method by which opinions may be analysed and criticised. They can be checked for coherence and revised if necessary. The procedure emphasises that information from answers to two opposing and relevant questions needs to be considered. It helps clarify exactly what questions need to be answered and how the questions should be formed. As the methodology develops, the scientist will move correspondingly towards a position in which a coherent summary of an inconclusive result can be given, ranging from very strong support for the prosecution (a large value of V) through irrelevance to very strong support for the defence (a small value of V, much less than 1). The methodology models variability in characteristics in such a way as to account for variation both between source and within source in such a way that the effect on the odds in favour of the ultimate issue can be measured on a continuous scale.

The Future

The role of probabilistic reasoning in the administration of justice is set to increase. There is a paradigm shift in this direction in forensic identification science (Saks and Koehler, 2005). This is aided by the increasing ability to collect, analyse and interpret data of relevance to the forensic context. The advent of DNA profiling in the 1980s acted as a catalyst for this shift. It illustrated to the courts the advantages of a rigorous probabilistic approach to evidence in which there was uncertainty. Identification by fingerprints is coming under increasing scrutiny, partly because of some high-profile misidentifications, and there is considerable research on statistical models for fingerprint characteristics from which probabilities of similarities may one day be able to be derived (e.g. Neumann et al., 2007).

Developments of graphical and visual methods of analysing evidence are also far advanced. Networks are being developed, where nodes represent separate items of evidence, edges represent links between items of evidence and conditional probability tables represent the strength of the links. Also, the time is not far off when decision theory will be able to offer assistance in the administration of justice through helping jurists consider more formally what the consequences of their decisions will be. There has already been some work in this area with the case, assessment and interpretation models developed by Cook et al. (1998a, b, 1999) and with the ideas of Taroni et al. (2005, 2006). Finally, work by Lark and Rawlins (2008) on an application of spatial statistics to the determination of the provenance of a soil sample through the creation of spatial databases also holds promise.

It is to be hoped that the ideas expressed in this chapter, the inspiration offered by the recent work on fingerprints and the opportunities offered by work on graphical models and decision theory will lead towards a more rigorous evaluation of evidence based on the characteristics of soil and on well-developed statistical analyses.

Acknowledgement Support is acknowledged of ESRC grant RES-000–23–0729 and EPSRC grant EP/C532627/1.

References

Aitken CGG and Taroni F (2004) Statistics and the evaluation of evidence for forensic scientists, John Wiley & Sons, Chichester.

Cook R, Evett IW, Jackson G, Jones PJ and Lambert JA (1998a) A model for case assessment and interpretation. Science and Justice 38:151–156.

Cook R, Evett IW, Jackson G, Jones PJ and Lambert JA (1998b) A hierarchy of propositions:deciding which level to address in casework. Science and Justice 38:231–239.

Cook R, Evett IW, Jackson G, Jones PJ and Lambert JA (1999) Case pre-assessment and review of a two-way transfer case. Science and Justice 39:103–122.

Darboux JG, Appell PE and Poincaré JH (1908) Examen critique des divers systœemes ou etudes graphologiques auxquels a donne lieu le bordereau. In: *L'affaire Drefus – La revision du procµes de Rennes – enquete de la chambre criminelle de la Cour de Cassation*. Ligue francaise des droits de l'homme et du citoyen, Paris, France:499–600.

Darroch J (1987) Probability and criminal trials; some comments prompted by the Splatt trial and The Royal Commission. The Professional Statistician 6:3–7.

Dudley RJ and Smalldon KW (1978) The comparison of distributional shapes with particular reference to a problem in forensic science. International Statistical Review 46:53–63.

Evett IW (1984) A quantitative theory for interpreting transfer evidence in criminal cases. Applied Statistics 33:25–32.

Evett IW (1991) Interpretation: a personal odyssey. In: The use of statistics in forensic science (Eds. CGG Aitken and DA Stoney), Ellis Horwood, Chichester:9–22.

Fairley WB and Mosteller W (1977) Statistics and public policy, Addison-Wesley, London:355–379.

Lark RM and Rawlins BG (2008) Can we predict the provenance of a soil sample for forensic purposes by reference to a spatial database? European Journal of Soil Science 59:1000–1006.

Meier P and Zabell S (1980) Benjamin Pierce and the Howland will. Journal of the American Statistical Association 75:497–506.

Neumann C, Champod C, Puch-Solis R, Egli N, Anthonioz A and Bromage-Griffths A (2007) Computation of likelihood ratios in fingerprint identification for configurations of any number of minutiae. Journal of Forensic Sciences 52:54–64.

Parker JB and Holford A (1968) Optimum test statistics with particular reference to a forensic science problem. Applied Statistics 17:237–251.

Pye K (2007) Geological and soil evidence, forensic applications, CRC, Boca Raton, FL.

Rawlins BG, Kemp SJ, Hodgkinson EH, Riding JB, Vane CH, Poulton C and Freeborough K (2006) Potential and pitfalls in establishing the provenance of earth-related samples in forensic identification. Journal of Forensic Sciences 51:832–845.

Saks MJ, Koehler JJ (2005) The coming paradigm shift in forensic identification science. Science 309:892:895.

Taroni F, Bozza S and Aitken CGG (2005) Decision analysis in forensic science. Journal of Forensic Sciences 50:894–905.

Taroni F, Aitken CGG, Garbolino P and Biedermann A (2006) Bayesian networks and probabilistic inference in forensic science, John Wiley & Sons, Chichester.

Thompson WC and Schumann EL (1987) Interpretation of statistical evidence in criminal trials. The prosecutor's fallacy and the defence attorney's fallacy. Law and Human Behaviour 11:167–187.

Chapter 4
Microbial Community Profiling for the Characterisation of Soil Evidence: Forensic Considerations

George F. Sensabaugh

Abstract Soil contains a very rich and diverse array of microbial species. The observation by soil scientists that soil samples from different locales possess different microbial species' profiles has suggested that this approach might have potential forensic utility for linking soil evidence samples to their sites of origin. This chapter outlines the biology and technology of microbial community profiling, particularly in relation to DNA analysis. Three research challenges are posed and discussed that must be addressed if this approach is to find a place in the forensic armamentarium: (i) it must be demonstrated that microbial population assemblages vary in such a way as to allow samples from a particular patch to be differentiated from samples deriving from other places; (ii) analytical approaches to microbial community profiling must be developed that combine discriminatory power, robustness and reliability; and (iii) statistical methods must be identified that provide objective measures for assessing the similarities and differences between samples.

Introduction

Soil is encountered often enough as evidence in crime investigations to make us wish we could better tap its potential to link persons and objects to specific places. In some cases, we may seek leads as to the likely origin of the soil so that we can focus investigative attention on a particular locale, to find, for example, where a body might be buried. This is a database question: we need to have a reference catalogue of the distinguishing characteristics of soils from different locales to be able to address this kind of question. A second and quite distinct question arises

G. F. Sensabaugh
School of Public Health, 50 University Hall, MC# 7360, University of California, Berkeley, CA USA
e-mail: sensaba@berkeley.edu

K. Ritz et al. (eds.), *Criminal and Environmental Soil Forensics*

when we have a soil sample in evidence, mud on a shoe for instance, and wish to determine whether it might have originated from a particular site, a patch of flowerbed for example. Here the question is quite specific: Is the evidence sample sufficiently similar to reference samples from the suspect source site and sufficiently different from samples collected at other sites to allow an inference that the evidence sample more likely originated from the suspect site? This is a comparison question and we cannot provide a meaningful answer unless we know what properties of soil might be useful for discriminating among these particular samples and what we mean by 'sufficiently similar' and 'sufficiently different'. This comparison question problem applies to many different types of forensic evidence though arguably the magnitude of the problem is greater for soil evidence than for most other categories of evidence.

To illustrate some of the salient aspects of the problem with soil comparisons, it is instructive to contrast these with human identification by DNA profiling. Human beings are discrete entities, each with a genetic profile established at conception and remaining constant through the life of the individual. We have good reason to believe, based on the principles of Mendelian genetics and population genetics, that every individual (identical sibs excepted) has a unique genetic profile. That the target of forensic interest, the individual, can be defined independently as a genetically unique entity is a key point. The principles of Mendelian genetics and population genetics also allow us to quantitatively model the degree of genetic similarity between individuals in families, in racial or ethnic groups, and in the population at large; importantly, these models make predictions that can be empirically validated by testing in real populations. Accordingly, we can readily determine which genetic markers behave as independent variables, estimate the discrimination power of genetic markers individually and in combination, and can project how many genetic markers need to be incorporated in a genetic profile to achieve a desired level of genetic discrimination. We are thus on solid theoretical and empirical ground when we make an estimate of the probability of a 'random match', the probability of having identified an individual unrelated to the true donor of the evidence sample who happens to possess an indistinguishable genetic profile.

An obvious and important difference between soil evidence comparison and human identification is that the target of forensic interest, our patch of flowerbed for example, is neither a discrete nor natural entity but rather is an artificially defined domain within a larger continuous space. The properties used to characterise the patch are unlikely to be uniform over its domain and so the patterns of variation within it must be mapped and compared to what is found in adjacent and more distant spaces. Moreover, the patterns of variation over a patch are expected to differ for each property measured and for any particular property may be continuous or discontinuous, course grained or fine. Accordingly, characterisation of a patch and its surrounding space may entail preparing an overlay of multiple maps, each representing the pattern of variation of a particular property. A second major difference between soil comparison and human identification is that the soil environment is dynamic and changes over time. As a result, it remains an open question whether meaningful comparisons can be made between a soil evidence sample deposited at the time of the crime and reference samples from the suspect site collected at a later point in time. A third difference is that the properties used to characterise soil differ substantially in kind, ranging from relatively simple features such as colour and pH to microscopic features relating to mineralogy and particle size distributions to detailed quantitative data on chemical trace element levels and stable isotope ratios. Although these properties are no doubt interrelated in some way, we do not yet have the keys to this knowledge. As a result, our capacity to make inferences from known data to the abstract case is limited. In other words, we lack sufficient knowledge about the basis for patterns of variation in soil to make the kind of quantitative statement that can be made with human DNA evidence regarding the likelihood that a particular patch can be distinguished from all others.

The objective of this chapter is to discuss these challenges as applied to microbial community profiling (MCP) as an approach to the analysis of soil evidence. MCP has been employed in the soil sciences since the mid-1990s as a tool for characterising the diversity of microbial species in soils. Its potential forensic application has generated several exploratory studies described in the forensic literature (Horswell et al. 2002; Heath and Saunders 2006; Moreno et al. 2006). Part of the attraction of MCP is that it examines the biological dimension of soils, a dimension that heretofore has received little forensic attention. MCP is predominantly based on the analysis of the DNA extracted from soil and in concept at least requires no technical expertise in microbiology – nothing has to be cultured. That MCP utilises DNA analysis no doubt contributes to the attraction; DNA analysis has provided spectacular benefits to human identification and hope springs eternal that it may bring comparable benefits to other categories of evidence. There is a practical side to the DNA connection as well: MCP can be done using the technical skills and facilities already present in the DNA unit of a crime laboratory, i.e. DNA extraction from test samples, DNA amplification by PCR, and separation of PCR products by high resolution gel or capillary electrophoresis. Thus, if MCP should prove to be of value with soil evidence, it could be implemented into current crime lab operations.

Microbial Community Profiling as an Approach to the Characterisation of Soil Diversity

Microbial Communities in Soils – An Overview

It has been long recognised that soils contain a rich variety of microorganisms. The term ‘microorganism’ is used here in the broad sense to include organisms in the two prokaryote domains, bacteria and archaea, and the microscopic members in the domain of eukaryotes – fungi, protozoa, and nematodes. Much of what we now know about the diversity of microbial species in soils has been gained through the characterisation of soil DNA using MCP methods. A key finding is that the average soil sample contains many more microbial species than can be detected by classical culture methods. Prokaryotes dominate the soil environment in terms of organism count and species diversity; a gram of soil may contain 10^6 to 10^{10} organisms representing as many as 50,000 different species (Torsvik et al. 2002; Fierer et al. 2007; Roesch et al. 2007). The most abundant bacterial phyla found in soils are Proteobacteria, Acidobacteria, Actinobacteria, Verrucomicrobia, Bacteroidetes, Chloroflexi, and Planctomycetes (Janssen 2006). Archaea may account for as much as 10% of the prokaryotic species content, depending on the soil type. Fungi are found in most soils, particularly in soils rich in organic material where they may be the dominant contributor to the soil microbial biomass (Ritz 2005; Ingham 2006). The filamentous fungi grow as branching thread-like strands (hyphae) of varying length. The branching mass of hyphae forms a mycelium that exists as a three dimensional network that intercalates within the micro-particulate structure of soil. Clonal mycelia can grow to great size, extending in some cases to a domain spanning several acres. The diversity of fungi in soils can be comparable to that observed for prokaryotes (Anderson and Cairney 2004; Fierer et al. 2007). Protozoa are single cell organisms that feed on bacteria and other organic material in soils (Ingham 2006). Among the protozoa, the flagellates tend to be the most abundant (10^4 to 10^8 organisms per gram of soil) followed by the amoebae (10^4 to 10^5 organisms per gram of soil); ciliates contribute relatively insignificant numbers (Adl and Coleman 2005; Esteban et al. 2006). Nematodes are non-segmented worms up to 1 mm in length; they feed on bacteria, archaea, fungi, protozoa, and other nematodes as well as plant and algal material. The abundance of nematodes depends on soil type but generally do not exceed several hundred per gram of soil (Ingham 2006).

Microbial Community DNA Profiling

The major fraction of the DNA extracted from soil, typically in the range of 1 to 100 μg DNA per gram of soil, is microbial in origin (Kuske et al. 2006). As noted above, MCP entails DNA extraction from soil samples, amplification of a target gene using the polymerase chain reaction (PCR), and separation of PCR products by high resolution gel or capillary electrophoresis; for a review of the variety of methods used for MCP see Nocker et al. (2007). The predominant target for most MCP methods is the gene region encoding the small subunit (SSU) ribosomal RNA (16S in prokaryotes, 18S in eukaryotes), a gene region present in all living organisms. The SSU rRNA genes contain conserved sequence segments that have evolved sufficiently slowly to allow deep phylogenetic relationships to be established among very distantly related species, even across the prokaryotic–eukaryotic divide; these data are the basis for the ‘tree of life’ (Pace 1997). Over 250,000 SSU rRNA gene sequences are archived in the Ribosomal Database Project and this database can be interrogated to identify the phylogenetic position of any unknown sequence (Cole et al. 2005). The conserved nature of the SSU rRNA sequences allows PCR primers to be designed to amplify targets with any level of species specificity desired, from the broadest scale (all bacteria, all fungi) to a limited species grouping within a genus (Blackwood et al. 2005). MCP based on SSU rRNA analysis thus can be 'tuned' to focus on whatever level of community structure is of interest. A second target within the ribosomal gene region sometimes employed in MCP is the intergenic spacer (IGS, also called the internal transcribed spacer, ITS) between the small subunit and large subunit rRNA genes. The IGS regions are much more variable in sequence than the flanking rRNA genes; they also exhibit length variation and both types of variation have been used to differentiate strains within a species (Ranjard et al. 2001). As with the SSU rRNA gene analysis, PCR primers can be anchored in conservative segments of the flanking small and large rRNA genes to achieve the degree of species specificity desired.

Figure 4.1 illustrates a microbial community profile output obtained using terminal restriction fragment length polymorphism (T-RFLP) analysis, probably the most widely used method for MCP; this figure is taken from the first paper to describe the potential forensic application of the MCP approach (Horswell et al. 2002). With T-RFLP, the SSU rRNA genes from the diverse microbial species present in the community DNA are amplified by PCR using primers tagged with fluorescent labels (Liu et al. 1997; Nocker et al. 2007). The PCR products are cut with restriction enzymes to produce restriction fragments which are then separated by gel or capillary electrophoresis; only the fluorescently labelled terminal fragments are detected. The rRNA genes from different microbial species differ in sequence and accordingly the size of the terminal fragments varies from species to species (Kent et al. 2003). The electrophoretic separation results in a profile consisting of an array of peaks in which peak positions represent the diverse microbial species present in the sample and the peak heights represent the relative abundance of those species. The T-RFLP method underestimates the full microbial diversity in a sample due to the resolution limits of the electrophoretic separation and redundancy in terminal restriction fragments sizes among some species groups. Nevertheless, it is possible to obtain profiles containing 100 or more resolvable peaks.

Other SSU rRNA-based techniques used for MCP include denaturing gradient gel electrophoresis, single strand conformation analysis, denaturating high performance liquid chromatography, hybridisation to species-specific probes in microarray formats, and mass sequencing (Brodie et al. 2006; Fierer et al. 2007; Nocker et al. 2007). The latter two approaches have the best capacity to resolve species among the MCP techniques and hence have the potential to provide more accurate estimates of microbial species diversity in a sample. The emergence of high through-put direct sequencing technologies with their potential to generate multiple thousands of sequences from a single sample in one pass provide the best yet assessments of species compositional diversity (Liu et al. 2007; Roesch et al. 2007). From a forensic perspective, however, the accuracy of the species diversity estimate is less important than the profile reproducibility in sample comparisons.

As previously noted, microbial DNA profiling detects the presence of unculturable organisms as well as those that can be cultured. Moreover, it does not distinguish between the DNA originating from living organisms, from dormant spores or cysts,

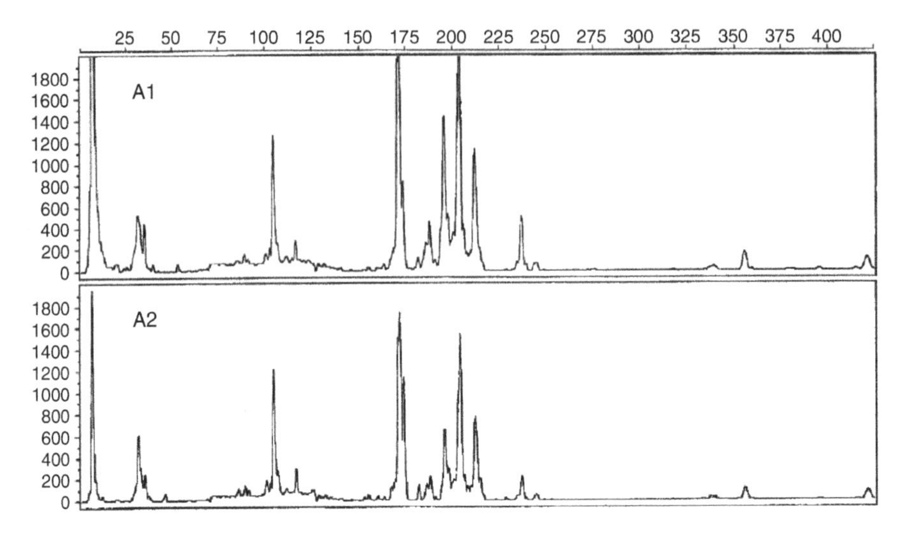

Fig. 4.1 Microbial community profiles generated by T-RFLP analysis in a simulated case example. Soil samples were collected from a shoe print impression in soil and from the shoe that made the impression. The profiles given by the shoe sample and by the impression sample are shown in the upper and lower panels respectively. Fragment sizes are indicated by the position of the peaks along the x-axis (in base pair units) and the abundance of each fragment is indicated by the peak heights on the y-axis. The Sorenson similarity index for the two profiles is 0.91. (From Horswell et al. (2002) with permission.)

or from dead organisms. This allows profiles to be obtained from years-old stored samples (Dolfing et al. 2004). Moreover, the size of soil samples appears to make little difference in the community profiles obtained provided the size exceeds minimum thresholds, greater than 125 mg for bacteria and greater than 1 g for fungi (Ranjard et al. 2003); the difference in minimum sample size likely reflects the different scales of homogeneity in soils.

Microbial Biogeography

MCP has contributed substantially over the last decade to advances in our understanding of the spatial and temporal diversification of microorganisms. There is accumulating evidence, much of it based on MCP studies, supporting the idea that microbial species vary in spatial and temporal distributions and that this variation is determined by the same sort of processes operating on plants and animals, i.e. variation in habitat modulated by historical contingency (Foissner 2006; Martiny et al. 2006). This challenges a long held view that the great abundance of individuals in microbial species and the ease of dispersal of small organisms results in ubiquitous globally distributed species (Finlay 2002); this view is encapsulated in the proposition "everything is everywhere – the environment selects"(Bass-Becking 1934). Clearly, variation in habitat affects microbial community structure; the habitat selects for an assemblage of adapted species. Habitat factors include soil type and structure (e.g. soil porosity, microaggregate stability), nutrient sources (organic material from plant and animal decomposition), and soil chemistry (pH and key nutrients) (Fierer and Jackson 2006). What molecular studies have shown is that local environmental and ecological history also play a role in determining patterns of variation. The microbial population structure at a site is shaped by its history of colonisation by new species, extinction of existing species, mutational change within species giving rise to strains with new adaptive potentials, and changes in the interaction dynamics between members of the community. Even within widely distributed genera and species, molecular studies have revealed localised genetic variation (Cho and Tiedje 2000; Whitaker et al. 2003; Davelos et al. 2004; Koch and Ekelund 2005); this variation suggests that adaptive change and/or genetic drift can occur at rates faster than individual organisms disperse. Thus the observed site-to-site variation in microbial community structure can be rationalised within the general framework of familiar population genetic and ecological theory, an important factor for gaining legal acceptance of MCP as a forensic tool.

Though the spatial and temporal distribution of microbial, plant, and animal species may be governed by common processes, there is a decided difference as to spatial scale. Plant and animal species can spread over extended ranges via active and/or passive dispersal mechanisms; these are well studied. The patterns of microbial dispersal, in contrast, are largely unknown. The motile protozoa and nematodes might actively migrate over short ranges and potentially can be dispersed passively over greater ranges. With fungi, we know some clonal mycelia can extend over large domains and that spore formers can spread by passive dispersal (Ritz 2005). Bacterial colonies in soil exist on the millimetre or sub-millimetre scale but clonally related colonies have been detected at kilometre distances, indicating that range of colonies and of clones are not identical (Grundmann 2004; Davelos et al. 2004). Spatial scale is obviously an important consideration for forensic application and needs more detailed study.

Research Challenges for Forensic Application

To achieve acceptance for forensic application, MCP must satisfy three broad and interconnected conditions.

1. It must be demonstrated that microbial population assemblages vary in such a way as to allow samples from a particular patch to be differentiated from samples deriving from other places.
2. Analytical approaches for MCP must be developed that combine discrimination power, robustness, and reliability.

3. Statistical methods must be identified that provide objective measures for assessing the similarities and differences between samples.

The interconnections of these questions are apparent: the patterns of variation we see in microbial communities will depend on the analytical approaches used; the analytical approach determines what sort of metrics can be used to look at the data, and the nature of the metrics determines the conceptualisation of patterns of variation. The following paragraphs outline some of the research challenges attendant to these conditions.

Characterisation of Microbial Population Variation in Space and Time

A primary challenge is the question of spatial scale. Consider the case of the mud on the shoe cited at the beginning of this piece. The patch of forensic interest, the flowerbed from which the mud might have originated, measures in metres; most patches of forensic interest, a particular field or a roadside pull-off, for example, would be on the same spatial scale. However, as noted in the preceding section, the changes in the composition of microbial populations are expected to occur over short distances, i.e. millimetres or centimetres. Although these differences in scale would seem to undermine the usefulness of MCP as a forensic tool, it is possible that the relative effect on the overall microbial community profile is small. Alternatively, it may be that an approach focused on particular species or species groups might circumvent the scale problem. It might be possible, for example, to abstract a set of site informative species from a representative sampling of the patch of forensic interest; this would entail a statistical characterisation of the profiles from the patch and from unrelated sites to identify the site informative species.

Assessing patterns of variation entails collection of samples from different sites and this in turn poses the question of sampling strategies. One might ask a big picture question: how many samples need to be collected from how many sites and how many different soil types to assess the diversity of microbial population assemblages in the world at large? Intuitively, one might feel confident about using MCP as a forensic tool if it were found that the technique had the capacity to individualise samples within a large collection of samples representing many sites and soil types. However this does not get at the relevant question for forensic comparisons. We want to know whether the evidence soil sample might have come from the particular suspect source flowerbed and not from some unrelated site, including sites that may be similar in soil type and environmental exposure such as other flowerbeds in the neighbourhood. To answer this question, we need to know more about the relationship between microbial population structure and other soil variables, specifically the extent to which other soil variables might shape microbial community structure. Put simply, we need to know whether it is likely that other flowerbeds in the neighbourhood might yield the same microbial community profile as our flowerbed of interest.

The above-mentioned considerations invoke the possibility that the forensic application of MCP may need to be tuned to fit particular local conditions and this in turn raises the question of designing a localised sampling strategy that includes both the patch of interest and other relevant sites. Clearly a research programme focused on evaluating the forensic applicability of MCP cannot cover all possible situations but it should provide models to guide analyses in real-life situations.

Geostatistical analysis provides a promising approach to modelling spatial patterns of variation in local settings (Goovaerts 1998). Geostatistical analysis provides statistical tools to assess correlations in properties between samples as a function of distance. This directly addresses the question of scale posed above and offers an approach to defining 'patches' as a function of feature variability. To illustrate, Franklin et al. (2002) determined by spatial autocorrelation analysis that the spatial extent of genetic similarity in microbial community structure along a riverbank was a patch about 35 cm in the horizontal dimension and 17 cm in the vertical dimension; notably, this is smaller than a typical patch of forensic interest. Geostatistical analysis can be used also to assess correlations between features in spatial terms and to predict patterns of feature variation in continuous space. These functions are useful for evaluating independence of variables and in identifying patterns of variation over large spaces (Franklin and Mills 2003; Ritz et al. 2004).

Development of Discriminating, Robust, and Reliable Profiling Systems

As noted previously, MCP can be used to survey species diversity at multiple levels, ranging from the broad base of all species within a domain or kingdom (e.g. all bacterial species, all fungal species) to a more limited base of the species within a large species group (e.g. actinobacteria) and to the narrow base of the species within a genus (e.g. all streptomycetes). Related methods can be used to distinguish strains within a species (Cho and Tiedje 2000; Whitaker et al. 2003; Davelos et al. 2004; Koch and Ekelund 2005). It is to be remembered, however, that the objective of forensic testing is to connect soils from geographically related sites and to distinguish soils from unrelated sites; the assessment of species diversity is a means to an end, not an end in itself. As knowledge about patterns of variation emerges, it will become apparent whether the profiling approach can be applied in a one-test-fits-all mode or needs to be tuned to fit particular soil types or geographic situations. This in turn will guide the choice of the most appropriate profiling approach.

The development of an analytic approach also needs to take into account that forensic evidence samples are unlike the typical samples collected in a soil science study. There are unknowns in the generation and history of the sample that may confound reliable analysis. For example, the forensic sample may originate from the soil surface or subsurface, it may be a mixture of soils from different sites, it may have spent various periods in the wet and/or dry states, it may be fresh or old, and it may have been exposed to potentially deleterious environmental conditions such as extreme temperatures and sunlight. The effects of these variables can be investigated in controlled studies with the goal of determining which have an effect on MCP outcomes and which do not. A further goal is to identify signals to warn when potentially confounding conditions are present. Though these investigations may appear formidable, studies such as these are part of the standard drill in the evaluation of other types of biological evidence.

Statistical Metrics for the Comparison of Sample Profiles

The principal hope of a good comparison metric is that it will maximise the probability of identifying samples that are truly related and of differentiating samples that are truly unrelated. Forensic application requires that the chance of false positives be known; the lower the chance of a false positive, the greater the evidentiary weight accorded to a positive result. It is also important to know the chance of false negatives, the chance that related samples would test out as different. As the chance of false negatives increases, the value of the test for excluding unrelated sources diminishes.

A number of statistical approaches have been used to assess relationships between samples in MCP studies. These include pairwise similarity measures, various forms of cluster analysis, and ordination methods such as principal components analysis (Fromin et al. 2002). In the best of all possible worlds, any of these techniques would draw the same bright line between the truly related and the truly unrelated and it would not matter which was used. Experience indicates otherwise, however. The various statistical approaches look at data in different ways and can reveal different, and sometimes inconsistent, facets of relationships. If different statistical approaches are found to give inconsistent pictures of the relationships among samples, the basis for these inconsistencies must be understood and accounted for. Otherwise, evidence based on MCP is open to challenge in court.

Conclusions

Many challenges remain to establish the applicability of MCP as a tool for the forensic analysis of soil evidence. Progress will depend on continued cooperation between the soil science community and the forensic community as the former advances understanding of soil ecology and biogeography and the latter addresses concerns specific to the forensic endeavour.

References

Adl S and Coleman D (2005). Dynamics of soil protozoa using a direct count method. Biology and Fertility of Soils 42:168–171.

Anderson IC and Cairney JWG (2004). Diversity and ecology of soil fungal communities: increased understanding through the application of molecular techniques. Environmental Microbiology 6:769–779.

Bass-Becking LGM (1934). Geobiologie of Inleiding to de Milieukunde. van Stockum & Zoon, The Hague.

Blackwood CB, Oaks A and Buyer JS (2005). Phylum and class specific PCR primers for general microbial community analysis. Applied and Environmental Microbiology 71:6193–6198.

Brodie EL, DeSantis TZ, Joyner DC, Baek SM, Larsen JT, Andersen GL, Hazen TC, Richardson PM, Herman DJ, Tokunaga TK, Wan JM and Firestone MK (2006). Application of a high density oligonucleotide microarray approach to study bacterial population dynamics during uranium reduction and reoxidation. Applied and Environmental Microbiology 72:6288–6298.

Cho JC and Tiedje JM (2000). Biogeography and degree of endemicity of fluorescent pseudomonas strains in soil. Applied and Environmental Microbiology 66:5448–5456.

Cole J, Chai B, Farris R, Wang Q, Kulam SA, McGarrell DM, Garrity GM and Tiedje JM (2005). The Ribosomal Database Project (RDP II): sequences and tools for high throughput rRNA analysis. Nucleic Acids Research 33:D294–D296.

Davelos AL, Xiao K, Samac DA, Martin AP and Kinkel LL (2004). Spatial variation in streptomyces genetic composition and diversity in a prairie soil. Microbial Ecology 48:601–612.

Dolfing J, Vos A, Bloem J and Kuikman PJ (2004). Microbial diversity in archived agricultural soils. The past as a guide to the future. In: Alterra rapport 916, Alterra, Wageningen, the Netherlands. *http://www.alterra.wur.nl/internet/modules/pub/PDFFiles/* Alterrararrporten/AlterraRapport916.pdf

Esteban GF, Clarke KJ, Olmo JL and Finlay BJ (2006). Soil protozoa – an intensive study of population dynamics and community structure in an upland grassland. Applied Soil Ecology 33:137–151.

Fierer N, Breitbart M, Nulton J, Salamon P, Lozupone C, Jones R, Robeson M, Edwards RA, Felts B, Rayhawk S, Knight R, Rohwer F and Jackson RB (2007). Metagenomic and small subunit rRNA analyses reveal the genetic diversity of bacteria, archaea, fungi, and viruses in soil. Applied and Environmental Microbiology 73:7059–7066.

Fierer N and Jackson RB (2006). The diversity and biogeography of soil bacterial communities. Proceedings National Academy Sciences 103:626–631.

Finlay BJ (2002). Global dispersal of free living microbial eukaryote species. Science 296:1061–1063.

Foissner W (2006). Biogeography and dispersal of micro organisms: a review emphasising protists. Acta Protozoologica 45:111–136.

Franklin RB, Blum LK, McComb AC and Mills AL (2002). A geostatistical analysis of small scale spatial variability in bacterial abundance and community structure in salt marsh creek bank sediments. FEMS Microbiology Ecology 42:71–80.

Franklin RB and Mills AL (2003). Multi scale variation in spatial heterogeneity for microbial community structure in an eastern Virginia agricultural field. FEMS Microbiology Ecology 44:335–346.

Fromin N, Hamelin J, Tarnawski S, Roesti D, Jourdain-Miserez K, Forestier N, Teyssier-Cuvelle S, Gillet F, Aragno M and Rossi P (2002). Statistical analysis of denaturing gel electrophoresis (DGGE) fingerprinting patterns. Environmental Microbiology 4:634–643.

Goovaerts P (1998). Geostatistical tools for characterizing the spatial variability of microbiological and physico chemical soil properties. Biology and Fertility of Soils 27:315–334.

Grundmann GL (2004). Spatial scales of soil bacterial diversity the size of a clone. FEMS Microbiology Ecology 48:119–127

Heath LE and Saunders VA (2006). Assessing the potential of bacterial DNA profiling for forensic soil comparisons. Journal of Forensic Sciences 51:1062–1068.

Horswell J CS, Maas EW, Martin TM, Sutherland KBW, Speir TW, Nogales B and Osborn AM (2002). Forensic comparison of soils by bacterial community DNA profiling. Journal of Forensic Sciences 47:350–353.

Ingham ER (2006). The Soil Biology Primer, Chapters 3–6. United States Department of Agriculture (http://soils.usda.gov/sqi/concepts/soil_biology). Accessed Dec. 1, 2007.

Janssen PH (2006). Identifying the dominant soil bacterial taxa in libraries of 16S rRNA and 16S rRNA genes. Applied and Environmental Microbiology 72:1719–1728.

Kent AD, Smith DJ, Benson BJ and Triplett EW (2003). Web based phylogenetic assignment tool for analysis of terminal restriction fragment length polymorphism profiles of microbial communities. Applied and Environmental Microbiology 69:6768–6776.

Koch TA and Ekelund F (2005). Strains of the heterotrophic flagellate Bodo designis from different environments vary considerably with respect to salinity preference and SSU rRNA gene composition. Protist 156:97–112.

Kuske CR, Barns SM, Grow CC and Merrill LJD (2006). Environmental survey for four pathogenic bacteria and closely related species using phylogenetic and functional genes. Journal of Forensic Sciences 51:548–558.

Liu W, Marsh T, Cheng H and Forney L (1997). Characterization of microbial diversity by determining terminal restriction fragment length polymorphisms of genes encoding 16S rRNA. Applied and Environmental Microbiology 63:4516–4522.

Liu Z, Lozupone C, Hamady M, Bushman FD and Knight R (2007). Short pyrosequencing reads suffice for accurate microbial community analysis. Nucleic Acids Research 35:120.

Martiny JBH, Bohannan BJM, Brown JH, Colwell RK, Fuhrman JA, Green JL, Horner-Devine MC, Kane M, Krumins JA, Kuske CR, Morin PJ, Naeem S, Øvreås L, Reysenbach A-L, Smith VH and Staley JT (2006). Microbial biogeography:putting microorganisms on the map. Nature Reviews Microbiology 4:102–112.

Moreno LI, Mills DK, Entry J, Sautter RT and Mathee K (2006). Microbial metagenome profiling using amplicon length heterogeneity polymerase chain reaction proves more effective than elemental analysis in discriminating soil specimens. Journal of Forensic Sciences 51:1315–1322.

Nocker A, Burr M and Camper A (2007). Genotypic microbial community profiling: a critical technical review. Microbial Ecology 54:276–289.

Pace NR (1997). A molecular view of microbial diversity and the biosphere. Science 276:734–740.

Ranjard L, Lejon DPH, Mougel C, Schehrer L, Merdinoglu D and Chaussod R (2003). Sampling strategy in molecular microbial ecology: influence of soil sample size on DNA fingerprinting analysis of fungal and bacterial communities. Environmental Microbiology 5:1111–1120.

Ranjard L, Poly F, Lata J C, Mougel C, Thioulouse J and Nazaret S (2001). Characterization of bacterial and fungal soil communities by automated ribosomal intergenic spacer analysis fingerprints: biological and methodological variability. Applied and Environmental Microbiology 67:4479–4487.

Ritz K (2005). Fungi. In: Encyclopedia of Soils in the Environment (Ed. D. Hillel), pp. 110–119. Elsevier Ltd, Oxford.

Ritz K, McNicol JW, Nunan N, Grayston S, Millard P, Atkinson D, Gollotte A, Habeshaw D, Boag B, Clegg CD, Griffiths BS, Wheatley RE, Glover LA, McCaig AE and Prosser JI (2004). Spatial structure in soil chemical and microbiological properties in an upland grassland. FEMS Microbiology Ecology 49:191–205.

Roesch LFW, Fulthorpe RR, Riva A, Casella G, Hadwin AKM, Kent AD, Daroub SH, Camargo FAO, Farmerie WG and Triplett EW (2007). Pyrosequencing enumerates and contrasts soil microbial diversity. ISME Journal 1:283–290.

Torsvik V, Ovreas L and Thingstad TF (2002). Prokaryotic diversity magnitude, dynamics, and controlling factors. Science 296:1064–1066.

Whitaker RJ, Grogan DW and Taylor JW (2003). Geographic barriers isolate endemic populations of hyperthermophilic archaea. Science 301:976–978.

Chapter 5
The Current Status of Forensic Soil Examination in the Russian Federation

Olga Gradusova and Ekaterina Nesterina

Abstract An overview of the current state of forensic soil examination in the Russian Federation is given, and the organisations involved reviewed. The main requirements for forensic soil examination are discussed, and methods of examination for trace amounts of urban soil presented. Four examples where the successful identification of soil in case work are described, where the presence of anthropogenic trace-evidence allowed both scenes of crime to be identified and evidence to be linked to suspects. It is shown that integrated multi-disciplinary studies are preferable to obtain more evidence-based conclusions in soil forensic investigations, where the association of diverse information strengthens evidence.

Introduction

In this chapter, we provide an overview of the work carried out by the Centre for Forensic Science in the Russian Federation, and focus via some case studies on aspects of soil forensic examination involving trace evidence.

A wide range of analytical techniques have been applied to the analysis of soil-based materials in forensic applications, including laser granulometry (Pye et al. 2007), elemental analysis using inductively coupled plasma spectrometry (Pye and Croft 2007), X-ray diffraction analysis (Ruffell and Wiltshire 2004) and electron microscopy coupled with energy dispersive X-ray analysis (Cengiz et al. 2004; Pye and Croft 2007; Pirrie et al., chapter 26). Automated data analysis programs of increasing sophistication are available for some of these analytical methods (Bogatyryev 2001). However, despite the power of these techniques, it is unlikely that a single approach provides sufficiently comprehensive forensic analysis, and that

O. Gradusova(✉) and E. Nesterina
Russian Federal Centre of Forensic Research, Building 2, Khokhlovsky Lane 13, 109028, Moscow, Russian Federation.
e-mail: bio_soil@rambler.ru

K. Ritz et al. (eds.), *Criminal and Environmental Soil Forensics*

no single technique is universally appropriate (Rawlins et al. 2006). The fact is that soils, whether natural or urban, are complex composite materials containing a great number of components of natural and manufactured origin, and thus a multidisciplinary study is preferable. Forensic experts have to apply appropriate analytical methods on a case-by-case basis; these always require an element of human perception and judgement, and increasingly rely on a strong awareness of the repertoire of techniques available. Clearly, there is no single analytical application or even a suite of techniques appropriate to all circumstances, and differences between legal systems and courts' views make such a concept untenable.

There is no doubt that statistical methods have utility in improving the basis of evidence. Statistics is a large discipline with a range of applications and developing research across the spectrum of sciences. Formal statistical methodology Bayesian theory and neural network approaches are being successfully applied in economics for making predictions (Adya and Collopy 1998) and in computer technologies, e.g. for automatically filtering spam emails (Yukun et al. 2004), and is now used in DNA and other forensic examinations (Aitken and Taroni 2004; Carresy et al.2004; Biedermann and Taroni 2006; Sjerps 2006; Aitken et al. 2008). However, there remain relatively few applications of formal statistical methodology to soil forensic examination, and a surprisingly small amount of literature on the subject.

The Russian Federation Perspective

The Russian Federal Centre of Forensic Science is the parent organisation in the system of the State Expert Institutions of the Ministry of Justice of the Russian Federation. Forensic examination status and forensic expert status (rights and charge) are stated in the Code of Criminal Procedure of the Federation, and the Federal Law covers expert activity within the Federation. The system of State Expert Institutions encompasses about 50 laboratories and centres, situated all over Russia. The basic purpose of the Centre is the realisation of forensic examinations with the aim of guaranteeing protection of human rights and freedom, and interests of the State. The main tasks of the Centre include: scientific and methodical work experts, in addition to making examinations; scientific research work (both singly and in collaboration with other research institutes); developing new and original forensic research methods; writing methodical manuals for use by forensic experts; review of experts' practice in Russia and globally; training and certification of forensic experts; review of experts' investigations with the purpose of improving the quality of their work.

The Centre was founded in 1962 with the establishment of two laboratories, one for actual casework applications, and the other for research. As of 2007 it consists of more than 20 laboratories offering services that cover about 100 different expert specialities. Within the Centre, the Laboratory of Forensic Soil and Biological Investigations is one of the oldest, founded in 1970. This laboratory necessarily collaborates closely with other laboratories. Initially, there were three expert specialities in the laboratory: forensic botany examination, forensic animal hair examination and forensic soil examination, but this was revised in 2007 to include mycology and DNA analysis of plants and animals. Botanical examination deals with identification and analysis of plant and plant particles, pollen grains and spores, diatoms and paper composition. Forensic examination of animal hair and bird feathers deals with identification of small fragments of such material. The greatest part of the examinations falls on soil forensic aspects which usually deal with investigations of high profile offences.

Soil forensics cases typically involve the examinations of soil layers and soil traces (of rural and urban origin), dust derived from building materials, and small mineral particulates of natural and technogenic origin.

In the period 1970–1980 a discipline of forensic examination theory (also termed *criminalistics*) developed markedly in the then Soviet Union, providing a formalisation of concepts and new direction that was previously absent. The first theoretical base was developed for 'traditional' forensic disciplines, such as handwriting examination, portrait examination, ballistics and fingerprint identification. Mitrychev (1976) developed a method of '*identification of the whole by parts*'. The main principles of expert identification theory in forensic examinations were well rehearsed in a seminal volume edited by Koldin (1996). Soil forensics experts began to collaborate with forensic theorists and soil forensic examinations were reviewed, summarised and analysed to provide a synthesised theory (Tjurikova 1976). The principle here is that in soil forensics, experts identify the scene of crime (the whole) by structured examination of the parts (e.g. soil layers from items). To achieve this, the expert undertakes a comparative study at both the broad, generic, level and in relation to particular differentiative or defining characteristics that may connect the scene of crime (burial, basement and garret of the house, garden or other places) and soil material or layers on associated items. The overall framework follows these basic tasks in relation to the criminalistic theory:

- *Identification tasks*: are the layers on the items under investigation derived from or composed of soil, and if so do they have generic and/or specific association with the scene of crime (burial or other place)? Are the soil layers on the items from this place?
- *Diagnostic tasks*: what is the provenance of a substance, e.g. what geographical region/location could the soil be from? These are often 'blind' cases in which only evidentiary soil samples are available.
- *Situational tasks*: what is the position/location of a soil sample on clothes (e.g. footwear) and what is its stratification? Does such location/position confirm the situation given in case reports? How was the soil deposited on the evidentiary item?
- It is rarely possible to be absolutely categorical in stating such identity or provenance, since it not tenable to find uniquely individualising characteristics for soils material in any place. Forensics reports therefore usually state evidence (*trace-forming object*) found on an *object-carrier* relates to a place or circumstance with the same characteristics of, for example, the soil cover. All courts in the Russian federation adopt such a formulation.

Very often, in forensic samples there is a mechanical mixture of soil micro-aggregates, mineral particles of natural and manufactured origin, plant particles and mammalian hair. For these applications, in soil cases we often apply a multidisciplinary approach in our laboratory. There is extensive experience spanning several decades with the participation of experts of different specialities. In every complex examination there is an expert-coordinator or the organiser, who understands the bases of all available analytical approaches and methods. We operate on the principle of involving a number of experts, with a balanced suite of methods, and apply principles of necessity and sufficiency. Analytical results are synthesised, with the generic and group characteristics are being stated. This combined approach is always more effective than separate examinations and the synthesis of diverse information strengthens the evidence in an investigation.

The General Approach to Soil Forensic Examinations

Forensic soil experts in Russia generally follow the defined guidelines in the manual "*Complex methods of forensic soil examination*" (VNIISE 1993), which are akin to those proposed by Sjerps (2006).

A forensic expert conducts examinations using special knowledge to answer questions posed by the case investigator. Although experts have the right to use any methods and procedures that they deem necessary, they must provide reasons for their choices in their examination reports. All forensic investigations begin with a careful study of case papers, especially reports from the scene of crime. Sometimes it is necessary to visit a crime scene to examine its physical environment (e.g. relief, character and homogeneity of soil cover and vegetation). Following this, the expert examines evidentiary items and chooses an analytical scheme. Finally, conclusions are made that answer the stated questions. As is well recognised in forensic science, all results are dependent upon the integral maintenance of the chain of custody of the evidence. All steps of an expert's examinations are depicted and defended in an examination report.

Forensic Examination of Trace Quantities Soils

Soil aggregates and the micro-aggregates of which they are formed are a unique product of soil formation (Khan et al. 2007). Their structure and position in a profile are defined by a generally consistent type of soil formation. Micro-aggregates are structural units which, as a rule, repeat in space and combine to make larger aggregates that associate to create soil genetic horizon. Micromorphological soil examination using thin sections derived from resin-embedded soil samples is widely used in soil science (e.g. Parfeneva and Yarilova 1977; Fitzpatrick 1993). But in a soil forensic examination, when there is often an insufficient amount of sample (typically $<100\,mg$), a preparation of thin sections leads to significant loss of analysable material. Possibilities of instrumental investigation of soil micro-aggregates are also limited.

In practice, trace amounts of soil extracted from shovels, footwear, automobile protectors and rugs usually constitutes a mechanical mixture of soil micro-aggregates (<2 mm) from different genetic horizons or strata from the surface layers of soil of different origins. Given such micro-scale material, comparative examinations of morphological singularities of soil micro-aggregates in via light stereomicroscopy can be particularly insightful. Pedological features which are traditionally used by soil scientists, such as colour, particle size-range distributions, proportion of gravel, mineralogy of mineral grains, mineraloid presence (including the orientation of clay particles), biological components (e.g. plant particles, diatoms, pollen, mycological fragments, etc.), and a complex of particles of technogenic origin, are also often pertinent. Such high-resolution examination of separate soil micro-aggregates often gives more objective information then the examination of homogenized or bulk samples.

Provenancing

When comparing soil samples, or when providing investigative information from soil examinations, it is often required to estimate the geographical provenance of a soil, and preferably to a narrow spatial location. The first work on localisation of a crime scene in forensic soil investigations in Russia was by Tjurikova (1980), who suggested that every unusual particle with individual characteristics (*viz.* 'exotic' particles defined by Rawlins et al. 2006) of natural or anthropogenic origin may be considered as important markers. However, from our practical experience, it is very hard to determine a boundary marker with precision. We consider that the most objective information in the forensic examination of small amounts of urban soils, or those with elevated amounts of manufactured materials arises from a morphological examination of small portions via light microscopy, in combination with chemical analysis using X-ray micro-probe fluorescence techniques and energy dispersive micro-probe X-ray analysis. It is also very desirable to estimate phase composition by means of synchrotron analysis, but access to such devices is restricted by their limited availability and high cost. Light microscopes equipped with digital cameras are particularly effective in the morphological examination of trace evidence, particularly when used in conjunction with scanning electron microscopy. The latter technique has a drawback in that sample colours are not visualised and there can be difficulty in interpretation of images.

Trace Evidence Investigations Involving Soil: Case Studies

We now present some case studies of successful examinations of urban soil traces using the techniques referred to above, particularly in a complementary fashion.

Case I: An Illegal Entry into a Dwelling

In a district of Moscow, a male person gained entry into a ground floor flat in a multi-storied home through a window when nobody was at home. When the female owner of the flat was entering the house upon returning home, she saw a man coming out of her flat. The woman began to shout and the malefactor ran upstairs. Neighbours, having heard the alarming cry, called the militia. The suspect was subsequently detained on the upper floors, but he denied the fact of illegal entry. He insisted that the door had already been open, and he only closed it upon exit.

To explore the possibility of illegal entry the following items were presented for forensic soil investigation: (i) comparative soil samples from the ground under the window, whence the suspect presumably gained entry to the flat; (ii) a soil substance found on a window sill; (iii) a sweater belonging to the occupant's of the flat which was on a chair near the window; (iv) the suspect's boots.

The amount of soil procured from the window and from the sweater were approximately 150 and 50 mg respectively, and the amount of soil from the suspect's boots (derived from near the heels) was approximately 15 mg.

The comparative soil samples were the first to be thoroughly examined. These samples were of a light loamy texture, and were rather homogeneous in colour and texture at both the inter- and intra-aggregate levels. This examination enabled us to determine the soil characteristics that we might expect in a very small volume of soil of this composition. A number of small, similarly sized sub-samples were then selected arbitrarily (*viz.* micro-aggregates) from the comparator samples and the evidential soils taken from the sill, the sweater, and the suspect's boots. Their colours and textures were compared within one field-of-view under a light microscope (Figure 5.1a). The soil structure of the sample taken from the sweater was not strongly damaged because of the spongy construction of the fabric. The soil structure of the samples taken from the boot was also relatively intact owing to their position. All of these samples had the same colour and texture. Soil substances associated with the objects under investigation were then washed with distilled water to remove the clay fraction and the resulting residues were examined further under the light microscope. It was established that the soil micro-aggregates contained the same complex of materials of natural origin including: ferriferrous concretions (Figure 5.1b), a small fragment of cormophyte moss (Figure 5.1c), material of technogenic or anthropogenic origin including distinctive glass fibre fragments (Figure 5.1d) and paint-like material of consistent colour (Figure 5.1e). In addition, some small ferromagnetic spheres were also found which were subproducts of welding processes and are generally more common in urban soils.

During the visit to the scene of the alleged crime it was observed that the majority of the surrounding ground area was covered with grass and crossed by asphalt footpaths, with the only wet place that contained cormophyte moss being near the house. It was also noted that the lower part of the house was painted with a very similar colour to that of the paint fragments observed in the micro-aggregate samples. Results of the investigation allowed us to conclude that the soil taken from the suspect's boots, from the sill, and the sweater were, with a very high degree of

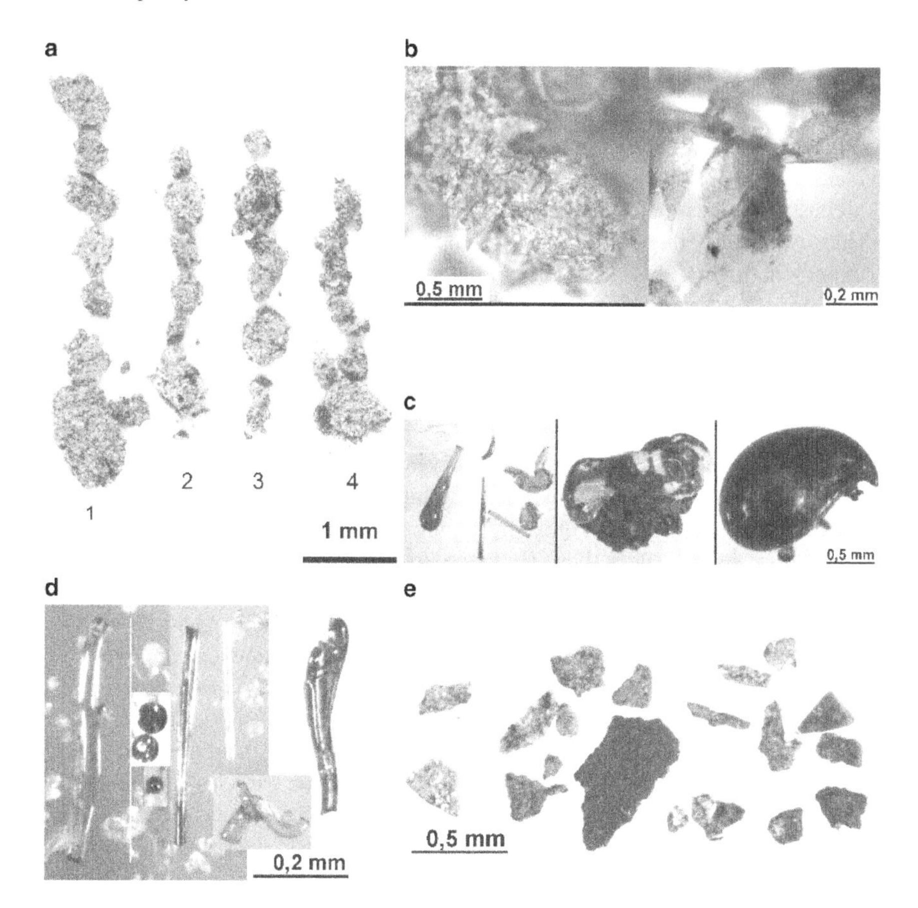

Fig. 5.1 Case study I. Soil micro-particles procured from site (a) Overview of material: 1 – comparative sample; 2 – window sill; 3 – sweater of the victim (owner of the flat), 4 – suspect's boots. (b) Fragment of cormophyte moss inside a soil micro-particle. (*left*) and extracted from it (*right*). (c) Glass fibre fragments extracted from comparator samples. (d) Glass fibre fragments extracted from washed soil traces, taken from the sill, the sweater and the boots respectively. (e) Micro-particles similar to paint fragments, taken from the comparative samples, the sill, the suspect's sweater and the suspect's boots *(see colour plate section for colour version)*

probability, from the place under the particular window in question, or from another place with soil that had the same features.

Case II: A Case of Assault with Intended Robbery

A girl was coming home from school at twilight. Passing a yard near her house she heard steps behind her. Somebody strongly pushed her in the back, and she fell down. Her assaulter began to strike her in the face demanding that she give him her mobile phone. Having duly got the phone the robber ran away. The girl began crying for help. The suspect was detained very soon thereafter, but he denied mugging her. There was no mobile phone found on his person, and it was later determined that the perpetrator would have had time to pass or sell the mobile phone to another person. Circumstantial evidence was therefore required to test the allegation that he was the perpetrator.

To test this allegation, comparator samples of soil procured from the location of the assault (as shown by the victim), and soil samples collected by investigators from the left and the right running shoes of the suspect were procured. During the visit to the scene of the crime, it was established that shortly before the assault, the region was subjected to construction for the replacement of insulation on heating pipes, and the soil in the vicinity was disturbed such that subsoil was present in the surface zones. An examination of comparator samples showed that the soil at the assault location was a clay, homogeneous in colour, and lacking in plant remains and humus. The amount of soil from the suspect's shoes (2.5 g) was sufficient to allow a full morphological and mineralogical examination including granulometric mass analysis. Small spherical particles coated with bitumen (Figure 5.2a) and having a very characteristic cellular structure and pearly sheen (Figure 5.2b) were present both in the comparator samples and in the soil taken from the suspect's shoes. The particles were extracted with chloroform and their chemical constitution determined using micro-probe X-ray fluorescence analysis. These particles were found to be a bituminous perlite, known to as a comparatively new heating insulation material. Hence it was established in comparative investigations that the soil substance procured from the suspect's shoes was identical to the soil of the comparator samples on the basis of a number of soil characteristics, including the same complex of technogenic particles (including bituminous perlite). Results of this investigation supported the conclusion that the soil taken from the suspect's shoes had essentially the same genetic and group characteristics associated with the scene of the crime, qualified that it could also have come from another place with soil having the same features.

Case III: A Case of Rape of a Teenager

A crime scene where the rape of a teenager occurred comprised a four-section basement of a house with a central heating system. There was a wet region therein, under some rusty water pipes. The items under investigation were the sport boots of the accused, the sport boots and sport trousers of the victim, and four comparator samples from the scene of the crime. Only a small mass (c. 30 mg) of trace material was available, comprising a mechanical mixture of soil micro-aggregates and micro-particles of natural and technogenic origin. The identification task here was to establish whether both individuals involved had been present at the site, via the similarity between the soil on the suspect's boots, the soil on the victim's boots and clothes, and the location. In this case, comparative examinations of soil micro-particles by light microscopy in combination with instrumental methods was very successful. It was established that the comparator sample was a grey-brown soil,

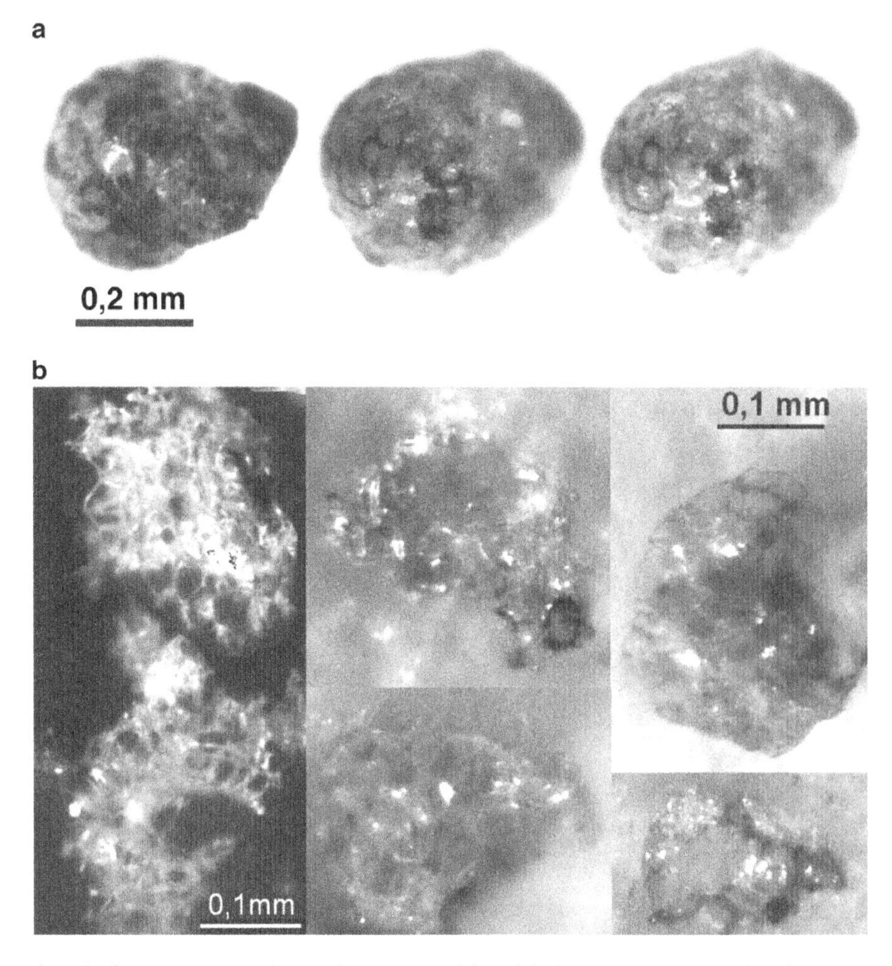

Fig. 5.2 Case study II. (a) Trace evidence comprising globules of bituminous perlite after several stages of washing in chloroform. (b) Inner structure of one such perlite particle *(see colour plate section for colour version)*

which included small fragments of glass fibres with iron hydroxide (Figure 5.3a), glass fibre-fragments from a heating pipe insulator (Figure 5.3b), aerosol ferromagnetic spheres from a calx material (Figure 5.3c), small fragments of bones (Figure 5.3d), and vivianite globules (Figure 5.3e). The vivianite was identified by micro X-ray fluorescence spectrometry (EAGLE-III) and electron microscope micro-probe energy dispersive X-ray analysis. Natural samples of vivianite mineral from a collection at the Museum of the Earth on the Moscow State University campus were used as standards. Vivianite formation is possible in anthropogenic conditions and may be found in buildings (Kazdym 2006). It was also observed that there were some soil micro-particles on the accused person's shoes of the same colour and texture as the soil component of the comparator samples. These micro-aggregates also contained small particles of iron hydroxide, glass fibres, aerosol balls, and a single vivianite globule.

Accordingly, integrated analyses of the mechanical mixture (microparticulate material from soil of contrasting nature here) enabled more objective information about the trace material to be obtained. Localised technogenic forming processes of Fe(III) hydroxide and vivianite mineral enabled the establishment to a high degree of certainty that the accused person was likely at the scene of the crime. The caveat is that the individuals could have been present in another basement of a house containing soil material of the same characteristics.

Case IV: A Murder of a Young Man

A decapitated corpse was found in a field about 75 km from Moscow. This case involved an attempt to determine the provenance of soil traces on his body. The items

Fig. 5.3 Case study III. (a) Soil micro-aggregates with embedded glass fibres. (b) Glass fibres with Fe (III) hydroxide. (c) Calx, the source of iron in vivianite formation. (d) Fragments of bones, the source of phosphorus in vivianite formation. (e) Vivianite globule in a soil aggregate (*left*) and extracted from it (*right*) *(see colour plate section for colour version)*

under investigation were four samples of soil, taken from different parts of the victim's body and five comparator samples from the place where the body was found. The presence of vivianite micro-globules in the stratifications (Figure 5.4a), taken from the corpse, suggested that the body had not come from the region where it had been found. The presence of vivianite mineral is not typical for natural soil cover sites in the Moscow suburbs. Glass microspheres were also found (Figure 5.4b) in the soil samples taken from the victim's body in contrast to the comparator samples. The morphological features of the glass particles were characteristic of fly ash from thermal power stations (Anshits et al. 2005; Consoly et al. 2007). In view of the variation in chemical composition of ash particles in the environs of thermal power stations (Vereshchagina et al. 2001), it was not possible to establish precisely from which thermal power station in or around Moscow these small particles were derived, even with the use of precise physical and chemical methods of analysis. However, taking into account that the body of the victim had been located southeast of Moscow, it was assumed that the most probable source of pollution by ashes was one of the largest thermal power stations located in the south or southeast of Moscow. Our laboratory's archives contained data about the presence of small particles of fly ash in the soils to the south of Moscow area. Furthermore, the locations of the pollution had been mapped. This allowed the conclusion that the provenance of the soil on the corpse was probably within 10–15 km from a source of such pollution. It transpired that the inspector who investigated the case then informed us that the last mobile phone calls made by the victim were from exactly the same region. This is an example of how anthropogenic particles can enable the more precise localisation of samples, but that additional information about the spatial location of sources that generate such material is required. The value of archive material and data is also well demonstrated in this case.

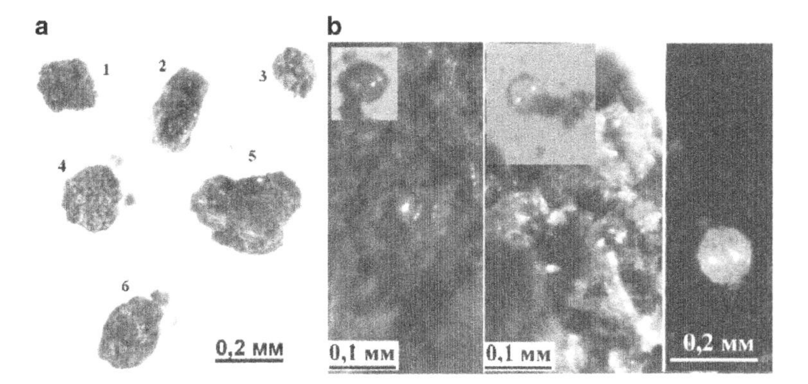

Fig. 5.4 Case study IV. Micro-particles procured from the corpse in this case. (a) Vivianite particles separated from soil traces obtained from the corpse. (b) Glass microspheres of fly ash inside (*left* two panels) and separated from (*right*) soil traces obtained from the corpse *(see colour plate section for colour version)*

Conclusions

This chapter demonstrates that an integrated analytical method to soil case examinations allows more objective information to be obtained about trace-forming materials in a soil forensic context. As it follows from our practice, the most objective information in forensic examination of small amounts of urban or highly technogenic soils is from a morphological examination of small to microscopic portions of soil, using light microscopy in combination with micro-probe fluorescence and X-ray analysis. These approaches can be used in both provenancing and comparator terms, providing an effective means of producing quality evidence for use in soil forensic science.

Acknowledgements We thank our colleagues Vladimir Sirotinkin and Sergey Sorokin in carrying out instrumental researches (micro-probe fluorescence and electron microscope microprobe energy dispersive X-ray analysis). Also we express our special thanks to Alastair Ruffell and Marianne Stamm for critical reading and helpful assistance in preparing this chapter.

References

Adya M and Collopy F (1998). How effective are neural networks at forecasting and prediction? A review and evaluation. Journal of Forecasting 17:481–495.

Aitken C and Taroni F (2004). Statistics and the Evaluation of Evidence for Forensic Scientists (Statistics in Practice). 2nd edition. Wiley, Chichester.

Aitken C, Biedermann A, Garbolino P and Taroni F (2008). Bayesian networks and probabilistic inference in forensic science (Statistics in Practice). Statistical Papers 49:393–394.

Anshits NN, Anshits AG, Bayukov OA, Salanov AN and Vereshchagina TA (2005).The nature of nanoparticles of crystalline phase in cenospheres and morphology of their shells. Glass Physics and Chemistry 31:306–315.

Biedermann A and Taroni F (2006). Bayesian networks and probabilistic reasoning about scientific evidence when there is a lack of data. Forensic Science International 57:163–167.

Bogatyryev VS (2001). Opportunities of statistical research of small parts for reconstruction of material conditions of material evidences being. In: Proceedings of Criminalistics of 21st Century, pp. 364–368. MVD, Moscow.

Carresy P, Causin V, Marigo A and Schiavone S (2004). Bayesian framework for the evaluation of fiber evidence in a double murder – a case report. Forensic Science International 141:151–170.

Cengiz S, Karaca AC, Çakir I, Üner HBand Sevendic A (2004). SEM-EDS analysis and discrimination of forensic soil. Forensic Science International 141:33–37.

Consoly NC, Heineck KS, Coop MR, Fonseca AV and Ferreira C (2007). Coal bottom ash as a geomaterial: influence of particle morphology on the behavior of granular materials. Soils and Foundations 47(2):361–373.

Fitzpatrick EA (1993). Soil Microscopy and Micromorphology. Wiley, Chichester.

Kazdym AA (2006). Technogenic minerals of cultural layers. In: A Technogenic Sediments of Ancient and Modern Urban Territories. Paleoecological Aspect (Ed. SA Nesmeyanov), pp. 116–120. Nauka, Moscow.

Khan KY, Pozdnyakov AI and Son BK (2007). Structure and stability of soil aggregates. Eurasian Soil Science 4:451–456.

Koldin VY (1996). Expert identification in forensic examinations. Theoretical basis. (Methodical manual for post-graduates, investigators and judges). RFCFR at the Ministry of Justice, Moscow.

Mitrychev VS (1976). Criminalistic identification of the whole in parts. In: Theory and Practice of Identification the Whole in Parts. Collection of Research Works 24:3–111.VNIISE at the Ministry of Justice, Moscow.

Parfeneva E and Yarilova E (1977). The Manual for Micro Morphological Researches in Soil Science. Nauka, Moscow.

Pye K and Blott SJ (2004). Particle size analysis of sediments, soils and related particulate materials for forensic purposes using laser granulometry. Forensic Science International 144:19–27.

Pye K and Croft DJ (2007). Forensic analysis of soil and sediment traces by scanning electron microscopy and energy-dispersive X-ray analysis: an experimental investigation. Forensic Science International 165:52–63.

Pye K, Blott SJ, Croft DJ and Witton SJ (2007). Discrimination between sediment and soil samples for forensic purposes using elemental data: An investigation of particle size effects. Forensic Science International 167:30–42.

Rawlins BG, Semon SJ, Hodgkinson EH, Riding JB, Vane KH, Poulton C and Freeborough K (2006). Potential and pitfalls in establishing the provenance of earth-related samples in forensic investigations. Journal of Forensic Sciences 51:832–844.

Ruffell A and Wiltshire P (2004). Conjunctive use of quantitative and qualitative X-ray diffraction analysis of soils and rocks for forensic analysis. Forensic Science International 145:13–23.

Sjerps M (2006). The role of statistics in forensic science casework and research. Problems of Forensic Science 55:82–90.

Tjurikova VV (1976). The conception of identification the whole in parts in soil forensic examination. In: Theory and Practice of Identification the Whole in Parts. Collection of Research Works 24:128–130.VNIISE at the Ministry of Justice, Moscow.

Tjurikova VV (1980). Local grounds as an object of criminalistic identification. In: Theoretical and Methodical Questions of Forensic Soil Examination. Collection of Research Works 47:13–27. VNIISE at the Ministry of Justice, Moscow.

Vereshchagina TA, Anshits NN, Zykova ID and Salanov AN (2001). Preparation of cenospheres of controlled composition from energy ashes and their properties. Chemistry for Sustainable Development 9:306–315.

VNIISE (1993). Complex methods of forensic soil examination in 3 parts. The methodical recommendations for experts, inspectors and judges. VNIISE at the Ministry of Justice, Moscow.

Yukun C, Xiaofeng L and Yunfeng L (2004). An e-mail filtering approach using neural network. In: Lecture Notes in Computer Science. Advances in Neural Networks (Ed. D Hutchison), pp. 688–694.Springer, Berlin.

Chapter 6
Characterisation and Discrimination of Urban Soils: Preliminary Results from the Soil Forensics University Network

Andrew R. Morrisson, Suzzanne M. McColl, Lorna A. Dawson and Mark J. Brewer

Abstract Soils from urban environments have a particular pertinence in a forensics context, yet relatively little is known about urban soils compared to those from non-urban regions. Collaboration between universities with forensic science teaching can provide opportunities to develop shared databases for potential use by forensic practitioners, such as the fibre database project hosted by Staffordshire University. The aim of this network project is to broaden the knowledge of soil characterisation and discrimination within the urban environment. In the context of soil evidence, a coordinated network of projects has been established as collaborative ventures across a range of universities and research organisations. A coordinated approach with a high degree of similarity in protocols and quality control between the participating organisations has been established. Between-city and within-city discrimination of common urban land-use vegetation (LUV) classes is being tested. Preliminary results for the parameter of soil colour over one year for sites in Scotland and north-west England are presented.

Introduction

Soil is a complex matrix composed of a unique combination of mineral, organic and molecular level signatures. As soil particles can readily adhere to, and transfer from, items such as clothing, shoes, vehicles and tools, they have the potential to

A.R. Morrisson (✉)
School of Life Sciences, The Robert Gordon University, Aberdeen AB25 1HG, UK
e-mail: a.morrisson@rgu.ac.uk

S.M. McColl
School of Biomolecular Sciences, Liverpool John Moores University, Liverpool L3 3AF, UK

L.A. Dawson
The Macaulay Institute, Craigiebuckler, Aberdeen AB15 8QH, UK

M.J. Brewer
Biomathematics and Statistics Scotland, The Macaulay Institute, Craigiebuckler, Aberdeen AB15 8QH, UK

be used as trace evidence, potentially linking or eliminating suspects to and from a crime scene. Such evidence can provide information of use in the intelligence phase of a police investigation, including the elimination of areas of enquiry, thus limiting areas of search, or also as evidence in a court of law as to the presence of a suspect or a vehicle at a particular location. To date, the application of soil evidence in criminal cases has been largely under-utilised.

A recent development in UK Higher Education has seen a large number of undergraduate forensic science degree courses being developed. Many of these have evolved from applied chemistry courses, and students study analytical science topics to a significant depth as part of the curriculum. Final year student projects often involve the application of these analytical skills in the forensic context. Therefore, opportunities exist for collaboration between universities to develop databases for use by forensic practitioners, such as the fibre database project hosted by Staffordshire University (2008). Within the context of soil information, a coordinated network of projects is being developed, as collaborative ventures across research organisations, such as the SoilFit project (Macaulay Institute 2008a), and the Soil Forensics University Network, SoilFUN (Macaulay Institute 2008b).

The national soil surveys for Scotland, England and Wales, and Northern Ireland reside with the Macaulay Institute, National Soil Resources Institute (NSRI) and Department of Agriculture and Rural Development Northern Ireland (DARDNI), respectively. Soil maps for these areas are published at scales of between 1:25,000 and 1:250,000, and include reports (memoirs) incorporating information about the soils and their associated landscapes and environments. For some areas, the data on soils were also interpreted to provide guidance about the suitability for specific land uses, such as agriculture or forestry. Soil maps at 1:250,000 exist for all of the UK. In Northern Ireland, a high-resolution survey named TELLUS was completed in 2007 by the Geological Survey of Northern Ireland (GSNI), providing full coverage and analytical data for a range of soil properties. In addition, national coverage for certain attributes of soils and all of the geology for the UK is held by the British Geological Survey (BGS). Underpinning these maps are field observations and interpreted aerial photographs. Soil profile pits are dug, described and samples taken according to agreed methodologies and protocols, and then the samples are analysed for a suite of chemical and physical properties.

Forensic examination of soil evidence conventionally considers various soil attributes such as colour, particle size, mineralogy and palynology (for example pollen, spores, and dinoflagellate cysts). Colour assessment can be used to eliminate soil samples from further evaluative comparison (Junger 1996). Mineralogical techniques define the general geological region of origin (Ruffell and Wiltshire 2004; Ruffell and Mckinley 2005), and palynology provides detailed site specific information relating to the likely vegetative history (Horrocks and Walsh 1999; Brown et al. 2002; Mildenhall et al. 2006). Such soil comparisons have a proven record but, due to requirements for specialised equipment and considerable expert knowledge, they are time-consuming, specialised and costly (Bryant et al. 1990). Of potential interest for forensic investigations are soil samples taken from surface layers (0 to 1 cm depth) from urban areas, which are likely to be of limited size (Rawlins et al. 2006).

The aim of the SoilFUN project is to broaden the knowledge of soils within the urban environment. It is necessary to establish a coordinated approach with a high degree of similarity in protocols and quality control (QC) between the participating organisations. In doing so, the SoilFUN partners are building a database of urban soils information from across the UK. Specifically it is hoped to address between city and within city discrimination of common urban land-use vegetation (LUV) classes. This chapter describes the background to the approach being taken, and presents preliminary results, for sites in Scotland and north-west England.

SoilFUN

Several soil parameters have been selected as having the potential to allow discrimination of samples taken from the soil surface of different land-use types found in urban areas. These were selected to reflect measurements for which appropriate equipment is likely to be available in university departments and accessible to undergraduate and postgraduate students for use in project work. In addition, they were considered suitable for small sized samples, which is important for application to forensic case work. The details of the parameters are shown in Table 6.1. Full sets of protocols including standard operating procedures (SOPs) have been developed for each test and are available from the authors.

Each individual project targets four LUV classes, each of which is represented at four sites across the urban area. Four replicate samples are taken at each site, giving a total of 64 samples per area. Urban soils are very diverse and challenging to classify, and due consideration must be given to the selection of urban soil LUV classes. Hollis (1990) described a hierarchical system of classification, in which a clear, high level, distinction is made between disturbed/non-disturbed soils and is likely to be most appropriate for urban situations. For the purposes of SoilFUN, it is acknowledged that urban soils fall into three broad categories encompassing a range of disturbance:

Table 6.1 Soil attributes and equipment required for the parameters described in the SoilFUN work programmes

Soil attribute	Equipment required
Visual inspection	None
Colour determination	Munsell Colour Charts/Spectrophotometer
Organic matter content	Muffle furnace
Wax biomarker characterisation (*n*-alkanes)	GC/GC-MS
Soil organic and mineral characterisation	ATR-FTIR

Where, GC/GC-MS is Gas Chromatography/Gas Chromatography-Mass Spectrometry, and ATR-FTIR is Attenuated Total Reflectance/Fourier Transform Infrared.

1. Semi-natural: Soils such as recreational woodlands, scrub-like green-space used for dog walking. Distinct O and A horizons may persist. These may be relatively non-disturbed, depending upon age.
2. Man-made – managed: Soils such as council managed flowerbeds and lawn areas. It is likely that these will have a man-made organic A-horizon.
3. Man-made – disturbed: Highly disturbed soils such as road sides, canal and rail embankments and wasteland or building sites, likely to contain foreign mineral material, and lack of an organic/A-horizon.

These three classes are common to all SoilFUN projects, and a fourth is selected that is relevant to a particular urban location. The selection of three common LUV classes allows the testing of between city discrimination. The fourth LUV is from one of the following: riverbanks/mudflats, canal embankments, residential gardens, playing fields, abandoned land/brownfield sites etc, depending upon the interests of the individual student/university. Further details of the three standard urban LUV and their relevance to particular types of crime is given in Table 6.2.

In order to apply relevant multivariate statistical approaches, it is important that the chosen LUV classes are represented at four different sites across an urban area i.e. four different lay-bys, four different managed flowerbeds or shrub borders within a city and four different areas of woodland. This provides information on the degree of similarity/dissimilarity of soil characteristics within a LUV class across a given urban area. Such information is currently unavailable for urban soil LUV classes. Maps of land cover or topographic details (e.g. from the Ordnance Survey) are first examined, informing the selection of four sites which reflect the LUV types in that city. The underlying geology of the area is considered by reference to geological maps and, if possible, samples taken which are distributed across the main bedrock types.

Site Replication

Previous work (Milton 2006) showed that four replicates were the minimum number required to obtain discrimination between garden sites in the city of Aberdeen. Based on this experience, four replicate samples were taken for the sites selected to represent an area that might be involved in a scene of crime in each LUV type. This provides information on the level of variability at a potential crime site.

The sampling approach differs between different LUV types. At the four roadside lay-by sites, the samples are taken at a standard distance from the road, in a parallel line with the road, 0.5 m apart. This is to represent a car pulling off the road and material adhering to the wheel arch.

In the flowerbed, four samples are taken in a transect at 0.5 m intervals, parallel to the edge of the border. This could reflect a zone where a struggle had taken place or where a body had been buried.

Table 6.2 Characteristics of the three standard urban LUV and their relevance to particular types of crime

Urban LUV	Type and level of disturbance	Description	Soil sampling	Potential crime relevance
Semi-natural: urban woodland pocket	For example, semi-natural unmanaged vegetation – low	Mixed deciduous woodland. In Aberdeen, these are largely dominated with beech, lightly interspersed with oak, elm, holly, rowan. Under-story tends to be either sparse grass or bracken	Sampling should be off trodden pathways by 10 m	Relevant to physical assault cases where victim is pulled off path and away from obvious view, or dumping or hiding activities
Man-made managed: flowerbed or shrub border	For example, managed parkland – medium	Border area in council managed city park. These tend to be planted with shrubs toward the back and bulbs at the front	Sampling should be at 20 cm from the front edge	Relevant to physical attacks and youth crime. Flowerbeds are often cut across to shorten exit route from scene of crime and avoid lit path areas
Man-made disturbed: roadside lay-by	For example, mineral man-made environment – high	These sites should be semi-official parking areas, where vehicles can pull completely off the road, and not just compacted soil at the roadside. Very mineral surface, used as parking to access woodland pockets and scrub areas. No vegetation coverage	Sampling across the length and breadth of the area that a vehicle may have driven over	Relevant to vehicular evidence and access to off the beaten track area. Near dumping or hiding areas

In the woodland, the samples are taken at a minimum of 10 m into the woodland (and at least 10 m from any path) 1 m from a tree base. This is to represent a site where a burial may have taken place. The species of tree is noted or a leaf collected for further identification if necessary. The four samples are taken 0.5 m apart, along a line radiating out from the tree base, the species of which is recorded.

Soil Characterisation

The characterisation of the soils is carried out using the SoilFUN prescribed parameters (Table 6.1). This starts with a visual inspection of the soil. The approximate colour, the colour distribution and uniformity of the sample is noted, the stone abundance (few 1–5%, slightly stony 6–15%, moderately stony 16–35%, very stony 36–70%, extremely stony >70%) and the root abundance (few, common, many, abundant) is described. The presence of any additional trace materials is noted (such as hair, fibres, paint chips, glass etc.) as these may have a potential forensic value and are recorded as examples of good practice.

For this study, soil colour was done either visually estimated, using Munsell colour charts, or measured by spectrophotometry, to provide notations in hue, value and chroma. Where mottles are present in a soil sample, the colour of the matrix and mottles are determined separately. The colour of the soil matrix is determined on a finely-ground dry sample, using the Munsell Colour System which gives notations in hue, value and chroma. Colour is a soil characteristic frequently used as a fast and simple means of differentiating soils from different locations, providing some information as to the mineral content (Viscara Rossel et al. 2006).

Where possible, only one observer and a new Munsell colour chart were used, to reduce variation introduced by different observers and change in colour chips (Islam et al. 2004; Sanchez-Maranon et al. 2005). Readings were taken both of air-dried and wet soil and then converted using the free-ware colour conversion software previously mentioned. In some horizons, particularly mottled horizons that have more than one colour, more care is required to distinguish one colour from another. In this case, both colours are recorded.

The organic matter content of each sample is determined by ashing to 105 °C. This "loss on ignition" procedure is a pre-requisite to the determination of *n*-alkanes in the soil. It is also an appropriate variable to use to differentiate organic from mineral soils.

The major part of the characterisation of the surface soils is the determination of the wax compounds which derive from the decomposition of the vegetation. The method used is an adaptation of that developed for use on plant material and faeces (Dove and Mayes 2006) and adapted for use on soil (Dawson et al. 2004). As with most other lipids, the common principle for analysing wax compounds in soil is a stepwise process involving solvent extraction, purification to separate the crude extract into particular lipid compound classes, followed by analysis of individual compounds by gas chromatography (GC) or gas chromatography/mass spectrometry (GCMS), which offers higher sensitivity and more definitive compound identification. There is a variety of classes of wax marker including *n*-alkanes, fatty acids, fatty alcohols etc. Each has the potential for discrimination, but the use of major plant *n*-alkanes with odd-numbered carbon chains in the range C21 to C35 show good potential for discrimination (Dawson et al. 2004).

Fourier transform infrared (FTIR) analysis can provide information both on the soil organic matter and on the mineral composition. By using an attenuated total reflectance (ATR) sampling accessory, spectra can be obtained rapidly without the need for complex sample preparation. Spectra are converted to 'Jcamp' format prior to storing on the database.

Other analysis techniques used by participating organisations include: soil pH, particle size analysis, elemental analysis by atomic spectroscopy, mineral analysis by X-ray powder diffraction, pollen characterisation and soil DNA.

Appropriate quality control (QC) in such a collaborative project is important. Two QC soils are provided for all attributes to ensure consistency of analysis. Additionally for the GC analysis, a mixed chemical standard containing a known mixture of *n*-alkane molecules is prescribed and provided to all partners. This allows the standardisation of GC data provided from the different analytical instruments available in each university department.

Statistical Analysis

Within individual projects, it will be possible to use appropriate statistical methods to determine differences between LUVs, and between site locations. Such comparisons may be made on the basis of individual and combined parameters. However, the overall aim of the SoilFUN project is to use the dataset as a whole, and to test the ability to discriminate between cities. In addition, information about variability at different LUV types will be used to inform thinking about the replication necessary across the range of site types. As the SoilFUN partnerships develop, it is anticipated that subsequent projects will return to the same sites to test year on year variation, and to test mock crime scene scenarios against the developing database.

Preliminary Results

As part of the SoilFUN project, data for different cities were collected by students from the Robert Gordon University, Aberdeen and Liverpool John Moores University. The urban areas used were were Aberdeen, Edinburgh, Glasgow and a region based around Burscough, a small town on the Lancashire plain. Details of the locations used and the LUV types sampled are given in Table 6.3.

In general four LUV types at each of four sites in each urban location were sampled. In the case of Edinburgh, time constraints for the student project only allowed three sites to be sampled from each LUV. Only the three LUV types common to each project are presented here; roadside lay-by, park flowerbed and woodland. Data presented in this chapter relate only to an initial comparison of colour data from a small number of cities.

Table 6.3 The locations used and the LUV types sampled for the initial phase of the SoilFUN Project

Location	LUV class	Site sampled
Aberdeen	Park flowerbed	Duthie Park; Seaton Park; Westburn Park; Hazelhead Park
	Roadside lay-by	Dobbie's Garden Centre; Queen's Links; Airport; Doonies Farm
	Woodland	Den Wood; Perwinnes Moss; Seaton Park; The Gramps
Edinburgh	Park flowerbed	Corstorphine; Braids; Heriot-Watt
	Roadside lay-by	Corstorphine; Braids; Heriot-Watt
	Woodland	Corstorphine; Braids; Heriot-Watt
Glasgow	Park flowerbed	Alexandra; Linn; Queens; Kelvingrove
	Roadside lay-by	Alexandra; Linn; Queens; Kelvingrove
	Woodland	Alexandra; Linn; Kelvingrove
Lancashire	Park flowerbed	Knowsley Road Park, Ormskirk; Junction Lane Park, Burscough; Memorial Park, Burscough; Coronation Park, Ormskirk
	Roadside lay-by	Flax Lane, Burscough; Blythe Lane, Lathom; Maltkiln Lane, Aughton; Vicarage Lane, Ormskirk
	Woodland	Ruff Woods, Ormskirk; Aughton Woods, Granville Park, Aughton; Rufford Park Wood, Rufford; Ashurst Beacon, Dalton

Examples of Preliminary Results from Samples from Scottish Cities and Lancashire

Statistical analysis of the colour data was performed using canonical variate analysis (CVA). This is a multivariate statistical method used when each observation in a dataset is associated with one of a number of groups. It works by finding linear combinations of the original variables which maximise the ratio of between-group to within-group variation, and is thus most useful for highlighting differences between the groups. In a similar manner to principal component analysis (PCA), it is possible to derive scatterplots of the canonical variate scores to provide a graphical tool for identifying differences and similarities between groups. Unlike with PCA, scaling is irrelevant in CVA since it is the ratio of between- and within-group variances which are examined.

Figure 6.1a shows a CVA plot for the complete data set from the four cities. A considerable amount of overlap can be observed between the groups. It appears that colour alone is unlikely to be enough to enable identification of a specific location of origin. The data for Edinburgh and Glasgow, coming from the same central belt location, were the most similar. The data from Aberdeen and Lancashire were most dissimilar. Although there was considerable variation, with further replicated sampling, it should be possible to ascertain predictions with a defined probability level via a statistical classification procedure; however, individual samples would not be classified accurately.

Figure 6.1b shows the results of a CVA of data from the sites within Scotland only, grouped by LUV, showing no clear discrimination between the three LUVs.

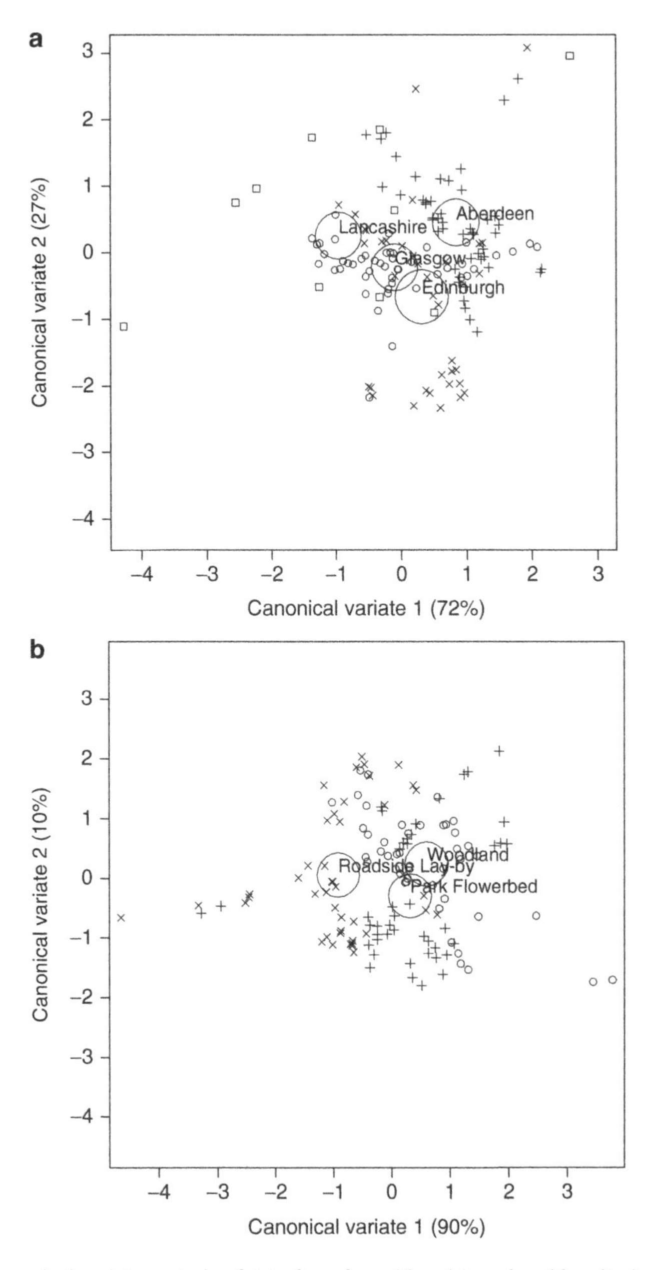

Fig. 6.1 Canonical variate analysis of data from four cities: (**a**) analysed by city including four cities sampled. Aberdeen (+), Edinburgh (X), Glasgow (O), Lancashire (□); (**b**) analysed by LUV type including Scottish cities only. Park Flowerbed (+), Roadside lay-by (X), Woodland (O). The circles show the 95% confidence levels around the mean values for each city.

However, there is a suggestion that the roadside lay-by category can be discriminated on the basis of colour alone. It is possible that data from a wider range of analytical methods would help improve discrimination between roadside lay-bys and the other two categories.

In Figure 6.2a, data aggregated across all sites from the city of Edinburgh alone is analysed, with again the roadside LUV being significantly separated from the other two classes. Figure 6.2b, also shows data from Edinburgh alone, but this time aggregated across the three LUV types (woodland, park flowerbeds and lay-bys). Here there does appear to be a useful degree of separation between some of the sites, although two of the Heriot-Watt sites and one of the Corstorphine sites overlap to a large extent. Figures 6.2c and 6.2d show results of a CVA applied to the woodland and park flowerbeds sites of Edinburgh respectively. It is clear that once city and LUV have been determined, discriminating between the sites is a much simpler task than when considering all cities or all LUVs together. This preliminary data analysis

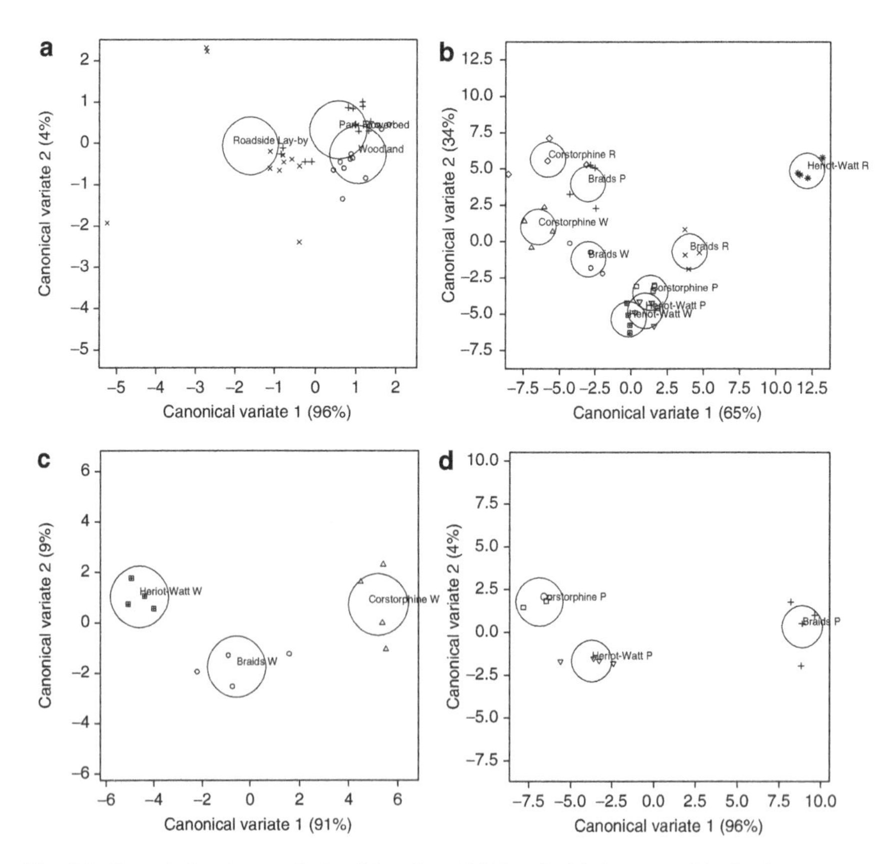

Fig. 6.2 Canonical variate analysis of data from Edinburgh. **(a)** Aggregated by land-use vegetation class, Park Flowerbed (+); Roadside Lay-by (×); Woodland (o). **(b)** Aggregated by site, R – roadside lay-by, W – woodland, P – park flowerbed. **(c)** Aggregated by woodland locations in the city. **(d)** Aggregated by park locations in the city

provides evidence that if intelligence suggests a crime has been committed in any of the three particular types of LUV studied here, one could potentially relate the location back to any of the individual site loci examined.

Conclusions

This collaborative project is in the early stages. Good collaboration with a number of Universities in the UK has been established, leading to the database being populated. As more Universities become involved, it is hoped that a significant public resource will be established, which will eventually be of use to forensic practitioners, particularly in respect to the intelligence phase of an investigation.

Preliminary results show that increasing the extent and availability of analytical information will support greater discrimination between samples from different sites. Colour may be useful for excluding or relating an unknown sample to a known type of land use within a certain city. It is likely that, combined with other indicators that reflect the geological basis of a city (e.g. FTIR/XRD), which may be more powerful in provenancing the individual cities, and with the inclusion of other indicators of organic matter characterisation (e.g. alkanes), site and LUV specific discrimination will be improved. When the more extensive datasets are available such hypotheses will be tested. Further information could potentially be gained through comparison of soil material to data contained within national (Dawson et al. 2007) and worldwide soil archives (Hallett et al. 2006), introducing a new resource to forensic investigation.

References

Brown AG, Smith A and Elmhurst O (2002). The combined use of pollen and soil analyses in a search and subsequent murder investigation. Journal of Forensic Sciences 47:614–618.

Bryant VM, Jones JG and Mildenhall DC (1990). Forensic palynology in the United States of America. Palynology 14:193–208.

Dawson LA, Miller DR and Towers W (2007). Dirty work: New uses for soil databases. http://vector1media.com/article/feature/dirty-work-new-uses-for-soil-databases/ Accessed 22/07/2008.

Dawson LA, Towers W, Mayes RW, Craig J, Vaisanen K and Waterhouse EC (2004). The use of plant hydrocarbon signatures in characterizing soil organic matter. In: Forensic Geoscience: Principles Techniques and Applications (Eds. K Pye and DJ Croft) 232:269–276. The Geological Society of London Special Publications, London.

Dove H and Mayes RW (2006). Protocol for the analysis of *n*-alkanes and other plant-wax compounds and for their use as markers for quantifying the nutrient supply of large mammalian herbivores. Nature Protocols 1:1680–1697. http://www.natureprotocols.com/2006/11/16/protocol_for_the_analysis_of_n.php.

Hallett SH, Bullock P and Baillie I. (2006). Towards a world soil survey archive and catalogue. Soil Use and Management 22:227–228.

Hollis JM (1990). The classification of soils in urban areas. In: Soils in the Urban Environment (Eds. P Bullock and PJ Gregory), pp. 5–27. Blackwell Scientific, London.

Horrocks M and Walsh KAJ (1999). Fine resolution of pollen patterns in limited space: Differentiating a crime scene and alibi scene seven meters apart. Journal of Forensic Sciences 44:417–420.

Islam K, McBratney AB and Singh B (2004). Estimation of soil colour from visible reflectance spectra. Supersoil 2004: 3rd Australian New Zealand Soil Conference, University of Sydney, Australia, 5th to 9th December 2004. www.regional.org.au/au/asssi/supersoil2004.

Junger EP (1996). Assessing the unique characteristics of close-proximity soil samples: Just how useful is soil evidence? Journal of Forensic Sciences 41:27–34.

Macaulay Institute (2008a). The SoilFit Project. http://www.macaulay.ac.uk/soilfit/. Accessed 01/07/2008.

Macaulay Institute (2008b). The SoilFun Project. http://www.macaulay.ac.uk/forensics/soilfun/. Accessed 01/07/2008.

Mildenhall DC, Wiltshire PEJ and Bryant VM (2006). Forensic palynology: Why do it and how it works. Forensic Science International 163:163–172.

Milton S (2006). The potential value of fatty acid markers as soil evidence. MSc thesis. The Robert Gordon University, Aberdeen, UK.

Rawlins BG, Kemp SJ, Hodgkinson EH, Riding JB, Vane CH, Poulton C and Freeborough K (2006). Potential pitfalls in establishing the provenance of earth-related samples in forensic investigations. Journal of Forensic Sciences 51:832–845.

Ruffell A and Mckinley J (2005). Forensic geoscience: Applications of geology, geomorphology and geophysics to criminal investigations. Earth-Science Reviews 69:235–247.

Ruffell A and Wiltshire P (2004). Conjunctive use of quantitative and qualitative X-ray diffraction analysis of soils and rocks for forensic analysis. Forensic Science International 145:13–23.

Sanchez-Maranon M, Huertas R and Melgosa M (2005). Colour Variation in standard soil-colour charts. Australian Journal of Soil Research 43:827–837.

Staffordshire University (2008). The Production of National Trace Evidence Databases. http://www.staffs.ac.uk/faculties/sciences/research/forensic_science/#production. Accessed 01/07/2008.

Viscara Rossel RAV, Minasny B, Roudier P and McBratney AB (2006). Colour space models for soil science. Geoderma 133:320–337.

Chapter 7
Environmental Considerations for Common Burial Site Selection After Pandemic Events

Anna Williams, Tracey Temple, Simon J. Pollard, Robert J.A. Jones and Karl Ritz

Abstract In light of the increasing threat of an avian flu pandemic in the UK, the Home Office have been investigating a range of methods for managing the potential problem of excess deaths that could exceed the capabilities of existing burial and funeral facilities. There is currently unprecedented pressure on the Government to find an environmentally, ethically, socially and economically sound solution to the problem of disposal of bodies. The use of common burials or mass interments has been mentioned as a possible means for disposal of the 'excess deaths' that may be manifest. This chapter examines the potential environmental considerations and consequences of the development and utilisation of such mass burials in both the short and long term. Structured risk management approaches, including source-pathway-receptor analysis of the potential hazards, are reviewed. Such research is informed by previous incidents such as the UK Foot and Mouth Crisis of 2001, where large numbers of animal carcasses were buried in mass graves that would be analogous to those after a human pandemic in terms of environmental impact. It also draws from previous environmental waste management research and strategies that are in place to mitigate the environmental impact of other large waste disposal mechanisms, such as landfill sites. Factors which should be considered when selecting a site for the purpose of constructing large common burial pits such as body decomposition, soil characteristics, the potential for groundwater contamination, vegetation and ecology, and the practicality of implementing contingency or mitigation measures are reviewed. Some recommendations are given for common grave site selection through analysis of soil characteristics, the application of soil databases, and how existing taphonomic knowledge may inform these issues.

A. Williams(✉) and T. Temple
Department of Applied Science, Security and Resilience, Defence Academy of the UK, Cranfield University, Shrivenham, SN6 8LA, UK
e-mail: a.williams@cranfield.ac.uk

S.J. Pollard
Sustainable Systems Department, School of Applied Science, Cranfield University, Cranfield, Bedfordshire MK43 0AL, UK

R.J.A. Jones and K. Ritz
National Soil Resources Institute, Natural Resources Department, School of Applied Sciences, Cranfield University, Cranfield, Bedfordshire, MK43 0AL, UK

K. Ritz et al. (eds.), *Criminal and Environmental Soil Forensics*

Introduction

This chapter explores the potential environmental effects following the predicted use of so-called 'common burials', as a means of managing the increased numbers of deaths over a short time during pandemics or mass disasters. We consider how the disciplines of soil forensics, environmental science and particularly taphonomic studies of human remains, might inform such considerations. Epidemics can occur on local, national or international scales. Contingency planners have to consider a wide range of limiting factors, such as the availability and area of land, and infrastructure resources. In the event of a pandemic, the UK Home Office (2007) suggests that common burials may be a solution to increased demands for cemetery space, suggesting that "large areas within the cemetery would need to be allocated for the common grave" and "it cannot be ruled out that new ground will need to be identified". We provide a general, broad overview of some of the issues worthy of consideration in any potential identification of suitable land for this purpose. In particular, we concentrate on factors such as body decomposition, soil characteristics, potential for groundwater contamination, vegetation and ecology that should be considered when selecting a site for the purpose of constructing large common burial pits, and the practicality of implementing contingency or mitigation measures.

Notwithstanding the personal and societal impacts of disease pandemics, wherever they occur, there remains a need to plan and account for the disposal of the dead in a respectful manner that preserves customs and protects public health and the environment.

It is imperative for such contingency plans to be prepared and in place in advance, as was found with the UK's experience of the Foot and Mouth Crisis of 2001. As Drummond (1999) iterated "there is little time to discuss the merits of one disposal method against another when disease has broken out and carcasses on the infected premises are starting to decompose".

The potential for burial grounds to pollute the environment, and thus the effective management of this risk, has been the subject of substantive research (Pacheco et al. 1991; Janaway 1997; Üçisik and Rushbrook 1998; Dent and Knight 1998; Young et al. 1999; Spongberg and Becks 2000a, b; Hart 2005). In general terms, when bodies and grave contents decompose, decomposition products are released to the wider environment. The human body (about 70 kg) is a complex matrix of organic (typically 17% w/w protein; 17% fat and 6% w/w carbohydrate) and inorganic (predominantly N, P, Ca, Na, S) constituents. Decomposition also poses a microbiological hazard. However, for cemeteries operating under normal conditions, there appears to be a growing consensus that the principal receptor of concern for the key pollutants (which are usually regarded as NO_3^- and $NH_4^+{}_{(aq)}$) is usually the groundwater below the site (Pacheco et al. 1991; Knight and Dent 1998; Lelliot 2002; Buss et al. 2003; Dent 2005; Pollard et al. 2008). Common burials pose threats similar to that of 'normal' cemeteries, but on a larger scale, and some of the potential hazards are due to the volume of remains and coffin contents interred, as well as the long-term disruption to the land. According to the Home Office (2007) "cemetery managers should plan for alternative ways of providing graves" and "... are likely to want to move to the provision of common graves, which would allow interments to be undertaken more quickly due to the more efficient mechanical preparation of the site".

Risk Management Frameworks

Decisions about the safe and responsible disposal of the dead must take account of the loading in common graves (i.e. the number of bodies), the preparation of bodies and burial practice, the full grave contents, and local environmental settings above- and below-ground in the context of a very wide range of environmental receptors of statutory and non-statutory status (Kim et al. 2008). A potentially useful approach to integrating the key issues in relation to management of common burials is through structured risk management (Figure 7.1). Such frameworks have been widely used in considering locations for animal carcass disposals in the UK in the event of disease outbreaks, and they provide a means of identifying, and thus informing, the management of key exposures to public and environmental health in the short, mid- and long term (Pollard et al. 2008). The application of risk-based decision-making may also refocus contingency plans towards the critical points of control that are essential to have well managed in the event of such large scale events (Pollard et al. 2008).

Successful risk management has to include appropriate co-ordination of the wide range of organisations involved. These disparate parties have to be in concordance and satisfied at every stage of the management of the disposal in common burial sites. Such issues were identified during the Foot and Mouth Crisis in 2001 (Anderson 2002).

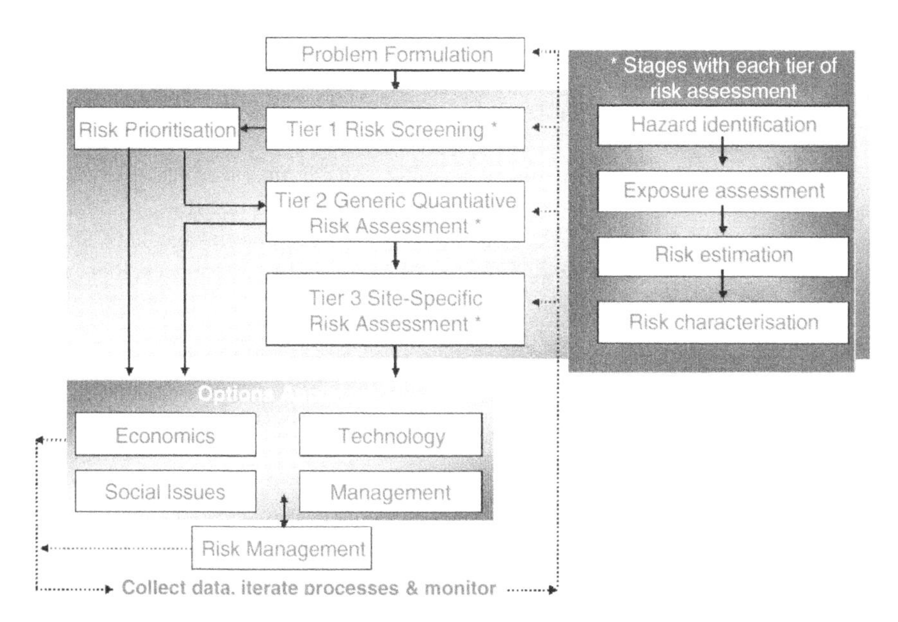

Fig. 7.1 A generalised framework for environmental risk management (after DETR, IEH, and Environment Agency 2000)

Table 7.1 Organisations in the UK with responsibility for environmental and public issues that might be raised with common burial site selection (After Anderson 2002)

Issue	Lead oganisation
Government policy and legislation affecting the environment	Department for Environment, Food and Rural Affairs (Defra)
	Scottish Environmental Protection Agency
	National Assembly for Wales Department of Health
Air pollution, i.e. noise, smoke, smells	Local authority
Environmental health	
Planning permission	Local planning authority
Contaminated land	Local authority (sometimes EA will have a lead role)
Supply of public drinking water	Local water company
Monitoring of quality of private drinking water supplies	Local authority
Sites of special scientific interest and nature reserves	English Nature
National parks	Appropriate National Park Authority
Rights of way and access to the countryside	Countryside Agency
Drinking water quality	Drinking Water Inspectorate

In the UK, such organisations have responsibility and jurisdiction over the different issues discussed in this chapter (Table 7.1).

For the risk assessment, which dominates the upper parts of Figure 7.1, we are concerned with understanding and managing the hazard and any potential receptor that it may affect. Conventional approaches to environmental risk assessment (Pollard et al. 2006) place a strong reliance on characterising these relationships and, if necessary, where potential risks are deemed significant changes may be required (Figure 7.2).

Understanding the source, pathway, receptor concept may then become critical in making informed decisions or appropriate mitigating actions. It is important to note that, in some situations, the source, pathway, receptor linkage may not be complete, for example where a direct receptor exists. In these instances, there may be no significant risk to consider.

Source

Body Decomposition

The products of decomposition from human burials can be divided into two broad categories: natural and synthetic. The decomposition products from a body include water and all the naturally occurring biological elements but these are rarely miner-

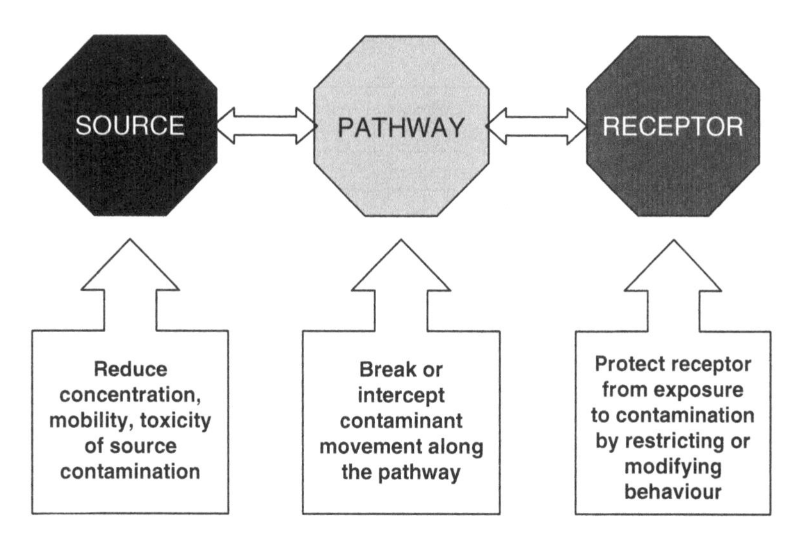

Fig. 7.2 Conventional approach to environmental risk assessment and risk management

alised to elemental form during decomposition and occur in a variety of compounds that may be more or less noxious to the environment or operators. Large quantities of ammonium are typically released from decomposing cadavers, as well as soluble organic nitrogenous compounds (c.f. Carter et al., Chapter 21). Such soluble compounds can be mobile and represent a significant threat to groundwater contamination. A putrefying body also produces gases (Table 7.2) and, in large concentrated burial pits, these could make a significant contribution to the surrounding atmosphere unless contingency measures were put in place. These include hazardous volatile amines such putrescine and cadaverine, which can pose a considerable threat to body handlers and post-mortem workers. They are members of the family of toxic ptomaines and can displace or compete with oxygen as inspired gases, especially in confined spaces (Haglund and Sorg 1996), and have a similar effect to carbon monoxide.

The toxicology of cadaverine and putrescine is little known (Safety Officer in Physical Chemistry University of Oxford 2005), but they are known to be harmful if ingested, inhaled or absorbed through skin. There have been some reports of such noxious gases accumulating in confined areas and threatening the safety of excavators (Canadian Content 2008). Methane is also given off by bodies undergoing putrefaction, as is hydrogen sulphide, which is highly poisonous, with toxicity similar to cyanide (Santarsiero et al. 2000). Ammonia is produced in the lungs of a cadaver soon after death but diffuses very quickly and would not pose a significant problem as a gas unless it was allowed to accumulate. Phosphine is also produced, which is highly flammable and toxic to animals undergoing oxidative respiration. It is used in pest control by fumigation because it can cause death at relatively low concentrations. Other gaseous products include mercaptans which, although highly odoriferous, are generally deemed benign (Santarsiero et al. 2000).

Table 7.2 Approximate relative exposure limits of some of the more common gaseous body decomposition products

Gaseous product of body decomposition	Short term exposure limits (ppm)[a]	Long term exposure limits (ppm)[b]
	Work exposure limit (WEL)	Work exposure limit (WEL)
Ammonia (NH_3)	35	25
Cadaverine ($NH_2(CH_2)_5NH_2$)	No OSHA, NIOSH, ACGIH data available	–
Similar compound: Butylamine ($C_4H_{11}N$) (UK)	5	
Carbon monoxide (CO)	200	30
Carbon dioxide (CO_2)	15,000	5,000
Hydrogen sulphide (H_2S)	10	5
Methane (CH_4)	–	10,000 (Switzerland – occupational exposure limit)
Phosphine (PH_3)	0.2	0.1
Putrescine ($NH_2(CH_2)_4NH_2$) Similar compound:	No OSHA, NIOSH, ACGIH data available	
Butylamine ($C_4H_{11}N$) (UK)	5	–

[a]Short term exposure limit usually takes place within 15 min.
[b]Long term exposure limit usually takes place within 8 h.
Source: Sigma-Aldrich Material Safety Data Sheets.

Micro-organisms that naturally inhabit the body are released as decomposition progresses, when the body cavity can no longer contain them. The bulk of microbial biomass in animals occurs in the digestive system, and is constituted of specialised gut-dwelling organisms that are unlikely to prevail in the soil environment. However, some groups are potential pathogens such as Enterobacteriaceae, Streptococci, Clostridia, Bacilli and Staphylococci (Üçisik and Rushbrook 1998; Morgan 2004) and, where soils are inundated with such microbes, their attenuation and decay rates may be relatively slow compared to low environmental loadings. In the context of a pandemic, considerably more risk would be associated with the causal organism(s) of the associated disease, whether soil adapted or not. Viruses are also not contained by the body and, in the case of an influenza pandemic, this would be especially pertinent. Whilst the entomology of body decay is a relatively well-studied process, the microbiology of such processes is still poorly understood. This situation is changing (c.f. Tibbett and Carter, Chapter 20, and other chapters herein). However, forensic taphonomy research to date has largely concentrated on decomposition of individual bodies, usually in isolation, and extrapolating such knowledge to large-scale circumstances represents a challenge.

The decomposition of synthetic or inorganic materials associated with burial also needs to be considered. Products included in synthetic clothing or in coffin materials, such as dyes, plastics and metals (including heavy metals such as mercury, lead, zinc and copper) may all contribute to potential contamination of the soil and groundwater, as could artificial products from the body such as silicone or formaldehyde used in funerary procedures (Kim et al. 2008).

Pathways

Burial Site Structure

The relationship between the source and pathways from such points will be governed by the constructional form of common burial sites. Clearly, there is much potential scope for the way such sites are engineered. Burial sites have been likened to 'dilute and disperse' municipal landfill sites (Mather 1989), and landfill technology, at least in terms of municipal and industrial waste disposal, is both a thoroughly-rehearsed and developing arena (e.g. Hamer 2003; Bagchi 2004; Velinni 2007; Renou et al. 2008). A review of such landfill systems is beyond the scope of this chapter, but appropriate formation and structuring of common burials is obviously an effective way to modulate the connection between source and pathways to potential receptors. This notwithstanding, there are particular social issues that would be associated with common burial of humans. Common burials share some characteristics of both mass graves and individual interments. The exact size and location of the common burials would be dependent on the particular spatial and topographic characteristics of the individual cemetery. UK contingency planners envisage that the common graves dug after a pandemic would allow for individual burial plots and individual or familial grave markers, possibly with plinths of earth between each burial. Burials would be in coffins, although these would be limited in type and size to ensure manufacturers could cope with demand (Home Office 2007). Mechanical or wooden shoring of the walls of the grave may be necessary, as is the requirement to ensure sufficient depth for future interment of family members together. However, depth and density of use needs to be limited to ensure that removal of remains for identification or re-interment at a later date would still be possible.

Soil Characteristics

Soil characteristics affect the potential of the soil environment to act as a filter of waste or any disused product buried within it. This applies both to gaseous and volatile emanations, as well as liquids and solutes arising from decomposing cadavers. For example, both the fundamental textural class (*viz.* proportions of sand, silt, clay) and the prevailing particle size distribution of the soil influences the hydro-geological processes that occur therein. Coarse-textured topsoil allows water to readily infiltrate and can wash away decomposition products from bodies and coffins relatively rapidly compared to fine-textured soils. Clay and organic matter may bind such compounds resulting in their sequestration for variable, including potentially long, periods. Also, the density of vegetation cover will affect the discharge of excess water and the time it takes for groundwater beneath a burial pit to recharge. This has an important effect on the process of decomposition, which can be retarded by excess water. The thickness of any unsaturated zone around a burial pit is significant, as it acts as a barrier to contamination if thick enough, and as a means for air to circulate, which in turn facilitates decomposition. Edaphic factors such as texture, pH, redox potential, salinity, organic matter and nutrient status will all influence the rate of decomposition (Tibbett and Carter, Chapter 20), and need to be factored into site selection criteria.

It will be necessary to create guidelines about what the intended outcomes of common burial are in terms of skeletonisation of cadavers. Some governmental legislation, for example in Italy (Santarsiero et al. 2000), gives conditions regarding the minimum time necessary to obtain complete skeletonisation of the corpse in burial grounds of a soil with given characteristics. This establishes a requirement for complete skeletonisation within a certain time frame in order to allow subsequent re-use of the burial ground. The choice of sites and soil characteristics could be tailored to encourage rapid decomposition, and ensure that complete decomposition occurred within a pre-determined timescale. This could be of benefit to a small island of limited land space such as the UK. Santarsiero et al. (2000) report that decomposition is accelerated in sandy, loamy soils, with permeability coefficients of more than 10^{-3} m/s, and retarded in waterlogged, impervious soils with permeability coefficients of less than 10^{-6} m/s. The rate of decomposition of buried remains can also be altered by the presence or absence of burial clothes, shrouds or vessels. The more separation there is between the body and the soil, through coffin, shroud or plastic sheeting for example, the greater the time for decomposition to occur.

It may be equally possible to deliberately decelerate the process of decomposition through the appropriate selection of soil characteristics and features, which may be desirable to allow an extended period of time for exhumations to occur for identification purposes, or for specific long-term research to be undertaken. This is an example of where taphonomic research could usefully inform such procedures.

Other geological and geographical features, such as slope and the amount of rainfall, will all affect its suitability as a site for a common burial. It is imperative that the geological characteristics of a potential burial site are analysed for the possibility of the right conditions for groundwater contamination, as "cemeteries are a potential risk to groundwater that can become a real risk if previous geological and hydrogeological studies are not consulted" (Pachecho et al. 1991). Indeed, larger-scale considerations such as these could be well informed by developments in catchment science and modelling tools used in catchment management.

Receptors

Burial Operatives

The potential range of noxious compounds and organisms to which mortuary, cemetery and other emergency staff may be exposed in mass burial circumstances are briefly reviewed above. Management strategies to mitigate such exposure at source might include stringent health and safety measures for people expected to be involved in the transportation and burial of bodies that take into account the potential exposure to harmful bio-toxic gases. To minimise any risk to health due to gaseous products, bodies would need to be buried and/or refrigerated as soon after death as possible. This reiterates the need for timely burial of bodies or adequate refrigerated storage before burial in order to retard the rate of decomposition (Home Office 2007). This was not achieved in all instances during the UK Foot and Mouth crisis of 2001, when some animals were not buried until up to seven days after culling (Anderson 2002), which allowed decomposition to progress to potentially levels which increased risk to operators (Haglund and Sorg 1996). Crucially, control of the production of potential toxicants and pollutants at source will also considerably reduce likely exposure to other receptors.

Groundwater

A large number of corpses in cemetery soil provide an obvious source of both organic and inorganic contamination (Spongberg and Becks 2000a, b). Through the decomposition process, many potential contaminants are released into the soil over a period of time. The fluids from decomposing bodies can percolate into underlying groundwater if non-leakproof coffins or no coffins are used (Freedman and Fleming 2003). Areas with a high water table and high annual rainfall are at most risk of groundwater contamination. There have been several historical cases of contamination of groundwater, for example those cited by Bouwer (1978). In Paris, water from sources close to cemeteries was found to have a 'sweetish taste and infected odour', and a higher incidence of typhoid fever in communities near cemeteries was reported in late 19th century Berlin. Pachecho et al. (1991) and Schraps (1972) found evidence of bacteria, ammonium and nitrite ions in a plume of decreasing concentration with distance from a cemetery, and indications of proteolytic and lipolytic bacteria in groundwater, accompanied by malodours (Pachecho et al. 1991). Groundwater is also vulnerable to contamination from inorganic compounds associated with burials and burial goods, such as formaldehyde from embalming practices, and arsenic, varnishes, preservatives, and metals including mercury, copper, zinc and steel. Metals leaching from burials may accumulate at depth.

It is possible to draw on past experience in order to inform best practice to avoid groundwater contamination. During the UK Foot and Mouth Crisis of 2001, the Environment Agency (EA) assessed the risk of groundwater contamination around the burial pits, and declared the risk low, due to the presence of a layer of non-porous boulder clay beneath the pits (Gray 2001). Although this was accepted at the time, it may have created another set of problems, such as 'ponding' or the accumulation of trapped leachate, which was unable to disperse into the soil. The long-term results of these preventative measures are as yet unclear.

Within the EA guidelines, it is stated that no permit is required for human cemeteries under water pollution or waste management legislation, although, if a

Table 7.3 Summary of EA groundwater protection policy and guidelines

Environment Agency Groundwater Protection Policy and Guidelines – Summary
No burials within zone 1 groundwater source protection zones around a spring, well and borehole
A minimum distance of 250 m from graves to wells, boreholes or springs used for water supply
A minimum distance of 30 m from graves to other springs or watercourses
A minimum distance of 10 m from graves to field drains
No burial into standing water and the base of the grave should be above the local water table

Source: www.environment-agency.gov.uk.

cemetery were to cause pollution of groundwater, or other controlled waters, it is possible that action could be undertaken under the Groundwater Regulations 1998. The EA Policy and Practice for the Protection of Groundwater (Environment Agency 2007) and supporting tools such as the Groundwater Vulnerability and Source Protection Zone Maps (Environment Agency 2007) could form the basis of an initial risk-screening exercise. It must be noted that the above points have not yet been incorporated into EA policy, and their groundwater protection policy and guidelines are currently under review, with consideration of the incorporation of specific guidelines for burials in the revision of the policy document (Table 7.3).

Adoption of these practices should avoid the need for site-specific assessments for the lower risk proposals (low burial rates and low groundwater vulnerability). In addition, these good practice measures, site-specific risk assessments are needed for higher burial rates and in areas where groundwater is deemed inherently more vulnerable.

Aboveground Ecology

It is well documented that the presence of buried remains has an effect on the growth of surface vegetation (Hunter et al. 1996; Hunter and Cox 2005). Disturbance to the soil affects the growth of vegetation on the soil by affecting the moisture retention ability of the soil and the air content. Disturbed soil is generally looser and more highly aerated, and can lead to increased vegetation growth on the surface above the burial, most likely arising as a consequence of accelerated nutrient release from decomposition. Deeper burials can tend to retard vegetation growth, as primary root structures are destroyed, which can lead to differential growth on the surface. Some species thrive on the conditions created by the grave, to the extent where they can be used as indicators of mass graves, for example *Artemisia vulgaris* (wormwood) in the Balkans (Hunter and Cox 2005). Differential vegetation growth can create strong visual effects that can be interpreted from aerial photography, and can be of use for the search and location of graves in a forensic context (c.f. Harrison and Donnelly, Chapter 13).

As discussed above, common burial sites are akin to landfill sites, and are often areas of land which are unique in that they tend to incorporate land previously disturbed either by historic quarrying or mineral extraction activities, or waste deposition of some nature. Therefore, as these sites tend not to be intensively managed, farmed or cropped, they often support and encourage a wide range of flora and fauna. Earthworms, soil-dwelling organisms and detritus-feeders often accumulate at the sites, as will larger animals, attracted by the gaseous and chemical products of decomposition. In Northern European countries, these scavengers tend to be birds, rodents, canines and felines but the type and size of scavengers attracted will depend on the geographical location and local fauna of the region of the burial ground.

Precautions would be advised to avoid either an influx of opportunist animals by means of fencing or deterrents, or to ensure that the arrival of new species as a result of the burial will not create an ecological imbalance to the detriment of existing species. The attraction of some species of flora or fauna may even be seen as an added benefit to the construction of a burial site. In the UK, the 1981 Wildlife and Countryside Act, the Habitats Directive, The Countryside Rights of Way Act and other wildlife legislation compels the landfill developer and operator to establish the presence or absence of protected species and undertake measures to ensure their conservation (Carver 2003).

The Guide for Burial Ground Managers (Webb 2005) and the Home Office paper Cemeteries and their Management (Wilson and Robson 2004) recognise the value of burial grounds as conservation areas and environmentally important sites. They both recommend training for managers to improve awareness of the ecological issues but, as yet, there are no legislative guidelines for cemeteries and burial grounds that are concerned with the restriction or monitoring of faunal activity. Pandemic contingency planners need to be aware of the potential for animal scavengers to be vectors of disease, which may render a contagious disease unpredictable and uncontainable.

Soil and Geological Databases

The sensible and timely selection of sites suitable for the construction of common burial pits is imperative. Soil and geological databases are potentially excellent tools for selecting sites with the appropriate characteristics for the requirements of a common burial ground. Soils data are available at a range of resolutions depending upon the country. The European Soil Database v.1.0 (Heineke et al. 1998) has been constructed from source material prepared and published at a scale of 1:1,000,000 (CEC 1985; King et al. 1994). The resulting soil data have been harmonised for the area of coverage, according to a standard international classification (FAO-UNESCO-ISRIC 1990), together with analytical data for standard profiles (Madsen and Jones 1995). The spatial component of this database comprises polygons, which define Soil Mapping Units (SMUs). These data are the only harmonised soil information available at a European level, but notably only characterise the upper 1.5 m of mantle. To characterise the deeper substrate, i.e. >2 m depth, requires access to geological databases. Again, these prevail for countries at differing spatial resolutions. In the UK, the highest resolution for which soils data has been mapped with *full coverage* as of 2008 is at a scale of 1:250,000 for England and Wales (Soil Survey Staff 1983), 1:250,000 for Scotland (Soil Survey of Scotland 1984) and 1:50,000 for Northern Ireland (Cruikshank 1997). For geological data, effectively full coverage of bedrock and superficials (i.e. surface horizons) is at 1:50,000 (British Geological Survey 2007). Particular sub-regions have been mapped at a larger cartographic scale, and sometimes at relatively high resolutions, but these are currently sporadic. There is evidence through recent experience during the UK Foot and Mouth Crisis of 2001 to suggest that the available databases relating to Wales were neither sufficiently complete nor of an appropriate spatial resolution to allow a reliable assessment for a mass burial site selection to be made at the time (Anderson 2002). This highlights the need for improvements to be made to existing soil and geological databases to ensure that the same shortfall is not encountered during future pandemics. Organisations involved in the production of such data are engaged in a process of continually increasing coverage and resolution, and with a total emphasis on digitised databases and affiliated information systems.

Site selection would also have to be informed by other considerations including prevailing meteorology, proximity to habitation, transport infrastructure, etc. Geographical information systems (GIS) are the most effective tools to combine such data in operationally useful ways, noting that such additional data are also available at different spatial scales and update cycles.

Conclusions and Projections

The discipline of soil forensics can both inform effective management of mass burial, and potentially be informed by the study of extant common burial sites or directed study of new sites in the event of their establishment. It is clear that any policy to construct common burials during pandemics or after a mass fatality incident should take into account the potential effect of such a construction on the environment, in particular, soil functions, potential for groundwater contamination and vegetation and ecology, and *vice versa*, the effect that changes in the soil environment can have on the decomposition process and management of such sites. In the future, detailed quantification of the changes to the soil, nearby groundwater and atmosphere found in association with concentrated burials is imperative, so that the exact nature of the contribution of the burials to environmental change can be understood. Site-specific data is needed from existing 'mass graves' from the Foot and Mouth epidemic or other similar animal outbreaks. This could be compared to data from archaeological burial grounds after pandemic or atrocity-related human mass graves. Assessment of the alternatives to burial that may be viable in a pandemic or mass disaster situation is also needed. These range from cremation, with its particular environmental implications, to bio-composting or disposal at sea, legislative frameworks notwithstanding. This will then allow balanced, informed comparisons to be made with the common burial option. In addition, whatever the science suggests as optimal, there are very significant economic, social and ethical considerations required which should be investigated further and incorporated into capacity or response planning.

Acknowledgements We thank Russell Lawley (British Geological Survey) for clarifications regarding UK geological databases, and Mark Kibblewhite for useful discussions in the preparation of this chapter.

References

Anderson I (Ed.) (2002). Foot and Mouth Disease 2001: Lessons To Be Learned Inquiry Report. The Stationery Office, London, UK.

Bagchi A (2004). Design of Landfills and Integrated Solid Waste Management. Wiley, Chichester, UK.

Bouwer H (1978). Groundwater Hydrology. McGraw-Hill, New York.

British Geological Survey (2007). Digital Geological Map of Great Britain 1:50,000 Scale (DiGMapGB-50) Data Version 4.16. British Geological Survey, Keyworth, Nottingham.

Buss S, Herbert A, Morgan P and Thornton S (2003). Review of ammonium attenuation in soil and groundwater. Quarterly Journal of Engineering Geology and Hydrogeology 37:347–359.

Canadian Content (2008). Rotting Corpses Forces Doomsday Group Out of Cave. 20 May 2008. http://www.canadiancontent.net/commtr/rotting-corpses-forces-doomsday-group-cave_913.html. Accessed 17/07/2008.

Carver SM (2003). The Protection of Ecology on Landfill Sites. In: Land Reclamation: Extending the Boundaries – Proceedings of the 7th International Conference (Eds. H Moore, H Fox and S Elliott). AA Balkema, UK.

CEC (1985). Soil Map of the European Communities, 1:1,000,000. Office for Official Publications of the European Communities, Luxembourg.

Cruikshank JG (Ed.) (1997). Soil and Environment: Northern Ireland. Department of Agriculture Northern Ireland and Queens University, Belfast, Northern Ireland.

Dent BB (2005). Vulnerability and the unsaturated zone – the case for cemeteries. In: Proceedings "Where Waters Meet", Joint Conference of the New Zealand Hydrological Society, International Association of Hydrogeologists Australian Chapter and New Zealand Soil Science Society, Auckland, 30 November–2 December, paper A13.

Dent BB and Knight MJK (1998). Cemeteries: A special kind of landfill. The context of their sustainable management. In: Groundwater: Sustainable Solutions, pp. 451–456. International Groundwater Conference, International Association of Hydrogeologists, Melbourne.

Drummond RD (1999). Notifiable Disease Preparedness within the State Veterinary Service. Ministry of Agriculture, Fisheries and Food, UK.

Environment Agency (2007). www.environment-agency.gov.uk. Accessed 04/12/2007.

FAO-UNESCO-ISRIC (1990). FAO-UNESCO Soil Map of the World: Revised Legend. World Soil Resources Report 60. FAO, Rome.

Freedman R and Fleming R (2003). Water Quality Impacts of Burying Livestock Mortalities. Ridgetown College, Ridgetown, Ontario.

Gray C (2001). Foot-and-Mouth Crisis: Army Digs Mass Grave to Bury 500,000. The Independent, 26 March 2001.

Haglund W and Sorg M (1996). Forensic Taphonomy: The Postmortem Fate of Human Remains. CRC, Boca Raton, FL.

Hamer G (2003). Solid waste treatment and disposal: effects on public health and environmental safety. Biotechnology Advances 22:71–79.

Hart A (2005). Ammonia shadow of my former self. Land Contamination and Reclamation 13(3):239–245.

Heineke HJ, Eckelmann W, Thomasson AJ, Jones RJA, Montanarella L and Buckley B (Eds.) (1998). Land Information Systems: Developments for Planning the Sustainable Use of Land Resources. European Soil Bureau Research Report No. 4, EUR 17729 EN. Office for Official Publications of the European Communities, Luxembourg.

Home Office (2007). Planning for a Possible Influenza Pandemic: A Framework for Planners Preparing to Manage Deaths. Crime Reduction and Community Safety Group, London, UK.

Hunter J and Cox M (2005). Forensic Archaeology: Advances in Theory and Practice. Routledge, London.

Hunter J, Roberts C and Martin A (1996). Studies in Crime: An Introduction to Forensic Archaeology. Routledge, London.

Janaway RC (1997). The decay of buried human remains and their associated materials. In: Studies in Crime: An Introduction to Forensic Archaeology (Eds. J Hunter, C Roberts and A Martin), pp. 58–85, Routledge, London.

Kim KH, Hall ML, Hart A and Pollard SJT (2008). A survey of green burial sites in England and Wales and an assessment of the feasibility of a groundwater vulnerability tool. Environmental Technology 29(1):1–12.

King D, Daroussin J and Tavernier R (1994). Development of a soil geographical database from the soil map of the European Communities. Catena 21:37–26.

Knight MJ and Dent BB (1998). Sustainability of waste and groundwater management systems. In: Groundwater: Sustainable Solutions, pp. 359–374. Proceedings of the Conference of the International Association of Hydrogeologists, Melbourne.

Lelliot M (2002). Hydrogeology, pollution and cemeteries. Teaching Earth Sciences, 27:68–73.

Madsen HB and Jones RJA (1995). Soil profile analytical database for the European Union. Danish Journal of Geography 95:49–57.

Mather JD (1989). The attenuation of the organic component of landfill leachate in the unsaturated zone: a review. Quarterly Journal of English Geology 22:241–6. Cited in Morgan O (2004).

Morgan O (2004). Infectious disease risks from dead bodies following natural disasters. Pan American Journal of Public Health 15(5):307–12.

Pachecho A, Mendes J, Martins T, Hassuda S and Kimmelmann A (1991). Cemeteries: A potential risk to groundwater. Water Science and Technology: A Journal of the International Association of Water Pollution 24(11):97–104.

Pollard SJT, Smith R, Longhurst PJ, Eduljee G and Hall D (2006). Recent developments in the application of risk analysis to waste technologies. Environment International 32:1010–1020.

Pollard SJT, Hickman GAW, Irving P, Hough RL, Gauntlett DM, Howson S, Hart A, Gayford P and Gent N (2008). Exposure assessment of carcass disposal options in the event of a notifiable exotic animal disease – methodology and application to avian influenza virus. Environmental Science Technology 42(9):3145–3154.

Renou S, Givaudan JG, Poulain S, Dirassouyan F and Moulin P (2008). Landfill leachate treatment: Review and opportunity. Journal of Hazardous Materials 150:468–493.

Safety Officer in Physical Chemistry, University of Oxford (2005). Material Safety Data Sheet for 1,5-diaminopentane (cadaverine). http://msds.chem.ox.ac.uk/. Accessed 17/07/2008.

Santarsiero A, Minelli L, Cutilli D and Cappiello G (2000). Hygienic aspects related to burial. Microchemical Journal 67:135–139.

Schraps W (1972). Die Bedeutung der Filtereigenschaften des Bodens fur die Anlage von Friedhofen. Mittcilungen Deutsche Bodenkundl. Gesellschaft, 16:225–259. Cited in Bouwer H (1978).

Sigma-Aldrich (2007). http://www.sigmaaldrich.com/Local/SA_Splash.html. Accessed 17/07/2008.

Soil Survey of Scotland (1984). Soil and Land Capability for Agriculture. Macaulay Institute for Soil Research, Aberdeen.

Soil Survey Staff (1983). The National Soil Map of England & Wales, 1:250,000 Scale (in Six Sheets). Ordnance Survey, Southampton.

Spongberg A and Becks P (2000a). Organic contamination in soils associated with cemeteries. Journal of Soil Contamination 9(2):87–97.

Spongberg A and Becks P (2000b). Inorganic soil contamination from cemetery leachate. Water Air and Soil Pollution117:313–27.

Üçisik A and Rushbrook P (1998). The Impact of Cemeteries on the Environment and Public Health: An Introductory Briefing. WHO Regional Office for Europe. Nancy Project Office.

Velinni AA (2007). Landfill Research Trends. Nova Science Publishers, Hauppauge New York.

Webb B (2005). Guide for Burial Ground Managers. Department of Constitutional Affairs, London.

Wilson B and Robson J (2004). Cemeteries and Their Management. Research Development and Statistics Directorate, Home Office, London.

Young CP, Blackmore KM, Reynolds P and Leavens A (1999). Pollution potential of cemeteries: Draft guidance. R&D Report, Environment Agency, Bristol.

Part II
Evidence

Chapter 8
A Systematic Approach to Soil Forensics: Criminal Case Studies Involving Transference from Crime Scene to Forensic Evidence

Rob W. Fitzpatrick, Mark D. Raven and Sean T. Forrester

Abstract Forensic soil science represents a newly-developed discipline of soil science, and has matured to the extent that well-defined questions and successful crime scene investigations can be answered in increasingly refined ways. This chapter considers two case studies and highlights the kinds of investigations that have been carried out on complex soil materials from shoes, vehicles and crime scenes by the Centre for Australian Forensic Soil Science (CAFSS). The two case examples are described in ways that show parallel approaches to more recent types of case investigations where soils as evidence are being applied with more certainty in criminal and environmental investigations. The history of forensic soil science and the importance of pedology and soil mineralogy are also briefly reviewed from a world perspective. The significance and relevance of established concepts and standard terminologies used in soil science but especially in pedology with practical relevance to forensic science are discussed. The systematic forensic soil examination approach described in this paper uses soil morphology (e.g. colour, consistency, texture and structure), mineralogy (X-ray powder diffraction) and chemistry (e.g. based primarily upon mid-infrared spectroscopy/diffuse reflectance infrared Fourier transform (DRIFT) analyses). Forensic soil characterisation usually combines the descriptive and analytical steps for rapid characterisation of whole soil samples for screening, and detailed characterisation and quantification of composite and individual soil particles after sample selection, size fractionation and detailed mineralogical and organic matter analyses using advanced analytical methods. X-ray powder diffraction methods are arguably the most significant for both qualitative and quantitative analyses of solid materials in forensic soil science. The two crime scene examples described in this paper use combined pedological (including field investigations), mineralogical and spectroscopic methods in the forensic comparison of small amounts of soil adhering to a suspect's shoe and carpet in a vehicle boot with control soil specimens. These case examples illustrate that forensic soil examination can be very complex because of the vast diversity and

---

R.W. Fitzpatrick(✉), M.D. Raven and S.T. Forrester
Centre for Australian Forensic Soil Science/CSIRO Land and Water, Private Bag No 2, Glen Osmond, South Australia
e-mail: Rob.Fitzpatrick@csiro.au

K. Ritz et al. (eds.), *Criminal and Environmental Soil Forensics*

heterogeneity of soil samples. The interpretation of soil forensic tests and methods is not equally applicable to all soils, and should also be made in the context of the forensic soil examination (e.g. the sieving of large amounts of stone and gravel from ASS samples to obtain a more representative sample to make comparisons).

Introduction

Forensic soil scientists (or forensic geologists) are more specifically concerned with soils that have been disturbed or moved (usually by human activity), sometimes comparing them to natural soils, or matching them with soil databases, to help locate the scene of crimes. Forensic soil scientists usually obtain soil samples from crime scenes and suspected control sites from which soil may have been transported by shoes, a vehicle or a shovel. Soil properties are diverse and it is this diversity which may enable forensic soil scientists to use soils with certainty as evidence in criminal and environmental investigations (e.g. see reviews by: Murray and Tedrow 1991, 1975; Murray 2004; Ruffell and McKinley 2004; Pye 2007; Dawson et al. 2008; Fitzpatrick 2008). Forensic soil science is a relatively new activity that is strongly 'method-orientated' because it is mostly a technique-driven activity in the multidisciplinary areas of pedology, geochemistry, mineralogy, molecular biology, geophysics, archaeology and forensic science. Consequently, it does not have a large number of past practitioners such as in the older forensic disciplines like chemistry and physics. Identification of soil differences using various morphological soil attributes (e.g. colour, consistency, texture and structure) on whole soil samples is the first step for using soil information to help police investigators at crime scenes. The second step is to discriminate soils using the discriminating power inherent in soil materials, especially in the use of: (i) powder X-ray diffraction techniques that provide qualitative and quantitative data in the mineralogical composition of samples; and (ii) spectroscopy (e.g. Fourier transform infrared spectroscopy-, which provides qualitative and semi-quantitative information about the soil organic matter (protenaceous, aliphatic, lipid, carboxyl and aromatic), mineral composition (smectite, kaolin and illite clays, quartz) and prediction of soil physicochemical properties using partial least-squares analysis (MIR-PLS) and inductively-coupled plasma spectroscopy mass spectroscopy providing elemental composition. Detailed soil characterisation usually requires a joint approach that combines the descriptive and analytical steps for: (i) rapid characterisation of whole soil samples for screening (Stage 1) and (ii) detailed characterisation and quantification of composite and individual soil particles after sample selection, size fractionation and detailed mineralogical and organic matter analyses using advanced analytical methods (Stage 2). The same principles apply to environmental investigations of polluted sites. This chapter has two principal objectives: (i) to summarise briefly some established concepts and standard terminologies used in pedology, mineralogy and geochemistry that have practical relevance to forensic science; (ii) to provide two brief case investigation examples of the use of pedological and related mineralogical and spectroscopic methods (XRD and

DRIFT analyses) in the forensic comparison of small amounts of soil adhering to a suspect's shoe and to a carpet on the floor in the boot of a vehicle with control soil specimens from a range of soils (i.e. inland acid sulphate soils in a wetland/river and an anthropogenic soil in a quarry).

Case Study 1: Hit-and-Run—Outline

This hit-and-run case will be used as a frequent example in this chapter. It involved two suspects that left the scene of a fatal car collision. One of the suspects was chased through the Adelaide suburbs at night and was later observed crossing the River Torrens. The suspect ran down the river bank, jumped into the river and onto the extended gravely and stony river bank then proceeded up the opposite river bank before disappearing into the adjacent parklands (Figure 8.1). Two control samples of gravelly acid sulphate soils (ASS) with sulphidic material from the alleged 'crime trail' were taken, located (i) on the stony and gravelly bank (CAFSS_027.5; Figure 8.1a, b, c); and (ii) in the river channel (subaqueous ASS; CAFSS 027.4; Figure 8.1a). A sufficient amount of fine-grained soil material was recovered from the control site samples by sieving the gravely (95% gravel and rock fragments with 5% clay and silt) samples through a 50 μm sieve (i.e. <50 μm fraction). Two additional 'alibi samples' were collected from alibi trails or scenes (20 m upstream) and upper river bank (CAFSS 027.3 and CAFSS 027.6 respectively; Figure 8.1a), to determine whether or not the suspect had been along the alleged 'crime trail'. The suspect was apprehended by police three hours later but denied ever running through this section of the river. A small amount of fine yellowish-grey soil was strongly adhered to the side and in the treads of the sole

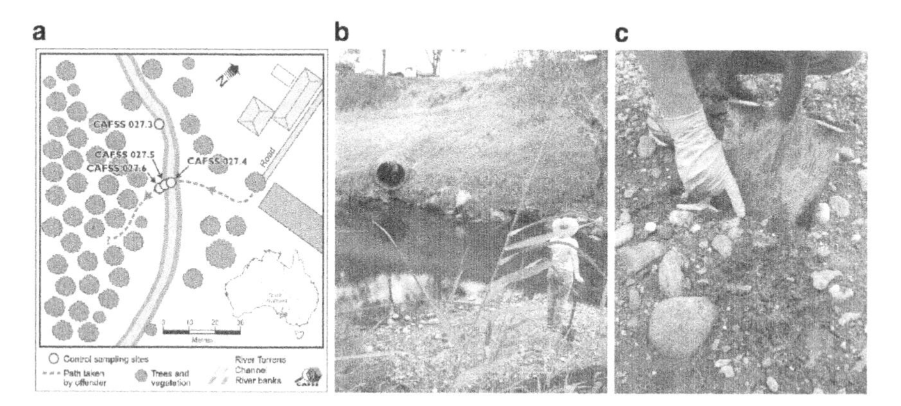

Fig. 8.1 The hit-and-run case study. **(a)** Schematic map showing locations, aspect, path taken by offender and sampling points with associated reference numbers. **(b)** Overview of river bank, person standing at point where Sample CAFSS_027.5 taken. **(c)** Close-up view of the soil surface near where a shoe impression matching the sole tread of the shoe worn by the offender *(see colour plate section for colour version)*

from the suspect's shoes (Figure 8.2a). A sufficient amount of the soil was recovered from the soles and sides of the shoes for forensic soil analyses by gently scraping the fine soil from the shoes using a plastic spatula (Figure 8.2b).

A control surface soil sample (0 to 3 cm) was taken where a shoe impression was located on the lower river bank (CAFSS 027.4; Figure 8.1) and where the suspect was seen to run (i.e. shoe imprint was similar to the sole tread of the shoe worn by the suspect). A second control soil sample (0 to 5 cm) was taken beneath 10 cm of water in the river channel one metre from the control sample site on the lower river bank. These two yellowish-grey to dark brownish-black samples are from acid sulphate soils (ASS) with sulphidic material, which comprises a mixture of 95% coarse gravel and stone fragments and only 5% clay and silt (<50 µm fraction).

Although the control ASS comprised 95% alluvial stone and coarse gravel with only 5% clay and silt, a sufficient amount of fine soil (<50 µm) was recovered by sieving. As shown in Figure 8.2e, this fine soil material closely resembles (colour and texture)

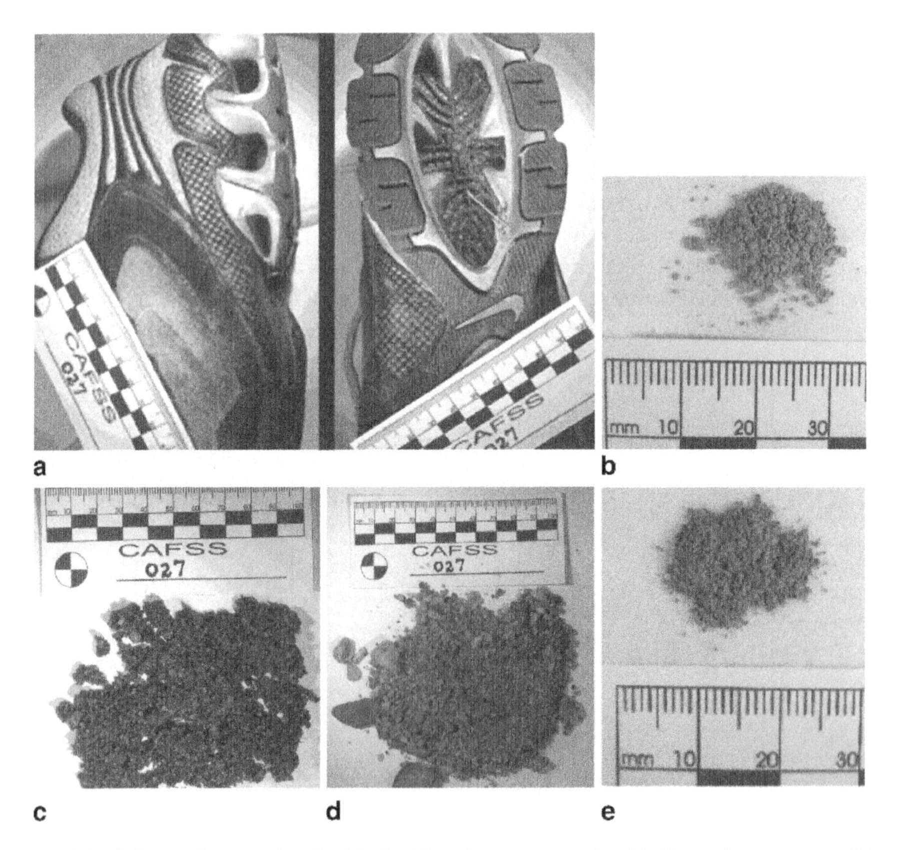

Fig. 8.2 Soil samples associated with the hit-and-run case study. **(a)** Shoes from suspect. **(b)** Sample scraped from the shoe. **(c)** Control soil specimen from the river channel. **(d)** Control soil specimen from the bank of river. **(e)** The <50 µm fraction separated from this sample *(see colour plate section for colour version)*

the fine soil material that was tightly trapped in grooves and treads in the rubber sole of the suspect's shoe (Figure 8.2b). Analyses of these two soil materials using soil morphological descriptors (e.g. Munsell Soil Color Charts 2000; Schoeneberger et al. 2002), microscopical, XRD and DRIFT methods indicated that the soil from the river bank and soil on the suspect's shoes were similar (see below). Two alibi samples were collected on the surface (0 to 3 cm) of: (i) a gravelly hydromorphic soil on the lower river bank, 20 m upstream (CAFSS 027.3; 1 m from the river edge) from the two control sites; and (ii) a non-gravelly alluvial soil on the upper river bank (CAFSS 027.6; 5 m from the river edge), to determine whether or not the suspect had run along the alleged crime trail shown in Figure 8.1 (soil analyses are not reported in this chapter).

Soil as a Powerful Contact Trace

Theory of Transfer of Materials from One Surface to Another as a Result of Contact

In 1910 the French scientist, Edmond Locard, inspired by the Adventures of Sherlock Holmes, postulated the fundamental principle on which forensic science and trace evidence is based, namely the 'Locard Exchange Principle' (Chisum and Turvey 2000), which states: "Whenever two objects come into physical contact – an exchange of materials takes place". When two things come in contact, physical components will be exchanged. For example, the exchange can take the form of soil material from a location transferring to shoes of a person who walked through a particular area. These types of transfers are referred to as primary transfers. Once a trace material has transferred, any subsequent movements of that material, in this case from shoes, are referred to as secondary transfers. These secondary transfer materials can also be significant in evaluating the nature and source(s) of contact. Hence, the surface of soils can provide information linking persons to crime scenes.

Aardahl (2003) lists six properties of the ideal trace evidence, *viz.* nearly invisible; is highly individualistic; has a high probability of transfer and retention; can be quickly collected, separated and concentrated; the merest traces are easily characterised; and is able to have computerised database capacity. In this context, Blackledge and Jones (2007) consider that 'glitter' (i.e. entirely manmade tiny pieces of Al foil or plastic with vapour-deposited Al layer) may be the ideal contact trace. Soil materials may be considered as approaching the ideal 'contact trace', and the following brief discussion considers how closely they fulfil the criteria of Aardahl.

Soil Is Highly Individualistic

Pedology has two broad purposes: to describe and classify, and to interpret soil differences with respect to their management or use requirements. An appropriate definition of pedology can be found in Wilding (1994) as "that component of earth science that quantifies the factors and processes of soil formation including the quality, extent, distribution, spatial variability and interpretation of soils from microscopic to megascopic scales". This definition introduces the words *extent, distribution, spatial variability and interpretation* in a general way. It is fair to presume though that such terms for the pedologist include primarily the descriptive aspects of the science – the field and laboratory descriptions of soil attributes such as presence and degree of development of particular soil features (e.g. soil colour and mottling) – and the interpretive aspects of those attributes (e.g. soil in relation to drainage class or wetness). This description and its interpretation can then be explained in relation to the forensic comparison of soils. In addressing the questions 'What is the soil like?' and 'Where does it come from?' (i.e. provenance determination), we are involved in studies relating to characterising and locating the sources of soils to make forensic comparisons.

The major question posed is how can soils be used to make accurate forensic comparisons when we know that soils are highly complex and that there are thousands of different soil types in existence? For example, according to the United States Department of Agriculture (USDA), which collects soil data at many different scales, there are over 50,000 different varieties of soil in the United States alone. Parent material, climate, organisms, and the amount of time it takes for these properties to interact, will vary worldwide.

There are two key issues which are especially important in forensic soil examination because the diversity of soil strongly depends on topography and climate, together with anthropogenic contaminants: (i) forensic soil examination can be complex because of the diversity and in-homogeneity of soil samples, i.e such diversity and complexity enable forensic examiners to distinguish between soil samples, which may appear similar to the untrained observer; (ii) a major problem in forensic soil examination is the limitation in the discrimination power of the standard and non-standard procedures and methods.

No standard forensic soil examination method exists. The main reasons for this are that examination of soil is concerned with detection of both: (i) naturally-occurring soils (e.g. minerals, organic matter, soil animals and included rock fragments), and (ii) anthropogenic soils that contain manufactured materials such as ions and fragments from different environments whose presence may impart soil with characteristics that will make it unique to a particular location (e.g. material from quarries, asphalt, brick fragments, cinders, objects containing lead from glass, hydrocarbons, paint chips and synthetic fertilisers with nitrate, phosphate, and sulphate). These anthropogenic properties make soil even more individualistic.

Soil Has a High Probability of Transfer and Retention

In general, soil usually has a strong capacity to transfer and stick, especially the fine fractions in soils (clay and silt size fractions) and organic matter. The larger quartz particles (e.g. >2 mm size fractions) have poor retention on clothes and shoes and carpets.

Fine soil material (e.g. their <50 to 100μm fractions) may often only occur in small quantities, as illustrated in the hit-and-run case above (cf also Fitzpatrick et al. 2007), where a remarkably small amount of fine soil was transferred from a gravelly and stony soil on a river bank (control site) to running shoes (forensic evidence items).

Soil Can Quickly be Collected, Separated and Concentrated

Although a suspect may be unaware that soil material, especially the fine fraction, has been transferred directly to the person (e.g. shoes) or surroundings (e.g. carpet in a suspects car), soil materials are easily located and collected when inspecting crime scenes or examining items of physical evidence (e.g. Figures 8.1 and 8.2). Traces of soil particles can easily and quickly be located directly using hand lenses or light microscopes.

Soil samples must be carefully collected and handled at the crime scene or control sites using the established approaches and then compared by a soil scientist with forensic science experience to ensure that the soil samples can be useful during an investigation. The size and type of samples to be taken are strongly dependent on the nature of the environment being investigated, especially the type of soil and nature of activity that may have taken place at the scene (e.g. if suspect footwear is heavily coated with mud on the uppers and the ground is wet and soft then the control sample should be collected to a depth of around 0 to 10 cm, see Figures 8.1 and 2). Subaqueous soils from the bottom of river channels, streams, ponds, lakes or dams can be obtained by pressing a plastic tube or container into the bed and removing it with a scooping action. In deeper water, samples can be taken using specialised sampling devices such as the Russian D-auger. If the soil is very hard and dry; and only the shoe tread was in contact with the soil, then collect to a depth of 0 to 0.5 cm – or thinner. It is critical to wear clean latex gloves but do not use talc powder in the gloves because the layer silicate mineral talc will contaminate the soil sample. Always use clean tools (e.g. shovel, trowel, artist's palette knife, which are made of stainless steel). Plastic spades and trowels generally lack the strength required to dig soils, especially for Australian soil conditions. Artist's palette knives are useful for sampling very thin layers surfaces of samples of mud or dust. Preferably place samples in rigid plastic containers, rather than polythene bags or paper bags (i.e. package must keep lumps intact). Do not use paper envelopes as they easily tear and leak. If soil is adhering to items of clothing or shoes, first air dry whole garment and then package whole intact sample and garment. If the soil is subaqueous or adhering in a wet/moist condition to tyres, vehicles, garments or shovels, either air dry then package or, as in the case of obvious sequential layers of soil being present, remove 'surface layer' and then air dry. Store dry samples at room temperature and ensure containers are sealed and take appropriate caution when storing and transporting. If biological material is attached, package using clean cardboard box/paper bags because samples are prone to rapid deterioration.

Several standard methods are available for quick separation and concentration of soil materials or particles, such as sieving (e.g. Figure 8.2), magnetic extraction and heavy mineral separation (e.g. Figure 8.3).

Soil Is Nearly Invisible

As described in the hit-and-run case study above, under typical viewing conditions by the naked eye we do not really see the yellow-brown colour of the fine 5% clay and silt (<50 μm fraction) fractions hidden in the gravelly soil (Figure 8.2d) until the sample is sieved and the fine fraction concentrated (Figure 8.2d and 8.2e). This is, for example, often unlike the more obvious bright transfer colours of blood, lipstick smears and paint. Hence, not being obviously aware of the presence of fine

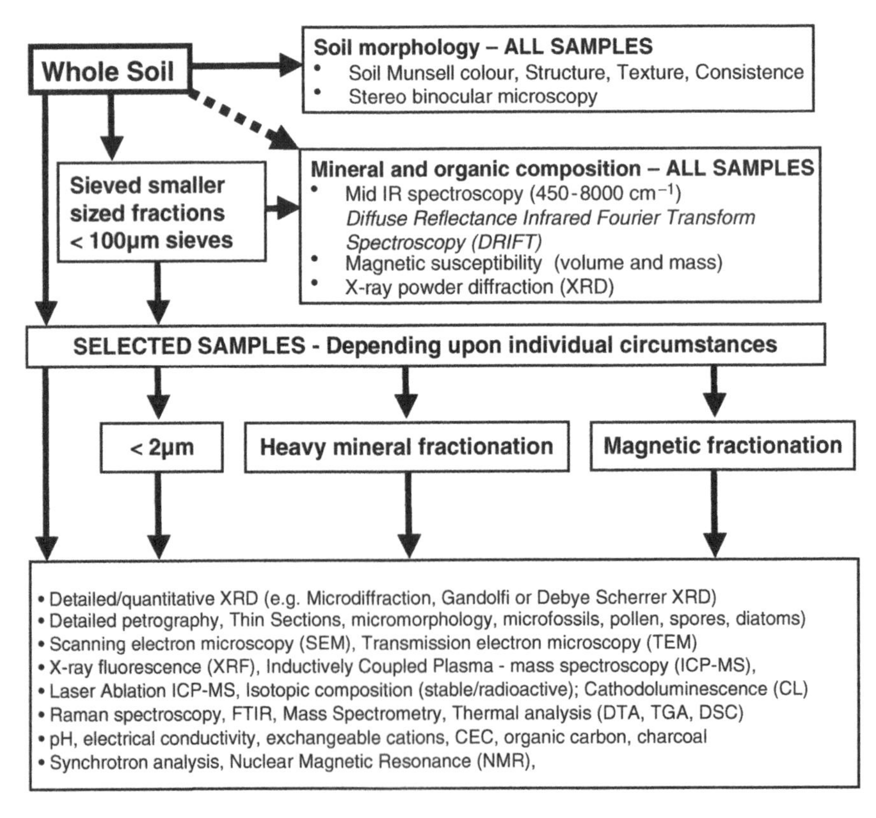

Fig. 8.3 A systematic approach to discriminate soils for forensic soil examinations where, FTIR is Fourier transform infrared spectroscopy, DTA is differential thermal analysis, TGA is thermogravimetric analysis, DSC is differential scanning calorimetry and CEC is Cation Exchange Capacity (modified from Fitzpatrick et al. 2006)

soil materials, especially when they impregnate vehicle carpeting, shoes or clothing, a suspect will often make little effort to remove soil materials.

Computerised Soil Databases: Capacity

Soil profiles and their horizons usually change across landscapes, and also change with depth in a soil at one location. In fact, soil samples taken at the surface may have entirely different characteristics and appearances from soil dug deeper in the soil profile. One common reason why soil horizons are different at depth is because there is mixing of organic material in the upper horizons, and weathering and leaching in the lower horizons.

Erosion, deposition, and other forms of disturbance might also affect the appearance of a soil profile at a particular location. For example, soils on alluvial flats with regular flooding often have clear sedimentary layers. Various soil-forming processes create and destroy layers and it is the balance between these competing processes that will determine how distinct layers are in a given soil. Some of the more common natural processes include the actions of soil fauna (e.g. worms, termites), and the depletion and accumulation of constituents including clay, organic matter and calcium carbonate. In contrast, the main anthropogenic soil-forming processes that destroy layers are excavation (e.g. ploughing and grave digging) and fertiliser applications.

The mapping of the surface and subsurface of both natural and anthropogenic soils provides crucial information as to the origin of a site's specific location, function, land degradation and management. In Australia (e.g. Johnston et al. 2003) and also in many developed countries in the world, soil data has been encoded into computer-compatible form. Hence, in Australia for example, a soil map can be produced by downloading information directly from the internet. The Australian Soil Resources Information System (ASRIS) database has compiled the best publicly available soil information available across Australian agencies into a national database of soil profile data, digital soil and land resources maps, and climate, terrain, and lithology datasets. Most datasets are thematic grids that cover the intensively-used land-use zones in Australia (Johnston et al. 2003). Hence, the first step when sampling across a region or wider area is to consult these existing/available soil maps of the region of interest in conjunction with, or with help from, experienced soil scientists. The areas of broadly similar soil type can then be identified as high priority areas for further sampling and comparative analyses, using morphological and analytical information. However, in the absence of obvious features, systematic sampling of the area should be conducted (cross-pattern to fully characterise the soil patterns in the area).

Easy-to-Characterise Soil Materials: Large and Trace Amounts

Soil materials are easy to characterise, especially by way of the following published historical examples, which demonstrate how large amounts of soil materials have been characterised using quick morphological and light optical methods to solve crime cases. On a Prussian railroad, in April 1856, a barrel which contained silver coins was found on arrival at its destination to have been emptied and refilled with sand. Prof Ehrenburg of Berlin acquired samples of sand from stations along railway lines and used a light microscope to match the sand to the station from which the sand must have come (Science and Art 1856). This is arguably the very first documented case where a forensic comparison of soils was used to help police solve a crime (Fitzpatrick 2008). Then in 1887 Sir Arthur Conan Doyle published several fictional cases involving Sherlock Holmes such as 'A Study in Scarlet' in Beeton's Christmas Annual of London where Holmes can: "Tell at a glance different soils from each other …has shown me splashes upon his trousers, and told me by their colour and consistence in what part of London he had received them". In 1891 in 'The Five Orange Pips', Holmes observed: 'chalk-rich soil' on boots. This clearly indicates that Conan Doyle was well aware of the key soil morphological properties (colour and consistence) and soil mineralogy (chalk) in forensic soil comparisons. Further, as documented by Murray and Tedrow (1975): "October 1904, a forensic scientist in Frankfurt, Germany named George Popp was asked to examine the evidence in a murder case where a seamstress named Eva Disch had been strangled in a bean field with her own scarf. George Popp successfully examined soil and dust from clothes for identification to solve this real criminalistic case".

Soil morphological descriptions follow strict conventions whereby a standard array of data is described in a sequence, and each term is defined according to both the United States Department of Agriculture (USDA) Field book for describing and sampling soils, Version 2.0 (Schoeneberger et al. 2002) and National standard systems (e.g. Australian Soil and Land Survey Field Handbook by McDonald et al. 1990). Soil morphological descriptors such as colour, consistency, structure, texture, segregations/ coarse fragments (charcoal, ironstone or carbonates) and abundance of roots/pores are the most useful properties to aid the identification of soil materials (e.g. Fitzpatrick et al. 2003) and to assess practical soil conditions (e.g. Fitzpatrick et al. 1999).

Examples of several standard methods and results from various analytical methods used in forensic soil examination will be discussed below (see also several reviews covering mainly forensic geology by several workers (Murray 2004; Ruffell and McKinley 2004; Pye 2007; Dawson et al. 2008)).

Common and Standardised Techniques Used by Forensic Soil Scientists

Evaluation of Degree of Similarity between Questioned Samples and Control Soil Samples

It is important to first define the word 'compare' because no two physical objects can ever, in a theoretical sense, be the same (Murray and Tedrow 1991). Similarly, a sample of soil or any other earth material cannot be said, in the absolute sense, to have come from the same single place. However, according to Murray and Tedrow (1991), it is possible to establish "with a high degree of probability that a sample was or was not derived from a given place". For example, a portion of the soil (or other earth material) could have been removed to another location during human activity. Pye (2007) summarises different schemes commonly used by various members of the Forensic Science Service to convey weight of evidence relating to forms of comparisons such as trace or DNA evidence. For example, he has developed 'verbal categories' ranging from 0 (no scientific evidence) to 10 (conclusive), with no statistical significance of the ranks implied. He also states that there is a long history of the use of numerical scales in the context of evidential and legal matters.

Approaches and Methods for Making Comparisons Between Soil Samples

Forensic soil scientists must first determine if uncommon and unusual particles, or unusual combinations of particles, occur in the soil samples and must then compare them with similar soil in a known location. To do this properly, the soil must be systematically described and characterised using standard soil testing methods to deduce whether a soil sample can be used as evidence (Figure 8.3).

Methods for characterising soils for a forensic comparison involve subdividing methods into two major steps, descriptive (morphological) and analytical (Figure 8.3). Detailed soil characterisation usually requires a joint approach that combines the descriptive and analytical steps in the following two stages:

Stage 1 – Rapid characterisation of composite soil particles in whole soil or bulk samples for screening of samples (Figure 8.3).

Stage 2 – Detailed characterisation and quantification of composite and individual soil particles following sample selection, size fractionation and detailed mineralogical and organic matter analyses using advanced analytical methods (Figure 8.3).

Stage 1: Initial Characterisation of Composite Soil Particles in Whole Samples for Screening

In the initial screening or comparison examination of whole soil samples, soil morphology, low magnification light microscopy, X-ray diffraction (XRD), Diffuse Reflectance Infrared Fourier Transform (DRIFT) spectroscopy, and magnetic susceptibility (volume and mass) are used to compare samples via bulk morphology, mineralogy and organic matter characterisation.

Soil Morphology

Pedological interpretation provides a visual, quick and non-destructive approach to screen and discriminate among many types of samples. Morphological soil indicators are arguably the most common and probably the simplest and it is for this reason that all samples are characterised first using the four key morphological descriptors of colour, consistency, texture and structure (Figure 8.3). In many respects, the soil resembles a sandwich with these easily observed characteristics and thickness, which conveys the concept of different soil layers with different properties. In soil samples from crime scenes and control sites in question where soil may have been transported, by vehicle, foot (e.g. Figures 8.1 and 8.2) or perhaps shovel, these four visual properties are important characteristics.

A checklist of six key soil morphological descriptors has been compiled from standard techniques used in soil science (e.g. Schoeneberger et al. 2002) for assessing the soil properties for forensic examinations. Observations of depth changes in various properties are recommended, *viz.* consistence, colour, texture, structure, segregations/coarse fragments (carbonates and ironstone), and abundance of roots in the different layers or horizons.

The use of petrography is a major and often precise method of studying and screening soils for discrimination in forensics (Figure 8.3). For example, nearly 50 common minerals (e.g. gypsum), as well as several less common minerals, can easily be seen by the naked eye, but using a hand lens or low power stereo-binocular microscope enables the forensic soil scientist to better detect mineral properties (e.g. particle shape and surface texture) and provide more accurate mineral identification. The petrographic microscope is also commonly available for studying thin sections of soil samples (resin impregnated), minerals and rocks. Thin sections of soil materials are mounted on a glass slide and viewed with the petrographic microscope under different incident light conditions through its special attachments (e.g. Stoops 2003). Morphological and petrographic descriptors are useful in assessing soil conditions because they involve rapid field and laboratory assessments. Other methods, such as more detailed mineralogy (see below) and geochemistry, are complex and more costly to carry out. In addition, they can be used in research to evaluate causes for variations in soil condition induced by weathering (that may range from recent, in the case of the formation of sulphuric materials from sulphidic materials in Acid Sulphate Soils, to thousands to millions or even billions of years), anthropogenic activities, land management, hydrology and weather conditions. In forensic soil science, a provenance examination or determination, also known as geographic sourcing, has developed to identify the origin of a sample by placing constraints on the environment from which the sample originated.

Mineral and Organic Matter Identification and Composition

Once a familiarity with the morphology of the materials has been achieved using visual and light microscopic methods, most of the mineralogical and organic matter components in a particular whole or bulk soil sample can be determined using the following three selected methods.

X-ray diffraction (XRD) methods are arguably the most significant for both qualitative and quantitative analyses of solid materials in forensic soil science. Extremely small sample quantities (e.g. up to a few tens of milligrams) as well as large quantities can be successfully analysed using XRD. The critical advantage of XRD methods in forensic soil science is based on the unique character of the diffraction patterns of crystalline and even poorly crystalline soil minerals. Elements and their oxides, polymorphic forms, and mixed crystals can be distinguished by non-destructive examinations. Part of the comparison involves identification of as many of the crystalline components as possible, either by reference to the ICDD Powder Diffraction File (Kugler 2003), or to a local collection of standard reference diffraction patterns (e.g. Rendle 2004), coupled with expert interpretation.

A new rapid mid-infrared spectroscopic method, diffuse reflectance infrared Fourier transform (DRIFT), coupled with chemometric approaches, specifically partial least-squares (MIR-PLS) modelling has been developed by Janik et al. (1998) as applied to soils to predict soil physicochemical properties and has been routinely applied to rapidly screen and compare crime scene samples (Figure 8.3). Principal components analysis (PCA), which models the spectral signatures alone, is also a powerful discriminatory tool, providing an objective method of comparing the mid-infrared spectra of the soil samples being examined when enough samples are available for the technique to be viable. The main advantages of DRIFT spectroscopy are that the analysis is non-destructive and can be rapidly applied, and that the mid-infrared portion of the electromagnetic spectrum is sensitive to organic materials, clay minerals, quartz, due to peaks at vibrational frequencies of the molecular functional groups (Van der Marel and Beutelspacher 1976; Nguyen et al. 1991; Janik and Skjemstad 1995; Janik et al. 1998). As such, this technique is a powerful qualitative tool, which can then be used semi-quantitatively to predict analytes of interest when combined with PLS (MIR-PLS) or other chemometric techniques (Stage 2).

Added to the above two rapid methods and techniques are the use of rapid mass and volume magnetic susceptibility methods, which should also always be used before moving to the more costly detailed methods, which require sample separation (Figure 8.3). Mineral magnetic techniques are a relatively recent development (post 1971) and have now become a very powerful and widely used research tool to characterise natural materials in landscapes (e.g. Thompson and Oldfield 1986).

Stage 2: Detailed Characterisation of Composite and Individual Soil Particles

X-ray Diffraction (XRD) Methods

In many soil forensic case investigations, the amount of soil available for analyses (e.g. on the sole of a shoe) may preclude routine bulk analyses. In such situations, it is best to use an XRD fitted with a system for analysis of extremely small samples (e.g. thin coatings or single particles of the order of 2 to 10 mg) loaded into thin glass capillaries. For analysis in a Gandolfi or Debye-Scherrer powder camera, extremely small specimens (e.g. single mineral particles and paint flakes) can be mounted on the end of glass fibres. Consequently, according to Kugler (2003), X-ray methods are often the only ones that will permit further differentiation of materials under laboratory conditions. According to Murray (2004), "Quantitative XRD could possibly revolutionise forensic soil examination". Methods such as XRD, XRF and DRIFT spectroscopy, whose results partially overlap, are used. These overlapping results confirm each other and give a secure result to the examination.

Scanning Electron Microscopes (SEM) and Transmission Electron Microscopes (TEM)

SEM-TEM are also frequently used to examine the morphology and chemical composition (via energy dispersive spectroscopy) of particles magnified to over 100,000 times their original size making them very useful for discrimination (e.g. Smale 1973). Soil minerals, fossils and pollen spores that occur in soils can be described and analysed in detail by SEM and TEM and are therefore very useful indicators when studying soil samples (e.g. Smale 1973; Pye 2007).

All these techniques and others listed in Figure 8.3 (e.g. heavy and magnetic mineral separations) in combination achieve reliable, definite and accurate results, and provide additional information about the mineralogical, chemical and physical properties of the suspected soil material.

Forensic Applications

Up to this point we have briefly discussed the (i) theory, (ii) significance and relevance of established concepts and standard terminologies used in pedology, (iii) laboratory analytical techniques commonly used, and (iv) systematic forensic soil examination approach, which uses soil morphology (e.g. colour, consistency, texture and structure), mineralogy (powder X-ray diffraction) and chemistry (e.g. based upon infrared spectroscopy analyses) to distinguish between soils associated with forensic examinations. The remainder of this chapter will discuss the ways in which this information can be applied advantageously in forensic casework using two case examples.

Case Study 1: Hit-and-Run – Analyses

The following soil analyses methods were required in the hit-and-run case, which was outlined above. The first step was to visually compare the questioned soil samples from the suspect's shoes (i.e. adhered soil scraped from the soles and sides of the running shoes shown in Figure 8.2) and control samples (i.e. soils shown in Figures 8.1 and 8.2). The control samples were obtained from sulphidic material (Soil Survey Staff 1999) in the acid sulphate soils located both in the river (subaqueous soil) and on the river bank where the suspect was seen to run and left a shoe impression, which was similar to the sole tread of the shoe worn by the suspect.

The visual comparison of the questioned samples from the shoe and control samples after sieving to obtain fine fractions (<50 μm) was conducted by eye and by low power stereo-binocular light microscopy. From these visual observations, it appeared that the fine fractions (<50 μm) from sulphidic material in the acid sulphate soils in both the river bank and in the channel samples had a similar yellow colour to the soil adhered to the shoe (Munsell Soil Color Charts 2000). Consequently, because the river bank sample contained over 95% coarse gravel and stones, a sub-sample was sieved using a 50 μm sieve to obtain a finer fraction (<50 μm). The fine soil fraction from the river bank and soil on the shoe had a remarkably similar colour (Munsell Soil Color) and mass magnetic susceptibility. Hence, in accordance with the systematic approach outlined in Figure 8.3, the third step was to check their mineralogical and chemical composition by using XRD and DRIFT analyses.

The XRD patterns that can be likened to finger print comparisons of the shoe (suspect) and ASS river bank (control) soil samples closely relate to each other (Figure 8.4). However, what is the significance of this close similarity in XRD patterns to the degree of similarity in terms of mineralogical composition? If the two soil samples, for example, contain only one crystalline component such as quartz (i.e. silicon dioxide), which is very common in soils, the significance of the similarity and its evidential value in terms of comparison criteria will be low. If, however, the two soils contain four or five crystalline mineral components, some of them unusual, then the degree of similarity will be considered to be high. In both cases, it was possible to evaluate the mineralogical data and formulate an opinion regarding the significance of the results obtained. The mineralogical compositions of the two samples are summarised in Table 8.1 and have a high degree of similarity because they both contain quartz, mica, albite, orthoclase, dolomite, chlorite, calcite, amphibole and kaolin. Relative proportions of the minerals are slightly different, likely due to the different distributions of particle sizes of the samples.

DRIFT analysis was conducted on the same samples after XRD analyses (Fitzpatrick et al. 2007). Electromagnetic energy in the mid-infrared range (4,000 to 500 cm^{-1}) is focused on the surface of the air-dried, finely ground soil samples (using an agate mortar). Some of the beam penetrates a small distance into the sample and is reflected back into the spectrometer where the spectrum is collected, the spectra are expressed in absorbance (A) units (where A = Log 1/Reflectance). Whilst the two samples are spectrally similar (Figure 8.5), they do differ slightly in the amount of aliphatic organic matter (Table 8.2), which is reflected in peaks centred on 2,850 and 2,930 cm^{-1} (i.e. because the shoe sample has a slightly higher organic carbon content). They are very similar with regards to clay mineralogy (kaolinite clay 3,690 to 3,620 cm^{-1}) and the amount of quartz (2,000 to 1,650 cm^{-1}) in the samples. A peak around 2,520 cm^{-1} also indicates the presence of a small amount of carbonate in both samples, with marginally more in the bank sample.

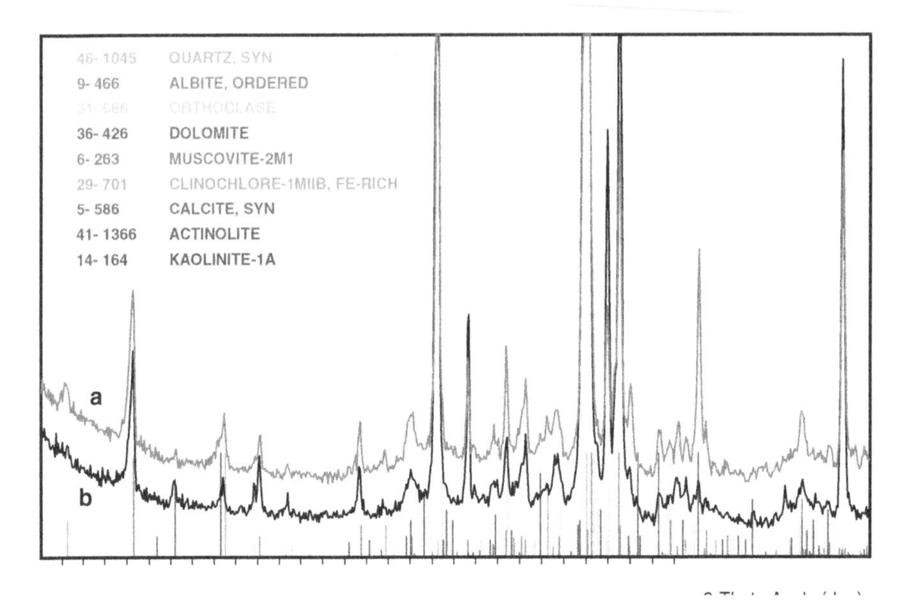

Fig. 8.4 Comparisons between X-ray diffraction (XRD) patterns of soil samples from the shoe (**b**) and river bank (<50 μm fraction) (**a**) in Figures 1 and 2. The <50 μm fraction was separated from the stony river bank soil by sieving through a 50 μm sieve. Shoe and river bank samples were both ground using an agate mortar and pestle before being lightly pressed into aluminium sample holders for XRD analysis. XRD patterns were recorded with a Philips PW1800 microprocessor-controlled diffractometer using Co Kα radiation, variable divergence slit and graphite monochromator (from Fitzpatrick et al. 2007) *(see colour plate section for colour version).*

Table 8.1 Summary of mineralogical composition from XRD analysis (from Fitzpatrick et al. 2007)

Soil samples	River bank[a]	Shoe[b]
Quartz	D	D
Mica	SD	M
Albite	M	M
Orthoclase	M	M
Dolomite	M	T
Chlorite	T	T
Calcite	T	T
Amphibole	T	T
Kaolin	T	T

D – Dominant (>60%), SD – Sub-dominant (20% to 60%), M – Minor (5% to 20%), T Trace (<5%).

[a]River bank sample (LRJ/ CAFSS 027.5) was sieved (<50 μm fraction).

[b]Shoe sample (CAFSS 027.0) was not sieved (i.e. approximately <50 μm).

These comparisons indicate that the two samples have a high degree of similarity and are most likely to have been derived from the same general location. In contrast, there is a lower degree of similarity with the two alibi soils samples (data not shown in this chapter) shown in Figure 8.1 and briefly described in the Case Study 1: Hit-and-Run Case outline above Box 1.

To conclude, sufficient soil morphological, mineralogical (XRD) and physicochemical (DRIFT and MIR-PLS) data was acquired on the two samples to be able to determine if they 'compare' or 'do not compare'. The soil from the shoe has a high degree of morphological, chemical and mineralogical similarity to the fine fraction (<50 µm) contained in the stony/gravelly soil on the river bank and in the river. Hence, the soil from the shoe is most likely sourced from the stony/

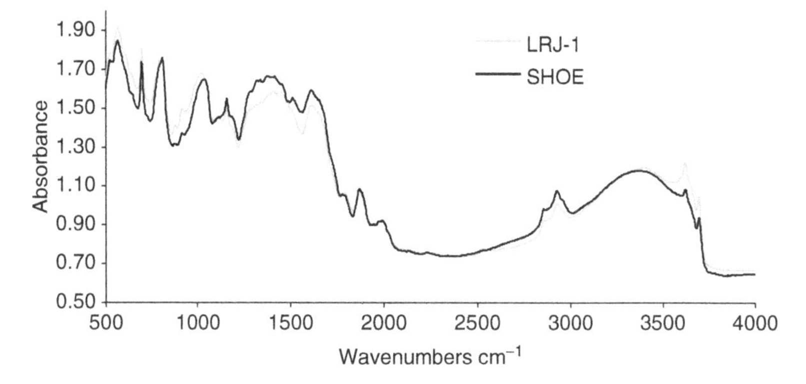

Fig. 8.5 Comparison of diffuse reflectance infrared Fourier transform (DRIFT) spectra between the yellow-brown soil on the shoe (black tone) and the <50 µm fraction in the stony soil from the river bank (grey tone). Shoe and river bank samples were both ground using an agate mortar and pestle (from Fitzpatrick et al. 2007)

Table 8.2 Predictions of charcoal (char), total organic carbon (TOC), pH ($CaCl_2$), calcium carbonate ($CaCO_3$), cation exchange capacity (CEC), clay, silt and sand contents from MIR-PLS analysis (Janik et al. 1998)

Sample	CAFSS	[a]Shoe CAFSS 027.0	[b]River bank CAFSS 027.5 LRJ-1
Char	%	0.2	0.2
TOC	%	6.1	3.4
pH	$CaCl_2$	4.9	5.5
$CaCO_3$	%	1.5	4.6
CEC	meq /100 g	17	15
Clay	%	6	12
Silt	%	27	35
Sand	%	68	53

[a]Shoe sample not sieved because it was already fine(i.e. approximately <50 µm).
[b]The river bank was sieved <50 µm fraction (LRJ-1/CAFSS 027.5).

gravelly soil on the river bank and in the river. Partly as a result of these analyses, the suspect was subsequently found guilty of 'Hit-and-Run' in the supreme court of South Australia.

Case Study 2: Abduction Case from a Quarry Near Stirling, South Australia

The same main soil analyses methods as in the previous case were undertaken for an abduction case (Raven and Fitzpatrick 2005). Once again, the first step was to visually compare the questioned soil sample from the suspect (i.e. adhered soil from the carpet in a vehicle boot) with control soil samples of quartz gravel and clay nodules from an anthropogenic soil in an abandoned quarry (Figure 8.6, Table 8.3).

The case involved the abduction, false imprisonment and life threats made against a male victim by two male offenders. The victim was abducted from the city and placed into the boot of the accused's vehicle before being driven to the quarry. He was allegedly pulled from the boot, threatened with a rock and drowning before being returned to the boot and later released. The accused's vehicle was seized by police and Forensic Science Centre examiners recovered soil material from hand and foot marks in the boot. Soil samples were also collected from the quarry where the offences allegedly took place for comparison.

Samples from the quarry and vehicle boot were also described and photographed under a stereo-microscope. The soil samples from the quarry and vehicle boot show similarities in colour (Table 8.3, similar Munsell Soil Color Hue – 10YR). The coarse and fine particles (i.e. $<50\,\mu m$ sieved samples) showed very similar morphological properties indicating they were likely taken from the same locality. The following grain properties (Table 8.3) also indicate that the two soil samples have similarities: (i) shape, degree of rounding/angularity and surface texture of quartz particles; (ii) colour of coatings on surfaces of the grains (very pale brown); (iii) colour and morphology of clay fragments/aggregates. The soil sample from the vehicle boot contained very small amounts of fibres (dark blue and back), which likely came from the black mat/carpet in the boot. The two XRD patterns (Figure 8.7) closely relate to each other (i.e. have remarkably close similarity). Mineralogy by XRD showed that both have dominant quartz, mica (muscovite), kaolinite with traces of orthoclase, albite, goethite, calcite and tourmaline (schorl).

Combined DRIFT spectral patterns are displayed in Figure 8.8. Samples AGU1 and AGU2 are very similar spectrally, with the main differences being that AGU2 contains slightly more organic matter (weak peaks near 2,850 to 2,930 cm^{-1}) and less quartz (peaks in the ranges 1,750 to 2,000 cm^{-1} and 450 to 1,300 cm^{-1}) than AGU1.

In summary, the soil samples from the quarry and vehicle boot show remarkably similar morphological, physical, chemical and mineralogical properties, viz. (i) soil Munsell colours (10YR Hue; Very pale brown [dry] and light yellowish brown [moist]) (Table 8.3); (ii) single grain (loose to powdery) with some clay, and medium to coarse sand, with no roots or detectable carbonate (Table 8.3); (iii) similar

Fig. 8.6 Abandoned quarry near Stirling, South Australia. Control soil was sampled in the surface layer (0 to 2 cm) of an anthropogenic soil (i.e. so-called man made soil that has recently formed in the floor of the quarry) between the tire tracks and metal construction (sieve) on the left hand side and at the base of the quarry *(see colour plate section for colour version)*

Table 8.3 Morphological description of materials collected from the mat/carpet on the floor of the boot of suspect's vehicle (forensic exhibit) and quarry (control sample)

Sample number	Brief description, analyses conducted	Munsell colour (dry) and (moist)	Colour; texture, segregations/coarse fragments and quartz grain shape and surface texture
AGU1	Material from quarry *XRD, DRIFT*	10YR 8/4 (dry)	Very pale brown. Single grain (loose to powdery) with some white clay fragments, and medium to coarse sand. No roots or detectable carbonate. Abundant (>75% grains) very angular with conchoidal fractures and smooth surface.
		10YR 6/4 (moist)	Light yellowish brown. As above.
AGU2	Material recovered from the mat/carpet on the floor of the boot of suspect's vehicle *XRD, DRIFT*	10YR 8/4 (dry)	Very pale brown. Single grain (loose to powdery) with some white clay fragments, and fine to medium sand. No roots or detectable carbonate. Trace of synthetic fibres. Abundant (>75% grains) very angular with conchoidal fractures and smooth surface.
		10YR 6/4 (moist)	Light yellowish brown. As above.

particle size distribution as expressed by percent of clay (30% to 31%), silt (60% to 65%) and sand (10% to 4%) (Table 8.4); (iv) calcium carbonate content is low (Table 8.4); (v) ESP (0%, Table 8.4); (vi) total organic carbon content (0.6 to 0.9, Table 8.4); (vii) pH (5) (Table 8.4); (viii) mineralogy by XRD: both have dominant quartz, mica (muscovite), kaolinite with traces of orthoclase, albite, goethite, calcite and tourmaline (schorl) (Figure 8.7).

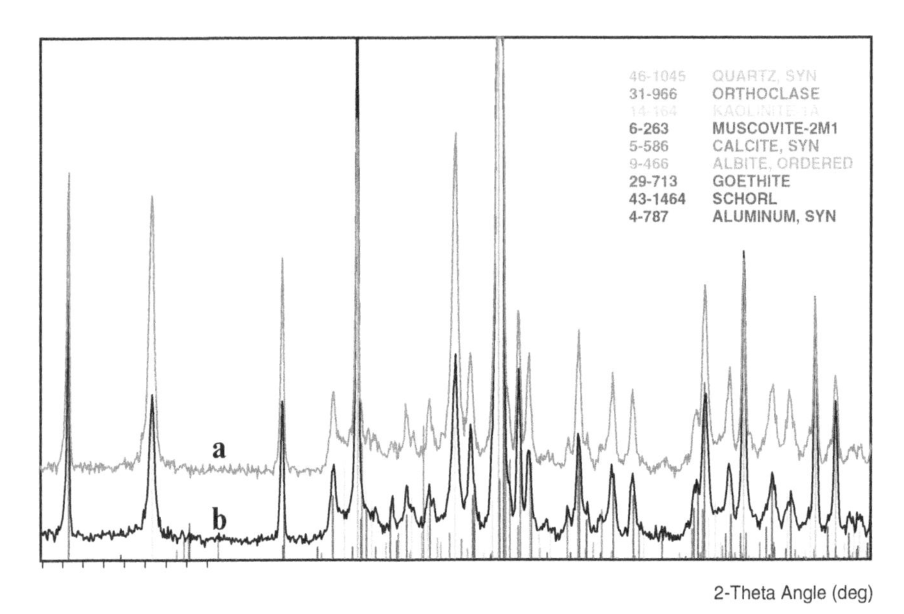

Fig. 8.7 Comparisons between representative X-ray diffraction (XRD) patterns of soil samples from the High Street quarry, Stirling (a, AGU1) and material recovered from the boot of the suspect's vehicle (b, AGU2). There are considerable similarities between the XRD patterns indicating that they are most likely to have been derived from the same location *(see colour plate section for colour version)*

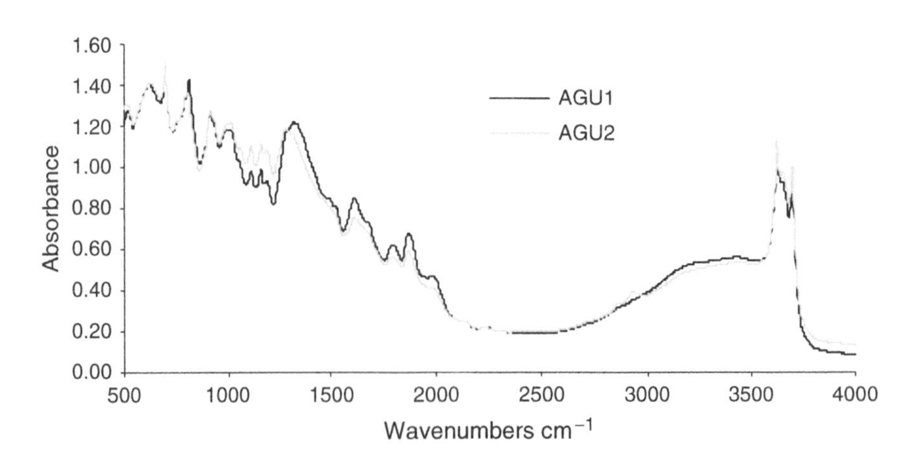

Fig. 8.8 Comparisons between representative diffuse reflectance infrared Fourier transform (DRIFT) spectra of <50 μm sieved samples of soil from the High Street quarry, Stirling (AGU1, black) and material recovered from the boot of the suspect's vehicle (AGU2, grey tone). There are considerable similarities between the DRIFT spectra indicating that they are most likely to have been derived from the same general location

Table 8.4 Predictions of total organic carbon (TOC), calcium carbonate ($CaCO_3$), pH (water), electrical conductivity (1:5 water), exchangeable sodium percent (ESP), clay, silt and sand contents from MIR-PLS analysis (Janik et al. 1998).

Sample	AGU1	AGU2
Locality	Quarry	Vehicle boot
TOC%	0.6	0.9
pH water	5	5
$CaCO_3$%	0	0.4
EC dS/m	0.7	0.6
ESP	0	0
Clay%	30	31
Silt%	60	65
Sand%	10	4

To summarise, sufficient soil morphological, mineralogical (XRD) and physicochemical (DRIFT) data was acquired on the soil samples from the quarry site (control soil sample) and vehicle boot (forensic exhibit containing an extremely small amount of soil material) to be able to determine if they 'compare' or 'do not compare'. Samples from the quarry have a high degree of morphological, chemical, mineralogical and organic matter composition similarity with the very small amount of soil collected from the carpet in the vehicle boot. Hence, samples collected from the quarry compare well with samples from the vehicle boot and are likely to have been sourced from within the same locality.

Two males were found guilty of abduction and assault at the supreme court in South Australia. This case was interesting in that it involved transfer of an extremely small amount of anthropogenic soil material from an abandoned quarry to the carpet in the boot of the suspect's car.

Conclusions

Soil materials are routinely encountered as evidence by police (physical evidence branch) for crime scene investigators and forensic staff. However, most forensic and physical evidence laboratories either do not accept or are unable to adequately characterise soil materials. The main reason for this is that morphological, mineralogical and spectroscopic analytical knowledge required to examine and interpret such soil evidence needs a large amount of training and expertise.

The two crime scenario examples illustrate the use of combined pedological, mineralogical and spectroscopic methods in the forensic comparison of transported soil samples to forensic evidence items (e.g. shoes or a carpet in a boot) with control soil samples from either the scene of the crime or a site traversed by the suspect in association with the crime. Forensic soil examination can be complex because of the diversity and in-homogeneity of soil samples. However, such diversity and complexity enables forensic examiners to distinguish between soils, which may appear to be similar. There is a general lack of expertise in this relatively new area among soil scientists. For research and practical application in this area to grow appreciably, it will need to be considered and taught as an integral part of both soil science and forensic science courses. Finally, an attempt should be made to develop and refine methodologies and approaches to develop a practical 'Soil forensic manual with soil kit for sampling, describing and interpreting soils' (Fitzpatrick 2008).

Acknowledgements Mr Adrian Beech for chemical analyses and Mr Stuart McClure for SEM analyses from several previous forensic soil investigations. Dr Jock Churchman, Mr Richard Merry and Dr Phil Slade for constructive editorial comments.

References

Aardahl K (2003). Evidential value of glitter particle trace evidence. Master's Thesis, National University, San Diego, CA, USA.

Blackledge RD and Jones EL (2007). All that glitters is gold! In: Forensic Analysis on the Cutting Edge: New Methods for Trace Evidence Analysis (Ed. RD Blackledge), pp. 1–32. Wiley, New York.

Chisum W and Turvey B (2000). Evidence dynamics: Locard's exchange principle and crime reconstruction. Journal of Behavioral Profiling 1:1–15.

Dawson LA, Campbell CD, Hillier S and Brewer MJ (2008). Methods of characterizing and fingerprinting soil for forensic application. In: Soil Analysis in Forensic Taphonomy: Chemical and Biological Effects of Buried Human Remains (Eds. M Tibbett and DO Carter), pp. 271–315. CRC, Boca Raton, FL.

Fitzpatrick RW (2008). Nature, distribution and origin of soil materials in the forensic comparison of soils. In: Soil Analysis in Forensic Taphonomy: Chemical and Biological Effects of Buried Human Remains (Eds. M Tibbett and DO Carter), pp. 1–28. CRC, Boca Raton, FL.

Fitzpatrick RW, McKenzie NJ and Maschmedt D (1999). Soil morphological indicators and their importance to soil fertility. In: Soil Analysis: An Interpretation Manual (Eds. K Peverell, LA Sparrow and DJ Reuter), pp. 55–69. CSIRO Publishing, Melbourne, Australia.

Fitzpatrick RW, Powell B, McKenzie NJ, Maschmedt DJ, Schoknecht N and Jacquier DW (2003). Demands on soil classification in Australia. In: Soil Classification: A Global Desk Reference (Eds. H Eswaran, T Rice, R Ahrens and BA Stewart), pp. 77–100. CRC, Boca Raton, FL.

Fitzpatrick RW, Raven MD and Forrester ST (2007). Investigation to determine if shoes seized by South Australia Police contain soil materials that compare with a control soil sample from the bank of the Torrens River, Adelaide. CSIRO Land and Water Client Report CAFSS_027. January 2007. Restricted Report.

Fitzpatrick RW, Raven M and McLaughlin MJ (2006). Forensic soil science: an overview with reference to case investigations and challenges. In: Proceedings of the First International Workshop on Criminal and Environmental Forensics (Ed. RW Fitzpatrick). Centre for Australian Forensic Soil Science, Perth, Australia. http://www.clw.csiro.au/cafss/

ICDD (2003). The *Powder Diffraction File*. International Center for *Diffraction* Data, 12 Campus Boulevard, Newton Square, PA 19073-3273, USA.

Janik LJ and Skjemstad JO (1995). Characterization and analysis of soils using mid-infrared partial least squares. II. Correlations with some laboratory data. Australian Journal of Soil Research 33:637–650.

Janik LJ, Merry RH and Skjemstad JO (1998). Can mid infrared diffuse reflectance analysis replace soil extractions? Australian Journal Experimental Agriculture 38:637–650.

Johnston RM, Barry SJ, Bleys E, Bui EN, Moran CJ, Simon DAP, Carlile P, McKenzie NJ, Henderson BL, Chapman G, Imhoff M, Maschmedt D, Howe D, Grose C and Schoknecht N (2003). ASRIS: the database. Australian Journal of Soil Research 41(6):1021–1036.

Kugler W (2003). X-ray diffraction analysis in the forensic science: the last resort in many criminal cases. JCPDS – International Centre for Diffraction Data 2003, Advances in X-ray Analysis, Volume 46. http://www.icdd.com/resources/axa/vol46/v46_01.pdf

McDonald RC, Isbell RF, Speight JG, Walker J and Hopkins MS (1990). Australian soil and land survey. Field Handbook. 2nd Edition. Inkata, Melbourne.

Munsell Soil Color Charts (2000). X-Rite, Incorporated and GretagMacbeth AG/LLC USA; 4300 44th Street SE; Grand Rapids, MI 49512 USA.

Murray RC (2004). Evidence from the Earth: Forensic Geology and Criminal Investigation. Mountain Press, Missoula, Montana.

Murray RC and Tedrow JCF (1975). Forensic Geology: Earth Sciences and Criminal Investigation (republished 1986). Rutgers University Press, New York.

Murray RC and Tedrow JCF (1991). Forensic Geology. Prentice Hall, Englewood Cliffs, NJ.

Nguyen TT, Janik LJ and Raupach M (1991). Diffuse reflectance infrared Fourier transform (DRIFT) spectroscopy in soil studies. Australian Journal of Soil Research 29:49–67.

Pye (2007). Geological and Soil Evidence: Forensic Applications. CRC, Boca Raton, FL.

Raven M and Fitzpatrick RW (2005). Comparison of soils from the High Street quarry, Stirling and material recovered from the boot of a vehicle submitted by South Australia Police: Uren case. CSIRO Land and Water Client Report CAFSS_007. June 2005. Restricted Report. 17 pp.

Rendle DF (2004). Database use in forensic analysis. Crystallography Reviews 10:23–28.

Ruffell A and McKinley J (2004). Forensic geoscience: applications of geology, geomorphology and geophysics to criminal investigations. Earth-Science Reviews 69:235–247.

Schoeneberger PJ, Wysocki DA, Benham EC and Broderson WD (Eds.) (2002). Field Book for Describing and Sampling Soils, Version 2.0. Natural Resources Conservation Service, National Soil Survey Center, Lincoln, NE.

Science and Art (1856). Curious use of the microscope. Scientific American 11:240.

Smale D (1973). The examination of paint flakes, glass and soils for forensic purposes, with special reference to the electron probe microanalysis. Journal of Forensic Sciences Society 13:5–15.

Soil Survey Staff (1999). Soil Taxonomy – a Basic System of Soil Classification for Making and Interpreting Soil Surveys. 2nd Edition. United States Department of Agriculture, Natural Resources Conservation Service, USA Agriculture Handbook No. 436. 869 pp.

Stoops G (2003). Guidelines for Analysis and Description of Soil and Regolith Thin Sections. Soil Science Society of America, Madison, WI.

Thompson R and Oldfield F (1986). Environmental Magnetism. Ch.2. Allen and Unwin, London.

Van der Marel HW and Beutelspacher H (1976). Atlas of Infrared Spectroscopy of Clay Minerals and Their Admixtures (Eds. HW van der Marel and H Beutelspacher). Elsevier Scientific, Amsterdam. 396 pp.

Wilding LP (1994). Factors of soil formation: contributions to pedology. In: Factors of Soil Formation: A Fiftieth Anniversary Retrospective. SSSA Special Publication 33, pp. 15–30. Soil Science Society of America, Madison, WI.

Chapter 9
Forensic Ecology, Botany, and Palynology: Some Aspects of Their Role in Criminal Investigation

Patricia E.J. Wiltshire

Abstract Ecology, botany, and palynology are now accepted as part of the armoury of forensic techniques. These disciplines have been tested in court and have provided evidence for contact of objects and places, location of clandestinely-disposed human remains and graves, estimating times of deposition of bodies, differentiating murder sites from deposition sites, and provenancing the origin of objects and materials. It is important that the forensic palynologist is a competent botanist and ecologist. Sadly, not all practitioners have this essential background and, therefore, produce work inadequate to withstand scrutiny in court. Palynology involves the identification of many classes of microscopic entities, the most important being pollen, plant spores, and fungal spores. The practitioner needs to be able to identify palynomorphs in damaged and decayed states and this requires experience and skill. However, identification is still the lowest level of palynological expertise, and interpretation of palynological assemblages requires knowledge of plant distribution, developmental responses, and phenology, as well as ecosystem structure and function. The forensic palynologist must also understand highly manipulated and artificial systems, and the complexities of taphonomic processes. There have been attempts to make forensic palynology 'more scientific' by the construction of test trials, the application of current statistical techniques, mathematical modelling, and reference to aerobiological data and pollen calendars. But these appear to be of limited use in the forensic context where outcomes are scrutinised in court. There is a high degree of heterogeneity and variability in palynological profiles, and every location is unique. It is impossible to achieve meaningful and forensically-useful databases of the palynological characteristics of places; predictive models will always be crude and unlikely to be of practical value. In spite of this, the experienced ecologist/palynologist has been able to identify places, demonstrate links between objects and places, estimate body deposition times, and differentiate

P.E.J. Wiltshire
Department of Geography and Environment, University of Aberdeen, Elphinstone Road, Aberdeen AB24 3UF, UK; and Department of Natural and Social Sciences, University of Gloucestershire, Swindon Road, Cheltenham Gl50 4AZ, UK.
e-mail: patricia.wiltshire1@btinternet.com

K. Ritz et al. (eds.), *Criminal and Environmental Soil Forensics*

pertinent from irrelevant places very successfully. Nevertheless, there has been no substitute for examination of every pertinent place, and every relevant exhibit in each criminal investigation.

Introduction

Ecology is the study of organisms together with their environments – the study of ecosystems. By its nature, palynology is a subdiscipline of botanical ecology and, to work in a forensic context, the palynologist must have a sound botanical and ecological training. In Britain, forensic palynology is an acknowledged aid to criminal investigation, providing valuable evidence in cases of murder, manslaughter, rape, and abduction. The body of literature for the discipline in peer-reviewed journals is relatively small and, although many reports and interpretive material are technically in the public domain, they are only accessible through court and police records.

The crime scenes which benefit from palynological help are invariably ecosystems themselves but they may be highly modified by human activities. Therefore, as well as the understanding of natural and semi-natural habitats, the forensic ecologist/palynologist must also appreciate the complexities of highly manipulated systems, such as gardens, parks, rubbish dumps, plantations, ponds, canals, roadsides verges, hedgerows and wasteland. Because of the breadth of the discipline, the forensic ecologist cannot be expert in every aspect of ecological science but, to be of use to a criminal investigator, the essential requirements are knowledge of soil, and of aquatic and terrestrial sediments.

Soils and sediments exhibit great variability in the origin of their parent materials, structure, and chemistry but it is important for the forensic ecologist and palynologist to realise that soil is particularly complex because of its dynamic nature. It provides a habitat where communities of organisms live and complete their life-cycles, and these organisms profoundly affect the chemistry of the inorganic matrix as well as any organic object or material present.

Most plants rely on soil as a source of mineral nutrition, water, and physical support. Depending on their responses to climate, microclimate, and their ecological tolerances and needs, the geographical distribution of plant species can reflect historical geography, and the patterning of soil types at local, regional, and national levels. Plant distribution is also profoundly affected by biotic factors – other plants, animals, micro-organisms and people. Again, the forensic ecologist and botanist must have an understanding of the factors underlying plant distribution, plant response to change, and to have a grasp of the variability created by human intervention. This is achieved by strengthening and modifying theoretical knowledge with extensive field experience.

Over time, organic components of soil will decompose to their constituent molecules. The speed of decomposition will depend on the communities of resident decomposer organisms, and their function depends largely on the physico-chemical nature of the soil itself. Palynomorphs are important organic particles in soils and sediments and, in recent years, these have provided valuable trace evidence in criminal investigation.

Originally, the term 'palynomorph' was used to describe pollen grains and plant spores. Over the years, however, the term has expanded to include: other microscopic plant remains such as trichomes (plant hairs and glands); fungal spores and other fungal bodies; diatoms; cyanobacteria; and microscopic animals such as testate amoebae, nematode eggs and mouth parts, mites, and other arthropod body parts. The palynologist needs to be able to identify many of these kinds of palynomorph, or seek additional expertise.

Applications of Ecology, Botany, and Palynology in Criminal Investigation

I have contributed ecological, botanical, and palynological evidence in over 200 criminal cases, and presented it for cross examination in court on many occasions. Table 9.1 lists a range of objects and matrices from which I have analysed thousands

Table 9.1 Items from which palynomorphs have been successfully retrieved

A-G	H-R	S-W
Ants' nests	Hair (living and deceased persons)	Sediment
Babies' dummies	Honey and other food	Shoes and trainers
Books and paper	Humus	Skin (living and deceased persons)
Boots	Lawns	Soft furnishings
Clothing	Leaves	Soil
Contents of the lower gut	Mosses	Stomach contents
Decaying plant material and compost heaps	Nasal passages of corpses (turbinate bones)	Stone walls and brickwork
Drug resins	Nylon tights and stockings	Swabs
Dust and dusty impressions on flooring or paper	Paving stones	Tea caddies
Fabrics	Petrol cans	Tools (spades, forks, hoes, rakes)
Faeces	Plant litter	Vacuum flasks
Fences and posts (wooden and metal)	Plant surfaces (leaves, stems, bark, fruits)	Vehicles
Finger nails (of living and deceased persons)	Plastic sheeting and seat covers	Vomit
Fodder	Pot plants	Walls
Fur	Roofs	Weapons
Furniture drawers	Ropes and baskets	Wild animal dens and setts
Ground vegetation		Wooden benches

of samples (see also Milne et al. 2005). There are many cases where ecology, botany, and palynology have successfully helped in: (i) linking objects and places (e.g. Wiltshire 2006a; Mildenhall, 2006); (ii) locating hidden human remains and provenancing of objects (Brown et al. 2002; Wiltshire 2005a); (iii) estimating temporal aspects of deposition of remains (Szibor et al. 1998; Wiltshire 2002a; 2003b); and, (iv) differentiating murder scenes from deposition sites (Wiltshire 2002b).

Knowledge of the anatomy of plants, animals, and other organisms helps in the identification of what victims have eaten or inhaled before death, and whether or not an object is of biological rather than manufactured origin (Wiltshire 2003a, 2004a; 2006b). An understanding of plant and fungal development, and the activity of scavenging animals, has given valuable information on the length of time a corpse has lain *in situ* or the length of time since an offender walked on vegetation (Hawksworth 2008a; Wiltshire 2007a, b). Knowledge of soil stratigraphy, coupled with plant development and distribution, has resulted in establishing the premeditated nature of a victim's grave (Wiltshire 2005b). Finally, exploitation of knowledge of plant and fungal taxonomy (Hawksworth 2008b; Wiltshire 2005c, 2008) has been used in assessing the potential of plants being involved in attempted murder by poisoning, or manslaughter through shamanism.

Therefore, it is clear that the identity, structure, chemistry, life-cycles, and growth responses of whole organisms play an important role in criminal investigation, and that even fragments of organisms provide valuable forensic evidence.

Palynology: The Background

The study of palynomorphs gave rise to the science of palynology, first coined by Hyde and Williams in the 1940s (Hyde and Williams 1944). Its derivation is from the Greek verb *palynein*, meaning 'to spread or sprinkle around'. Hyde and Williams were aeropalynologists, concerned with airborne allergens, and had little interest in soil palynomorphs. Their work involved trapping airborne particles and identifying temporal sequences of anthesis (pollen release) for the construction of 'pollen calendars'.

The discipline of palynology is now over 100 years old. The founder of *modern* pollen analysis was Swedish geologist, Lennart von Post, but the subject was developed and promoted by fellow Swedish botanists Rutger Sernander and Gustaf Lagerheim. The first major work in the subject was published by Erdtman (1921).

The first recorded cases of palynology being used as a forensic tool were described by Erdtman (1969). Although applied occasionally (e.g. Frei 1979; Nowicke and Meselson 1984), it has only been used more routinely in the last 15 years or so. Mildenhall pioneered the techniques in New Zealand, Bryant in the United States (Mildenhall 1982; Bryant et al. 1990; Bryant and Mildenhall 1998), and I have developed forensic palynology, forensic ecology, and forensic botany in the British Isles.

Palynomorphs and Their Identification

The range of palynomorphs requiring identification in criminal investigation has grown, and every attempt should be made to identify anything of apparent significance in a palynological preparation. That said, the most abundant and frequently encountered palynomorphs are pollen grains and plant spores. There are many web sites and publications providing pictures, diagrams, and descriptions of pollen and spore structures to facilitate their identification (e.g. Moore et al. 1991; Beug 2004). However, there is no substitute for authenticated reference material. Any unknown pollen grain or spore must be compared with actual, accurately identified material and every attempt made to obtain a prepared or fresh specimen. Serious misidentifications have been made by those relying solely on pictures. While this is regrettable in any area of palynological study, it could have dire consequences in forensic investigation. Keys and pictures should only be used as guides; final identification must involve critical comparative examination of actual palynomorphs under the microscope.

Pollen and plant spores are identified by their shape, size, outer wall (exine) structure, surface sculpturing, and the type, number, and arrangement of apertures. To achieve the highest resolution of identification, it is essential to remove the inner part of the grain so that only the exine remains, involving use of toxic and corrosive acids. It results in the dissolution of background humic material, cellulose, and silica, and only structures which are resistant to the treatment will be retained (Moore et al. 1991). These include the outer walls of pollen and plants spores which are composed of sporopollenin, a very robust polymer. Fungal and arthropod remains, composed of chitin, and some testate amoebae, will also be left after treatment. Although not so important for the identification of some non-botanical palynomorphs, the chemical processing of pollen and plant spores is critical for precise identification. Some sculpturing features of the palynomorph are small (e.g. 0.5 μm), needing observation under phase contrast microscopy at ×1000 or more.

The resolution in pollen and spore identification is variable. In some plant families, taxa can be identified to species (e.g. *Plantago lanceolata* – ribwort plantain; *Sanguisorba officinalis* – greater burnet; *Centaurea scabiosa* – greater knapweed). Others can only be identified reliably to genus (e.g. *Quercus* – oaks; *Salix* – willows; *Polygala* – milkworts; *Aesculus* – horse chestnuts), while some can only be identified to family (Cupressaceae – cypress; Chenopodiaceae/Amaranthaceae – goosefoot family; Poaceae – grasses). In some families such as the Rosaceae, taxa can be identified to species (*Rubus chamaemorus* – cloudberry; *Agrimonia eupatorium* – agrimony), to genus (*Geum* – wood avens and water avens), to type (*Potentilla* – type which includes three genera), and to groups of genera and species whose morphologies merge one into another and so are difficult to differentiate reliably (e.g. many *Prunus* – cherry/plum/peach/almond species). It is also difficult to differentiate between pollen taxa such as *Rosa* (roses), *Rubus* (brambles and others), *Sorbus* (rowan and others), *Crataegus* (hawthorns), and some species of *Prunus* are in the rose-bramble-hawthorn group. As introduced garden species and cultivars are important in the forensic context, particular care must be taken in identification of these groups, but they can also provide surprisingly distinctive markers.

Identification will be relatively crude in the absence of chemical processing. If the inner part of the pollen grain, and the soil matrix (or other material), are not removed from the background, only identification to family may be possible. Another source of error is in the identification of taxa which are impossible to differentiate by standard techniques. Knowing what is possible requires considerable experience of pollen and spores from of a wide range of species. Reference to text books will not always resolve the problem of inter-generic and inter-species variation. Further, many fossil spores (from the Mesozoic to Caenozoic eras), and other remains, find their way into palynological preparations; and these have been known to be wrongly identified as modern taxa. This is particularly the case for Pteridophyta (ferns and allies), sometimes identified as *Sphagnum* moss spores, or as modern fern spores. Fossil fungal remains are also retrieved from exhibits and considered modern. Fungal spores have been misidentified as pollen!

Many areas of palynological investigation (e.g. palaeoecology, melissopalynology) can achieve clean samples with palynomorphs in good condition (Figure 9.1a), but forensic palynology often involves examination of very poor samples (Figure 9.1b). The samples may be laden with cellulosic debris, fly-ash, soot, and other materials which can obscure the view of the palynomorph on the slide. The palynomorphs themselves might also be crushed, crumpled, broken, and partially decayed. In the case of forensic samples, there must be no positive opinion given unless the analyst can demonstrate criteria for identification which are robust enough for legal challenge.

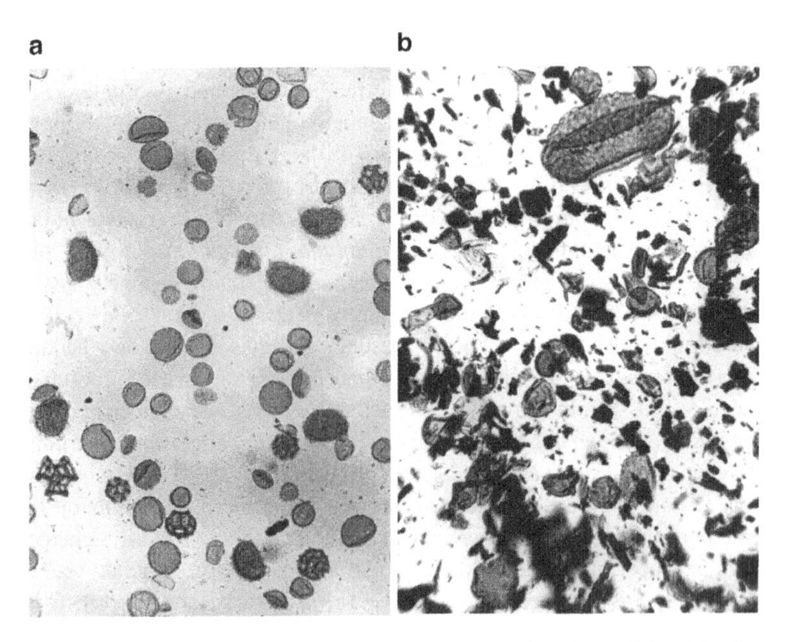

Fig. 9.1 Examples of how palynological samples can appear when viewed through the light microscope. (**a**) 'Clean' sample with palynomorphs in good condition. (**b**) Typical sample from a forensic context

Scanning electron microscopy (SEM) and other sophisticated microscopical methods (e.g. confocal electron microscopy) have little practical application for routine forensic palynological work. Although, on occasion, a sample might consist of a single taxon, identified by SEM, this is rare. SEM presents a picture of only the outer surface of the pollen grain or spore. In the case of pollen, it is often the elements making up the outer wall (exine) seen in section, that are pivotal for precise identification. These details are more easily differentiated by high-powered, bright field and phase contrast microscopy. In forensic samples, it is usually necessary to identify and count hundreds of palynomorphs found in a prepared slide and, although it is technically feasible using SEM (Jones and Bryant 2007), the experienced palynologist does not need SEM for identification. If only SEM micrographs are available, identification can be impossible on some occasions.

Palynomorphs as Trace Evidence

Locard's Principle ('every contact leaves a trace') is known to every police detective (White 2004). As outlined above, palynomorphs, especially pollen and spores, are excellent proxy indicators of place. Offenders walk on soil, mud, or vegetation (short and tall); they have been known to hide in, or walk through, hedges, lean against buildings, trees, and posts, or sit on seats. Important evidence has been retrieved from very many objects and matrices and some of these are shown in Table 9.1. If palynomorphs are transferred from a place to an offender, a victim, or any object, they can be retrieved. The transferred assemblage can then be evaluated in terms of the likelihood of the offender, or victim, or object having contacted the specific place (Wiltshire 2004c).

Pollen grains have evolved for sticking to the female part of the plant and, unlike fibres (which are readily shed from clothing and other objects), will embed into fabrics and small interstices in footwear and other objects; pollen and spores are not easily removed. They are held firmly by their surface sculpturing and by static charges, and are not easily shed, even from clothing and footwear that have been subjected to washing in a machine (Wiltshire 1997). This quality of tenacious adherence makes them very valuable as trace evidence and indicators of places or specific surfaces. The value and advantages of palynology to forensic investigation are obvious and the discipline proven to be effective. However, it is not simple and the practitioner may need to apply caveats to any conclusion.

Caveats and Limitations

It is important to be aware of caveats and limitations in forensic palynology as the outcome of the work could result in someone losing their liberty. It is not an academic exercise. All environments are highly variable, particularly those outside buildings.

Even run-down inner-city estates can have surfaces yielding distinctive palynological profiles with areas of bare soil, lawns, weedy cracks in pavements and corners, and some vegetation. But great care is required in planning the sampling strategy, the refinement of preparation, the resolution of palynomorph identification, and the interpretation of the volume of data gathered.

One of the most commonly-used arguments against the expert witness palynologist is that the observed profile could have been picked up from 'anywhere'. Another common challenge is that the observed profile had accumulated through repeated contact with a variety of palyniferous surfaces, each contributing to a profile that, collectively, happens to be characteristic of the crime scene. The reply would need to stress the improbability of this happening. Experience has shown that, while every place will yield a unique profile, some places are similar to others in various degrees. Considerable credibility as to the uniqueness of an assemblage is achieved when palynomorphs that are generally rare (in palynological terms) are found in the profile.

Databases and Statistical Analyses

It might be considered that a reference database could be compiled of palynological profiles of oak woodlands, grasslands, roadside verges, or any other kind of recognisable environment. This might be feasible in the broadest sense but only at a level where the resolution would never be sufficiently high to be of value in criminal investigation. Databases for populations of palynomorph species can be constructed but not for assemblages from places.

This can be exemplified by Hampshire murder case (Wiltshire and Black 2006) where a man was garrotted beneath a beech tree (*Fagus sylvatica*) and placed, face-down, in a shallow grave immediately adjacent to where he was killed. The area around the grave also had holly (*Ilex aquifolium*), oak (*Quercus robur*), birch (*Betula pendula*), pine (*Pinus*), honeysuckle (*Lonicera periclymenum*), and bramble (*Rubus fruticosus*), with a ground flora dominated by, amongst others, bluebell (*Hyacinthoides non-scripta*), wind-flower (*Anemone nemorosa*), and bracken (*Pteridium aquilinum*). The palynomorph assemblage in the comparator samples was complex; many other taxa (some relatively uncommon in palynological assemblages) were recorded in addition to the most obvious plants growing in the vicinity of the grave. The assemblages retrieved from footwear and a vehicle belonging to two suspects were very similar to that in the comparator samples from the crime scene. Included in the assemblage were abundant conidia (asexual spores) of *Triposporium elegans*. Although found on a range of woody species, this fungus is very common on the cupules of beech fruits.

An inevitable argument likely to have been presented by Defence Counsel, and which needed addressing, was that the assemblage could have 'come from any woodland in Hampshire and Sussex'. Investigation revealed 14 woodlands that could, conceivably have had the same species composition as the crime scene. Each was extensively field-walked with the purpose of finding an environment which might offer a similar palynological assemblage to that from the crime scene. Three were found, sampled, and analysed. None resembled the crime scene closely, although all the dominant species were growing in the three places. Furthermore, only one yielded *Triposporium elegans*, even though all three were thickly strewn with beech cupules; this fungal spore helped eliminate two of the sites.

Pollen and spore reference material in a good collection will contain examples taken from different anthers/sporangia, different plants, and different places at different times for each species; and to be competent, a palynologist requires access to an authenticated and comprehensive reference collection of pollen, plant spores, fungal spores, and other microscopic entities. Although there have been occasions where a single palynomorph taxon has been useful in a criminal case (Mildenhall 2006), it is rare for one palynomorph or single palynomorph taxon to be of evidential value. More usually, the evidence consists of palynological *assemblages* comprised of up to 200 or more different palynomorphs in varying abundances. For many classes of trace evidence (e.g. fibres, glass, brick, paper, paint, and ink), although each is huge, their populations are finite; given time and resources, comprehensive reference collections for these types of evidence could be constructed for statistical comparison. To date this has proved impossible for whole palynological profiles.

Some attempt has been made to invoke Bayes' theorem to palynological data (Horrocks and Walsh 1998). Unfortunately, there were some false assumptions in their study, and the 'cases' appeared to be highly theoretical, simple, scenarios constructed to demonstrate their hypotheses. If full datasets from real cases had been used, the conclusions drawn may have been different. The Bayesian approach is important philosophically and is useful for framing the presentation of conclusions but, at the present time, it is difficult to see how it can be applied to the volume of data accrued from real cases. Multivariate analysis has, on occasion, proved useful in dealing with data where there are high numbers of exhibits and comparator samples (Wiltshire 2002c, 2004a; Riding et al. 2007), but only for guidance. Conclusions, and professional opinion presented in court, must be based on botanical and palynological criteria, skill, and experience.

Taphonomy may be defined as 'The sum of all the factors that influence whether a palynomorph (pollen, spore, or other microscopic entity) will be found at a specific place at a specific time' (Wiltshire 2006a). The taphonomic processes influencing the pattern of palynomorph deposition in any environment are numerous. Unlike populations of fibres, glass and other man-made materials, those of palynological profiles are infinite. The many variables that affect the accumulation of palynomorphs in soils, on surfaces, and on items such as footwear, vehicles, and clothing, preclude routine statistical techniques from making useful contributions to forensic palynological studies.

If two sampling points are spatially close, there will be a higher likelihood of them being similar than if they were widely separated. Similarity diminishes with distance although, under some circumstances, samples in close proximity to each other can be distinctly different. My interpretation of many thousands of samples has shown that every place will yield a unique palynological profile, and that even each sampling point within a site will have its own special characteristics. To gain a palynological 'picture of place' requires the analysis of many comparator samples to build up a matrix of profiles, each one contributing to the bigger picture. Such a bigger picture will be unique for that place.

The complexity of palynological taphonomy makes the discipline an exquisite tool in the hands of the experienced forensic palynologist. However, it can be potentially misleading and suspect when those with inappropriate, or insufficient, experience are involved, and the caveats and constraints are not properly addressed.

Taphonomic Considerations

Some of the major taphonomic factors affecting palynological profiles have been briefly reviewed by Mildenhall et al. (2006). They note that of importance is the level of pollen and spore production, and the way these entities are dispersed.

A spore produced by a fungus, moss, or fern will germinate and form a new organism if conditions are amenable. Pollen produced by conifers and flowering plants is carried to female stigmas to effect fertilisation and production of seeds. Both spores and pollen are carried by vectors, mostly wind, insects, and rain-splash. Those pollen and spores, which achieve dispersal into turbulent air and get carried up and away from the parent plant, form the 'airspora'; this will eventually fall as 'pollen rain' onto surfaces.

In general, wind-pollinated plants produce large amounts of well-dispersed pollen, while insect-pollinated ones produce relatively small amounts of poorly-dispersed pollen. Pollen derived from wind-pollinated trees and shrubs, and tall, wind-pollinated herbs are common components in the airspora, and are often over-represented. But crime scenes are often dominated by insect- or self-pollinated plants. Invariably their pollen simply falls in a halo on the ground around the parent plant, or is released only when the plant dies and falls to the ground. Pollen from these plants, as well as many fungal spores (including those of lichens), may never be found in the airspora, and contribute little to the pollen rain, but they may be the most abundant on an item of footwear. The wearer may have walked through organic debris or meadow vegetation, and abundant pollen from the airspora at the time of the offence may form only a small proportion of the profile relevant to the criminal investigation.

Some palynomorph taxa are thus exceptionally useful markers of the place and, if there were many such plants at a specific location, that place would be endowed with a characteristic palynological signature which would be difficult to replicate artificially. Analysis of many samples at numerous crime scenes has demonstrated that whatever the mode of pollination, the major part of the pollen load of any plant falls close to the parent.

Many palynologists are engaged in the reconstruction of past environments, vegetation change through time, and past land-use; this involves the analysis of cores of mire or lake sediments, or buried soils. Few carry out independent and extensive analysis of multiple surface samples at either local or regional level. Most depend on the researches of other palynologists interested in the taphonomic problems associated with dispersal and fall-out. The latter includes the Pollen Monitoring Programme, a research initiative sponsored by the International Quaternary Association (INQUA) (Hicks et al., 2001; Tinsley, 2001; Barkenow et al., 2007). Another initiative, POLLANDCAL (Pollen-Landuse Calibrations), involves many palynologists collecting modern pollen data and using sophisticated statistical techniques to generate predictive models (Sugita et al., 1999; Eklöf et al., 2004; Bunting et al., 2005; Bunting and Middleton, 2005; Soepboer et al., 2007).

As with earlier studies on pollen productivity and dispersal, the work of the PMP and POLLANDCAL is, in the main, concerned with obtaining information to enhance interpretation of palaeoecological profiles. It aims to interpret past vegetation by extrapolation of results from patterns of modern pollen deposition to profiles obtained from ancient deposits.

Airspora studies depend largely on the use of pollen traps (see Caulton et al. 1995; Levetin et al. 2000; Wiltshire 2006a). Although there are many trapping sites, the areas covered are, nevertheless, insignificant in comparison to the total land surface available for deposition. Extrapolating modern data to the past requires a leap of faith since there may be no past homologue or analogue of modern vegetation patterning. The situation is less tenable when data from pollen traps are applied to more distant sites, and inappropriate when applied to ancient sites, hundreds of miles from the sampling locations. Irrespective of the sophistication of statistical modelling, the nature of the pollen trapping site is critical. There may be common elements in vegetation composition between the sampling site and a palaeoecological profile, but the chance of them having been truly representative of the ecology of another is remote.

The study of airborne palynomorphs is fraught with difficulty and has been a focus of debate for many years. There is considerable variation in the pollen rain at any one place at any one time, and this variation is reflected in fall-out patterns onto surfaces. The variation will depend on pollen production and dispersal characteristics, the presence of physical and geographical barriers, the structure and mass of the palynomorph affecting sedimentation rates, and many other factors. There are classic models constructed to explain the observed heterogeneity in the airspora (Tauber 1965, 1967), and large-scale and small-scale modern pollen studies which have led to the construction of predictive models, often quoted by palynologists working on the reconstruction of past vegetation and land-use (Davies 1967; Jacobson and Bradshaw 1981; Prentice 1985).

A search on the internet provides large numbers of papers, written over the last 40 to 50 years or so, describing and examining these taphonomic phenomena, as well as more recent research projects and papers. There is little doubt that data collected through the PMP and POLLANDCAL, and specific studies involving dispersal from individual species, are not sufficiently refined for use in forensic studies. Inevitably, their investigations concentrate either on palynomorphs, which become airborne and contribute to the airspora, or are geared towards plant/vector inter-relationships. The former are mostly trees, shrubs, and wind-pollinated herbs which have high pollen production and good dispersal, and the latter related to crop plants which rely on insect pollination.

There is some danger in adhering to some of the conventional wisdoms in palynology where models are based on the work of a limited number of researchers in a limited range of scenarios. For example, it is often assumed that some taxa such as *Pinus* (pine) exhibit long distance transport. Some grains are capable of being transported many miles from source but, if there are physical obstacles between the source and accumulating surface, the pollen may be deposited only very locally. In every plant, most of its pollen or spore production will fall near the parent. Enigmatically, there have been cases where a prolific pollen producer such as pine registered less than 2% of the total pollen sum even though mature trees were within 10 m of the sampling site (Wiltshire unpubl.). In other instances, insect-pollinated plants such as *Aesculus hippocastanum* (horse chestnut), where pollen production is thought to be low and dispersal poor, have achieved the same values as *Quercus* (oak), a more prolific pollen producer, several hundred metres away from a mixed stand of horse chestnut and oak in a public park (Wiltshire unpubl.).

At a crime scene in Brierley Hill, near Birmingham, several samples from the vicinity of a young mature, fruiting tree of *Fagus sylvatica* (beech) failed to yield any beech pollen (Wiltshire 2003c). A similar situation was observed in a case in South Wales (Wiltshire 2004c) where a murder victim was buried on a hillside, dominated by *Picea sitchensis* (sitka spruce). Analysis of the surface soils around the grave showed that spruce pollen hardly registered in the comparator samples, but that pine pollen was very well represented. In the lower fill of the grave-fill, the assemblage was *dominated* by pine. The spruce trees were very large and it might be expected that their pollen would swamp the surface. However, very few of the trees had reached sexual maturity. A single, small but mature, pine tree about 100 m from the site was the source of the surface pine pollen. That in the deeper profile was enigmatic, but could have represented the vegetation before the spruce was planted about 40 years earlier. Without understanding such systems, or by not visiting the crime scene, the palynologist might well have assumed pine rather than spruce woodland to be associated with the suspects.

These few examples demonstrate the danger of adhering strictly to simplistic scenarios. Further examples of the dangers of relying on airspora data in forensic investigation are found in Wiltshire (2006a).

Source of Trace Evidence

Pollen and spores falling at any one time will be mixed with pollen previously accumulated on the surfaces. Plants (both insect and wind-pollinated) colonising new ground will also contribute to pre-existing assemblages. This means that time is important in forensic sampling. A natural/semi-natural habitat such as a woodland might yield very similar profiles for many years, but there could be drastic changes if the environment were a manipulated one, such as a plantation or garden, even within short periods. Further, any object contacting a palyniferous surface will receive only a fragment of the pollen rain that had accumulated on it over time, and a fragment of the biological signature of the habitat as a whole. This is why it is essential for the palynologist to select target locations within a crime scene.

Although trace evidence is transferred to the belongings of offenders when they contact soil and sediments, there have been cases where soil has not played any role in investigations (Wiltshire 1997, 2007b). Many surfaces are completely vegetated, or covered in deep leaf litter, such that soil may not be contacted by footwear, clothing, or tools, and vehicles. Plant surfaces, plant litter, humus, and compost can yield dust and, perhaps, the fine fraction of the soil through rain splash and wind action, but the most important particulates will be biological. The investigator must be aware that footwear, the outside of vehicles, and digging implements might be irrelevant to a case and the main source of evidence would be the clothing on the upper body, with no trace of soil.

The palynological profile from any crime scene is built up from multiples of comparator samples; it is, therefore, composed of a pattern of fragments. It also follows that, to gain a workable picture of the place, the larger the sampling area, and the greater the number of samples obtained, the closer the results will be to the actual profile (even though that is unknowable in detail). An offender contacting a crime scene will pick up only a fragment of the crime scene's palynological profile. If the trace evidence is then secondarily transferred to, say, a vehicle, only a third-order fragment will be retrieved. Palynological interpretation is, therefore, complex and requires visualisation skills as well as an understanding of the complex taphonomy underlying assemblages.

In spite of all the caveats that apply, the assemblages distinctive enough to establish convincing links between items, places, and vehicles have been repeatedly demonstrated. As previously stated, palynological samples obtained from exhibits are fragmentary in nature. For links between them and crime scenes to be acceptable to the Court, there needs to be either: (a) a highly complex assemblage where there are many points of similarity between place and object, or (b) some unusual or rare component or components.

Mixed Samples: Fabrics

Garments worn repeatedly for considerable periods will pick up palynomorphs from various places, so any retrieved assemblage will be mixed. They are transferred easily from palyniferous surfaces, but few seem to be picked up from air. Except where there is obvious soiling, it is impossible to separate various depositional events by sampling. But, unlike footwear, most items of clothing generally have limited contact with soil, vegetation, and other intensely palyniferous surfaces. As with footwear and vehicles, sufficient comparator samples are needed to be able to eliminate sources other than the crime scene but, if the assemblage accumulated from the crime scene is sufficiently distinctive, multiple deposition need not be an insurmountable problem.

A complication with fabric is that an offender may already have had soil on clothing before committing the offence, or after visiting the burial site. Palynomorphs from the crime scene can then be superimposed on the pre-existing soil marks. In one case there was an apparent conflict of evidence where a soil scientist and palynologist were not aware of each others' roles (Wiltshire 2001b). Soil on the suspect's sweatshirt was 'innocent', and was derived from deep sub-soil accumulated during the digging of garage foundations; analysis of the soil from the excavation showed it to contain no palynomorphs. While wearing the soiled clothing, the offender buried the victim near a hedge in a pasture. Before the grave was dug, there was little exposed soil in the meadow but, importantly, the offender picked up spores and pollen from tall vegetation on the path to and from the area around the grave site on his soiled sweatshirt. The palynomorph assemblage on the garment was similar to that at and around the deposition site. Thus, if only the soil evidence had been taken, there would have been no link between the garment and the burial site. This case provided a salutary lesson to investigators; the soil analyst and palynologist should work together to gain the deepest level of understanding from the respective data.

It is one of the strengths of palynology that pollen grains and plant spores will embed themselves in fabriucs such that they can be retrieved from exhibits even after being put through the washing machine (Wiltshire 1997).

Mixed Samples: Footwear

Footwear presents another complex of problems and challenges. Usually, samples are taken from specific areas within the crime scene, known to have been walked upon by a suspect, so that they can be compared with palynological assemblages on the footwear. An offender will have had to contact the edge (and inside) of a grave during digging, and a rape victim might be able to locate the exact places trodden by her attacker; samples should be taken from such identified locations. Any footprints or depressions in soil and mud are obvious targets, but these are usually seized by the police for casting and foot mark analysis. It is now standard practice to scrape away the deposit at the interface of the underneath of the cast and the adhering soil layer to obtain the most relevant comparator sample. Even if the offender accumulated layers of soil/mud from elsewhere prior, or subsequent, to the offence, the mixed profile on the footwear should contain some of the trace evidence retrieved from the cast. The palynologist then has to differentiate the relevant profile from the irrelevant one.

There has been some attempt to make forensic palynology more 'scientific' by setting up hypothetical crime scenes and testing outcomes from various kinds of contact. Such studies are useful exercises and, for the objects used in the experiment, or places tested, the results might be valuable. However, some results presented, should be considered to be preliminary in view of the low pollen counts in each case, and the limited number of treatments within the trials (e.g. Riding et al. 2007). For different sets of footwear exposed to the same palyniferous surfaces, or other footwear exposed to other palyniferous surfaces, the outcomes might be very different. It is dangerous to formulate predictive models, or form firm conclusions, based on relatively few test items, in a few test scenarios, with low pollen counts. Footwear invariably accumulates multiple depositions of palynomorphs. It is, therefore, often necessary to count many hundreds (sometimes thousands) of pollen and spores to achieve an assemblage large enough to allow the differentiation of the crime scene from other places where palynomorphs may have been transferred to the footwear. If footwear yielded sparse palynomorphs, low counts might still be useful, but only if there were some very distinctive components present in the assemblage.

My analysis of thousands of items of footwear, including wellington boots, baby's bootees and Gucci court shoes, has shown that variation is so great that general models are unlikely to be attainable goals. There are many variables associated with the palyniferous material itself (soil, sediment, leaf litter, vegetation), but there are others which can affect the deposition and removal of palynomorphs. These include the materials making up the footwear, the gait and wear patterns of the wearer, the weight of the wearer, and even ambient weather conditions (and hence the wetness of the surface). For criminal investigation, in every case, it is important that microscopic analysis of footwear and other items is carried out so that they can be compared directly with comparator samples from the crime scene and other pertinent places. In the forensic context, a model is never likely to provide adequate information for prediction of events or outcomes.

By its very nature, at any one time, footwear will have a palynomorph load accumulated from a range of different places. Depending on the frequency and pattern of wear, trace evidence will continually be gained and lost. In my experience relatively few pollen grains are picked up from paved or metalled surfaces, although spores can be transferred from lichens growing on hard-standing. Palynomorphs can also be picked up from pavements and gravel paths on which decomposed and decomposing plant litter have accumulated. Invariably, the most significant assemblages of palynomorphs of all kinds are picked up from bare soil, mud, leaf litter, organic debris, and vegetation. Day-to-day, most people do not usually wear muddy or soiled shoes, and any noticeable accumulations of soil or mud on footwear are obvious targets for sampling in criminal investigations. These have proved useful in linking footwear with crime scenes and other sites but, more commonly, footwear from suspects in criminal cases is relatively clean or only slightly soiled. If there were several obvious depositions of soil/mud, then every attempt should be made to analyse each one separately. However, even when sampling is meticulous, it is rare for perfectly uncontaminated samples to be obtained; the palynological preparations will contain mixed assemblages.

In some cases, the pertinent layer of material on a shoe can be beneath subsequent, superficial layers of soil and mud, and it becomes impossible to differentiate them physically. Here, the situation is similar to that of the relatively clean shoe where the whole item must be sampled. There is no substitute for counting many hundreds of palynomorphs (or as many as possible) and comparing them with the crime scene as well as other pertinent places for elimination purposes.

A complicated case was that of R vs Anthia (Wiltshire 2004d) concurrently investigated by two police forces. Because of the *modus operandi*, the two forces suspected that the same man was responsible for at least six attacks in Hertfordshire and London. I visited the six crime scenes, which varied from woodland, roadside verge hedges, golf courses, and wooded areas of parks. I was given the suspect's clothing, items from his vehicle, and several items of footwear. There were convincing palynological similarities between one of the Hertfordshire sites and the upper clothing, car seat, and one pair of footwear. The palynology of another pair of boots yielded very similar assemblages to another Hertfordshire crime scene as well as one of the London sites. These two crime scenes were sufficiently distinctive that both could be recognised in the mixed assemblage on the boots. The defendant was convicted and received 10 life sentences. Approaching 20,000 palynomorphs were counted in this case, and it was the exceedingly high resolution of the analysis which allowed the various crime scenes to be differentiated.

It is now standard practice to analyse each item in a pair of footwear separately. In many cases, both feet pick up similar palynological assemblages and it may be thought unnecessary to do separate analysis. However, there have been at least two cases where each shoe differed, and the results were pivotal to interpretation of the cases. In a drugs-related case (Wiltshire 2001a), an informant claimed that, although he had stood on an area of hard-standing at the edge of a woodland where a grave had been dug, he had not entered the scene and remained standing on an area of muddy concrete about 30m from the actual grave site. His statement needed verification. His shoes were analysed and the palynological assemblages on both of them showed that he had, indeed, picked up woodland palynomorphs, and the trees and shrubs were the same as those at the crime scene. But, only one of his feet yielded the assemblage characteristic of the gravesite itself; one foot had picked up components of *Alnus* (alder), *Quercus* (oak), *Pinus* (pine), *Corylus* (hazel), and other woody taxa and ferns. The other had the same assemblage and few grains of *Hyacinthoides* (bluebell). The floor of the woodland was carpeted with bluebells and *Anemone* (wind flower), and there were *Rhododendron* bushes next to the grave. If he had walked to the grave site, he would not have been able to avoid picking up *larger* amounts of bluebell pollen, that of *Anemone* and, possibly, *Rhododendron* on both feet. It would appear that he had picked up the woodland palynomorphs from the muddy concrete but there was no evidence that he had walked into the woodland with *both* feet. Only one yielded bluebell and this had probably been carried in soil to the concrete on the grave-diggers' feet. If both items of footwear had been amalgamated, the case for his non-involvement would have been weaker.

Evidence in a murder case in Greater Manchester was also enhanced by separate analysis of shoes (Wiltshire 2003d). Palynologically, this was a very complex case and resulted in my being cross-examined continuously for five days, and the accused being given a life sentence. The naked body of a woman, who had spurned her lover, was found lying on a path at a local woodland beauty spot. She had been beaten and there were foot marks on her face. One aspect of the case was to confirm a statement that events witnessed in the yard of the local public house might have been relevant. The victim had yellow stains on her jacket sleeve; the stains were composed of *Forsythia* pollen and green algae. The palynological assemblage from the suspect's shoes was shown to have a strong similarity to the actual crime scene a few hundred metres away, but one had a large number of grains of fenestrate *Lactuceae* (dandelion-like) pollen. The other shoe had only a couple of grains of dandelion-like pollen. A visit to the public house yard showed a concrete fence covered in green algae, with a flowering *Forsythia* bush growing over its top. There was a very narrow verge along the fence, dominated by dense growths of *Taraxacum officinale* (dandelion). Extensive searches and sampling of the local area failed to find anywhere which would offer such an assemblage of plants and palynomorphs in close proximity to one another. It was suggested that the suspect had started abusing the victim in the yard, pushed her against the fence, and whilst doing so, stepped on the verge and the dandelions with one foot. This was difficult for the accused to deny.

Mixed Samples: Digging Implements and Vehicles

Whenever buried remains are found, key exhibits will include digging implements. Unless a spade or shovel was bought for the criminal activity, it could have a palimpsest of soil layers distributed heterogeneously over the blade. Obtaining appropriate samples from such an item can be fraught with difficulty, and the best option might be to attempt a multiple sampling strategy. Again, a mixed sample will ensue and the skill of the palynologist can be severely tested in such cases. A soil-laden spade might also be laid on the ground where there is no bare soil but only a close cover of vegetation. Palynomorphs from the turf could dominate any number of profiles previously accumulated. Again, it is the rare or unusual assemblage of palynomorphs, or even a very rare palynomorph, that might indicate a link with a specific place. It is unlikely to be formed with widespread, common taxa.

In the case of the murder of Joanne Nelson, the 'Valentine Girl' (Wiltshire 2005a), her lover killed her but forgot where he had placed her body. His statement to the police was incoherent and they were anxious to find her remains. From palynological analysis of his vehicle, footwear, and a garden fork, I was able to eliminate his own, or his parents' garden, as being the source of the critical palynomorph assemblage. There were distinct similarities between the profiles from the car, one pair of shoes, and the garden fork and, from them, I was able to envisage the kind of place her body had been deposited and the vegetation of the place in question. The fork had not been used to bury her body but to cover it with woody, forest-floor litter. The signature of that litter was super-imposed on the pre-existing mud. Again, the soil on the exhibit was irrelevant to the investigation, and there was no soil from the crime scene on the fork. If soil analysis alone had been carried out, the girl would never have been located. Police found her body very quickly from the provision of an accurate description of the site. The assemblage of palynomorphs was an unusual one because of the nature of the Forestry Commission plantings; it also included spores of *Polypodium* (polypody) fern which is very uncommon in the area.

Seasonality and Temporal Interpretation

The time of an offence, or activities surrounding criminal activity, are often important aspects of police intelligence. Palynology has, on occasion, been used to confirm the temporal aspects of cases (Wiltshire, *sub-judice* cases ongoing).

The timing of anthesis (pollen release), especially of wind-pollinated plants, is critically important to those involved in studying allergy. Pollen and fungal spores are important allergens, and considerable effort is focused on pollen calendars. Such calendars are produced, and information exchanged, by various institutes, universities, and hospitals in many countries (Hyde 1969; Michel et al. 1976; O'Rourke 1990). The pollen calendars give start and finishing times, and the periods of peak release for pollen and spores for selected species. However, such calendars have limited use for forensic work because of the frequent, and sometimes extreme, variation in pollen release times from region to region.

In Britain there is a network of 10 pollen monitoring stations and 11 stations which monitor only grass pollen; the pollen stations are situated in towns in lowland areas, so their results may be unrepresentative of much of the British airspora. Even at a local level, 'pollen calendars' can never be precise – certainly, they cannot offer the precision required for forensic investigation. If the pollen calendar for a specific place were known in detail, there might be some application for interpretation of data relating to that place; but, because of specificity and inherent variability, the uses of seasonal records are very limited. They can only be used in the crudest way. For example, Montali et al. (2006) showed that pollen retrieved from corpses could not be related to the local pollen calendars because of the degree of variability in local conditions. However, they concluded that they could differentiate between winter/spring, spring/summer, and summer/autumn. To be sufficiently convincing to be useful in the forensic context, a great deal more work would be necessary and the phenomenon of residuality addressed. Depending on the environment, palynomorphs can remain *in situ* for very long periods. A soil sample might contain pollen accumulated over decades, and it can remain on foliage and bark for more than one year (Adam et al. 1967; Groeneman-van Waateringe 1998). Any pollen assemblage transferred during day-to-day activity would inevitably contain palynomorphs from a number of seasons. To rely on pollen calendar data for estimating seasonality in forensic work is imprudent.

Conclusions

In natural ecosystems, organisms occupy niches that may be narrow or wide. Some have a wide geographical distribution and others a narrow one. This is useful for predicting the nature of places from which palynomorphs were transferred to offenders. However, ecosystems are rarely natural in the true sense, with enormous environmental manipulation wherever people have had influence. From place to place, and sample to sample, palynological profiles are characterised by their variability and uniqueness. It follows that models for pollen dispersal and pattern of fall-out, which might aid the palaeoecologist in the interpretation of past environments where human intervention was minimal, will be of limited use in forensic case work. The amount of variation inherent in any system makes it impossible to construct databases of palynological profiles that could be used with confidence in criminal investigation and preparation of court statements. Every sample from every crime scene, and every assemblage retrieved from every exhibit, will be unique and will need independent evaluation. Predictions of origin of any organic particulate can only be crude, and there is no substitute for detailed analysis of the crime scene, other places pertinent to the investigation, and the objects that are thought to have had contact with them.

Worldwide, palynology is an under-used resource for criminal investigation. This is due, in part, to the perennial dearth of competent palynologists who possess not only comprehensive botanical knowledge, ecological training, and appropriate and extensive field experience, but who can cope with the rigours of cross-examination in the courts.

Acknowledgements Grateful thanks are due to Judy Webb, Julia Newberry, and Peter Murphy for the excellence of their contribution to my case work over the years. I would also like to thank David L. Hawksworth for assistance in the finalisation of this contribution.

References

Adam DP, Ferguson CW and LaMarch VC Jr. (1967). Enclosed bark as a pollen trap. Science 157:1067–1068.

Barnekow L, Loader NJ, Hicks S, Froyd CA and Goslar T (2007). Strong correlation between summer temperature and pollen accumulation rates for *Pinus sylvestris, Picea abies* and *Betula* spp. in a high resolution record from northern Sweden. Journal if Quaternary Science 222:653–658.

Beug H-J (2004). Leitfaden der Pollenbestimmung für Mitteleuropa und angrenzende Gebiete. Verlag Dr. Friedrich Pfeil, München.

Brown AG, Smith A and Elmhurst O (2002). The combined use of pollen and soil analyses in a search and subsequent murder investigation. Journal of Forensic Sciences 47:614–618.

Bryant VM and Mildenhall DC (1998). Forensic palynology: a new way to catch crooks. In: New Developments in Palynomorph Sampling, Extraction, and Analysis (Eds. VM Bryant and JH Wrenn). American Association of Stratigraphic Palynologists Foundation. Contributions Series Number 33:145–155.

Bryant VM, Mildenhall DC and Jones JG (1990). Forensic palynology in the United States of America. Palynology 14:193–208.

Bunting MJ, Armitage R, Binney H and Waller M (2005). Estimates of 'relative pollen productivity' and 'relevant source area of pollen' for major tree taxa in two Norfolk (UK) woodlands. The Holocene 15:459–465.

Bunting MJ and Middleton R (2005). Modelling pollen dispersal and deposition using HUMPOL software, including simulating windroses and irregular lakes. Review of Palaeobotany and Palynology 134:185–196.

Caulton E, Lacey M, Allitt U, Crosby R, Emberlin J and Hirst J (1995). Airborne pollens and spores: a guide to trapping and counting. The British Aerobiology Federation, Worcester.

Davies MB (1967). Late-glacial climate in northern United States: a comparison of New England and the Great Lakes Region. In: Quaternary Palaeoecology (Eds. EJ Cushing, HE Wright), pp. 11–43. Yale University Press, New Haven.
Eklöf M, Broström A, Gaillard M-J and Pilesjö P (2004). OPENLAND3: a computer program to estimate plant abundance around pollen sampling sites from vegetation maps: a necessary step for calculation of pollen productivity estimates. Review of Palaeobotany and Palynology 132:67–77.
Erdtman G (1921). Pollenanalytische Untersuchungen von Torfmooren und marinen Sedimenten in Südwest-Schweden. Arkiv för Botanik 17(10):1–173.
Erdtman G (1969). Handbook of Palynology. Munksgaard, Copenhagen.
Frei M (1979). Plant species of pollen samples from the Shroud. Appendix to the Turin Shroud, I. Wilson. Penguin Books, London, Appendix E.
Groeneman-van Waateringe W (1998). Bark as a natural pollen trap. Review of Palynology and Palaeoecology 103:289–294.
Hawksworth DL (2008a). Estimation of post-mortem interval and time of deposition from fungal growth on a corpse: *Sub-judice* case: Hertfordshire Constabulary.
Hawksworth DL (2008b). Examination of fungal remains: Operation Jalap: Avon and Somerset Police.
Hicks S, Tinsley H, Huusko A, Jensen C, Hattesstrand M, Gerasimides A and Kvavadze E (2001). Some comments on spatial variation in arboreal pollen deposition: first records from the Pollen Monitoring Programme (PMP). Review of Palaeobotany and Palynology 117:183–194.
Horrocks M and Walsh AJ (1998). Forensic palynology: assessing the value of the evidence. Review of Palaeobotany and Palynology 103:69–74.
Hyde HA (1969). Aeropalynology in Britain – an outline. New Phytologist 86(3):579–590.
Jacobson GL and Bradshaw EHW (1981). The selection of sites for palaeovegetational studies. Quaternary Research 16:80–96.
Jones GD and Bryant VM Jr. (2007). A comparison of pollen counts: light versus scanning electron microscopy. Grana 46:20–33.
Levetin E, Rogers CA and Hall SA (2000). Comparison of pollen sampling with a Burkard spore trap and a Tauber trap in a warm temperate climate. Grana 39:294–302.
Michel FB, Cour P, Lyne Q and Marty JP (1976). Qualitative and quantitative comparison of pollen calendars for plain and mountain areas. Clinical and Experimental Allergy 6(4):383–393.
Mildenhall DC (1982). Forensic palynology. Geological Society of New Zealand Newsletter 58:25.
Mildenhall DC (2006). *Hypericum* pollen determines the presence of burglars at the scene of a crime: an example of forensic palynology. Forensic Science International 163:231–235.
Mildenhall DC, Wiltshire PEJ and Bryant VM Jr (2006). Forensic palynology: why do it and how it works: Editorial. Forensic Science International 163:163–172.
Milne LA, Bryant VM Jr and Mildenhall DC (2005). Forensic palynology. In: Forensic Botany: Principles and Applications to Criminal Casework (Ed. HM Coyle). CRC, Boca Raton, FL.
Montali E, Mercuri AM, Trevisan Grandi G and Accorsi CA (2006). Towards a "crime pollen calendar" – pollen analysis on corpses throughout one year. Forensic Science International 163:211–223.
Moore PD, Webb JA and Collinson ME (1991). Pollen Analysis (2nd Ed). Blackwell Scientific, Oxford.
Nowicke JW and Meselson J (1984). Yellow rain – a palynological analysis. Nature 309:205–206.
O'Rourke MK (1990). Comparative pollen calendars from Tucson, Arizona: Durham vs Burkard Samplers. Aerobiologia 6:136–140.
Prentice IC (1985). Pollen representation, source area, and basin size: toward a unified theory of pollen analysis. Quaternary Research 23:76–86.
Riding JB, Rawlins BG and Coley KH (2007). Changes in soil pollen assemblages on footwear worn at different sites. Palynology 31:135–151.

Soepboer W, Sugita S, Lotter AF, Van Leeuwen JFN and ver der Knaap WO (2007). Pollen productivity estimates for quantitative reconstruction of vegetation cover on the Swiss Plateau. The Holocene 17(1):65–77.
Sugita S, Gaillard MJ and Broström A (1999). Landscape openness and pollen records: a simulation approach. The Holocene 9:409–421.
Szibor R, Schubert C, Schöning R, Krause D and Wendt U (1998). Pollen analysis reveals murder season. Nature 395:449–450.
Tauber H (1965). Differential pollen dispersion and the interpretation of pollen diagrams. Geological Survey Denmark, Series II 89:1–69.
Tauber H (1967). Differential pollen dispersion and filtration. In: Quaternary Palaeoecology (Eds. EJ Cushing and HE Wright), pp. 131–141. Yale University Press, New Haven.
Tinsley H (2001). Modern pollen deposition in traps on a transect cross an anthropogenic tree-line on Exmoor, south-west England: a note summarizing the first three years data. Review of Palaeobotany and Palynology 117: 153–159.
White PC (Ed.) (2004). Crime Scene to Court: The Essentials of Forensic Science. The Royal Society of Chemistry, London.
Wiltshire PEJ (1997). Operation Gratis. Hertfordshire Constabulary.
Wiltshire PEJ (2001a). Operation Alfalfa. Hertfordshire Constabulary.
Wiltshire PEJ (2001b). Palynological analysis of soils and exhibits. Operation Maple. Sussex Police.
Wiltshire PEJ (2002a). Field evaluation of deposition site: Operation Ruby. Surrey Police.
Wiltshire PEJ (2002b). Operation Bracken. Norfolk Constabulary.
Wiltshire PEJ (2002c). Murder of Baby Carrie. Police Service Northern Ireland.
Wiltshire PEJ (2003a). Report on post-mortem oesophageal sections: Operation Lara Cumbria Police.
Wiltshire PEJ (2003b). Field evaluation of the corpse and deposition site at Noak Hill, Romford, Essex: Operation Woolhope. Metropolitan Police.
Wiltshire PEJ (2003c). Operation Pomerol. West Midlands Police.
Wiltshire PEJ (2003d). Murder of Karen Doubleday. Greater Manchester Police.
Wiltshire PEJ (2004a). Microscopical analysis of material retrieved from lung tissue: The death of William Pettener. North Wales Police.
Wiltshire PEJ (2004b). The murder of Alan McCullough. Police Services Northern Ireland.
Wiltshire PEJ (2004c). R v J. Thomas and A. W. Tilley. South Wales Police.
Wiltshire PEJ (2004d). R. vs Anthia Hertfordshire Constabulary and Metropolitan Police.
Wiltshire PEJ (2005a). Palynological contribution to the finding the remains of Joanne Nelson. Humberside Police.
Wiltshire PEJ (2005b). Preliminary evaluation of crime scene and grave deposits. Operation Relator. Hertfordshire Constabulary.
Wiltshire PEJ (2005c). Comments on potential toxicity of a range of plants listed by suspect Brian David Lawrence. Thames Valley Police.
Wiltshire PEJ (2006a). Consideration of some taphonomic variables of relevance to forensic palynological investigation in the UK. Forensic Science International 163:173–182.
Wiltshire PEJ (2006b). Microscopical analysis of putative vomit, stomach contents, and commercial food items: Operation Ultra. Bedfordshire Police.
Wiltshire PEJ (2007a). Report on environmental evidence obtained from the cadaver and the place where she was found: Operation Salute Suffolk Police.
Wiltshire PEJ (2007b). The murder of Barrie Horrell. Gwent Police.
Wiltshire PEJ (2008). Operation Jalap. Avon and Somerset Police.
Wiltshire PEJ and Black S (2006). The cribriform approach to the retrieval of palynological evidence from the turbinates of murder victims. Forensic Science International 163:224–230.

Chapter 10
Sediment and Soil Environmental Forensics: What Do We Know?

Stephen M. Mudge

Abstract Environmental forensic practitioners have been identifying the source of contamination in the environment for many years. In some cases, their results have been used successfully to prosecute offenders. The use of DNA in criminal cases has led to a considerable advance in the ability of the scientist to identify the source of materials; in environmental forensics the same may be true, but the 'DNA' may be the chemical fingerprint or the biological assemblage. Legislation dictates that the correct source of compounds in the environment is identified or the source apportioned between potentially responsible parties. The implementation of the Environmental Liability Directive in the European Union may require remediation of damaged sites to their baseline condition but those conditions may not be quite so easy to define. Chemical composition of multi-compound materials such as oil may enable fingerprints to be developed based on the internal relationship between groups of chemicals: this may be considered as its 'DNA' and enable it to be tracked in the environment. Chemicals have a whole range of water solubilities and hence mobilities in the environment. In some cases, the partitioning between the solid and solution phase is dependent on the prevailing environmental conditions, such as pH or salinity. These compounds, known as hydrophobic ionogenic organic compounds, include Triclosan, a widely used anti-microbial agent in domestic products. In highly variable environments such as estuaries, this compound may not be where you initially expect it to be. In a similar manner, the biota that make up an assemblage may provide a unique signature for that environment or those conditions; this may also be tracked in the environment. In a study using the meiofauna of surface sediments of a lagoon in Portugal, the sewage-signature based on the species assemblage near known discharge sites was able to indicate the position of previously unrecognised discharges to the lagoon. The ability to use these data may also rely on multivariate statistical methods that enable signatures to be defined and then quantifiably extracted from environmental data. These methods have become

S.M. Mudge(✉)
School of Ocean Sciences, Bangor University, Menai Bridge, LL59 5AB, UK
e-mail: s.m.mudge@bangor.ac.uk

K. Ritz et al. (eds.), *Criminal and Environmental Soil Forensics*

relatively simple to use with the development of dedicated computer programs, but there is still the need to ensure that the data are appropriate for the method and *vice versa*. Over-reliance on statistical results may be undesirable in court presentations but, if used together with a whole range of methods, may help explain the situation to the lay judiciary. Stable isotopes have provided considerable advances in source identification and apportionment. In one example, lead isotopes were able to show that in the Conwy Estuary, North Wales, there was no simple single source (e.g. mine tailings) and that multiple sources contributed to the receptor site at the mouth of the estuary. As with any other branch of forensic science, the ability to identify the source of materials is paramount and, in the case of environmental forensics, methods based on a whole range of chemical, biological and statistical data have become the 'DNA' of source apportionment.

Introduction

There have been many advances in forensic science in the past few decades including the wide usage of DNA to identify the potential culprit or to at least rule persons out of an enquiry. This goes back to Edmond Locard's famous phrase that "every contact leaves a trace" and this is true outside of criminal forensics too. In the field of environmental forensics (EF), there has been the application of a wide range of existing technologies to environmental crime. In many ways, these methods provide the same function as DNA analysis does in crimes against the person. The purpose is primarily the same – to identify the source of contaminating materials in the environment. However, there is another aspect particularly related to EF and that is source apportionment; how much of the contamination came from each potentially responsible party?

In the UK, much of the post-1995 legislation has been driven by European Directives although it is likely that most prosecutions for environmental crime are still under the Water Resources Act 1991 c.57 (Evans pers. comm.). One of the key pieces of legislation that will increase the activity of EF practitioners across Europe is the Environmental Liability Directive (2004/35/CE) which requires polluters to prevent imminent contamination or to remediate, reimburse or replace contaminated sites. In the UK, this directive is being enacted through The Environmental Damage (Prevention and Remediation) Regulations 2008. These Regulations cover aspects of damage to both water (linked to the Water Framework Directive) and land. In the case of land, 'environmental damage' is said to have occurred if there is a significant risk to human health. As part of the requirement to remediate, one of the potentially contentious aspects will be determining background or baseline conditions when these data were not available at the time of damage occurring.

The key focus of EF must be the identification of the source/s of contamination in the environment but a case must be built around three aspects of environmental science. A schematic of such an investigation can be seen in Figure 10.1. There needs to be at least one identifiable source of the contaminant available and much

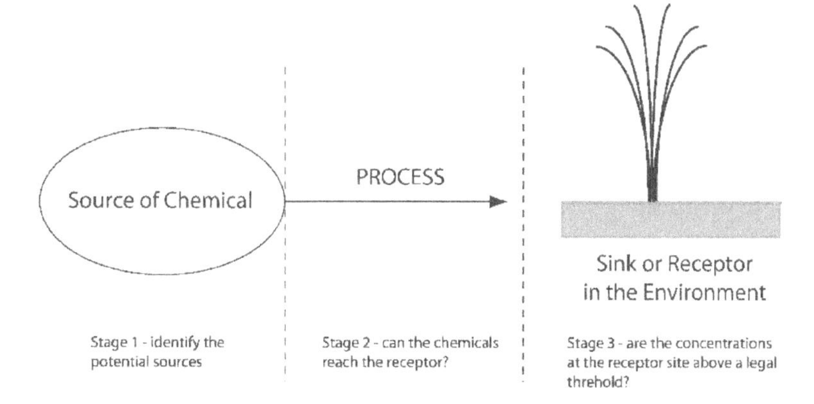

Fig. 10.1 Schematic for the simple case of movement of a contaminant from a single source to receptor. In reality, there may be multiple potential sources, multiple processes by which they are trans-located and multiple potential receptors

work revolves around correctly identifying the chemistry of source materials. Secondly, there must be an identifiable process by which it might be transferred from the source or sources to the site of contamination (a receptor or sink). There may be sources within range of the receptor but if there is no mechanism by which the chemicals can reach the receptor, they must be ruled out. Finally, the concentration at the receptor must be above a threshold value and some form of damage must have occurred. Figure 10.1 presents a simplistic case of a single source, an obvious process and a receptor of the chemical. However, this is rarely the case as there are often several potential sources, multiple routes by which chemicals may travel from one location to another and a whole host of receptors which may be impacted differentially. Examples of these may be seen in Mudge (2008).

For an individual or industry to be a potentially responsible party, there needs to be both a source of the contaminant at their facility and a process (groundwater flow, atmospheric dispersion route, etc.) by which the material could reach a site remote from the source. The second aspect of this is crucial as there may be some contaminants (e.g. oil) which have a wide range of potential sources in the environment, but not all would have routes to the receptor site. In this case, it will be necessary to determine which of these potentially responsible parties have contributed to the contamination. The complex chemical nature of crude oil does hold the answers, however, in that the wide range of trace components such as steranes, terpanes and diamonoids (Wang and Fingas 2003) act as a signature and may be considered 'the chemical DNA' of the oil. Complex multivariate statistical methods such as Principal Component Analysis (PCA) and the Projection to Latent Structures by means of Partial Least Squares (PLS) do assist in determining the dissimilarity/similarity between potential sources and receptor samples (Mudge 2007). Care must be taken, however, in the presentation of statistics in court, and presentation without degrading the method is critical. It is important that the correct procedures for any statistical method are followed with due regard to limitations such as: sparse data, zeros in the dataset and non-normal distributions. Over reliance on these methods should be avoided and they should just be one component in a 'total raft' and one should never apply statistical methods blindly without thinking about their appropriateness in a given situation.

As statistical software packages have become easier to use, there is the potential to misuse statistical methods (wrong data for the technique or wrong technique for the data). Conclusions drawn from inappropriately applied statistical methods may lead to misleading results and reports being rejected in court and cases lost or inappropriately won.

Sometimes, conclusions are drawn from datasets with many missing values. Since some software programs can assign a system missing value to these cases, artefacts may arise indicating 'correlation' where is does not really exist. Similar problems can arise with zeros in data (Mudge 2007); are they truly zeros or are they really indicative of the concentration being below the limit of detection for the analytical method? It may be correctly argued that there were no beetles in the soil sample (i.e. a true zero) but in the case of, say, dioxins, if we had a better analytical method, we may detect them (i.e. not a true zero). One way of accounting for these differences is to adopt a value of half of the limit of detection for any analyte below the limit of detection. However, if there were lots of these, artefacts may again arise in the statistical results.

Many statistical methods assume normally distributed data. While it is not always necessary, clearer interpretation can be seen with such data. In the case of PCA (Mudge 2007), substantial improvements may be seen. Several methods exist to 'normalise' the data and the most common method with chemical data is taking the logarithm of the number, although with biological data square root transformations are often applied (Mudge 2007). However, in all cases where statistics are used, *Beware of the Black Box*!

Example Cases

Lead (Pb) in the Conwy Estuary, UK

An example where intrinsic chemical properties are particularly important is the use of stable isotopes in fingerprinting the chemicals of the source materials. Stable isotopes of carbon (^{13}C), hydrogen (^{2}H), oxygen (^{18}O) and nitrogen (^{15}N) are often used with organic matter (Philp 2007), although there are several other geochemically-significant isotopes, such as strontium (Montgomery et al. 2007). One such example is the multiple isotopes of lead that are formed from the radiogenic decay of heavier elements. Natural lead is comprised of four isotopes (^{204}Pb (1.4%), ^{206}Pb (24.1%), ^{207}Pb (22.1%), ^{208}Pb (52.4%)), although the proportions vary according to the initial heavy radionuclides present in the rocks, as the $^{206,\,207}$ and ^{208}Pb originate from ^{238}U, ^{235}U and ^{232}Th, respectively. The relative proportions of each can also be considered as a signature of the different sources within a catchment and for each anthropogenic source (Ghazi and Millette 2006).

The Conwy Estuary in North Wales, UK has received lead from many sources over the years including lead added to petrol (Farmer et al. 2006), used as flashing in roof construction and, in this catchment in particular, from lead mining. Lead was mined in the Gwydyr Forest region until 1954 (Gao and Bradshaw 1995) and remediation of the tailings was performed in 1977 to 1978, to prevent excessive runoff of metal rich materials into the River Conwy (Shu and Bradshaw 1995). In an unpublished study to determine the source of lead in the sediments of the lower estuary, samples were collected in 2006 from several soils, sediments and waters around the catchment. Raw metal concentrations exceeded the Environmental Quality Standard (EQS) of 45 µg/g or 90 µg/l in sediments/soils and waters at various points across the catchment (Figure 10.2) and the relationship with grain size data (Figure 10.3) shows the surface sediment lead content across

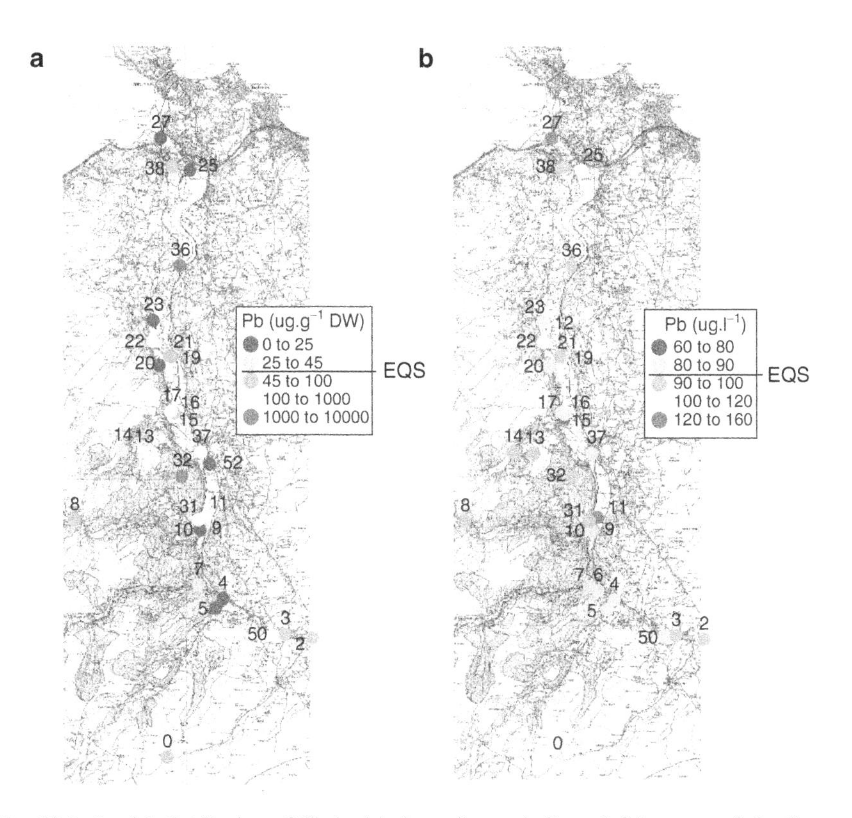

Fig. 10.2 Spatial distribution of Pb in (**a**) the sediments/soils and (**b**) waters of the Conwy Catchment. EQS = Environmental Quality Standard. These data do not show a simple gradient away from a single source and potentially imply multiple sources of Pb in this region. Background map images Crown Copyright/Database Right 2008. An Ordnance Survey/(Datacentre) supplied service *(see colour plate section for colour version)*

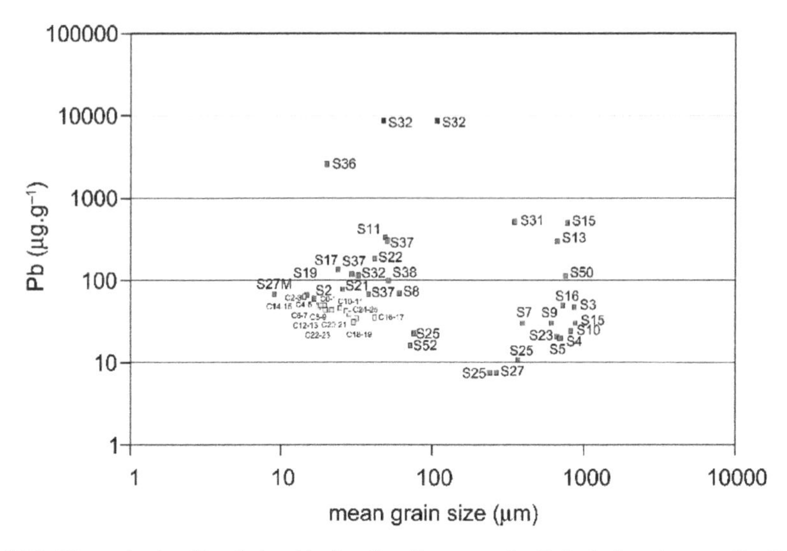

Fig. 10.3 The grain size–Pb relationship for all sediments and soils including the core. Replicate analyses are included and both axes are shown in their logarithm form. These data do not describe the typical high concentrations in the small grain size sediments due to multiple sources and equilibria processes

the system and in the core taken near Site 36. In the case of contaminants associated with sediments, it is usually the case that the concentration of trace constituents increases as the mean grain diameter decreases. This is due to the increase in the grain surface area which increases as a square function in relation to the diameter of the grains. Given time, in most systems, the contaminants reach an equilibrium condition with the sediments such that a plot of a measure of grain surface area (e.g. % <63 μm or Total Organic Carbon (TOC) content, a proxy for surface area) has a well defined, sometimes linear, relationship with total concentration of the contaminant (Mudge et al. 2001).

Figure 10.3 shows there is no simple relationship between the mean grain size of the sediments/soils and the measured Pb concentration. It might be expected that the concentration would be greatest in the sediments with a small mean grain size (10 μm) and decrease as the mean grain size increased toward the sands (1,000 μm). In this system, the lead is not in equilibrium with the grain size, unlike many marine systems (Mudge et al. 2001) and samples taken near the mine tailings are significantly enriched in lead and do not exhibit a clear gradient away from this source. The core samples also have similar grain sizes and cluster relatively closely together in Figure 10.3. From these two figures (Figures 10.2 and 10.3), it is not possible to unambiguously identify which of the potential sources is contributing to the pollution in the sediments of the lower Conwy Estuary. Analysis of the stable isotopes by ICP-MS, however, can provide an insight into the signature at each source (Geffen et al. 1998; Ghazi and Millette 2006). In this case, the sediment data describe a range of values, although the natural geologic source at Llyn Conwy (Site 0) is different to that from the Pen-y-Parc mine (Site 32 on Figure 10.4).

The ratios in the core samples taken from the mid-reaches of the estuary cluster around the mine tailing ratio but the marine samples at sites 25 and 27 are intermediate between the non-mining source and mine tailings indicating that not all lead at Conwy may be attributed to a single source. The core profile (inset to Figure 10.4) shows how the concentrations have decreased to below the EQS and shows the improvement with time. This also indicates the remedial action to stabilise the mine tailings undertaken in 1977 to 1978 has largely been successful in reducing the flux of lead to the estuary. In this case, the stable isotope profile of the lead has acted as the 'DNA' for its identification in the environment.

Using Changes in the Biological Community

In some cases, the potential source is not continuous or the chemicals are water soluble and may have dissipated to concentrations below the analytical limits of detection before samples are taken (Hewitt and Mudge 2004). It is also possible that the discharges may have taken place at some time in the past and they have degraded to other compounds or also simply been washed away. In cases such as these, the biology may provide a useful tool to determine if discharges have in fact occurred. It is well known that biological communities or assemblages

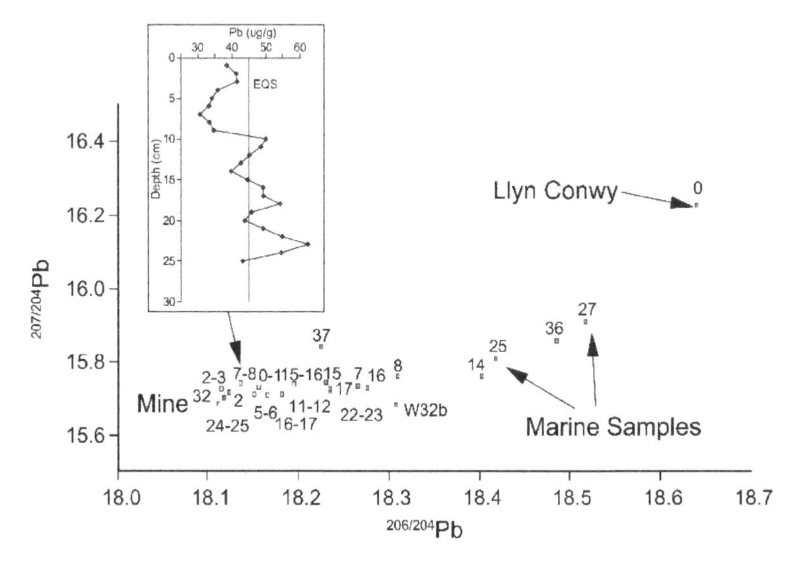

Fig. 10.4 The $^{207/204}$Pb versus $^{206/204}$Pb cross-plot which may be used to indicate sources of Pb. The concentration profile down the core taken near site 36 is shown as an inset. The EQS at 45 µg/g shows how the environmental concentrations have decreased with time.

change depending upon the stress they are under and this may be used to classify waters (Borja et al. 2003, 2007). The classification of transitional waters under the Water Framework Directive is partly reliant on this approach to enable comparison between sites and to monitor improvement (Borja et al. 2007). This approach may also be used in an EF context to identify changes in the wake of contamination events. For example, the meiofauna in lagoonal sediments successfully pointed to the location of undocumented sewage discharges (Hewitt and Mudge 2004), although the macrofauna were less sensitive and, therefore, less diagnostic (Hopkins and Mudge 2004; Chenery and Mudge 2005). In these particular cases, a multivariate signature was developed based on the biological assemblage at sites that were known to be impacted by sewage and this 'fingerprint' was used to identify other locations that had similar species compositions or the spread of the sewage derived products (Figure 10.5). These methods are not as sensitive as the chemical methods (Mudge and Duce 2005) but do provide an indicator in areas where chemicals are not appropriate or simply not there. Therefore, the community structure or assemblage may be acting as the DNA to identify that contamination has taken place.

Figure 10.5 shows the predictable variance in each sediment sample based on four signatures developed from the meiofaunal assemblage (Hewitt and Mudge 2004). The sewage signature explained significant amounts of variance in samples that were not thought to be close to such discharges. However, re-examination in the light of these results indicated small surface water drains that were impacting the area.

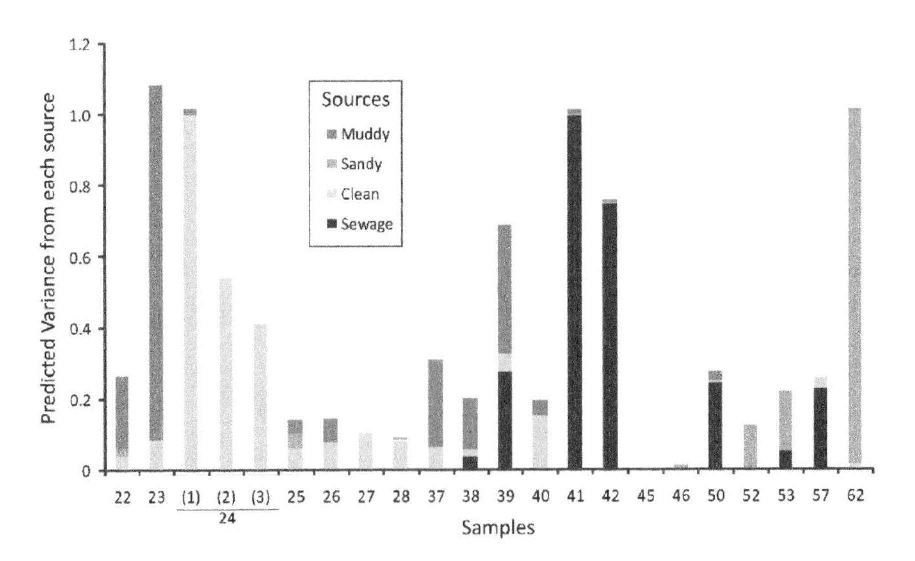

Fig. 10.5 The predictable variance in each sediment sample based on four signatures developed from the meiofaunal assemblage (Hewitt and Mudge 2004). The Sewage signature explained significant amounts of variance in samples that were not thought to be close to such discharges. However, re-examination in the light of these results indicated small surface water drains that were impacting the area

Water Solubility

Some chemicals are water soluble and will be rapidly washed out from soils and may not leave a trace for future EF practitioners to come along and find. Examples of this include detergents (e.g. nonyl phenol polyethoxylate) but, by thinking laterally, their passage may be determined by what was *not* there rather than what *was* there. In soils, for instance, some of the complex, low molecular weight, non-water soluble phenols may be washed out by the detergent leaving an inverted or negative plume (Figure 10.6). Soil analyses taken across a plume of a detergent spill may show the depletion in a particular set of chemicals as these are washed out and removed with the detergent.

Some chemicals such as Triclosan which are used as an antimicrobial agent in many domestic and hygiene products (Yazdankhah et al. 2006) pass through sewage treatment systems and eventually reach the environment (Singer et al. 2002). The key factor for this and several other chemicals is that they are hydrophobic ionogenic organic compounds (HIOC) that may change their preference for the solid or solution phases in the environment depending on the environmental conditions (Jafvert et al. 1990). In this case, Triclosan has a pK_a of 8.1 which is similar to the pH of seawater (7.8 to 8.1) and passage of the compound through estuaries may

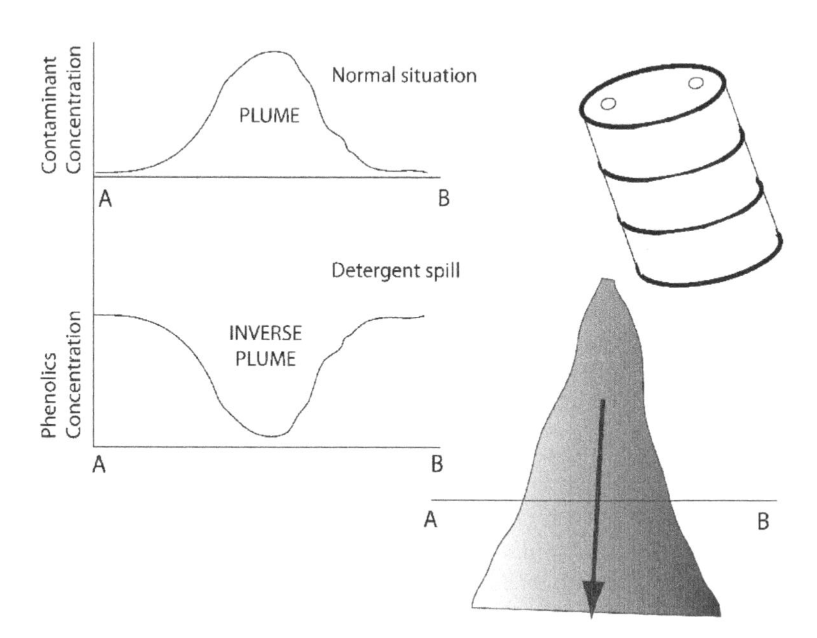

Fig. 10.6 Potential outcomes with two different spilled chemicals. In the normal situation with a low water solubility product, residuals are left in the soil showing the plume axis. In the case of a detergent, the material itself has been washed away and has removed some of the soil phenolic compounds as well leaving an inverse plume – what is not there!

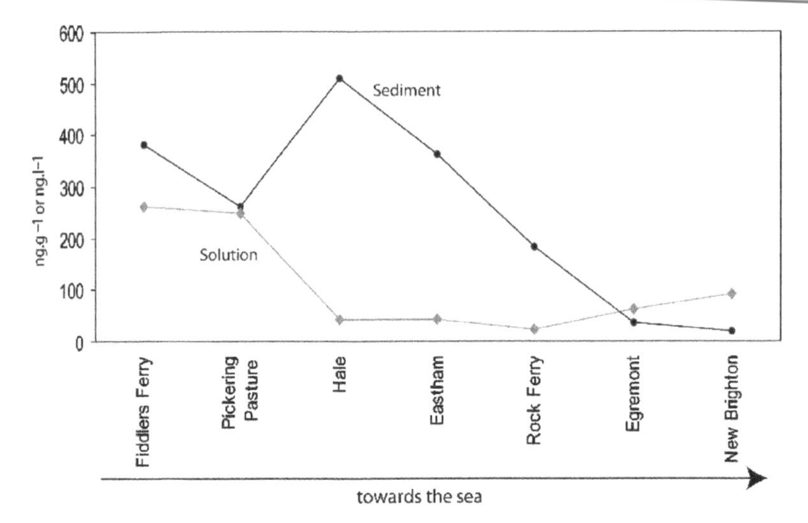

Fig. 10.7 The partitioning of Triclosan between the solid and solution phases down the Mersey Estuary (data from Veenstra 2005)

lead to association with the solid phase rather than the solution phase in freshwaters. In a study of the fate of Triclosan in such waters (Veenstra 2005), data were collected from the Mersey Estuary, NW England.

Figure 10.7 shows the concentrations of Triclosan in the solid and solution phases down the Mersey Estuary; as salinity and pH increase (seawater has a pH of ~8.1 compared to freshwater of 6 or below), the Triclosan becomes preferentially bound to the sediments rather than remaining in the solution phase. Therefore, when collecting samples in an EF investigation, it is important to know about the environmental behaviour of the chemicals involved so that samples are collected from the appropriate phase. In this case, water sampling alone in the lower reaches would miss detection of the chemical in the sediments and underestimate its presence in the environment.

Conclusions

EF has come a long way in the past decade and is now a recognised profession and scientific discipline. The practises of EF are akin to those used in criminal cases by the various forensic science practitioners around the world. Key aspects include identification of the source of contaminants in the environment and, under new legislation in the EU (Environmental Liability Directive and its national Regulations) or existing USA legislation (CERCLA – The Comprehensive Environmental Response, Compensation and Liability Act), apportioning responsibility or blame where necessary. What is needed is a method similar to DNA analysis which will unambiguously rule potential sources in or out of an investigation. A case will always have more chance of success in court if the results of several different types of analysis all point to the same answer. Reliance on a single approach is not to be recommended and the importance of ruling out the alternatives can be as powerful as identifying the correct source.

Several tools exist that can give us a signature (stable isotopes, multivariate statistics, etc.), although care needs to be exercised when presenting such results to courts to ensure understanding by the lay judiciary. In the case of the Pb isotopes in the Conwy Estuary, the data immediately indicates the system is much more complicated than a single source – process – receptor model, and the ability to use stable isotopes here has enhanced our understanding of the system enormously. However, not all chemicals have such properties and care also needs to be taken when tracking compounds, as they may not be where they initially could be expected to be due to their intrinsic nature (water solubility, HIOC, etc.). In other cases, emissions may be fugitive, in the past or ephemeral and, unless a sampler is there at the right time, nothing will be collected for analysis. In such cases, the biological assemblage may provide a measure of the prior presence of chemicals that have since degraded or moved on.

The nature of environmental forensics is such that practitioners need to understand the behaviour of the contaminant they are tracing, the environment it is in, and the associated interactions, as well as how to construct a scientifically and legally defensible case to allow prosecution or defend against inappropriate action. This fusion of disciplines is exciting and likely to grow in demand as we strive to protect our environment.

Acknowledgements The investigation into lead in the Conwy Estuary presented in this chapter was collected by EF students J. Adams, R. Ironmonger, P. Lloyd, A. McInnes, L. Waterman, W. Williams and H. Woods. We would like to thank Dr. Bill Perkins for the use of the ICP-MS for the lead isotope analysis.

References

Borja A, Muxika I and Franco J (2003). The application of a Marine Biotic Index to different impact sources affecting soft-bottom benthic communities along European coasts. Marine Pollution Bulletin 46:835–845.

Borja A, Josefson AB, Miles A, Muxika I, Olsgard F, Phillips G, Rodriguez JG and Rygg B (2007). An approach to the intercalibration of benthic ecological status assessment in the North Atlantic ecoregion, according to the European Water Framework Directive. Marine Pollution Bulletin 55:42–52.

Chenery AM and Mudge SM (2005). Detecting anthropogenic stress in an ecosystem: 3. Mesoscale variability and biotic indices. Environmental Forensics 6:371–384.

Farmer JG, MacKenzie AB and Moody GH (2006). Human teeth as historical biomonitors of environmental and dietary lead: some lessons from isotopic studies of 19th and 20th century archival material. Environmental Geochemistry and Health 28:421–430.

Gao Y and Bradshaw AD (1995). The containment of toxic wastes 2. Metal movement in leachate and drainage at Parc Lead-Zinc Mine, North Wales. Environmental Pollution 90:379–382.

Geffen AJ, Pearce NJG and Perkins WT (1998). Metal concentrations in fish otoliths in relation to body composition after laboratory exposure to mercury and lead. Marine Ecology-Progress Series 165:235–245.

Ghazi AM and Millette JR (2006). Lead. In: Environmental Forensics. Contaminant Specific Guide (Eds. RD Morrison and BL Murphy), pp. 55–79. Academic, Burlington, MA.
Hewitt EJ and Mudge SM (2004). Detecting anthropogenic stress in an ecosystem: 1. Meiofauna in a sewage gradient. Environmental Forensics 5:155–170.
Hopkins FE and Mudge SM (2004). Detecting anthropogenic stress in an ecosystem: 2. Macrofauna in a sewage gradient. Environmental Forensics 5:213–223.
Jafvert CT, Westall JC, Grieder E and Schwarzenbach RP (1990). Distribution of hydrophobic ionogenic organic-compounds between octanol and water – organic-acids. Environmental Science and Technology 24:1795–1803.
Montgomery J, Evans JA and Cooper RE (2007). Resolving archaeological populations with Sr-isotope mixing models. Applied Geochemistry 22:1502–1514.
Mudge SM (2007). Multivariate statistics in environmental forensics. Environmental Forensics 8:155–163.
Mudge SM (2008). Environmental forensics and the importance of source identification. In: Issues in Environmental Science and Technology (Ed. R Harrison). RSC, Cambridge.
Mudge SM and Duce CE (2005). Identifying the source, transport path and sinks of sewage derived organic matter. Environmental Pollution 136:209–220.
Mudge SM, Assinder DJ and Russell AT (2001). Micro-scale variability of contaminants in surface sediments: the implications for sampling. R&D Technical Report P3-057/TR, Environment Agency:88.
Philp RP (2007). The emergence of stable isotopes in environmental and forensic geochemistry studies: a review. Environmental Chemistry Letters 5:57–66.
Shu JM and Bradshaw AD (1995). The containment of toxic wastes. 1. Long-term metal movement in soils over a covered metalliferous waste heap at Parc Lead-Zinc Mine, North Wales. Environmental Pollution 90:371–377.
Singer H, Muller S, Tixier C and Pillonel L (2002). Triclosan: occurrence and fate of a widely used biocide in the aquatic environment: field measurements in wastewater treatment plants, surface waters, and lake sediments. Environmental Science and Technology 36:4998–5004.
Veenstra S (2005). The occurrence of Triclosan in the marine environment. MSc Dissertation, Department of Ocean Sciences, Bangor University.
Wang Z and Fingas MF (2003). Development of oil hydrocarbon fingerprinting and identification techniques. Marine Pollution Bulletin 47:423–452.
Yazdankhah SP, Scheie AA, Hoiby EA, Lunestad B-T, Heir E, Fotland TO, Naterstad K and Kruse H (2006). Triclosan and antimicrobial resistance in bacteria: an overview. Microbial Drug Resistance 12:83–90.

Chapter 11
Petrography and Geochemical Analysis for the Forensic Assessment of Concrete Damage

Isabel Fernandes, Maarten A.T.M. Broekmans and Fernando Noronha

Abstract Concrete deterioration was recognised in the early 1900s and at the time was considered a natural consequence of aging. Since then, a number of different damage mechanisms have been identified, compromising performance and reducing service life. Proper identification of primary and secondary causes of deterioration is essential to determine correct rehabilitation strategies, and to prevent future damage. Results from such assessments have been used to decide disputes and warranty claims especially, in recent structures, from a forensic perspective. In older structures, such data are typically used to plan maintenance and rehabilitation and, in general, to provide workers with the required expertise and know-how to prevent the use of materials known to be deleterious, or mixes proven to have a poor performance in new structures. Reduced concrete performance can be assessed by a number of standard methods to produce data on, amongst others, compressive/tensile strength, water infiltration depth, total porosity, permeability, and chloride content. However, besides the characterisation of the actual performance of the material, it is necessary to identify the *cause* of deterioration, for which analytical methods, based in geological techniques, have proven to be powerful and versatile, notably petrographic microscopy and geochemistry. Petrography can be applied on plane sections from extracted drill cores, as well as on thin sections under an optical microscope using polarised light. Polished sections can be used for analysis by microprobe (EMPA), including element mapping. Appropriately prepared thin sections enable identification and assessment of the spatial distribution of micro-structural features, including capillary porosity. Petrographic data on modal content of coarse and fine constituents, and rock types and minerals present in concrete, are essential for correct interpretation of geochemical assessment of bulk 'whole rock' concrete using methods such as XRF or ICP.

I. Fernandes(✉) and F. Noronha
Department and Centre of Geology, Faculty of Science, University of Porto, Rua do Campo Alegre 687, 4169-007 PORTO, Portugal
e-mail: ifernand@fc.up.pt

M.A.T.M. Broekmans
Geological Survey of Norway, Department of Industrial Minerals and Ores
N-7491 Trondheim, Norway

K. Ritz et al. (eds.), *Criminal and Environmental Soil Forensics*

Issues which can prove challenging using conventional bulk testing methods can be more easily resolved using geological methods, especially when forensic issues are involved. This chapter presents the applications of these techniques in case studies, illustrating the potential of petrography, combined with geochemical analysis where applicable, to resolve the cause of deterioration where traditional methods failed.

Introduction

Modern concrete consists of coarse and fine aggregate rock particles cemented together by hydrated ordinary Portland cement (OPC) as an inorganic binder. This is often with supplementary mineral materials such as pozzolanas or fillers, to enhance concrete properties (e.g. Taylor 1997; Hewlett 1998; Neville 1999). To minimise transport expenses, aggregate is typically sourced close to the structure, resulting in a wide lithological variety of aggregate materials applied in concrete. Adding further to this variation, concrete mix designs are customised to meet requirements for their particular application (e.g. Fookes et al. 1993). Mix design thus specifies aggregate grading, amount and type of cement, water/cement ratio and use of additives (e.g. plasticiser, accelerator, air entraining agent). Finally, when a concrete mix of a given composition is being poured and compacted, imperfections leading to local inhomogeneity and anisotropy are virtually inevitable.

During their service life, concrete structures can be exposed to variable meteorological conditions (wet/dry seasons), in various climates (hot, cold, arid, wet). In addition, concrete structures may be exposed to aggressive chemicals, such as sodium chloride in seawater or as deicer, or alternative deicers, fertiliser, distilled water, acid or caustic solutions. The net sum of these exposure conditions may result in particularly unfavourable combinations leading to damage by untimely degradation, compromising designed service life.

This chapter presents a short review of geological methods to study concrete, notably petrographic analysis and bulk 'whole rock' geochemistry. Given the scope of this volume, the chapter focuses on petrography and geochemical analysis as methods for the forensic assessment of damaged concrete, as illustrated by a few case studies. While typically perceived as dull and bulk, concrete can be delicate too if studied with proper tools.

Concrete Petrography

History and Background

Transcribed from its Greek origin, the concept of 'petrography' can be described as 'the art of describing a rock'. A comprehensive petrographic description includes information on mineral modal content in percentage volume (vol%), grain size, grain shape, grain contacts, as well as the spatial arrangement of these grains, i.e. rock fabric/texture, and rock structure. Thus, for concrete as a composite material, a petrographic description would have to include data on the mineral modal content of aggregate constituents, coarse and fine, and the cement paste, as well as textural data of the bulk concrete and the cement paste itself. In practice, however, concrete petrography is understood to be virtually synonymous with 'thin section analysis'. Case studies on petrography, at macro- and micro-scales, show this to be a misconception, with authors such as Laugesen (1999) providing an overview of 'petrographic features' at different scales of observation.

The first rock thin section was made in 1849 by Sir Henry Clifton Sorby (Humphries 1992). Due to the relatively soft and friable cement paste being ground away between the much harder aggregate grains, the first successful thin section studies of concrete did not appear before 1925–1935. It was noted that some paste constituents altered during sample preparation (St John et al. 1998). These and other problems have been resolved by improved equipment, and by consistently following strict procedures, as elaborated below.

Concrete Deterioration

Concrete deterioration is the generic denominator for a multitude of decay processes, either resulting from natural wear and tear by exposure to environmental conditions, or untimely degradation from poor design, poor construction practice, or wrong or excessive use. The layman term 'concrete rot' is more or less synonymous with 'steel reinforcement corrosion'. However, this chapter focuses on deterioration of the stony part of this construction material.

Since the early 1900s when the first petrographic studies on concrete decay were published (e.g. St John et al. 1998), many different causes of deterioration have been recognised. Concrete durability can be negatively influenced by poor workmanship during building and construction, but also by factors affecting the concrete rock itself, so that the cementitious binder, the aggregate, or both of the concrete rock may be altered.

For a diversity of mechanisms, petrography is able to identify the cause of deterioration effectively and at comparatively low cost (St John et al. 1998), including alkali-aggregate reaction (AAR), sulphate attack (DEF, TSA), leaching and other fluid interaction, fire damage, ordinary carbonation, and also poor workmanship. The most common form of alkali-aggregate reaction (AAR) is alkali-silica reaction (ASR). This is a chemical reaction between hydroxyl ions in the pore water with reactive silica in the aggregate, mostly quartz but also amorphous or poorly crystalline polymorphs.

Other reactive aggregate constituents, including volcanic glass or other silicates (SLEASSR) and carbonate rock (ACR), are also well known. The reaction with silica produces a hygroscopic and hydraulic silica gel that upon reaction with water causes expansion. This leads to penetrative cracking, loss of strength, and a substantially reduced service life of the structure. The incubation time before damage becomes manifest is typically several decades, and rehabilitation is temporary as ASR is an inherent quality of a concrete containing reactive aggregate. The modal composition of aggregate, in terms of mineral and rock constituents, is considered paramount to determine reaction potential (Berra et al. 2003). On a global basis, AAR costs approximately 2 billion EUR annually for the assessment, maintenance and/or replacement of affected structures.

Sulphate attack is caused by chemical alteration of the cement paste by dissolved SO_4 (Skalny et al. 2002). Sulphate may originate externally from various resources, or internally from the sulphate added to control the time for setting. There are two main mechanisms: the delayed ettringite formation (DEF), and thaumasite sulphate attack (TSA). Incipient sulphate attack manifests itself in ettringite clusters in the paste, and more advanced attacks by closely interspaced *en-echelon* cracks, filled with ettringite and/or minor gypsum. Finally, the process leads to extensive cracking, expansion and progressive loss of concrete strength, with the alteration of the paste composition and formation of gypsum ($CaSO_4 \cdot 2H_2O$), ettringite $[Ca_3Al(OH)_6 \cdot 12H_2O]_2|(SO_4)_3 \cdot 2H_2O$ and/or thaumasite $[Ca_3Si(OH)_6 \cdot 12H_2O]_2|(SO_4)_3 \cdot (CO_3)_2$, often associated with popcorn-calcite deposition (Hagelia et al. 2003).

Hydrated cement paste exposed to excessive fluid may decompose, depending on porosity/permeability (i.e. 'accessibility'), the amount of fluid, and its relative pH. The paste gets depleted, becomes soft and friable, and insoluble aggregate grains are often observed to protrude from the corroded surface.

Exposure of concrete to high temperatures effectively dehydrates the hydrated cement paste, its extent depending on the highest temperature reached and exposure time. The concrete may show discoloration (a faint 'brick red') due to oxidation of iron, delamination, and eventually strength loss and complete disintegration. In addition, volatiles may enter the concrete, e.g. chlorine vapour from burning PVC, and if the surface gets shock-cooled by water to quench the fire, the thermal stress may cause explosive spalling.

Carbonation is an acid–base neutralisation reaction, with atmospheric CO_2 initially reacting with portlandite $Ca[OH]_2$ in partially water-saturated paste to form $CaCO_3$ and H_2O. After depletion of the paste for portlandite, silicate paste minerals will decompose and become carbonated as well. As a result, pH drops beyond the value at which it passivates the steel reinforcement from corroding (~pH 10). The resulting rebar corrosion is colloquially known as concrete rot, but has its origin in alteration of the cement paste (e.g. MacLeod et al. 1990).

All of the deterioration mechanisms outlined above comprise chemical reactions. For any reaction to continue, access to unreacted, virgin material is essential. Thus, permeability through connected porosity and/or cracks greatly facilitates attack by providing an easy path for reaction constituents to meet. During building and construction, good workmanship, combined with skill and experience, will generally produce concrete with minimal cracking or other discontinuities, making its surface contiguous and impermeable, and the bulk material beneath it dense, homogenous, and isotropic, without segregation of its coarse and fine grained constituents. Conversely, poor workmanship might produce a cracked surface facilitating infiltration of aggressive agents. Such cracks may rapidly spread through the permeable interior, following pre-existing bleeding pathways and open paste-aggregate interfaces, created due to excessive compaction, and along semi-plastic cracks caused by the premature demoulding of the inadequately set cement paste.

Sample Extraction and Preparation for Petrography

The total price of a concrete core comprises costs for extraction, transport, handling, storage, and preparation, and may also include peripheral expenses for road closing, and scaffolding. Calculated per kilogram, a core may be worth its own weight in gold, quite literally. Thus, precautions taken to minimise introduction of artifacts at any instance prior to analysis, including core extraction, are worthwhile. Broekmans (2002) outlines experiential procedures for core extraction and handling, later included in Dutch national guidelines (CUR-Recommendation 102, 2005).

Concrete damage is typically associated with some form of cracking at different size scales. As a first step in sample preparation to visualise cracks and defects, concrete cores are impregnated with epoxy resin containing 'Hudson Yellow' fluorescent dye. A suitable epoxy has low viscosity, but in particular a high capillary affinity for concrete. An extracted core is carefully exsiccated to open capillary porosity. After drying, the core is evacuated and completely immersed in epoxy, after which the vacuum is released. The atmospheric pressure then presses the epoxy deeper into the concrete. This procedure is described in detail in the Danish national standard DS423.39 (Danish Standards Association 2002a). An example of an impregnated core is given in Figure 11.1. Traditional thin sectioning uses carborundum (SiC) slurry on a rotating cast-iron disc but is inadequate for concrete. Grains of worn-off abrasive surfaces roll between the lapping surface and the specimen, milling away at the softest parts in a process known as 'undercutting'. Using a bonded abrasive significantly reduces such undercutting, while an increase in the difference in hardness between abrasive and specimen reduces the relative difference in hardness between individual constituents within the specimen. Replacing traditional carborundum slurry with bonded diamond (e.g. Baumann 1957) combines all of the above.

Even with bonded-diamond lapping equipment, the quality of thin sections may be less than desired, either because the paste is still too fragile to withstand even careful lapping, or due to chemical alteration during specimen preparation. Impregnating the concrete specimen with a high-capillary fluorescent epoxy (as described above for complete cores) reinforces the friable paste, while reducing accessible surface and thus limiting alteration. Detailed procedures for the preparation of concrete thin sections are described in Danish national standard DS423.40 (Danish Standards Association 2002b).

Impregnation can be combined with other specialty techniques, for example, to produce fluorescent thin sections of regular thickness of hydrophobic material such

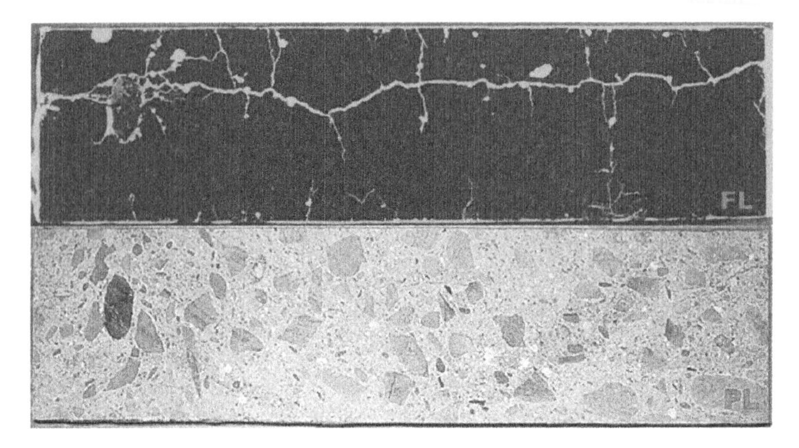

Fig. 11.1 Impregnated full core in fluorescence (FL), compared with the same core in plain light (PL). At >1 mm width, the crack is difficult to discern in plain light, with off-shoots and thinner cracks along the side essentially hidden. Core dimensions: 100 × 400 mm *(see colour plate section for colour version)*

as pure rock salt or salt-impregnated concrete (using an anhydrous coolant), or of sticky and ductile bituminous concrete (Broekmans 2007).

Petrographic Assessment of Concrete

The Hudson Yellow fluorescent dye dissolved in the epoxy has an optimum excitation wavelength in blue light near 485 nm (i.e. not ultraviolet), and an emission wavelength at 515 nm in the green part of the visible spectrum. In practice, 'black light' FL-tubes (UVA, 400 to 320 nm) induce powerful fluorescence on full cores, with impregnated cracks, pores, voids etc. standing out bright yellowish-green against a dark background (Figure 11.1). Impregnated thin sections containing the same fluorescent epoxy can be studied in a regular petrographic polarising microscope. Optimum fluorescence is achieved with regular halogen illumination and a custom filter set, allowing only blue light to pass on to the specimen for excitation, while only passing the green fluorescence emission to the observer (St John et al. 1998).

In addition to all the petrographic features that can be observed with regular thin sections, fluorescence-impregnated thin sections enable the operator to study and document capillary porosity, micro-crack fabric, paste-aggregate bonding, and a range of other features that otherwise remain hidden (St John et al. 1998). Using concrete thin sections of uniform thickness (usually 20 μm), the degree of fluorescence enables calibrated *in situ* assessment of the water/cement ratio (Elsen et al. 1995; Jakobsen et al. 2003). More importantly, the spatial distribution of features can be studied, revealing local variations in water/cement ratio, hydration degree, paste dispersion and packing density.

Concrete constituents can be quantified by grid counting of coarse aggregate in longitudinal core sections, and fine aggregate and cement paste in thin sections. Thus, the modal composition of the concrete mix can be reliably determined, essential for bulk geochemical assessment, as discussed further below (e.g. Broekmans 2002; 2006).

Scanning Electron Microscopy and Electron Microprobe

If greater detail is required than is achievable with optical microscopy, polished thin sections can be assessed by Scanning Electron Microscope (SEM), possibly equipped with an energy dispersive spectrometer (EDS; e.g. Fernandes et al. 2004, 2007). SEM provides detail in the finest grained aggregate types like greywacke or basalt, or on morphology of clinker hydration products (e.g. Taylor 1997), whereas qualitative chemical analysis by EDS can be very helpful to identify phases. SEM-EDS has also been used to determine phases in damaged concrete (e.g. Knudsen and Thaulow 1975; Fernandes 2005, 2007). In addition, back-scattered electron imaging can provide useful information about cement and clinker phases (e.g. Famy et al. 2002) and element maps can be made showing the spatial distribution of chemical species. For quantitative chemical analysis of phases in concrete as required for modelling and calculations, assessment using Electron Microprobe (EMPA) using calibrated standards is more appropriate (Potts et al. 1995).

Geochemical Assessment of Concrete

Sample Size Requirements and Comminution

While structural features are a major issue with petrographic analysis, they are much less relevant for geochemical assessment of concrete. Instead, sample size and representativeness are of key importance but, in most cases, are simply omitted or overestimated. If a sample is to be treated as representative for bulk concrete, its minimum size is defined by the nominal grain size of the coarse aggregate. Minimum representative sample sizes for coarse aggregate are listed in ASTM D75-97 (1997, Table 11.1 therein). For simple and straight-forward 1:2:3 concrete, containing 25 mm coarse aggregate, one representative sample would have to weigh a minimum of 108 kg, or about 40 L in volume (Broekmans 2006). This minimum size poses practical problems with respect to sample extraction, transport, storage and handling. Therefore, the practical size of concrete cores is often limited to 100 × 400 mm and often even less, or about 8% of the minimum representative size. Thus, claims that 'samples are representative for bulk concrete' should be regarded with scepticism.

Subsequently, bulk sample material must be prepared for analysis. Typically, procedures for whole rock analysis require only a few grams of material, whence bulk samples are comminuted and split in iterative steps. After initial comminution in a jaw crusher, deteriorated concrete with a friable paste often consists of coarse chunks of aggregate and fine cement dust, which may be challenging to split into representative subsamples (e.g. Gy 1979), requiring additional comminution steps (Broekmans 2006).

Whole Rock Concrete: Main Elements

Concrete is a material of variable composition, being the net sum of aggregate (mostly siliceous, occasionally carbonaceous) and binder (essentially calcic), with possible contributions from the environment to which the material has been exposed. Thus, to analyse whole rock concrete chemically in a comprehensive manner, a set of complementary methods is required, as outlined below.

The main element composition of bulk concrete is conveniently assessed by X-ray fluorescence (XRF). In a typical procedure, a weighed amount of pulverised sample material is digested in excess Li-borate flux at high temperature, to: (1) minimise matrix effects from the original material, and (2) homogenise the material (e.g. Kristmann 1977; Odler et al. 1981). The tablet produced by the digestion is fed into an XRF instrument and analysed, typically for main rock-forming elements Na, K, Ca, Mg, Mn, Cr, Fe, Al, Si, Ti, P, and possibly others depending on instrument type and set-up. Results are typically presented as percentage weight (wt%) oxide.

Alternatively, whole rock composition can be assessed by inductively coupled plasma–atomic emission spectrometry (ICP-AES) after digestion in excess acid. The pH of healthy OPC concrete is about 14, and somewhat lower (pH ~13) if also containing blast furnace slag, fly ash, microsilica or other pozzolanas (Bijen 1996), which requires to be taken into account when determining the amount of 'excess acid'. As silica and most silicates are only sparingly soluble in common strong acids, the acid mixture used to digest the material greatly affects the analytical result. By selecting a solvent that consumes the paste but leaves the aggregate unaffected, it is possible to chemically separate concrete constituents using selective digestion. Petrography is indispensable for the purpose of selecting a suitable acid solvent to discriminate between aggregate and paste, based on mineral content.

Concrete contains a substantial amount of volatile species not assessed by XRF or ICP, including hydration water bound by the paste, possibly CO_3^{2-} from carbonation of the concrete cover by uptake of atmospheric CO_2, from carbonatic aggregate, or from pulverised limestone added to self-compacting concrete. Hydration water and CO_3^{2-} can be assessed in bulk by gravimetric measurement of loss on ignition (LOI) at 1000 °C, at which temperature hydrated paste gets devolatilised, whereas most carbonates decompose at temperatures over 800 °C. Thus, the value for bulk-LOI comprises both H_2O and CO_2.

Total carbon content (total-C) can be analysed by ignition at 1450 °C and spectrometric analysis of released volatiles (e.g. Leco), which also gives total-S. Whilst total-S is recalculated to SO_3 and added to the main element oxides, total-C is recalculated in terms of CO_2 and can be subtracted from bulk-LOI to give the wt% of H_2O (Figg and Bowden 1971; Figg 1989). Thus, taking the analysis by XRF and the volatiles by LOI supplied by total-S and total-C, from Leco, provides description of the main elements of which bulk concrete is composed, with a net sum total of approximately 100 wt% (Broekmans and Jansen 1998; Broekmans 2002, 2006).

Concrete In-Situ: Electron Microprobe Analysis (EMPA)

If reliable quantitative data are required, for instance to verify compositional zoning, or for numerical modelling, data may be acquired from carefully polished thin sections using an electron microprobe instrument (EPMA). While derived from a SEM, a microprobe is specially constructed for optimum electron beam stability, accurate stage positioning, on-the-fly calibration against known standards, and to accommodate multiple detectors, notably wavelength-dispersive spectrometry (WDS) in addition to EDS (e.g. Potts et al. 1995). During an analysis, the electron beam switches between precisely known standards within the high vacuum chamber, and the selected spot on the specimen. Using data acquired for these standards, raw data can be amended for atomic number (Z), absorption (A), and fluorescence (F) factors, dubbed ZAF-correction. Thus, 10 to 12 elements from Be to U can be quantified per analysis, with detection limits, under favourable conditions, down to ppm-level. The size of the analytical spot can be reduced to 1 µm, but the droplet-shaped volume actually analysed underneath extends much deeper into the material (Wong and Buenfeld 2006).

Practical Examples of Concrete Deterioration

General

Some case studies of concrete deterioration are presented below which illustrate some of the problems which can arise at different stages in its structural life, from pre-production to well-matured. Using petrography and geochemical analysis as outlined above, it has been possible to pinpoint the primary causes of damage, and design effective remedial strategies. The examples in this section demonstrate the opportunities that petrography and geochemistry offer for the forensic assessment of concrete damage.

Pre-concrete: Pre-construction Assessment of Aggregate

Aggregate materials comprise up to two-thirds of the volume of bulk concrete. Properties such as particle size, grading, shape, impurities or contaminants are important for the quality and durability of the final concrete (e.g. Fookes 1980). In geology, rocks are traditionally classified according to origin (i.e. igneous, sedimentary and metamorphic), mineral (modal) content, grain fabric, texture and structure (e.g. Gjelle and Sigmond 1995). The classification of rocks in terms of alkali-reactivity potential mainly focuses on the quality of the main reactive component of quartz (α-SiO_2), or other silica polymorphs (e.g. Broekmans 2002). Features that have been attributed to alkali-reactivity potential include sub-graining and grain-size reduction (Grattan-Bellew 1992; Shayan and Grimstad 2006), and the available surface area of quartz (Wigum 1995), among many others, but few seem to apply universally. According to ASTM C295-98 (2002), potentially alkali-silica reactive silica includes quartz, and its polymorphs cristobalite and tridymite, fine grained varieties chert/flint and chalcedony, opal, various types of volcanic and industrial glasses, as well as a range of rock including igneous, sedimentary, and metamorphic types.

The procedure in RILEM AAR-1 (2003; see www.RILEM.net) describes petrographic assessment using point counting to determine the vol% of rock types present, followed by their classification as potentially alkali-reactive, ambiguous, or innocuous according to a specified list (see Wigum 2000). The total of both potentially reactive and ambiguous rock types is not to exceed a given limit in vol%. If it does, it is advised that the material be rejected for use in concrete or applied under special measures.

To illustrate the importance of aggregate characterisation prior to construction, an example is shown regarding alluvial material from a deposit in the north of Portugal, which has been widely applied as concrete aggregate. A recent petrographic analysis revealed a texture previously described as potentially reactive. Figure 11.2 shows two different sandstone particles. On the left particles are shown with serrate contacts between adjacent quartz grains from diagenetic compaction-deformation, and on the right particles with ‘accentuated grain boundaries’ (Broekmans and Jansen 1998), supposedly due to neogenic clay minerals lining grain boundaries. The open pore system between the fine-grained quartz grains allows ingress of the highly alkaline pore fluid (pH ~ 14) from the surrounding Portland cement paste.

To validate the performance of this material in practice, the concrete from a bridge containing this aggregate was assessed by thin section petrography. The concrete proved to be extensively cracked by deleterious alkali-silica reaction, including gel extrusion (Figure 11.3). Following the findings of this study, it is expected that damage will be identified in other structures with the same aggregate in the future. Had this aggregate material been assessed conform RILEM AAR-1 prior to its application, this damage could have been largely avoided.

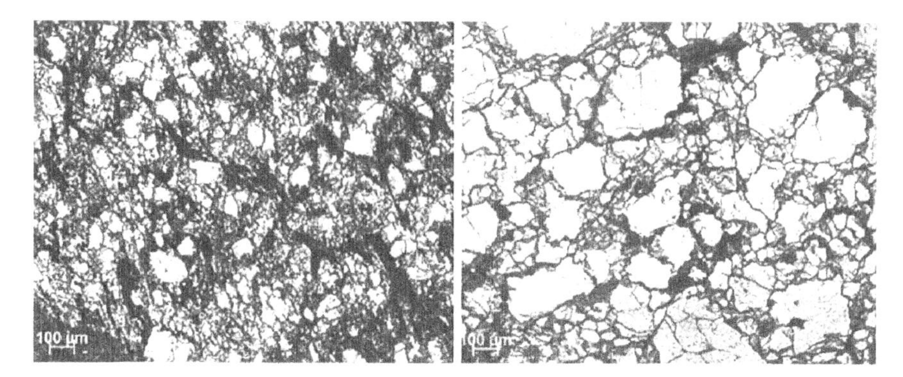

Fig. 11.2 Aggregate particles from an alluvial deposit containing potentially reactive textures: (*left*) siltstone with grain size reduction from deformation upon compaction; (*right*) sandstone with interstitial neogenic clay decorating grain boundaries *(see colour plate section for colour version)*

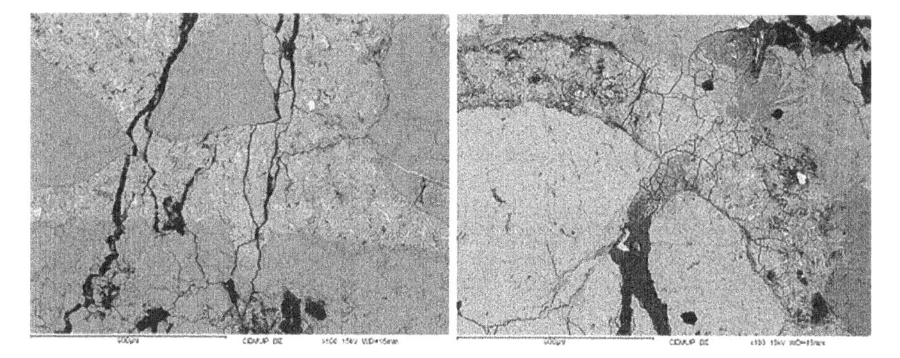

Fig. 11.3 SEM images showing cracks and extruded ASR gel in concrete containing aggregates from the alluvial deposit described in the text, now classified as alkali-reactive

Very Young Concrete: Delamination in an Industrial Floor

Shortly after the construction of an industrial floor had been completed, it was noticed that large areas of the floor surface sounded hollow. Upon inquiry, the contractor blamed the damage on a higher work load than that for which the floor had been designed (e.g. the use of forklift trucks). This argument was rejected because even areas of the floor which had not been used at all had the same damage. Cores were extracted from areas sounding hollow as well as those which were solid, and a number of plane and thin sections were prepared conform the Danish standards referred to above. Impregnation-fluorescence petrography on plane sections revealed that both hollow-sounding and solid concrete was delaminated at 7 to 10 mm below the surface (Figure 11.4). Detailed assessment of the delamination surface in thin sections revealed features recognised as 'sedimentary mass-transport' at a micro-scale,

Fig. 11.4 Impregnated core, showing delamination at 7 to 12 mm from the top surface. The wavy lower core face is due to pouring on plastic foil on sand bed. Core dimension: 100 × 170 mm

known to form in a semi-plastic state, i.e. before the concrete was fully set and hardened. In summary, the delamination resulted from the construction process before actual use, and was attributable to over-eager efforts to produce a smooth and dense finish. This conclusion led to the warranty being settled to the advantage of the floor's owner. More extensive details are given in Broekmans (2004).

Young Concrete: Spalling of a Sedimentation Basin Wall

After only one year of service the wall surface of a circular sedimentation basin in a municipal wastewater treatment installation showed severe spalling, exposing all aggregate. A bottom scraper drive wheel was being run along the top of the wall. To prevent slippage under winter conditions, urea was applied as a pre-icing agent. The installation design, building and construction were all supervised by the same engineering consultancy, which proposed urea as 'the smart choice' (*ipse dixit*), avoiding reinforcement corrosion, as urea is chloride-free.

An initial assessment attributed 'poor concrete quality' to an excessive cement content of 540 kg m^{-3}, compared with 300 kg m^{-3} required to conform to specification. The contractor was accused of not following specifications, convicted, and required to rehabilitate the structure at their own expense. The contractor considered that the concrete complied with the specifications set, so arbitration was required.

Using grid and point counting on plane and thin sections, the second assessment demonstrated unequivocally that modal contents in vol% of coarse and fine aggregate versus binder were as specified. Thin section petrography then revealed substantial carbonate present in the coarse aggregate, either as calcite $CaCO_3$ or dolomite $CaMg(CO_3)_2$ in some of the particles, but also as belemnite fragments (a fossil related to the modern squid). As calcite can be stained by Alizarin-R after etching in dilute acid (Warne 1962), non-carbonatic aggregate would stand out as holes in the easily etched and coloured cement matrix (Figure 11.5).

According to the appropriate standard, the cement content of hardened concrete is to be assessed by dissolving a weighed amount of pulverised bulk concrete in excess acid, weighing the dry residue, and obtaining the cement content by balance, thereby assuming the aggregate as insoluble. However, with substantial acid-soluble carbonate in the aggregate, the weight of insoluble residue is lower and consequently the cement content found to be greater. In this particular case, the identification of the nature of aggregates by petrography unambiguously identified the traditional acid dissolution method as unreliable for this type of aggregate, releasing the contractor from his initial conviction.

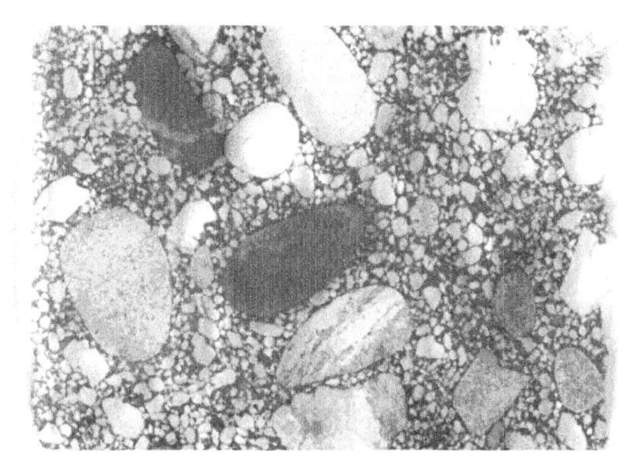

Fig. 11.5 Alizarin-R stained thin section, revealing a number of stained coarse aggregate particles obviously containing carbonate. Actual size: 26 × 35 mm *(see colour plate section for colour version)*

Concrete During Service: Sulphate Attack of a Portuguese Bridge

A concrete bridge located in the centre of Portugal containing a beam with extensive vertical fracturing which, potentially, could lead to local failure of the deck, was initially suspected of deleterious AAR. Cores were carefully extracted and thin sections produced using the Danish standards described above. Petrographic analysis revealed that the fractures originated from sulphate attack, not AAR. Ettringite was identified in closely interspaced *en-echelon* cracks and on the paste-aggregate interface, as well as in isolated clusters throughout the paste. Qualitative chemical analysis by SEM/EDS on polished thin sections identified this material as ettringite (Figure 11.6).

The abundance of ettringite and crack frequency increased from the concrete structure's top surface towards its foundation in contact with soil. Investigation of the geological setting and lithology of the soil substrate revealed that the structure was situated in an area formerly known for gypsum mining. Further investigation revealed that the sulphate attack could be attributed to groundwater infiltrating the foundation. In this case, the damage could have been avoided had local geology been appraised.

Untimely End-of-Life: Excessive Wear in Bituminous Concrete

Mainly used in road pavements, the aggregate in bituminous concrete is bonded by an organic binder, offering a degree of ductility as well as self-repairing qualities. The bitumen binder often consists of natural asphalt, mainly from Trinidad, partly or wholly replaced by epoxy, poly-urethane, or other synthetic binders. The formulation of bituminous concrete can be made to comply with a broad range of specifications,

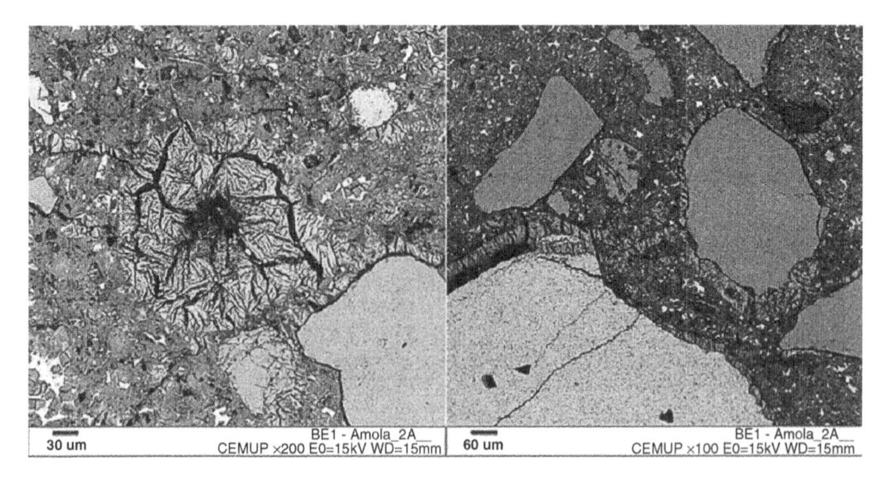

Fig. 11.6 Ettringite crystals filling an air void in the cement paste (*left*), and on the interface between aggregate and cement paste (*right*)

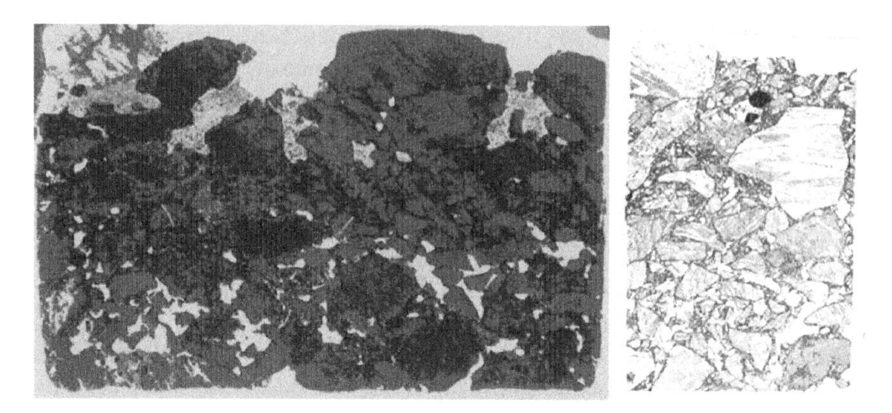

Fig. 11.7 Image of impregnated plane section in fluorescence (*left*), and thin section of bituminous concrete in plain light (*right*). Note the surface roughness and the pockets with debris. Core dimensions: 75 × 45 mm, thin section dimensions: 30 × 45 mm

by customising binder properties, choice of aggregate material and particle shape, and/or by adapting mix proportions.

In Trondheim, Norway, the use of studded tyres during the winter (from November to April) causes excessive wear of the road surface. For low traffic volumes, an inexpensive but not long-lasting greenstone is used, whereas the busiest roads are paved with a very hard and durable, and hence expensive, jasper. Extensive laboratory testing of a local cataclastic rock suggests it has an intermediate durability at a far more attractive price, presenting an interesting option for use by designers. However, in practice, the cataclasite appears to wear faster in surfaces with blended aggregate and, surprisingly, faster than greenstone (Broekmans 2007), as demonstrated by the heavy pitting of the road surface, as well as shattered car windshields.

A study of plane and thin sections reveals that deterioration cracks in the cataclasite follow welded pre-existing cracks in the original cataclastic fabric, with the rock behaving anisotropically. In contrast, cracks in jasper or greenstone do not seem to have a preferred orientation inherited from a pre-existing fabric. Both seem to behave isotropically (Broekmans 2007; Figure 11.7). Unless there is excellent orientation control (for example for roofing slate), the relevant mechanical anisotropy of a rock must be regarded with great care.

Summary and Conclusions

Well-established geological techniques for the assessment and analysis of natural rocks and minerals can also be applied for the forensic investigation of concrete. Petrographic analysis is not restricted to polarised light microscopy, but spans size scales ranging from a whole structure to very small details in a thin section. Initial problems with sample preparation have been resolved, and methods for the preparation of specimens for petrography are described in national guidelines. High quality specimens of even the most difficult materials can now be prepared almost routinely, using impregnation techniques with fluorescent epoxy.

Standard optical petrography provides data on structure, texture/fabric, as well as mineral content and composition of both aggregate and paste constituents. Grid and point counting and analysis of plane and thin sections provide reliable data on proportions of the mix of concrete in percentage volume, which can be used to assess data from chemical analysis. In addition, fluorescence petrography can reveal the presence of cracks, capillary porosity and permeability, and their spatial distribution, and can be used to quantitatively determine the water/cement ratio. The finest detail can be observed using a SEM.

Many types of concrete deterioration involve chemical reactions, which can be studied with the aid of geochemical assay. As bulk concrete contains volatile components in addition to solid matter, complementary methods are required for comprehensive analysis. Reliable results have been obtained using XRF, ICP, LOI, possibly in combination with other methods. Sample size and representativity, comminution, splitting and sample digestion have been proven crucial issues, all of which may compromise data reliability if not properly taken care of. Quantitative *in-situ* chemical analysis of individual grains is possible using an electron microprobe instruments (EMPA).

As a method, petrography is an indispensable 'intake assessment' method prior to use in further analysis, and is the most cost-effective method in the evaluation of causes of deterioration and the validation of results obtained by chemical analysis.

References

ASTM C295-98 (2002). Standard Guide for Petrographic Examination of Aggregates for Concrete. American Society for Testing and Materials, Philadelphia.

ASTM D75-97 (1997). Standard Practice for Sampling Aggregates. American Society for Testing and Materials, Philadelphia.

Baumann HN Jr (1957). Preparation of petrographic sections with bonded diamond wheels. American Mineralogist 42:416–421.

Berra M, Mangialardi T, and Paolini AE (2003). Alkali-silica reactive criteria for concrete aggregates. Materials and Structures 38:373–380.

Bijen J (1996). Blast Furnace Slag Cement. Association of the Netherlands Cement Industry, The Netherlands, 62 pp.

Broekmans MATM (2002). The alkali-silica reaction: mineralogical and geochemical aspects of some Dutch concretes and Norwegian mylonites. PhD Thesis, Utrecht University. Geologica Ultraiectina 217.

Broekmans MATM (2004). Microscale sedimentary transport phenomena reveal the origin of delamination in an industrial floor. Special Issue 29, Materials Characterization (53/2–4):233–241.

Broekmans MATM (2006). Sample representativity: effects of size and preparation on geochemical analysis. In: Marc-André Bérubé Symposium on Alkali-aggregate Reactivity in Concrete

(Ed. B Fournier), pp. 1–19. 8th CANMET/ACI International Conference on Recent Advances in Concrete Technology, Montréal, Canada.
Broekmans MATM (2007). Failure of greenstone, jasper and cataclasite aggregate in bituminous concrete due to studded tyres: similarities and differences. Special Issue 31, Materials Characterization 58/11–12:1171–1182.
Broekmans MATM and Jansen JBH (1998). Silica dissolution in impure sandstone: application to concrete. In: Proceedings of the Conference on Geochemical Engineering: Current Applications and Future Trends (Eds. SP Vriend and JJP Zijlstra). Special Volume, Journal of Geochemical Exploration 62:311–318.
CUR-Recommendation 102 (2005). Inspection and assessment of concrete structures in which ASR is suspected or has been confirmed. *www.cur.nl/upload/documents/CUR-Recommendation%20102%20versie%2009-05-08%20.pdf* Centre for Civil Engineering Research and Codes, Gouda.
Danish Standards Association (2002a). Testing of concrete – Hardened concrete – Production of fluorescence impregnated plane sections (in Danish). DS 423.39.
Danish Standards Association (2002b). Testing of concrete – Hardened concrete – Production of fluorescence impregnated thin sections (in Danish). DS 423.40.
Elsen J, Lens N, Aarre T, Quenard D, and Smolej V (1995). Determination of the w/c ratio of hardened cement paste and concrete samples on thin sections using automated image analysis techniques. Cement and Concrete Research 25:827–834.
Famy C, Scrivener KL and Crumbie AK (2002). What causes differences of C-S-H gel gray levels in backscattered electron images? Cement and Concrete Research 32:1465–1471.
Fernandes I (2005). Petrographic, physical and chemical characterisation of granitic aggregates for concrete. Case studies (in Portuguese). PhD Thesis, Universidade do Porto.
Fernandes I (2007). Composition of alkali-silica gel related to its location in concrete. In: Proceedings of the 11th Euroseminar on Microscopy Applied to Building Materials (Eds. I Fernandes, A Guedes, MA Ribeiro, F Noronha and M Teles), Porto, CD-ROM.
Fernandes I, Noronha F and Teles M (2004). Microscopic analysis of alkali-aggregate reaction products in a 50-year-old concrete. Special Issue 29, Materials Characterization 53/2–4:295–306.
Fernandes I, Noronha F and Teles M (2007). Examination of the concrete from an old Portuguese dam. Texture and composition of alkali-silica gel. Special Issue 31, Materials Characterization 58/11–12:1160–1170.
Figg JW (1989). Analysis of hardened concrete – a guide to tests, procedures and interpretation of results. A report of a joint working party of the Concrete Society and Society of Chemical Industry. The Concrete Society, London, Technical Report 32.
Figg JW and Bowden SR (1971). The Analysis of Concretes. Building Research Establishment, Her Majesty's Stationery Office, London.
Fookes PG (1980). An introduction to the influence of natural aggregates on the performance and durability of concrete. Quarterly Journal of Engineering Geology 13:207–229.
Fookes PG, Stoner JR and Mackintosh J (1993). Great man-made river project, Libya, phase I: A case study on the influence of climate and geology on concrete technology. Quarterly Journal of Engineering Geology 26:25–60.
Gjelle S and Sigmond EMO (1995). Bergartsklassifikasjon og kartfremstilling. With Norwegian-English and vv. glossary. Norges geologiske undersøkelse, Skrifter 113.
Grattan-Bellew PE (1992). Microcrystalline quartz, undulatory extinction and the alkali-silica reaction. In: Proceedings of the 9th International Conference on Alkali-Aggregate Reaction in Concrete 1 (Ed. A Poole), London. Published by The Concrete Society, Slough, England pp. 383–394.
Gy PM (1979). Sampling of particulate materials, theory and practice. Developments in Geomathematics 4. Elsevier Scientific, Amsterdam.
Hagelia P, Sibbick RG, Crammond NJ and Larsen CK (2003). Thaumasite and secondary calcite in some Norwegian concretes. Cement and Concrete Composites 25:1131–1140.
Hewlett PC (1998). Lea's Chemistry of Cement and Concrete. 4th edition. Arnold, London.

Humphries DW (1992). The preparation of thin sections of rocks, minerals and ceramics. Microscopy Handbooks 24, Royal Microscopical Society, Oxford Science.
Jakobsen UH, Brown DR, Comeau RJ, and Henriksen JHH (2003). Fluorescent epoxy impregnated thin sections prepared for a Round Robin test on w/c determination. In: Proceedings of the 9th Euroseminar on Microscopy of Building Materials (Eds. MATM Broekmans, V Jensen and B Brattli), Norway: CD-ROM.
Knudsen T and Thaulow N (1975). Quantitative microanalyses of alkali-silica gel in concrete. Cement and Concrete Research 5:443–454.
Kristmann M (1977). Portland cement clinker: mineralogical and chemical investigations. Part I: microscopy, X-ray fluorescence and X-ray diffraction. Cement and Concrete Research 7:649–658.
Laugesen P (1999). Concrete and its constituents. In: Proceedings of 7th Euroseminar on Microscopy of Building Materials (Eds. HS Pietersen, JA Larbi and Janssen HHA), The Netherlands, pp. 7–16.
MacLeod G, Hall AJ, and Fallick AE (1990). An applied mineralogical investigation of concrete degradation in a major concrete road bridge. Mineralogical Magazine 54:637–644.
Neville AM (1999). Properties of Concrete. 4th edition. Pearson Education, Essex.
Odler I, Abdul-Maula S, Nüdling P, and Richter T (1981). Über die mineralogische und oxidische Zusammensetzung industrieller Portlandzementklinker. Zement-Kalk-Gips 34(9):445–449.
Potts PJ, Bowles JFW, Reed SJB and Cave MR (1995). Microprobe Techniques in the Earth Sciences (Eds. PJ Potts, JFW Bowles, SJB Reed, and MR Cave), The Mineralogical Society Series 6.
RILEM (2003). AAR-1 – Detection of potential alkali-reactivity of aggregates – petrographic method. Materials and Structures 36:480–496.
Shayan A and Grimstad J (2006). Deterioration of concrete in a hydroelectric concrete gravity dam and its characterisation. Cement and Concrete Research 36:371–383.
Skalny J, Marchand J, and Odler I (2002). Sulphate attack on concrete. Modern Concrete Technology Series 10. E and FN Spon, London.
St John DA, Poole AB, and Sims I (1998). Concrete Petrography: A Handbook of Investigative Techniques. Arnold, London.
Taylor HFW (1997). Cement Chemistry. 2nd edition. Thomas Telford, London.
Warne SStJ (1962). A quick field or laboratory staining scheme for the differentiation of major carbonate minerals. Journal of Sedimentary Petrology 32/1:29–38.
Wigum BJ (1995). Alkali-aggregate reactions in concrete: properties, classification and testing of Norwegian cataclastic rocks. Doctor Ingeniør Thesis, Norwegian University of Science and Technology, Trondheim.
Wigum BJ (2000). "Normin2000" - A Norwegian AAR research program. In: Proceedings of the 11th International Conference on Alkali-Aggregate Reaction in Concrete (Eds. MA Bérubé, B Fournier and B Durand), Montreal, Canada, pp. 523–531.
Wong HS and Buenfeld NR (2006). Monte Carlo simulation of electron-solid interactions in cement-based materials. Cement and Concrete Research 36:1076–1082.

Chapter 12
Tracing Soil and Groundwater Pollution with Electromagnetic Profiling and Geo-Electrical Investigations

Kristine Martens, and Kristine Walraevens

Abstract Geophysical investigation is used to differentiate lithological units. In addition to this, geophysics is also useful for the detection, definition and monitoring of pollution in the ground, on the condition that the pollution, due to spills, leakage or illegal discharges causes a significant difference in conductivity/resistivity. The methodology of geophysical investigation in environmental issues will be illustrated. The first presented method is the electromagnetic profiling method, which measures the lateral variation in ground conductivity using a transmitter and a receiver. At the transmitter coil, a time-varying electromagnetic field is induced by an alternating current. This field interacts with the ground, proportional to ground conductivity. The resulting field is measured and recorded by the receiver. The background conductivity needs to be defined along a profile in a non-polluted zone. Subsequently, all the collected data can be used to plot the lateral variation in conductivity. Areas with higher conductivity reflect in most cases pollution from which the source can be traced. A second method is geo-electrical tomography, which is a combination of resistivity profiling and sounding where a large number of electrodes are placed at a constant distance along a line. During each measurement, the electrical potential caused by a current sent into the soil by two current electrodes is measured between the two potential electrodes. By automatically addressing a combination of four electrodes, and increasing the distance between the electrodes, the depth of penetration increases. Considering the resistivity of the corresponding lithology, pollution along the profile can be delimited vertically and horizontally. Finally, the investigation of the conductivity (resistivity) carried out with borehole loggings delivers information on the vertical distribution of the conductivity in the groundwater reservoir, resulting in the vertical delimitation of pollution close to the borehole. A great advantage for these methods applied to soil pollution consists in the avoidance of direct contact with the pollution, resulting in a reduction of health risks. These methods are non-destructive and fairly fast investigations are possible. They will be illustrated by case studies where the results are validated based on the analyses of soil and groundwater samples.

K. Martens(✉)
Laboratory of Applied Geology and Hydrogeology, Gent University, Krijgslaan 281 – S8, B-9000 Gent, Belgium
e-mail: kristine.martens@ugent.be

K. Ritz et al. (eds.), *Criminal and Environmental Soil Forensics*

Introduction

Geophysical investigation is used to differentiate lithological units. In addition, geophysics is also useful to detect, define and monitor pollution in the ground on the condition that the pollution, due to spills, leakage or illegal discharges, causes a significant difference to conductivity/resistivity. The methodology of geophysical investigation in environmental studies is not new (Stewart and Bretnall 1986; Hoekstra et al. 1992; Mack 1993; Williams et al. 1993), particularly in studies dealing with landfills (Kayabali et al. 1998; Nobes et al. 2000; Monteiro Santos et al. 2006; Kaya et al. 2007). However, other approaches of groundwater flow and groundwater analysis are rarely combined with the interpretation of geophysical methods. Although an integrated approach is recommended in the literature (Greenhouse and Harris 1983), geophysical methods are all too often applied as stand-alone approaches. Even more surprisingly, geophysical investigation is relatively unknown in soil forensics, although it offers in many cases great perspectives. As stated in Greenhouse and Slaine (1986), it should be used prior to, and in combination with, other investigation methods (i.e. drilling, soil and water sampling). This article gives a brief overview of the methodology of electromagnetic and geo-electrical prospection, together with examples of how it can be used in soil forensics.

Methodology

Electromagnetic Profiling

The first method presented is electromagnetic profiling, where the lateral variation in ground conductivity can be measured along several lines. The measurements are carried out with two coils: a transmitter and a receiver, connected with a cable. At the transmitter coil, a time-varying electromagnetic field is induced by an alternating current; depending on the specific conductivity of the ground, the frequency of the signal and the distance between the coils. The resulting field is measured and, after processing, the conductivity (or its reciprocal: the resistivity) can immediately be recorded by the receiver.

The penetration depth depends on the distance between the transmitter and the receiver. For homogeneous subsoils, the exploration depth for the horizontal dipole mode (coils are placed vertically) is 0.75 times the intercoil spacing, while for the vertical dipole mode (coils are placed horizontally) it reaches up to 1.5 times the intercoil spacing (McNeill 1980). The selection of the intercoil spacing (S) depends on the expected geology and the probable depth of (the source of) the pollution. Using the EM34-3 instrument, three intercoil spacings are applicable: 10 20 or 40 m. It is recommended to use three intercoil spacings to observe the variation in conductivity with depth. In addition to this, it is advised to keep the distance small between two successive measurements in order to trace the pollution. The orientation of the profiles is recommended to be based on the type of the suspected source of pollution. Dealing with an elongated source (pipe line, stream), the profiles in a first stage need to be parallel to the elongated source in order to detect propagation of pollution from the source. In the second phase, at each anomaly additional profiles perpendicular to the original profile need to be executed in order to delineate the extent of pollution from the source. In the case of single point pollution, a network of individual lines is suggested.

For the interpretation of the gathered data, the natural background conductivity first needs to be derived from a profile located in a non-polluted area. Secondly, the measured values can be presented on a map resulting in isolines of the apparent conductivity from which the source and the lateral extension of the pollution can be deduced.

The implementation of electromagnetic profiles is a useful tool to rapidly gather information on the lateral variation in ground conductivity. This method has the added advantage that direct contact with the pollution is avoided and that it can also be carried out on paved ground. Nevertheless, underground pipes, fences, and electric wire might cause interference.

Geo-Electrical Tomography

The second method is the application of surface resistivity prospection in its modern form: geo-electrical tomography. It is a combination of resistivity profiling and sounding, where a large number of electrodes are placed at a constant distance along a line. During each measurement, the electrical potential caused by a current sent into the soil by two current electrodes is measured between the two potential electrodes, from which the resistivity (in ohms) can be calculated. By automatically addressing a combination of four electrodes, and increasing the distance between the electrodes, the penetration depth increases. The instructions to define the orientation of the profiles are similar to those explained for the electromagnetic profiling method. Based on the lithology and its corresponding natural background resistivity, the pollution along the profile can be delimited vertically and horizontally.

The main limitation to this method is that the electrodes of the geo-electrical tomography are required to penetrate into the soil and a good conduction with the soil is necessary. Because of this, the method is not applicable in paved areas. The method is relatively fast, but requires more time and effort than the electromagnetic profiling method. The other mentioned advantages of electromagnetic profiling are that they are non-destructive, fast, accurate and with no direct contact made with the pollution. Compared to the electromagnetical profiling method, they are lateral, depth variations of resistivity are investigated, and the apparent measured resistivities can be interpreted into true formation resistivities.

Borehole Logging

The investigation of the conductivity (resistivity) carried out with borehole logging delivers information on the vertical distribution on the conductivity (resistivity) in the groundwater reservoir. Changes in lithology result in an increase or decrease of the conductivity (resistivity) which leads to the deduction of the structure of the groundwater reservoir. Generally, the coarser the sediment, the lower the conductivity (Williams et al. 1993). Unfortunately, when the reservoir is saturated with polluted or salt groundwater, it is impossible to deduce the structure. In such a case, the formation conductivity (resistivity) will provide information on the degree of pollution/salinisation, while lithological information has to be obtained using another method. Therefore, the professional probes for measurements include a unit to measure the natural-gamma radiation which is indispensable to distinguish the influences of the lithology and the pollution on the formation conductivity (resistivity). The natural-gamma radiation provides consistent information about the lithology, while in the case of pollution; the delimitation of the pollution will rely on the conductivity (resistivity).

Resistivity well logging is performed in open boreholes with four electrodes arranged in a normal array; downhole are placed one current electrode and one potential electrode, while the second current and potential electrodes are placed at an 'infinite' distance. The measuring principle is the same as for surface resistivity prospection. Borehole logging of the conductivity measured by means of electromagnetic induction is based on the same principle as the electromagnetic profiling method: a transmitter and a receiver are assembled on the axis of the probe in a vertical mode. This method has the supplementary advantage that it is applicable in plastic cased piezometers, which is not the case for resistivity measurements.

Application in Soil and Groundwater Pollution Studies

Detection and Delimitation of the Pollution

Particularly in industrial areas, there is a reasonable risk that the groundwater reservoir is polluted accidentally due to spills or leakages. Unfortunately, some pollution incidents are deliberately caused (illegal dumping or discharges) and it will be more difficult to trace the source of pollution when the discharge is underground. It is a controversial issue as to identifying the owner of the waste but initially the source of the pollution needs to be traced. Therefore, it is advisable to implement the electromagnetic profiling method because of its fast exploration capacities. Based on these results, areas with increased conductivity can be deduced and correspond most likely with pollution sites. In order to analyse the anomalies in more detail, it is recommended to use geo-electrical tomography along selected profiles. This will result in the vertical delimitation and the quantification of the true resistivity, from which the degree of pollution can be deduced.

Some results are presented here from geophysical investigations to illustrate how a source can be traced and how to perform the delimitation of the pollution.

Site 1

A former storage of salt used for road de-icing was located at site 1. These activities stopped several years ago, and the remaining storage and buildings have since been removed. The plot is now unused but the adjacent plots are in use for agriculture. The principal aim of the geophysical investigation was to find and delimitate the pollution by using the electromagnetic profiling method; with second phase geo-electrical tomography. The extension of the pollution has been correlated with the available analysed water samples at the water table. Figure 12.1 gives the location of the different profiles for the electromagnetic profiling and the geo-electrical tomography.

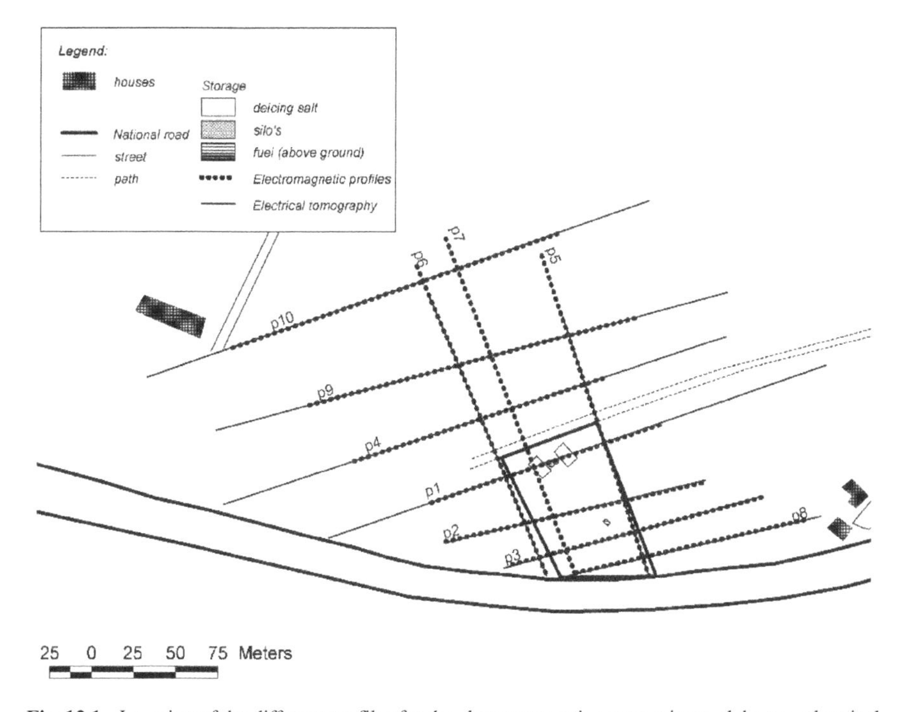

Fig. 12.1 Location of the different profiles for the electromagnetic prospection and the geo-electrical investigation

Electromagnetic Profiling

Along several lines, electromagnetic prospection (horizontal dipole mode) was performed by means of the EM34-3 instrument. With increasing intercoil spacing (S), the exploration depth increases. Along each profile the intercoil spacing was 10, 20 and 40 m. Exploration depths of ca. 7.5, 15 and 30 m were reached. For S = 10 m, the unsaturated zone up to (and including) the water table was reached. With a distance of S = 40 m, the entire groundwater reservoir was investigated.

From the negligible variation of the measured values for each intercoil spacing along profile 10 (Figure 12.2), the natural background conductivity for this study area is deduced. A conductivity above 20 mS/m indicated the occurrence of pollution for S = 10 m and S = 20 m. For S = 40 m, the natural background conductivity was below 30 mS/m. At the beginning of the profile for S = 40 m lower values were noticed, which are due to the presence of high voltage cables. Some metal accessories for cultivating chicory and a fence are responsible for the lower measured conductivity in the middle of the profile for S = 10 m.

Along profile 1 (Figure 12.3), through the central part of the study area, the highest conductivities (>60 mS/m) are measured with spacing S = 10 m. The increase in conductivity is a measure of the presence of the pollution, in this case, a dissolved salt. The absence of a significant increase in conductivity with S = 40 m is probably related to the absence of the pollution at greater depth. If the pollution had occurred up to the base of the groundwater reservoir, an increase in conductivity would be expected, which is not the case. So, it is suggested that the pollution only occurs in the unsaturated soil and in the upper part of the groundwater reservoir.

Once the measured values of the conductivity gathered with one intercoil spacing are plotted on a map, isolines of the conductivity can be drawn, and the maximum extension of the pollution can be deduced. This is illustrated on Figures 12.4 and 12.5 for the intercoil spacings of S = 10 and S = 40 m. Both figures indicate two sources of pollution. One source is located at the former storage of salt, which is expected, while a second one occurs in the southern part of the area, close to the

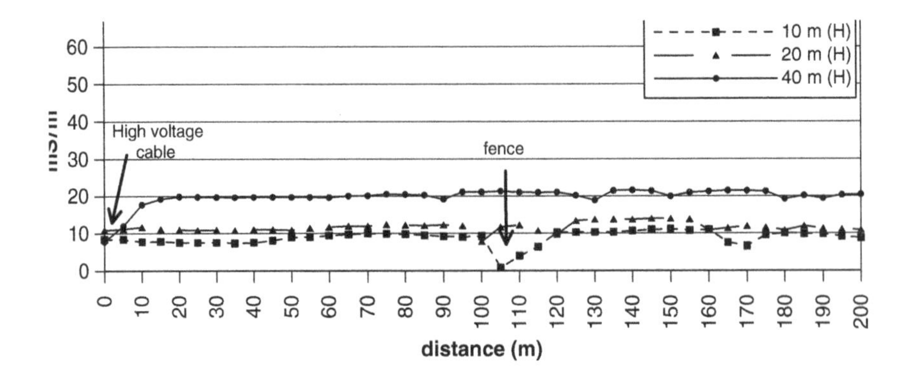

Fig. 12.2 Electromagnetic measurements with intercoil spacing of 10, 20 and 40 m along P10 (SW-NE oriented)

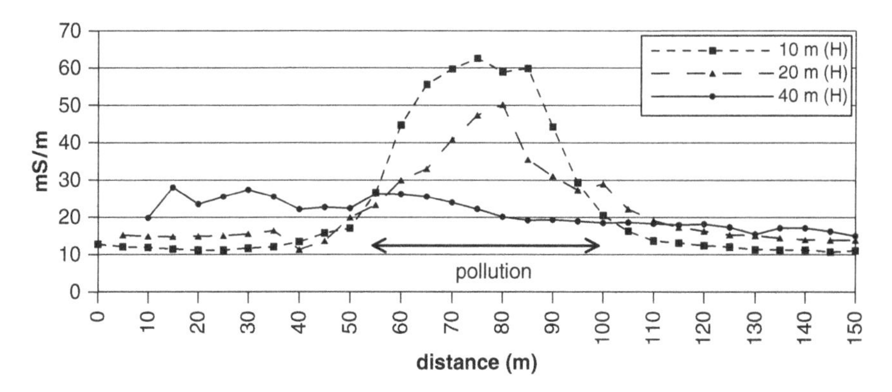

Fig. 12.3 Electromagnetic measurements with intercoil spacing of 10, 20 and 40 m along P1 (SW-NE oriented)

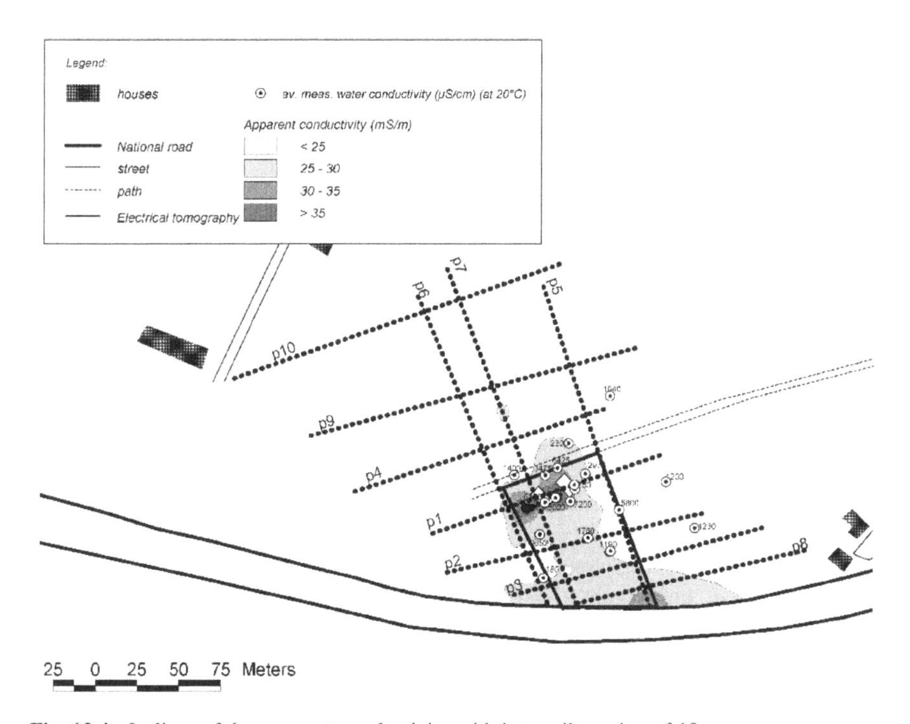

Fig. 12.4 Isolines of the apparent conductivity with intercoil spacing of 10 m

street. There are no significant indications that this pollution plume is due to storage or discharge, but rather due to transport of the de-icing salt.

Comparing Figure 12.4 to Figure 12.5, a displacement to the northwest of the sources can be observed, which is related to the groundwater flow (Martens et al. 2003). For S = 40 m, the highest conductivity can be observed northwest of the former storage, indicating that there the pollution probably has reached the base of the groundwater reservoir, due to gravity-driven flow, salt water being denser than fresh water.

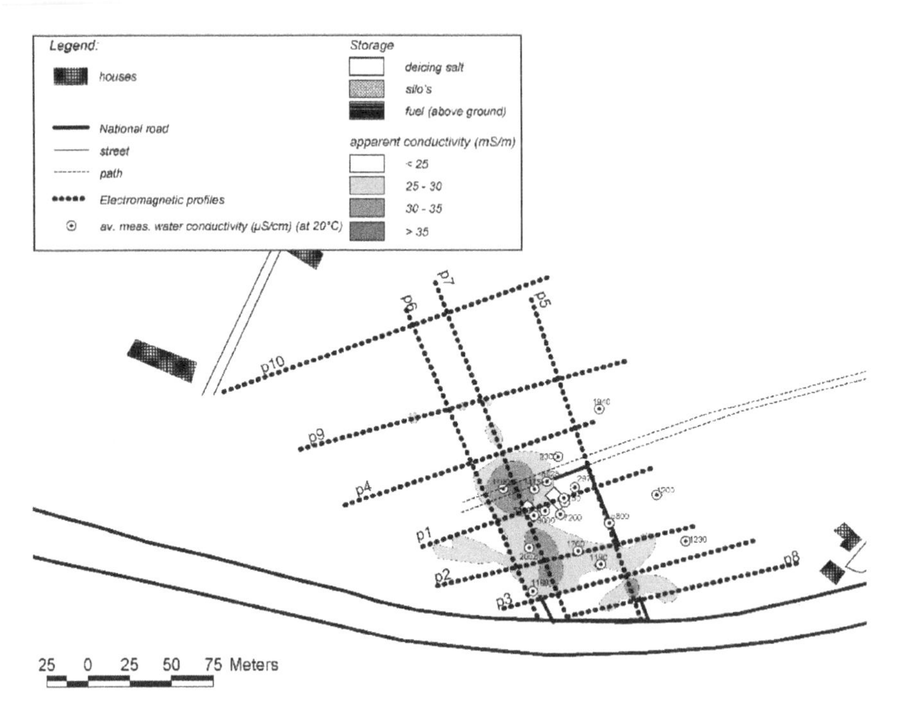

Fig. 12.5 Isolines of the apparent conductivity with intercoil spacing of 40 m

Geo-Electrical Tomography

Along the same profiles as with the electromagnetic profiling method, the apparent resistivity of the groundwater reservoir has been measured by means of the multi-electrode system. The measurements were done with the Wenner array. The distance between the electrodes is 5 or 10 m, depending on the available length along the profile. By automatically addressing a combination of four electrodes from the available 32 electrodes, the total length of the profiles is 155 or 310 m. The interpretations of the electrical tomography, by means of the program RES2DINV (Loke and Barker 1996; Loke 2002), permit characterisation of the subsoil both in a vertical and horizontal direction (Figure 12.6).

Profile 1, at the central part of the study area, has been carried out with both distances between the electrodes (5 and 10 m) retaining the same centre. The measurements carried out with a distance of 5 m between the electrodes, gave a detailed result of the upper part of the groundwater reservoir, even though the base of the groundwater reservoir at a depth of 31 m could not be delimited. The latter is deduced from the results of the tomography with a distance of 10 m.

Profile 10, presented in Figure 12.6, is representative of the reference situation. The natural background resistivity of the three different entities can be defined and is also marked on the profiles: the unsaturated zone (a resistivity of minimum 90 ohm), a phreatic groundwater reservoir where the resistivity decreases with depth, and an

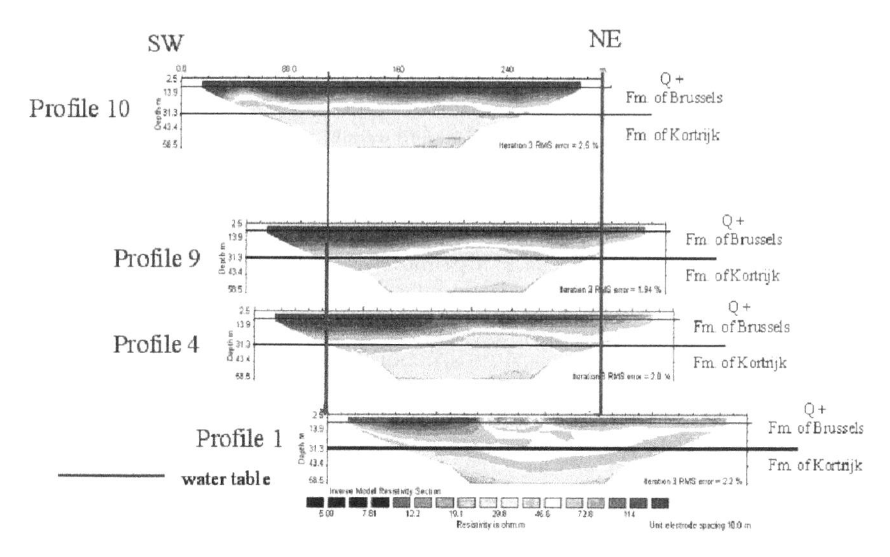

Fig. 12.6 Correlation between four parallel profiles (distance of the electrodes = 10 m) from profile 1 to profile 10

aquitard (maximum 30 ohm). The groundwater reservoir, consisting of fine to coarse sand with cemented banks, has an average resistivity of 70 ohm and is delimited at a depth of about 31 m with sandy clay which has a resistivity of less than 30 ohm.

The pollution in the groundwater reservoir causes a decrease of the resistivity as noticeable in profile 1: a resistivity of less than 30 ohm in the unsaturated zone and in the upper part of the groundwater reservoir. From profile 1 to profile 10, along with groundwater flow, it is clear that the pollution has moved from the unsaturated zone towards the base of the groundwater reservoir (profiles 4 and 9). These results confirm the prediction made by the EM-profiling and are additional information compared with the results of former studies, in which only the top of the groundwater reservoir has been examined.

Validation

During previous studies, the conductivity of the groundwater at the water table has been measured several times. The average measured conductivity at the water table is plotted in Figures 12.4 and 12.5 together with the results of the electromagnetic investigations. Unfortunately, there are no water samples available from the deeper part of the aquifer. The horizontal extension of the pollution derived with the electromagnetic investigation with intercoil spacing of 10 m, corresponds very well with the observed conductivity at the water table in the piezometers. The highest values are situated in the vicinity of the former salt storage area. In the southern part, the results of the electromagnetic profiling and the geo-electrical investigations by means of the multi-electrode system cannot be verified because of the absence of piezometers.

Conclusions

Electromagnetic profiling and geo-electrical tomography in this study area revealed two sources of pollution close to the street, and a displacement of the pollution with the groundwater flow up to the basement of the groundwater reservoir. In the former studies using the traditional method of soil and groundwater investigation, whereby shallow piezometers were installed and analysed without a preceding geophysical survey, only one source of pollution was traced and only the pollution at the top of the groundwater reservoir was detected. This case study illustrates that it is highly recommended to investigate also the migration towards the base of the groundwater reservoir. Geophysical prospection has been very useful in reaching this objective.

Site 2

This example is situated at an industrial site. Extended geophysical and hydrogeological investigations have been carried out, but in this chapter only the results of the borehole loggings will be quoted as an example illustrating their advantages.

Borehole loggings were performed in several borings, but only two are illustrated here. The resistivity measured by means of the short normal (SN) and long normal (LN) resistivity log and the natural-gamma radiation log for GB2 and GB1 are given in Figures 12.7 and 8 respectively. Both the natural gamma radiation and the resistivity loggings show the contrasting lithologies for GB2, located in the unpolluted area. At boring GB1, located in the polluted area, from the top to the bottom, the resistivity is very low (<10 ohm) and the variation in resistivity is hardly noticeable, suggesting either a thick homogeneous silt or clay layer, or a totally contaminated aquifer. The natural gamma radiation log of GB1 indicates different lithological units which are similar to the ones observed at GB2. It can thus be deduced that, at GB1, the whole groundwater reservoir is equally polluted. Pollution is clearly expressed by the low resistivities (Mack 1993; Williams et al. 1993; Walraevens et al. 1997).

The interpretation of resistivity measures always should be done in combination with the interpretation of natural gamma radiation (Walraevens et al. 1997) which is also illustrated by this example.

Site 3

The third case-study is concerned with a dump site in a former sand quarry. The dump was licensed to receive materials, which should not influence the subsoil and the groundwater. It was suspected that, in the eighties, 500,000 tonnes of heavily polluted chemical waste was dumped illegally on the site. The pollution was suspected

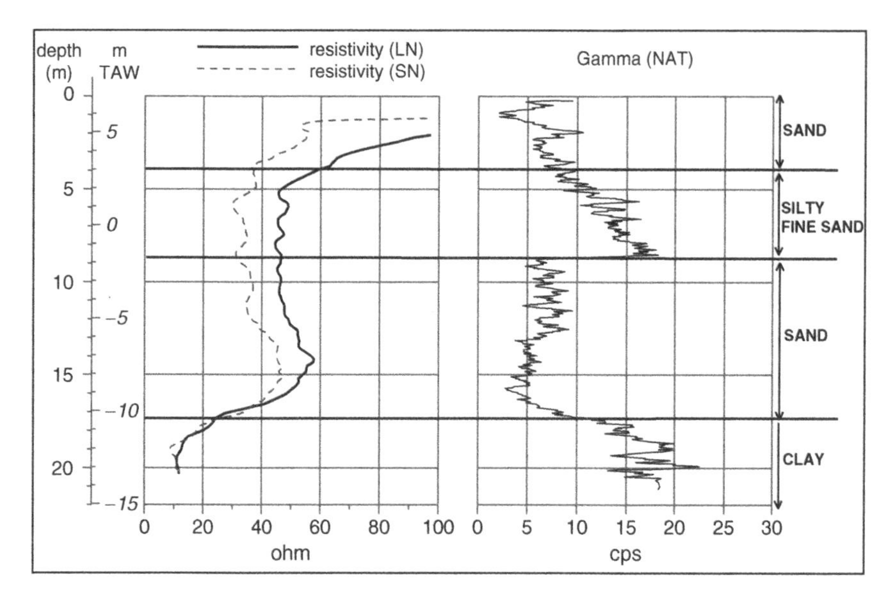

Fig. 12.7 Results of well logging in bore hole GB2

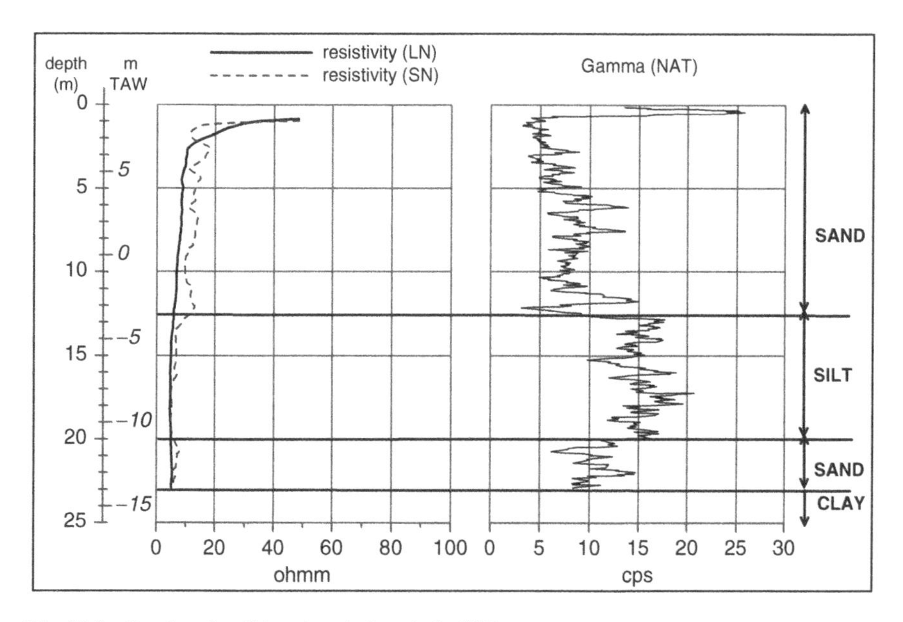

Fig. 12.8 Results of well loggings in bore hole GB1

to have left the dump and to have migrated with groundwater flow towards a rivulet. Along this rivulet, the most densely inhabited street of the village was situated. Allegations were made that cancer deaths and genetic disorders in people living in the street were caused by the effects of the illegally dumped waste.

The case was taken to court and, in the framework of an expert inquiry, an electromagnetic survey with an EM34-3 was performed, accompanied by well logging in new and existing drill holes. Figure 12.9 shows the survey lines and the apparent conductivity contour map derived from the measurements. From this map, it can be concluded that the background apparent conductivity values are less than 15 mS/m. Within the dump site, marked increases in apparent conductivity (up to over 150 mS/m) can be deduced, which indicate the presence of the important thicknesses of waste material. These high conductivities are contributed by the type of the waste on one side (e.g. metal scoria) and by the presence of leachate on the other hand. The central lower conductivities (60 to 90 mS/cm) correspond to the former quarry road. Based on the results of the contour conductivity map, the location of three new boreholes (B7, B8 and B9) (Figure 12.9) executed through the base of the dump has been deduced. This allowed confirmation that the license was not complied with: instead of the prescribed thickness of a minimum of one metre of clay material, in two drillings half a metre of loam was found instead (taken from the Quaternary loam cover found locally); in one drilling there was no lower-permeability liner at all, and the waste was directly overlying the highly permeable sand. Moreover, the EM mapping allowed a conclusion to be made that the high-conductivity leachate had left the dump at several positions, travelling to the west, in the direction of the rivulet. Further down flow, however, the pollution plume could not be followed, due to difficult terrain conditions (up to 20 m of unsaturated zone in medium to coarse sand; strongly sloping topography) and anthropogenic disturbing factors (power lines, electrical cables, pipelines and metal fences).

Fig. 12.9 Ground conductivity at a subsurface industrial dump site, mapped based on electromagnetic prospection with S = 20 m (Adapted from Walraevens et al. 2005)

Monitoring of Remediation

Once the pollution had been determined and the remediation commenced, it was recommended to monitor the evolution of the remediation. Also for this purpose, geophysical methods can be very usefully applied. If the pollution has a significant impact on the conductivity (resistivity), the evolution of the remediation will be noticeable. The conductivity (resistivity) should slowly return to the natural background conductivity. The monitoring should consist of measurements at different time steps during the period of remediation. The approach to use geophysical investigations in monitoring remediation has just begun; it is recommended to focus more on such methods, as the results are promising.

Conclusions

Geophysical investigation is a useful tool to trace and delimitate pollution. A great advantage of surface geophysical methods applied in soil pollution studies is avoidance of direct contact with the pollution, resulting in a reduction of associated health risks. These methods are non-destructive and fairly rapid. The interpretation of the geophysical results should be integrated in a standard hydrogeological investigation. The electromagnetic profiling method and the geo-electrical tomography make it possible to locate the pollution and deduce the likely source of it. Also conductivity/resistivity measurements in boreholes are useful to delimit the pollution in the vicinity of the borehole as long as these results are used in combination with natural gamma ray logging. The case studies shown here illustrate that geophysical investigation should be done prior to drilling activities for soil sampling and the installation of piezometers for groundwater sampling. Once the source of pollution and the extension are defined by geophysical investigation, the installation of piezometers will be more effective. Next, the pollutant causing the contamination should be identified, which needs to be done by analysing soil and/or groundwater. This might provide a 'fingerprint,' allowing the offender to be traced. Geophysical investigation may, in this sense, serve as a tool to confirm or to reject allegations.

References

Greenhouse JP and Harris RD (1983). Migration of contaminants at a landfill: a case study 7.DC, VLF, and inductive resistivity surveys. Journal of Hydrology 63:177–197.

Greenhouse JP and Slaine DD (1986). Geophysical modeling and mapping of contaminated groundwater around three waste disposal sites in southern Ontario. Canadian Geotechnology Journal 23:372–384.

Hoekstra P, Lahti R, Hild J, Bates CR and Phillips D (1992). Case histories of shallow time domain electromagnetics in environmental site assessment. Ground Water Monitoring Review (12)4:110–117.

Kaya M., Özürlan G and Sengül E (2007). Delineation of soil and groundwater contamination using geophysical methods at a waste disposal site in Canakkale, Turkey. Environmental Monitoring and Assessment 135:441–446.

Kayabali K, Yüksel FA and Yeken T (1998). Integrated use of hydrochemistry and resistivity methods in groundwater contamination caused by a recently closed solid waste site. Environmental Geology 36:227–234.

Loke MH (2002). RES2DINV ver.3.50. Rapid 2-D resistivity and IP inversion using the least-squares method. Wenner (α, β, γ), dipole–dipole, inline pole–pole, pole–dipole, equatorial dipole–dipole, Schlumberger and non-conventional arrays. On land, underwater and cross-borehole surveys. Geoelectrical Imaging 2-D and 3-D. Geotomo Software. Malaysia. 115p.

Loke MH and Barker RD (1996). Rapid least-squares inversion of apparent resistivity pseudosections using a quasi-Newton method. Geophysical Prospecting 44:31–152.

Mack T J (1993). Detection of contaminant plumes by borehole geophysical logging. Ground Water Monitoring Review 13:107–114.

Martens K, Beeuwsaert E and Walraevens K (2003). Geo-electrical tomography in the framework of soil investigation (in Dutch). Laboratory for Applied Geology and Hydrogeology. Ghent University. p.36 + annexes.

McNeil JD (1980). Electromagnetic terrain conductivity measurement at low induction numbers. Technical Note TN-6. Geonics Ltd., Ontario, Canada.

Monteiro Santos FA, Mateus A, Figueiras J and Gonçalves MA (2006). Mapping groundwater contamination around a landfill facility using the VLF-EM method – A case study. Journal of Applied Geophysics 60:115–125.

Nobes DC, Armstrong MJ and Close ME (2000). Delineation of a landfill leachate plume and flow channels in coastal sands near Christchurch, New Zealand, using a shallow electromagnetic survey method. Hydrogeology Journal 8:328–336.

Stewart M and Bretnall R (1986). Interpretation of VLF resistivity data for ground water contamination surveys. Ground Water Monitoring Review 6:71–75.

Walraevens K, Beeuwsaert E and De Breuck W (1997). Geophysical methods for prospecting industrial pollution: a case study. European Journal of Environmental and Engineering Geophysics 2:95–108.

Walraevens K, Coetsiers M and Martens K (2005). Large-scale mapping of soil and groundwater pollution to quantify pollution spreading. In: Soil and Sediment Remediation. Mechanisms, Technologies and Applications (Eds. P Lens, T Grotenhuis, G Malina and H Tabak), pp. 37–48. Integrated Environmental Technology Series. IWA, London.

Williams JH, Lapham WW and Barringer TH (1993). Application of electromagnetic logging to contamination investigations in glacial sand-and-gravel aquifers. Ground Water Monitoring Reviews 13:129–138.

Geoforensics

Chapter 13
Locating Concealed Homicide Victims: Developing the Role of Geoforensics

Mark Harrison and Laurance J. Donnelly

Abstract Historically police searches for homicide victims' graves have been undertaken by the use of large numbers of police, military and public volunteers, conducting visual or manual probe line searches covering formalised gridded sectored areas. Speculative digging of large areas of ground has also been employed with variable success. Mineral exploration geologists, engineering geologists and geohazards specialists traditionally investigate the ground using a range of methods and techniques. Before such ground investigations are undertaken, a conceptual geological model of the ground is developed. This provides information on, for example: tectonic setting, stratigraphy, lithology, structure, hydrogeology, hydrology, groundwater, hydrochemistry, superficial deposits, principal soil types, depth to bedrock, nature of bedrock interface, engineering and physical properties of the ground, geomorphological processes, mining, past land use, current land use, geological hazards and man's influences. In a similar way, the properties of a buried or concealed body may also be determined and how in particular these have influenced the geology. This provides estimates of the target's age, size, and geometry, expected depth of burial, time and duration of burial, state of preservation or decomposition, physical, chemical, hydrogeological and geotechnical variations compared to the surrounding ground. An understating of the undisturbed (pre-burial) and disturbed (post-burial) geology and the target (body and associated objects) properties are crucial before the correct search strategy and choice of instrumentation may be decided, and the optimum method of deployment identified. These may include

M. Harrison
National Policing Improvement Agency, Wyboston Lakes,
Great North Road, Wyboston, Bedfordshire MK443AL and School of Ocean and Earth Science,
University of Southampton, Waterfront Campus,
European Way, Southampton SO14 3ZH, UK

L.J. Donnelly(✉)
Halcrow Group Ltd., Deanway Technology Centre, Wilmslow Road, Handforth,
Cheshire, SK9 3FB, UK
e-mail: DonnellyLJ@Halcrow.com

K. Ritz et al. (eds.), *Criminal and Environmental Soil Forensics*

geophysics, geochemistry, satellite imagery, air photo interpretation and invasive methods (such as auguring, drilling, trial pitting and trenching). Geological investigative techniques are applicable to law enforcement searches, since the underlying search philosophy, concepts and principles are similar. That is, there is a buried/concealed 'object' or 'target' desirable to be found. The most important services a geologist can give the police and a law enforcement search strategist are: the production of a geological model of a potential grave site, an understanding of the geological and geomorphological processes, the characterisation and understanding of the origin, source and properties of the soils, rocks and target (body), and a choice of detecting methods. For the geologist (and other subject matter experts) to be effectively incorporated into a search team, he/she must be an effective communicator of complex geological (scientific) terminology, recognise the limitations of his/her skills and capabilities and be aware of the boundaries and interface with other subject matter forensic experts. The principal objective of this chapter is to describe the effective and efficient processes to locate concealed victims of homicide. It also seeks to show how the combined skills and expertise of law enforcement and geoforensic search specialists enable the ground in the vicinity of homicide graves to be better understood and more professionally searched.

Introduction

Searching for Homicide Graves: Law Enforcement Perspective

There are four core functions that can describe all law enforcement activity, namely investigation, interview, search and recovery. Law enforcement personnel throughout the world are usually well-trained and experienced in these functions. However, outwith the UK search techniques are less established, with little or no uniform training and framework models. This chapter is primarily concerned with 'search' and to a lesser degree 'recovery'.

To the casual observer, forensic technicians, or Crime Scene Investigators (CSI) as popularised by the media, appear to fulfil the role of search. However, on closer scrutiny they are trained to undertake the role of examining the crime scene and collecting evidence found. For instance, where a forensic technician uses a light source in a dwelling to detect blood, or the dusting of a window for fingerprints, these 'could' be described as 'searches', but are perhaps more appropriately 'examinations'. In these examples, the investigation is conducted at crime scenes to examine and recover evidence with forensic integrity to enable laboratory analysis. It is the 'search-to-locate' aspect of law enforcement that presents the greatest challenge. Without locating the evidence, or the victim's body, rarely will a conviction succeed and justice be served however satisfactorily the investigation has been conducted, and even if a confession has been obtained from the suspect. With regard to relatives and friends of a victim, the search to locate, and the recovery of, their loved one is the most important function that law enforcement can perform, with a conviction being of secondary concern.

A recognised UK law enforcement definition of search is defined as 'The application and management of systematic procedures and appropriate detection equipment to locate specified targets' (Harrison et al. 2006). It is the skill of 'looking' for a specific object and the art of 'finding'. Search requires the application of systems, combined with appropriate expertise, an understanding of the ground conditions (i.e. the geology) and the deployment of detection equipment, to locate a specific item.

In a law enforcement context, searching is for the prevention and detection of crime, and location of missing persons and objects. A search can be described as 'offensive/detective' or 'defensive/protective'. An offensive search is when evidence is sought where an event/crime has occurred, the aim of which would be to locate a specific item, obtain evidence and/or intelligence, or to restrict a suspect/offender's ability to continue criminal activities. A defensive search is proactive and preventative, to enable freedom of access and movement by the public, confirming that no evidence, item, or person is present within an area or location.

Law enforcement searching has five core objectives:

1. *Locating evidence to support a prosecution*: Following discovery of a crime, search activity will require appropriate legal authority and documentation, forensic awareness and continuity of evidence to assist in prosecuting those acting outside the law.
2. *Gathering intelligence*: As a by-product, search activity can gather intelligence that is useful for law enforcement. It may reveal an offender's *modus operandi*, assisting investigators in detecting further crimes.
3. *Depriving the criminal of their resources and opportunity*: Locating criminal resources and assets frustrates the offender's plans and limits opportunities to progress their intentions.
4. *Locating missing persons*: Law enforcement is responsible for co-ordinating searches for people that go missing, either in a non suspicious or suspicious context.
5. *Protecting potential targets*: A defensive/protective search may be part of a pre-planned event, providing security for a specified group, individual at an event, or for the protection of the event itself. Other instances may include safeguarding disparate actions from each other and specific operations designed to protect society, or certain sections identified as being at particular risk.

Searching for Homicide Graves: The Geological Perspective

Geoforensics (known also as forensic geology and forensic geoscience) is a specialist branch of geology that is concerned with the applications of geological sciences to criminal (domestic, international terrorist, humanitarian, environmental,fraudulent) investigations of what happened, where and when it occurred, and how and why it took place. Forensic geologists, therefore, need not limit their investigations to law enforcement (e.g. Murray 2004; Pye and Croft 2004; Ruffell and Mckinley 2008).

For the purposes of this chapter forensic geoscientists may be broadly divided into the two following principal fields, depending on their skills, expertise and capabilities:

1. Laboratory-based geoscientists, including geochemists, mineralogists, petrologists, micro-palaeontologists and isotope specialists. These provide physical (trace) evidence for use in court, assist investigations providing intelligence or identifying the location of a crime scene. Geoscientists also assist the police link an offender (or object) to the scene, link the victim to an offender, provide intelligence and information to assist and investigation and provide information capable of being used as physical evidence to implicate or eliminate a potential offender.
2. Field-based geoscientists, whose skills in exploration (e.g. geophysics, geochemistry, geomorphology, hydrogeology, environmental geology, remote sensing and geotechnics) are used to locate homicide victim's graves, weapons and other buried or concealed objects, and possibly assist with the recovery of such objects.

As this paper is primarily concerned with the search for victims of homicide, it is the field-based geoscientist who will be considered further. The most valuable contribution a geologist may provide for an investigator or search strategist is a detailed and thorough understanding of the ground conditions, that is 'to get the geology right'. In doing so, geologists provide an understanding of places of possible burial, diggability and geomorphological processes, which could have influenced selection of a grave site, or changes that have taken place at the grave site since burial (such as erosion and weathering), and to the identify methods and techniques suitable to locate the grave (Donnelly 2006).

The Search for a Grave

Historical Overview

Traditional police methods of finding graves often involves large-scale gridded areas with personnel 'finger-tip/line searches' and 'trial-and-error' excavations. These are inefficient, labour intensive, may destroy evidence and ignore subtle ground (geological) disturbances. They may be supported by non-specialists, such as public volunteers and local interest groups. However, the effectiveness and cost of large numbers of public volunteers, or military infantry forming lines and walking the grids across open land or woodland, must be weighed against their effectiveness in locating a concealed sub-surface burial (Boyd 1979).

Search Objectives

The search for a homicide victim's grave is one for, rather than of, the crime scene. The aim is to progress the investigation by locating the victim using an offensive/

detective search procedure to obtain evidence for a prosecution, gaining further intelligence, and locating the remains of the victim. The objective of the search is not to recover the victim.

The principle asset of any search is an experienced, trained, searcher working to a pre-prepared plan using recognised, proven techniques and capabilities. They can call upon a range of other assets to ensure that the search is effective and efficient. The choice of assets is crucial and often can only be undertaken following consultation with multidisciplinary search specialists. Financial, technical and/or logistical constraints or availability of specialist resource/equipment will be limiting factors which require careful consideration. The search strategist will decide on the most cost effective way to achieve the minimum standard (resolution) required for a high probability of search success. A pragmatic balance is required between a minimum acceptable standard and minimal expenditure when conducting a search.

Search Scale

Homicide searching occurs across a range of scales, geographical settings and timeframes, from small individual dwellings to extensive tracts of remote moorland. Some searches may take an afternoon whereas others may take several weeks and, in some instances, years. In general, searching tends to evolve from the macro-sized to the micro and from the non-invasive to the invasive. The choice of techniques and search methodology needs to consider the preservation of the crime scene and reduce the possibility for any cross contamination and destruction of evidence. The size of the grave relative to the size of the search area will also influence the choice of search technique with respect to detection resolution. Conventional geological exploration and some site investigation techniques and methodologies are applicable to these situations.

Search Philosophy

Searches for victims of homicide and missing persons may be categorised into search and rescue, scenario based, feature focused, intelligence led and systematic Standard Operating Procedure (SOP). Each is discussed further below.

Search and Rescue

During search and rescue, or 'search to rescue', for missing persons, subjective calculations are made in selecting an area to search. These rely on a group discussion and a map, to set the radius of the search area, usually from the last place the missing person was seen or the last known location of evidence associated with that person. Emphasis is placed on the correlation of time and distance. The general hypothesis is that, if the person is mobile, they will travel further as each hour passes; therefore the search area can be geographically large. The search managers will split the area into sectors, the size of which are determined by the terrain, resources and time available, and subsequently submit each sector to a subjective calculation of the probability of the person being sought in each sector. The vulnerability of the 'search and rescue' process is the subjectivity of the calculations being made.

Search and rescue techniques should be employed for persons believed to be lost, or injured, and are active in their self-discovery and recovery (i.e. the person wishes to be found). It may be an appropriate and proportionate response to deploy such techniques during initial phases of an investigation for a missing person. However, once this proves negative, or further information is provided (e.g. a possible suicide or homicide), scenario-based searching should be utilised with its appropriate focus on victimology and offender behaviour profiling.

Scenario-Based Searching

The process of scenario-based searching differs significantly from 'search and rescue' theory. Scenario-based searches are conducted using available intelligence and behavioural information, requiring the investigator to generate hypotheses that could account for the disappearance of the victim and the body disposal method chosen by the offender. By 'profiling' victim and offender behaviours, credible options for sites for disposal of a body can be determined. Forensic clinical psychologists and appropriately trained law enforcement personnel can assist the search strategist in understanding the capability, capacity, motive and resourcefulness of the offender with respect to the disposal of the victim's body.

Often the challenge for investigators is whether the disappearance is voluntary or not, and whether the person has been a victim of crime. Often a suspect will allege a victim has, for example, left for a better life elsewhere or committed suicide. Completing a full victimology profile of the missing person, and comparison with datasets and research relevant to suicide scenarios, the investigator can establish a degree of predictability as to the method of death. Factors such as the likely distance travelled, or suitability of a location for suicide or body disposal can be considered by the search strategist in their deliberations. The benefit of such an approach is that specific hypotheses regarding the person's disappearance are tested following a logical process, removing 'investigator's gut feeling' and challenging the search strategist to select the most appropriate resources to detect the person being sought. For example, if such a profile suggested that the person was more likely to commit suicide by drowning, the focus should be in the lake in the search area, and not woodland. The process assists the investigation gravitate to scenarios in an escalating and proportionate manner.

The search for a homicide victim differs significantly from traditional search and rescue techniques. The victim is not mobile, so there is no correlation of time and distance, nor will they be seeking their own discovery. An offender has chosen a location which may have no relevance to the victim, so the victim's last position may not be relevant. Furthermore, the offender may have stated they were the last to see the missing person in an attempt to deflect the search away from the victim's true location. Therefore, a search based around the victims last movements and local knowledge may be problematic, and the offender's selection of location for body disposal would be specific to their state of mind and knowledge of an area.

The scenario-based searching theory is based on physical 'features' of the ground, whereas traditional search and rescue theory focuses on 'area' searching. Careful consideration of all available intelligence, followed by detailed planning and management of the search, will enable the most appropriate choice of resources and search methodology to be identified.

Feature Focused

Feature searching enables the identification of physical landmarks that can be easily relocated by the offender, such as a dark rock amongst an area of light coloured rocks, a waterfall, or a characteristic weathered rock formation. The offender may have a particular reason to identify a landmark, for instance, a lake that is accessible and well known to him/her. Other landmarks may be chosen for their ease of access, relocation purposes or concealment. The focus is on searching key features within the search sector relevant to the search scenario, rather than searching the whole area within the sector.

Intelligence Led

No search undertaken by law enforcement should be of a speculative nature. It should be based on the currently known case facts, or intelligence that enables hypotheses generation. This will ensure that searches are based on logic and resources and finances can be justified. Furthermore, the deployment of all geophysical and other search techniques should be based on a sound scientific understanding of the geology and ground conditions.

Systematic Standard Operating Procedures

Standard Operating Procedures (SOPs) should be applied to all search techniques. These provide assurance of consistency throughout the area searched and enable peer or independent reviews to be made on the qualitative aspects of any search. The SOPs should be in written and descriptive form, and are part of the overall search strategy and documentation.

Behavioural and Geographical Characterisation Models Associated with Homicide Disposal

Attributes of Homicide Victim Disposal

An offender will choose the most suitable deposition site for a victim, key features of which are:

- *The principle of least effort*: The offender does the minimum needed to achieve their aim of concealment, prioritising possible locations for the bodies' disposal that afford the least effort, which may not always be the nearest location available to the offender. There are 'push-pull' factors to be considered regarding the benefits of the disposal site outweighing the additional distance to be travelled and thereby the risk of detection.
- *Familiar location*: The location chosen for the grave is likely to be previously known to, and visited by, the offender (i.e. we go where we know). The offender will have considered a plausible explanation for their presence at the location, should they be disturbed. There will be easy-to-navigate access and escape routes and it will be navigable in darkness, deposition being mainly a night time activity, although any return visits are more commonly in daylight. There will be clear, permanent, identifiable features/markers, as in choosing a site an offender selects permanent and prominent features by which to mark or navigate to the disposal site, for example, a large conspicuous boulder. There may be primary, secondary and tertiary markers.
- *Concealment*: The location will have a low witness potential to preclude observation and provide concealment from view, for example, in a ditch or gully. This is to enable the act of disposal to be concealed (in cover).

Body Disposal Site Survey

A body disposal site survey should be conducted prior to detailed searches. The geographical extent of the body disposal site should be cognisant of the access point and physical markers selected. Within the area being assessed, significant physical features should be identified as prominent, permanent markers for specific consideration and, where possible, enable 'feature based searching' to improve efficiency, such as searching around a single tree in a ploughed field. For such a site survey on land, it is essential that there is a hypothesis with regards the likely disposal method of the body. For example, has the body has been buried with clothing and artefacts? The survey considers the viability of burial with respect to the conceptual geological model and in particular wherever practicable, involves a diggability survey.

Development of a Conceptual Geological Model

As a concealed burial takes place in the ground, a conceptual geological model can be developed to provide information on a range of issues. Similarly, the properties of the target (e.g. a body or associated objects) may be determined and how these have been influenced by the surrounding ground. The conceptual geological model includes estimates of the target's age, size, and geometry, expected depth of burial, time and duration of burial, physical, chemical, hydrogeological and geotechnical variations compared to the surrounding ground (Donnelly 2002a). This model can then be used to identify the correct search strategy, appropriate choice of instrumentation, and optimum method of deployment, in collaboration with intelligence, victimology assessments and behavioural profiling information provided by law enforcement officers.

A conceptual model of a potential burial site summarises what is likely to be found and the condition of the target. Conceptual geological models are developed at the beginning of a search. It is a model to be tested, revised and tested again until it can be verified (at discovery) or proven otherwise and therefore abandoned (Figure 13.1). This model is based on the geologist's experience and from conducting investigations in comparable settings. High quality geological information

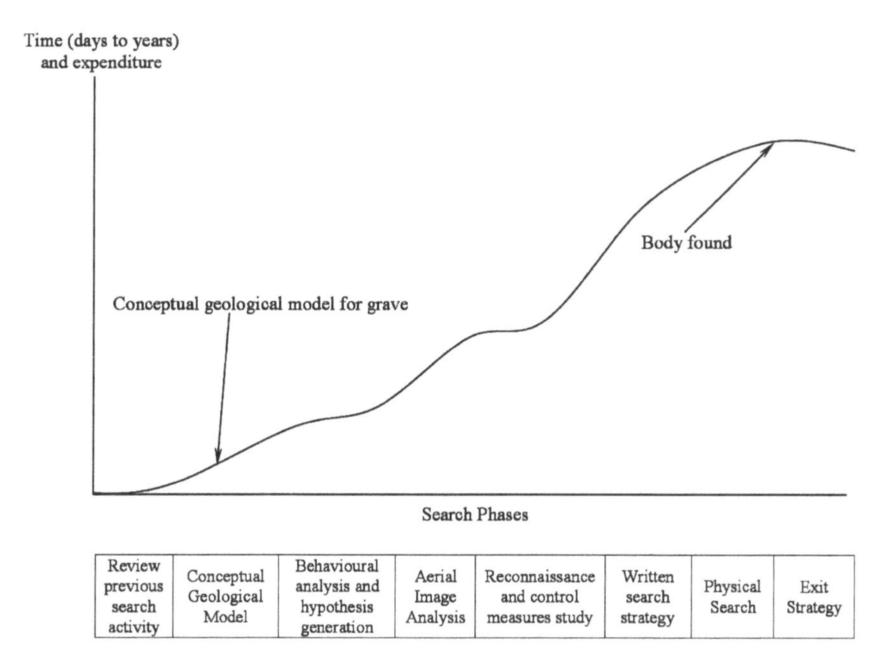

Fig. 13.1 A conceptual geological model for a grave is developed at the start of a search. This schematic diagram demonstrates the phases and time/expenditure in a search to locate a homicide victim, adopted and modified after a typical mineral exploration search programme. The duration, axiz values and morphology of the 'curve' will vary for each search. Modified from Donnelly 2002a.

(such as published geological maps, memoirs, papers and technical reports) will support a search effort with a high level of assurance. A weak geological model for the grave site will introduce uncertainty into the search, no matter how precise and accurate the subsequent exploration (search) techniques.

The development of a geological model for a victim of homicide, or a grave, requires a specific understanding of the natural (geological) ground conditions and how these have been influenced by the activities of the offender (e.g. digging, and subsequent reinstatement of the disturbed ground). At any one location there are likely to be a number of interactive, dynamic, active surface geological processes, which have affected the rocks, soil, groundwater and topography. These processes were active long before burial took place and are likely to have continued in the time which has passed since (Donnelly 2002a, b, c, 2003) (Figure 13.2).

To find a buried object it is important to understand the expected geology in the vicinity of the target. The geological model assists in predicting possible ground conditions and facilitates the choice of appropriate instruments for locating the target, thus reducing the risk of a common mistake made by law enforcement officers. That is, the deployment of a technique on the basis of its success at a previous

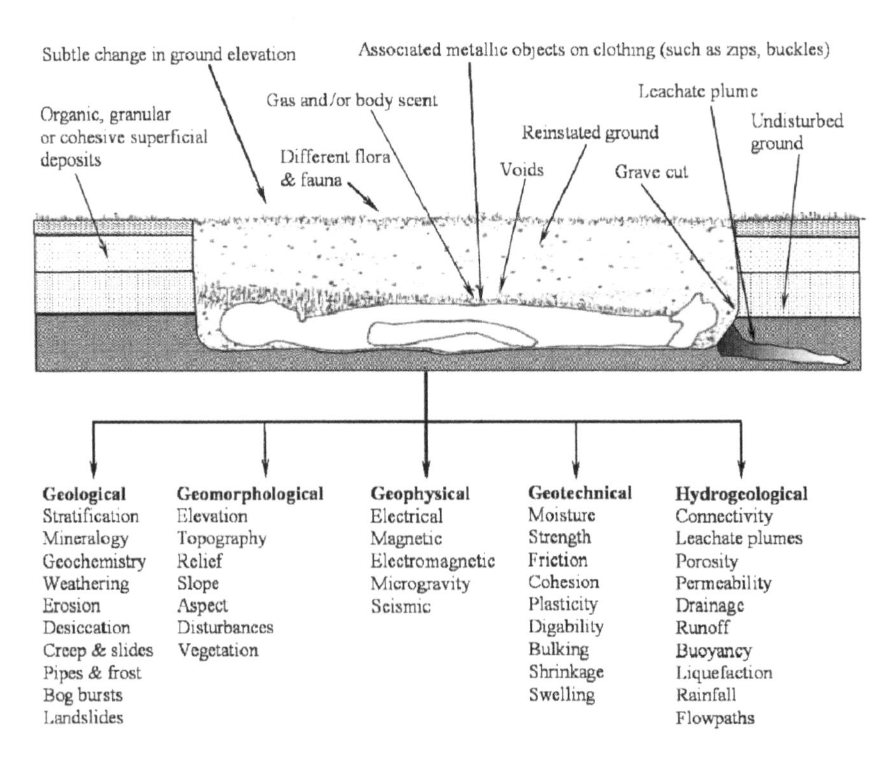

Fig. 13.2 An idealised conceptual geological model for a shallow homicide grave. The geological, geomorphological, geophysical, geotechnical and hydrogeology properties of the body, reinstated ground and undisturbed ground may change after burial. This type of model may assist in determining the most suitable suite of assets for conducting a search. This may include for example the deployment of geophysical surveys and specially trained cadaver dogs (after Donnelly 2002a, 2008)

search area (where the geology may have been different), without due consideration of the ground conditions and an assessment of the suitability of the technique.

The development of the geological model requires an initial desk study to obtain a general appreciation of rock and soil types, the influence of groundwater, accompanied by a reconnaissance (walk-over) survey to inspect and observe the ground conditions. Crime scene visits are crucial to make detailed, expert, observations of ground disturbance. This enables the geoscientist to decide if the observations are caused by natural events, biological activity or man-induced.

No single geological model suits all types of search area, and there is no single approach to producing a geological model, as each homicide case and search area will have unique characteristics. Due to the inherent complexity and variability of the ground (and grave sites) the actual geological conditions will only become apparent at the time of its excavation. However, the strength of the geological model is in providing an understanding of the geological and man-made processes which influenced the grave and the ground in its immediate vicinity. This provides a rational, scientific and objective basis for the deployment of measured and proportionate suitable search methodologies. The progressive accumulation of police intelligence and geological information allows the geological model for the grave site to evolve over the duration of the investigation.

For most of the United Kingdom, geological and pedological (soil) maps are available, generally at scales from 1:10,000 to 1:50,000 but not of sufficient detail to provide site specific information for search areas. Detailed geological and geomorphological mapping is therefore a pre-requisite before a site search. As well as recording the principal stratigraphy, superficial deposits and soils, these need to include accurate plotting and descriptions of topographic features, such as areas of ground disturbances (e.g. seeps, springs, compression, fissures, scarps, deposited debris, changes in slope profile).

Geological maps may also be used to reduce the geographical confinement of a search area. For example, during a search in Northern England, intelligence suggested the homicide victim was located in a grave with a sandy bottom. As the rocks which outcrop beneath the soil cover consisted of either coarse sandstone or shale, the areas where shale outcropped were eliminated as unlikely places to conceal the grave, as only weathered sand could occur beneath the soil cover where the sandstone outcropped. This geological information reduced the search area substantially, enabling resources to be focused in areas where sandstone occurred at shallow depths beneath the soil profile.

Diggability and Excavatability Survey

The method of excavation in rock is determined primarily by geology, mainly the engineering properties of the rock mass including the presence, geometry and type of discontinuities (such as joints, faults, bedding planes and schistocity), joint spacing and orientation, strength of the rock, presence and properties of fault gouge, groundwater, type and intensity of weathering and erosion. Typical methods of excavation in rock are digging or ripping (with the use of mechanical excavators) and blasting (with the use of explosives). In homicide searches such approaches to excavation may need to be considered if, for example, body disposal has taken place in a quarry, mine, pipeline trench or beneath concrete foundations. The excavatability of rock may be determined by a number of geotechnical methods involving assessments of the engineering properties of the rock mass (McCann and Fenning 1995).

More commonly, body disposal takes place in soil, superficial deposits, or softer rocks (such as shales and mud rocks). The ease with which the soil can be dug (i.e. its diggability) and placed back into grave (or reinstated) is of critical importance. The offender is likely to choose a site where the soil is sufficiently thick, and can be quickly dug then reinstated, with no or little surface indication that digging has taken. The diggability of soil depends on its geological properties such as intact strength, bulk density, natural water content, depth, weathering, proximity of the underlying bedrock, slope angle, groundwater, surface water, vegetation, stability of the walls upon excavation, bulking and swelling of the soil as well as the method of digging. There is no generally accepted quantitative measure of diggability, rather it is determined by *in situ* testing.

In situ diggability tests may be easily performed, before the main phase of the detailed survey (usually at the reconnaissance stage), involving either probing or digging using tools similar to those to which the offender is believed to have had access (Ruffell 2005a). This provides the opportunity to inspect soil structure and/or weathered bedrock and associated superficial deposits to determine whether it is granular (sand rich), cohesive (clay rich) or organic (peat). These observations are important as they have implications on the efficiency of burial and preservation of human remains depending on the time since burial (Cornwell 2002).

When soil is excavated it may increase in its bulk due to the decrease in density per unit volume and loss of inter-particulate moisture. The amount of bulking can be pre-determined, as in excess it may result in a slight increase in ground elevation, or soil distribution around the grave. Subsequent subsidence, or settlement, of the reinstated ground may cause a depression which could become filled with standing water or vegetation. Any mineralogical and geochemical differences in the sub-soil profile should be noted as these differences may be of value during exploration of the grave. For example, any excess sub-soil, distributed around the grave, may have characteristics (such as colour, consistency, grain size, structure, mineralogy) which could make it distinguishable form the rock, soil or vegetation on the ground surface.

A diggability survey will:

- Provide geological information on the soil/rock types as different geological processes operate in different places. For example, salt evaporation and the deposition of evaporite salts (mainly in arid climates), or the deposition of iron-pan deposits on a peat moorland may act as ground surface indicators if they are exposed or disturbed.
- Demonstrate the level of difficulty, and time required, for a shallow excavation, and the effective depth which can be achieved. Where there are bedrock outcrops

at, or close to, the surface, there may be insufficient soil cover for complete concealment, or the ground may contain perched water tables which inhibit digging and burial. Soil strength is also significant. If, for example, an excavation is dug on a beach above the inter-tidal zone, the excavation will be easily diggable due to the fine, dry sand. However, the sidewalls of the excavation may be susceptible to collapse, or 'running' conditions, which may hinder the burial of a corpse.
- Demonstrate how effectively the soil can be reinstated and what visible topographical features may exist, to indicate possible presence of a grave. These may include, for example, scarps, subsidence, fissures, ground compression or standing water, all characteristic signatures which may aid detection. However, these features need to be considered in conjunction with other geomorphological processes or biological activity which may generate similar features on the ground surface.
- Provide a prediction of the length of time it would take an offender to dispose of the body, which may be relevant where investigation has narrowed the time available for body deposition.

Once the assessment of diggability is complete a wider environmental assessment may be required, to include the surface vegetation to identify anomalous growth which may be accounted for by a buried human cadaver. It will also identify vegetation that has not been disturbed, and thus not the scene of any burial. Advanced surveys of this nature benefit from the accompaniment of, for example, an environmental geologist, palynologist or botanist.

Geophysical Investigations

Geophysical exploration and prospecting methods are widely used in mineral exploration, site investigation (CIRIA 2002; The Geological Society of London 1988; MacDougall et al. 2002; McCann et al. 1997) and archaeology (Gaffney et al. 2002). In the USA and UK, geophysical instruments have been successfully used for forensic purposes (Donnelly 2004; Fenning and Donnelly 2004; Hunter and Cox 2005; Killam 2004; Ruffell 2005b).

Several types of geophysical instruments are available to assist searches for missing and buried objects. The types of survey, methodology and interpretation depend upon several complex factors including: types of target buried (e.g. human remains, money, explosives, weapons); geological conditions; anticipated depth of burial; age of burial; and, experience and skill of the geoscientist. Geophysical surveys can be carried out by one person, or a team, be non-invasive or invasive, and may take hours to weeks to complete.

The assessment of geological data and maps, intelligence provided from the law enforcement officers, and a reconnaissance walk-over survey will be required before an approach is recommended for the full geophysical survey. Typical such surveys usually operate on a linear traverse, or a grid, with readings every 0.1 to 1.0 m, determined by the size and dimensions of the buried target, and prevailing topography and ground conditions. The use of automated data logging systems, computerised data processing, airborne surveys, and graphics technology combined with global positioning systems (GPS) have made it possible for relatively large areas of land to be investigated in a single day. The use of geophysical methods does not preclude the use of cadaver detecting canines and manual search methods; rather, they are complementary and supportive techniques.

The comparison of a geophysical interpretation with directly observed geological data is known as 'ground-truthing'. This enables the survey results to be extrapolated across areas where little, or no, ground truth information is available.

Geophysical Survey Types

Geophysical investigations are predominantly non-invasive for their operation and therefore adhere to the law enforcement preference of moving proportionately from the non-invasive to the invasive in forensic searches. This minimises evidential contamination and damage. The data obtained provides measures of vertical and lateral variation of the physical properties of the ground, such as electrical conductivity, microgravity, magnetic and electromagnetic properties of the geological materials or buried objects at a site. The conceptual geological model of the ground is used as a basis for data interpretation.

There are some common limitations which make search areas challenging for geophysical surveys. Principal amongst these are: presence of man-made metallic objects such as overhead power lines, utilities, sewers, gas mains, buildings, reinforced concrete and fences (i.e. geophysical noise); steep or irregular topography; human activities (such as digging, tipping, building, construction, farming and mining); electrical interferences (e.g. mobile phones, laptop computers, machinery, and power cables); seasonal variations in weather; access and logistical problems (e.g. trees, dense vegetation and areas of flooding); and public, media and compromised missions.

Geophysical Instruments

Geoscientists can adapt traditional methods and techniques, such as those used in mineral exploration, geohazard investigations and geotechnical engineering site investigations, to help locate graves and other buried objects. An overview of the main techniques follows.

Magnetic

The naked human body has virtually no associated magnetic anomaly and, when buried, is very unlikely to be detected by a magnetic survey. However, clothing may contain objects that can be detected such as metal buttons, zip fasteners, shoe eyelets and belt buckles, whilst pockets may contain spectacles, keys, coins, pens and other ferrous or non-ferrous objects.

Resistivity

Contrasts in the electrical resistivity between a target and its surroundings can be delineated using depth-sounding and profiling techniques, involving insertion of four steel electrodes into the ground and measuring vertical and horizontal variation in resistivity. Multi-electrode arrays have been utilised which involve up to 80 electrodes, producing a resistivity cross-section or image of the subsurface (Clark 1996; Noël 1992).

Induced Polarisation

Induced polarisation is commonly used in disseminated metallic ore prospecting. The induced polarisation effect is a transient voltage, observed after current flow ceases in a resistivity array. Aspinall and Lynam (1970) report use of the method in archaeological surveying, finding it slower and less effective than resistivity surveys.

Self-Potential

This is a naturally occurring ground potential due to electrochemical reactions between different rocks and ground water levels and flow. It is a simple and inexpensive technique that utilises two non-polarising ground electrodes and a millivoltmeter. Typically, it has been used in the location of metallic sulphides but is used increasingly in the mapping of geological boundaries and cavity features.

Electromagnetic (Conductivity)

An effective and rapid surveying alternative to resistivity profiling, the electromagnetic inductive conductivity profiling method (electrical conductivity is the reciprocal of electrical resistivity) allows continuous recording of the subsurface conductivity at a walking pace. Two instrument types are available; one measuring conductivity to about 1.5 m subsurface, the other to approximately 7 m (Clark 1996; Frohlic and Lancaster 1986).

Ground Penetrating Radar

Ground penetrating radar (GPR) operates in a frequency range of 25 MHz to 2 GHz, and identifies shallow physical anomalies in the ground. GPR applications received publicity in searches for buried murder victims, such as those in Cromwell Street, Gloucestershire (so-called ‘Fred West Murders’), and locating a buried cache of ransom money (kidnapped estate agent Stephanie Slater, 1992). The approach is very effective for locating buried graves which are lined 'cavity' structures (Ruffell 2006). Experiments using GPR at test sited with buried pigs, and buried bodies at the Forensic Anthropology Unit ('The Body Farm'), University of Tennessee, indicate that it can be used to locate such buried objects. Miller (2002) used GPR to investigate the effects of buried, decomposing, human body targets over a period of time. He showed that changes in GPR anomaly response related to stages of body decomposition. However, in experiments, the GPR operator knows the target location, and often in uniform ground, so not replicating real world search scenarios. Mellett (1992, 1996) discusses eight different GPR searches for human body remains in which only one was successfully located. In this case other evidence had been used to considerably reduce the search area; the victim was wearing synthetic clothing, and buried at a depth of 0.5 m. GPR surveys over a concrete floor may detect anomalies due to a void caused by compaction of underlying disturbed soil, or the space occupied by the body (Hammon et al. 2000; Hilderbrand et al. 2002). Therefore, the method may be more successful for indirectly locating buried bodies, through delineation of the change in physical properties of disturbed soil overlying the cadaver. The skills required include those of the instrument operator and the interpreter of the data collected (Ruffell 2005c; Ruffell et al. 2004).

Metal Detectors

These instruments use pulse induction, or time domain, principles and are one-man, portable, hand-held scanning devices with an audible signal or meter output. More sophisticated units embody interchangeable search heads and may be used to locate metal coins to a depth of 0.5 m, whilst larger targets such as metal spades may be located at depths of 1 m.

Airborne Geophysical Surveys

Airborne geophysical surveys may be deployed from a light aircraft, helicopter or un-manned low flying aircraft (drone). Typical types of airborne geophysical surveys are magnetic, electromagnetic, gravity and radiometric. Unmanned systems may be used at low altitude, or in hostile environments, where it would be too dangerous for manned flying. However, these attract public attention and may not be suitable for covert operations. Survey data are usually acquired in a grid pattern (e.g. flight paths), the interpretation of which requires an understanding of the geology and target.

Geophysical Equipment Exploration Platforms

Multiple geophysical sensors may be simultaneously deployed to provide a faster and more effective way to survey large areas of land. Instruments are mounted on a mobile platform, and towed around the search area on a 4 × 4 vehicle or tractor.

Positional data is provided by on-board GPS and compass. Although these techniques are yet to be proven in geoforensics, there exists potential where large tracts of ground require to be surveyed.

Seismic and Microgravity

Seismic (Hildebrand et al. 2002) and microgravity (Emsley and Bishop 1997) methods have been tested for use in forensic investigations but currently have limited application. However, microgravity has potential use in conjunction with other geophysical methods to locate relatively large underground voids such as vaults, caves or shallow mine workings.

Hydrochemical and Geochemical Investigations

An understanding of the hydrogeology in the vicinity of a grave site is required. This influences surface flow paths, run-off, groundwater flows, preservation of human remains, generation of leachate plumes from decomposing human remains, migration of body scent (gas/vapour), and the deployment of the most suitable non-invasive geophysical and geochemical techniques.

Leachate plumes can also increase the geographic footprint of geochemical and geophysical signatures, thereby enhancing the possibility of detection. The direction and distance of a plume will depend on factors such as ground permeability, geological structures, lithology, rainfall, and time elapsed since burial (Figure 13.3). Compared to exploration geophysical methods, hydrochemical and geochemical search techniques are less well developed.

The Design and Phased Implementation of a Search Strategy for a Homicide Grave

The type of search adopted for a victim's grave varies depending on resources available, expected location of the grave (on land or in water), intelligence, victimology assessments, behavioural profiling of the offender and geology. Well-organised searches follow an established pattern which begins with a 'desk study' and 'review' of all available information and intelligence, enabling the development of a conceptual geological model for the grave. The search ends with location of the grave and discovery of the victim. The body is subsequently recovered and the scene may be subject to further investigations. The generalised sequence is summarised below and in Figure 13.1.

- *Review (desk study)*: All available intelligence, geological and other information concerning the site is collated and analysed.

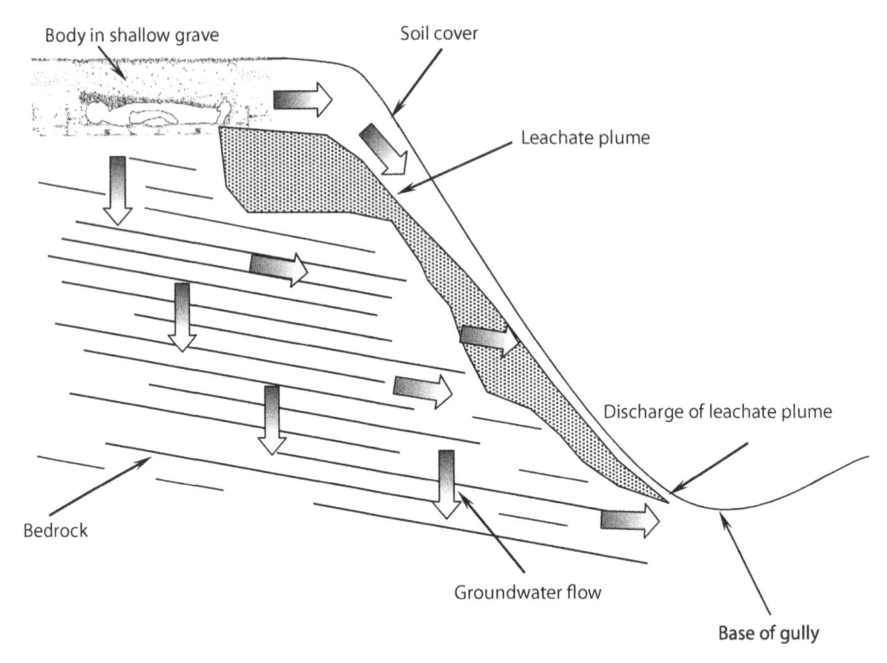

Fig. 13.3 Schematic model to illustrate leachate plumes generated from decomposing human remains. The presence of a leachate plume may increase the target area (after Donnelly 2002a)

- *Development of a conceptual geological model*: Produce a model to identify likely ground conditions, geology and general characteristics at, and in the vicinity of, a grave.
- *Behavioural analysis (offender profiling) and hypotheses generation*: A behavioural profile of a suspect or victim may provide information on the most likely scenario a suspect would have used to abduct, murder and dispose of the body. A comparison of the profile to confidential law enforcement body disposal databases then follows. Other predictive models may include missing persons and suicide search models, weaponry/drug concealment models and body deposition site assessment models. These will result in the development of the likely search scenario hypotheses and predictions of the expected place and condition of the victim.
- *Aerial image analysis*: Current and historical aerial and satellite imagery, preferably before the victim's disappearance, for identifying locations consistent with the suspect's behavioural profile parameters and identify ground disturbances and suitable conditions for the choice of the grave location.
- *Reconnaissance site visits*: A detailed 'walk-over' survey to obtain an appreciation of the search area (i.e. placing the desk study into context) and identify technical, logistical or access constraints (e.g. overhead power lines, which may influence choice of geophysical techniques). Identifying possible grave sites using attributes associated with homicide victim disposal, assessment and interpretation of topographic and geological maps. A diggability survey to assess the viability of

burial and, where applicable, undertake control measures to ensure the most appropriate method of detection or equipment within the target terrain.
- *Development of a written search strategy*: Producing a document to support search decisions. Include relevant information uncovered in the desk-based study, a likely scenario of disposal, informed by behavioural analysis, and the most appropriate way to search the target area based on in the field reconnaissance. Detailing the SOPs for any search assets their likely search duration and costs.
- *Search, review and continue*: Conducted under the supervision of a law enforcement officer qualified in search management procedures.
- *Development of a search strategy exit*: A written report detailing all the searches actioned, including any associated mapping and photography, and documentation of any objects found. Conclude that all investigative facts and intelligence are exhausted and recommend cessation of search.

Communication Between Law Enforcement Officers and Geoscientists During Homicide Investigations

Communication is a social skill, not a technical one, for the transfer of data and information. The most effective method of communication is the use of clear, simple, unambiguous, non-technical language. Visual material can facilitate effective communication, especially to a non-technical audience and professionals with little or no knowledge of geology. The transfer of knowledge, to be effective, needs to be delivered with confidence and consistency. The good communicator must also be a good listener, using silence, reflection, paraphrasing and non-verbal behaviour. If possible, there should be feedback from the targeted audience.

The search for homicide graves may involve teams of multi-disciplinary experts such as geoscientists, anthropologists, botanists, cadaver-detecting canines, remote sensing aerial assets, behavioural profilers, clinical psychologists and military personnel. These are co-ordinated and managed by a police search specialist or Senior Investigating Officer (SIO) (Donnelly 2008) (Figure 13.4). Law enforcement officers will, through experience, begin to understand where and when the geoscientist can be most effectively deployed to help assist with the investigation and, equally importantly, where the geoscientists' experience finishes. For victim recovery, the geoscientist would recommend the services of other specialists such as an anthropologist, archaeologist or pathologist. This clear and effective communication is of critical importance during searches for victims of homicide.

Conclusions

The search for homicide victims has evolved over the past decade, from traditional methods, involving the deployment of large numbers of non-specialist volunteers, to the design and implementation of search strategies which utilise targeted geological

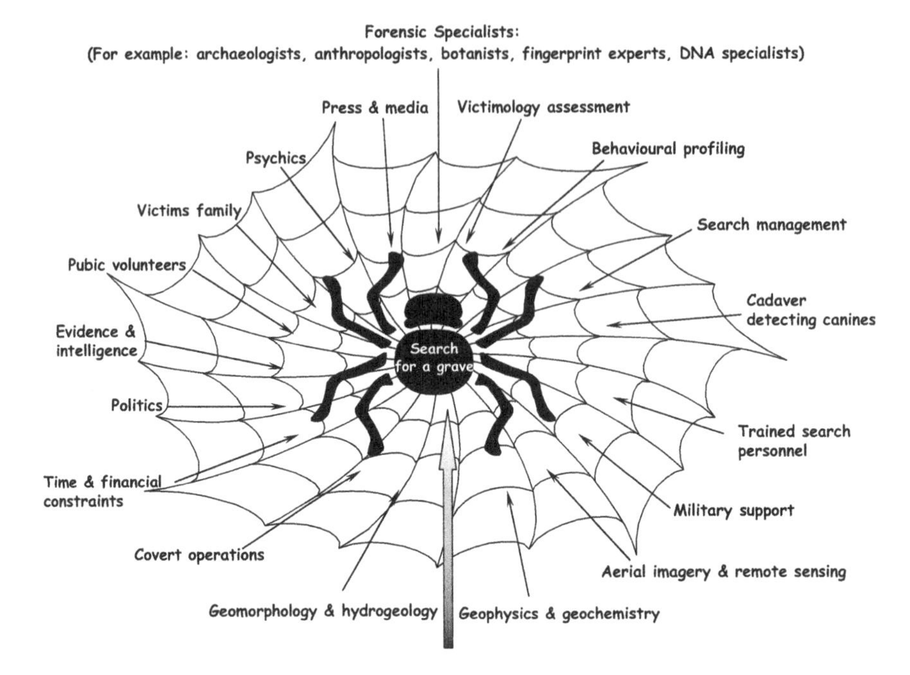

Fig. 13.4 The introduction of a geologist to a complex, multi-disciplinary police search team must be carefully coordinated and properly managed. The geologist must be able to effectively communicate with the other subject matter experts, be aware of his/her limitations and understand the role and capabilities of other experts (after Donnelly 2002a, 2008)

techniques, originally designed to assist with mineral exploration and geotechnical site investigation. This paper demonstrates how geoscientists can be proactive in offering targeted support in the 'search-to-locate' phase of a homicide investigation. It shows, to law enforcement officers and search strategists, a clearer application of geoforensics, and where and how they can benefit and positively influence search strategies and their implementation.

Historically, geoscientists have rarely been involved with homicide searches, yet they can 'read the ground' and understand geological, geomorphological and hydrogeological processes that affected the ground before, at the time of, and following, body disposal. This discipline assists in identifying potential search locations when incorporated with, for example; clinical psychological behavioural profiling and victimology assessments.

Traditional search techniques, which include the use of large numbers of volunteers, line searches and trial-and-error excavations, are resource intensive, cost prohibitive, and often non-productive and destructive. Construction of a conceptual geological model, and use of geophysical methods, provides cost-effective approaches for the location of homicide graves.

A conceptual geological model of the ground may be developed by the geoscientist to provide information on the target's age, size, geometry, expected depth of burial, time and duration of burial, physical, chemical, hydrogeological and geotechnical variations compared to surrounding ground. This information can be used to identify the correct search strategy, appropriate choice of instrumentation, and the optimum method of deployment. To carry out these operations successfully, the main challenges are not technical but those of communication.

Geoscientists must recognise how their expertise and capabilities fit into the broader homicide investigation. For most geoscientists involved in searching for victims of homicide, their input may begin to reduce once the target (body) has been located, and the investigation moves into victim recovery and crime scene investigation. At that stage it is important for a hand-over to forensic practitioners in related sciences, such as forensic archaeology and anthropology for the victim recovery and post-recovery analysis phases.

The investigations which underpinned this chapter have demonstrated, for the first time, the value of the combined efforts of geoscientists (experienced in exploration and ground investigation) and law enforcement officers (experienced in homicide searching). For these types of approaches to be more widely applied, both nationally and internationally, one of the main future challenges is standardising search criteria, techniques, and methodologies, and accrediting professional standards.

The law enforcement search strategist, facilitating the application of behavioural and geological sciences, can ensure geoforensic practitioners add value to a homicide investigation. This assists the search strategist in translating behavioural analysis of offender and victim profiles into physical search action by access to the correct interpretation of ground conditions and most appropriate detection instrumentation.

Acknowledgements The authors would like to acknowledge the National Policing Improvement Agency (NPIA), The University of Southampton and Halcrow Group Ltd. Any views expressed in the paper are those of the authors and not necessarily those of these organisations.

References

Aspinall A and Lynam JT (1970). An induced polarization instrument for the detection of near surface features. Prospezioni Archeologiche 5:67–75.

Boyd RM (1979). Buried body cases. FBI Law Enforcement Bulletin 48:1–7.

CIRIA (2002). Geophysics in Engineering Investigations. Construction Industry Research and Information Association (CIRIA) Report C562.

Clark A (1996). Seeing Beneath the Soil: Prospecting Methods in Archaeology. Routledge, London.

Cornwell P (2002). The Body Farm. Time Warner Paperbacks, New York, USA.

Donnelly LJ (2002a). How forensic geology helps solve crime. Record of presentation on Forensic Geology and the Moors Murders to the House of commons, Westminster Palace, on 12th March 2002, with contributions from J.R. Hunter and B. Simpson. All-Party Parliamentary Group for Earth Sciences. Britis Geological Survery & International Mining Consultants.

Donnelly LJ (2002b). Finding the silent witness: how forensic geology helps solve crimes. All-Party Parliamentary Group for Earth Science. The Geological Society of London. Geoscientist 12:24.

Donnelly LJ (2002c). Finding the silent witness. The Geological Society of London. Geoscientist 12:16–17.

Donnelly LJ (2003). The applications of forensic geology to help the police solve crimes. European geologist. Journal of the European Federation of Geologists 16:8–12.
Donnelly LJ (2004). Forensic geology; the discovery of spades on Saddleworth Moor. Geology Today 20:42.
Donnelly LJ (2006). The inaugural meeting of The Geological Society of London Forensic Geoscience Group. Geoscientists at Crime Scenes. Forensic Geoscience Group Meeting. The Geological Society of London, Burlington House, 20 December 2006.
Donnelly L.J. (2008). Communication in geology: A personal perspective and lessons from volcanic, mining, exploration, geotechnical, police and geoforensic investigations. In: Liverman D.G.E, Pereira C.P. & Marker B. (eds) Communicating Environmental Geoscience. Geological Society, London, Special Publication 350: 107–121
Emsley SJ and Bishop I (1997). Application of the microgravity technique to cavity location in the investigation for major civil engineering works. Modern Geophysics in Engineering Geology. The Geological Society Engineering Group Special Publication 12:183–192.
Fenning PJ and Donnelly LJ (2004). Geophysical techniques for forensic investigations. The Geological Society of London Special Publication 232:11–20.
Frohlic B and Lancaster WJ (1986). Electromagnetic surveying in current Middle Eastern archaeology: application and evaluation. Geophysics 51:1414–1425.
Gaffney C, Gater J and Ovenden S (2002). The use of geophysical techniques in archaeological evaluation. Institute of Field Archaeologists, University of Reading, Papers 6.
Hammon WS, McMechan GA and Xiaoxian Z (2000). Forensic GPR-finite-difference simulation of responses from buried remains. Journal of Applied Geophysics 45:171–186.
Harrison M, Hedges C and Sims C (2006). Practice Advice on Search Management and Procedures (Eds). National Policing Improvement Agency (NPIA).
Hildebrand JA, Wiggins SM, Henkart PC and Conyers LB (2002). Comparison of seismic reflection and ground penetrating radar imaging at the controlled archaeological test site, Champaign, Illinois. Archaeological Prospection 9:9–21.
Hunter J and Cox M (2005). Forensic Archaeology Advances in Theory and Practice. Taylor & Francis Oxon, UK
Killam EW (2004). The Detection of Human Remains. Charles C. Thomas. Springfield, IL.
MacDougall KA, Fenning PJ, Cooke DA, Preston H, Brown A, Hazzard J and Smith T (2002). Non-intrusive investigation techniques for groundwater pollution studies. Research & Development Technical Report P2–178/TR/1. Environment Agency, Bristol.
McCann DM, Culshaw MG and Fenning PJ (1997). Setting the standard for geophysical surveys in site investigation. In: Modern Geophysics in Engineering Geology (Eds. DM McCann, M Eddleston, PJ Fenning and GM Reeves). The Geological Society of London Engineering Geology Special Publication 12:3–34.
McCann DM and Fenning PJ (1995). Estimation of rippability and excavation conditions from seismic velocity measurements. In: Engineering Geology of Construction (Eds. M Eddleston, S Walthall, JC Cripps and MG Culshaw). The Geological Society of London Special Publication 110:335–343.
Mellett JS (1992). Location of human remains with ground penetrating radar. Fourth International Conference on GPR. Geological Survey of Finland Special Paper 16:359–365.
Mellett JS (1996). GPR in forensic and archaeological work: hits and misses. Proceedings of Application of Geophysics to Engineering and Environmental Problems. SAGEEP Conference 1991 Environmental and Engineering Geophysical Society. Co., USA, 487–491.
Miller M (2002). Coupling Ground Penetrating Radar Applications with Continually Changing Decomposing Human Targets: An Effort to Enhance Search Strategies of Buried Human Remains. MA thesis, University of Tennessee, Knoxville.
Murray RC (2004). Evidence from the Earth. Forensic Geology and Criminal Investigations. Mountain Press, Missoula, MT.
Noël M (1992). Multielectrode resistivity tomography for imaging archaeology. In: Geoprospection in the Archaeological Landscape (Ed. R Spoerry), pp. 89–99. Oxbow Monograph 18, Oxbow Books, Oxford.

Pye K and Croft DJ (2004). Forensic Geoscience: Principles Techniques and Applications (Eds. K Pye and DJ Croft). The Geological Society of London Special Publication.
Ruffell A (2005a). Forensic geoscience. Geology Today 22:69–72.
Ruffell A (2005b). Burial location using cheap and reliable quantitative probe measurements. Diversity in Forensic Anthropology, Special Publication of Forensic Science International. Forensic Science International 151:207–211
Ruffell A (2005c). Searching for the I.R.A. Disappeared: ground-penetrating radar investigation of a churchyard burial site, Northern Ireland. Journal of Forensic Sciences 50:414–424.
Ruffell A (2006). Freshwater under water scene of crime reconstruction using freshwater ground-penetrating radar (GPR) in the search for an amputated leg and jet-ski, Northern Ireland. Science and Justice 46:133–145.
Ruffell A, Geraghty L, Brown C and Barton K (2004). Ground-penetrating radar facies as an aid to sequence stratigraphic analysis: application to the archaeology of Clonmacnoise Castle, Ireland. Archaeological Prospection 11:1–16.
Ruffell A and Mckinley, J (2008). Geoforensics. Wiley-Blackwell, Chichester, UK.
The Geological Society of London (1988). Engineering Group working party report on engineering geophysics. The Working Party Report on Engineering Geophysics.

Chapter 14
Geological Trace Evidence: Forensic and Legal Perspectives

Antoinette Keaney, Alastair Ruffell and Jennifer McKinley

Abstract A case study is presented in which exclusionary and comparative analysis of soil from locations associated with a murder and on clothing from the victim and suspect proved important in the investigation of what happened. The soil on clothing comprised 1 to 2 mm diameter 'specks' and thus were considered trace evidence: how to handle such quantities of soil is considered in the context of a review of the history of trace evidence in legal cases. The criminal mind is becoming increasingly familiar with clean-up procedures following a criminal event. Police officers and forensic scientists are being faced with scenes of crime that exhibit smaller and smaller amounts of trace evidence available for analysis. Therefore the techniques which are carried out upon recoverable evidence are crucial – as the amount of material available for investigation is diminishing. Investigative research into the analysis of trace mud splashes on different substrate materials attempts to utilise the amount of information which can be obtained at this micro scale. Analysis of these splashes becomes increasingly important when a number of analytical techniques are used to establish their origin: in light of the limited material available non-destructive techniques are therefore top priority in this type of investigative research. This chapter aims to highlight ideas that should be considered during an investigation, information that can be obtained through non-destructive analysis and to provide some guidance as to limitations of analysis techniques.

Introduction

This chapter aims to highlight the importance of testing the combined scientific, cost and time effectiveness of non-destructive analysis of soil, rock dust and other crystalline materials as a rapid screening method for further forensic based scientific analysis.

A. Keaney (✉), A. Ruffell and J. McKinley.
School of Geography, Archaeology and Palaeoecology, Queen's University Belfast, Belfast BT7 1NN, Northern Ireland, UK.
e-mail: *akeaney04@qub.ac.uk*

K. Ritz et al. (eds.), *Criminal and Environmental Soil Forensics*

The importance of such testing can be seen using an example of a case in which analysis of soil recovered from a crime scene proved to be a crucial part of an investigation. In this case study, on the 22nd September 2003, a 16 year-old female (here named 'Mandy' for convenience) left her home in West Belfast to 'hang out' with friends at a chip shop. She departed at 02:00 in a car, driven by a friend, a 17 year-old member of the travelling community of West Belfast. She never arrived home. Concerned for her welfare, the 17 year-old took two friends for a drive to look for the missing girl. They arrived at an abandoned graveyard where they were horrified to find her mutilated body in a ditch. In trying to resuscitate her, and running back to the car, the 17 year-old's clothes became mud-spattered. As a witness these clothes were seized by the police. The victim's clothes were also examined: the base of her trousers was mud-stained, however the mud did not appear visually similar to the graveyard scene. The additional two witnesses stated that the 17 year-old man was wearing different clothes on the night of Mandy's disappearance to those already seized by police. The clothes relating to the day of the victim's disappearance were subsequently seized and were found to have very minor mud staining. These mud stains appeared more similar to those from the victims' trousers than from that of the graveyard – the apparent crime scene. Scanning electron microscope (SEM) analysis was carried out on evidence from the suspect, the victim, the graveyard and the surrounding area that indicated a better comparison between the suspect's 'clean' clothing, the victim's jeans and mud from a nearby quarry, than the graveyard in which the victim had been 'found'. Unfortunately not enough material was available from the suspect's clothing to provide a statistically robust link. Although convicted using other evidence (witness reports; CCTV footage), the account of what happened, and where, could have been better established using geological trace evidence. The question which arises from this case study, pertinent to this research is: could more information on the suspect's 'clean' clothing have been obtained using non-destructive methods before SEM?

This case provides an example of how crucial the analysis techniques are which are carried out upon these small amounts of sample. In this case only a small amount of material was available and, once SEM had been carried out upon the sample, no other technique could be used. In this chapter, experimental studies are discussed to illustrate the importance and uses of non-destructive analysis of trace materials, using this case as a basis for the investigation into analysis techniques, in order to provide information regarding a multi-proxy approach to an investigation to ensure that the maximum amount of information can be obtained from a minimum amount of sample.

Background

This chapter aims to identify and bring forward ideas that should be considered during an investigation, highlight the amount of information that can be obtained from a sample and provide some guidance as to the limitations of the analysis, as well as the best way in which samples must be preserved in order for them to provide accurate information. An example of this can be seen in relation to an investigation in which geological material may provide a crucial piece of evidence but the integrity of the sample may have been compromised by the entrance into a crime scene by many officers. A recent case which may have suffered as a result of this intrusion is the investigation in Portugal of missing toddler Madeleine McCann. Chief Inspector Olegario de Sousa, working on the case, stated that vital forensic clues may have been destroyed as a result of numerous people entering into her room and looking for her immediately after her disappearance. “At the very worst they would have destroyed all the evidence. This could prove fatal for the investigation” (BBC 2007).

“The most critical phase in the majority of criminal investigations is the preliminary investigation. The decisions made, the responsibilities assumed and the tasks performed apply to a wide variety of crimes and must be part of every investigators repertoire” (Bennett and Hess 2004). The implications for this have already been seen in relation to the collection of evidence from a crime scene. Another set of procedures has to be applied when a sample is brought into the laboratory for investigation. When faced with small (milligram) amounts of sample, the investigator has to decide which professional would be best suited to its analysis. Would microbiology be the most important factor in the sample in relation to the investigation or would the mineralogy hold the key to the suspect involved? These questions prove critical when the choice of what type of analysis will be carried out upon the samples, as this will most likely result in their complete destruction.

Division or Homogenisation of a Sample?

The division or homogenisation of samples can also cause problems. Division of samples could be perceived by a jury to be unrepresentative, as the division of the substance may result in each separate section containing different results. Alternatively, an important piece of information may be derived from one sample but its discovery is missed due to the decision to send it for inappropriate analysis that fails to detect it. All of the above result in loss of sample or sample integrity, leaving no way to reanalyse the material. Homogenisation of a sample can also prove to be detrimental to an investigation. “If analysis, no matter how detailed, complex and data producing requires that the sample be homogenized (by grinding or solution or both), then the associated results, be they positive or negative, cannot be tested using these techniques and thus the authenticity of any association is gravely compromised” (Morgan et al. 2006).

Screening Methods

It is at the preliminary stages of an investigation that a screening method may be used in order to aid intelligent decisions being made about the best way to deal with the samples obtained from a crime scene. Rapid and cost-effective analysis at the beginning of an investigation could prove critical in relation to the next steps to be taken, a practical application of this being a potential decision to return to a scene of crime before substantial changes (people/vehicle movement; soil disturbance; changes in weather) have taken place. Introducing a non-destructive screening method might affect the decision whether to follow analytical steps for organic or mineralogical materials. Also, when sample size is limited and is not large enough to be divided into two or more parts to enable different analysis to be carried out upon them, the use of non-destructive methods would allow the maximum amount of information to be obtained from a limited amount of sample without compromising it for future investigation. "It must be accepted that it is better to use fewer techniques and to reproduce the results through a series of different experimental runs than to use many techniques with no repeated checks" (Morgan and Bull 2007). These authors then go on to state that "It is also conventional and wise in forensic analysis, for at least half the sample to be left in its original state in order that other scientists may be able to check the findings or undertake other analysis" (Morgan and Bull 2007).

The potential use of soil in criminal investigations has seen a rapid expansion throughout the twentieth and into the twenty-first century. It is important however that in the recent development of this discipline certain restrictions and hindrances in the techniques used to analyse the soil are taken into consideration. "Forensic geology is concerned with the application of geological data and techniques in relation to issues which may come before a court of law … probably the most widely recognized application of forensic geology is the use of geological materials as trace evidence which can be of value in linking a suspect to a crime scene" (Pye and Blott 2004). Increasingly, the amount of available material obtained from a scene of crime is becoming smaller, as was the situation presented in the case study, and forensic geology provides a particularly beneficial route for investigation to be carried out on these types of samples as "The value of geological evidence results from the almost unlimited number of rock, mineral, soil and related material lines combined with our ability to use instruments that characterize these materials" (Murray 2004).

When an examiner is faced with a soil sample there are set guidelines to be followed, such as those laid out by Palenik (2000) and Murray (2004), as to the procedure which should be followed for an investigation, depending on the amount of material which has been collected. Increasingly, however, police officers and forensic scientists are being faced with smaller and smaller amounts of material evidence available for forensic purposes. These can be as small as 1 mm diameter 'specks' of soil or a coating of dust. With such amounts of material it is often impossible to determine what these 'specks' may be, a fundamental first step in choosing an analytical method. Here we arbitrarily consider 'trace' as being too small to carry out numerous standard destructive laboratory analyses. As the amount of material available for investigation is diminishing in many situations, not enough sample is present to carry out the tried and tested sequence of analysis. "If only small amounts of mud or particulates are present, there is usually a need to preserve as much evidence as possible for re-examination. Consequently non-destructive tests, or those which are minimally destructive, are preferable to destructive tests which require a relatively large amount of sample material" (Pye and Blott 2004). The amount of evidence obtained can be used to establish numerous aspects relating to crime scene investigation, alibi veracity, crime scene attendance, burial location and many more. Thus the implications of this type of investigative research can clearly be seen. Non-destructive analysis provides the investigator with information about the evidence that can inform fundamental decisions about which destructive tests would then be most appropriate. This could ensure that the sample is not destroyed unnecessarily or the validity of the sample becoming compromised as a result of the analysis technique applied. A key aspect underlying such work is to ensure that accurate, effective analysis is carried out to a standard that is validated by the current justice system, therefore ensuring an increase in the quality of evidence available resulting in more suitable convictions and/or aiding the release of innocent suspects.

Soil samples collected in the field generally allow the scientist to be in control of the amount of sample taken and the procedures which will be carried out upon the sample in order to ensure that the most effective analysis is utilised to answer the research question being investigated. In non-forensic laboratory conditions when laboratory technicians are dealing with soil for experimental purposes, the preservation of the soil is never really an issue; the examiner can use destructive techniques on it to try and find the best approach to extract the information contained in the soil. The role of a forensic geologist is somewhat different in comparison. Investigators receive an amount of soil which, in some unfortunate cases, they may not even have been in charge of collecting, and it is up to the individual investigator to choose which technique is most applicable for each particular sample. The investigator has least control over materials associated with suspects and victims – as mud specks, dust coatings, etc. were inadvertently 'collected' by the person, not the investigator. In fact, quite often, the suspect has gone to strenuous lengths to ensure they do not 'collect' or preserve material from the scene of a crime. The criminal mind is becoming increasingly familiar with clean-up procedures following criminal activity, gloves are worn to hide fingerprint evidence, masks are worn to avoid identification by CCTV cameras, clothing may be washed or destroyed after committal of a crime. However, it remains very difficult for a suspect to cover their shoes or the bottoms of their trousers without it being considered to be suspicious to potential eye witnesses, "during an everyday routine it is normal to see an individual wearing gloves, but it is not normal to see individuals wearing protection over their shoes" (Hilderbrand 1999). The analysis of these important items and samples obtained during an investigation is critical. One of the main benefits of trace evidence for use in analysis is that the suspects themselves may not be aware that they have collected this evidence on their person, or on their belongings, or left it behind them at a crime scene. Although a comprehensive clean-up operation may have been employed, there is a possibility crucial evidence may still be found and analysed (Ruffell and McKinley 2008). Consequently, one of the biggest issues which examiners may face is the amount of sample which is placed in front of them and the important questions which are being asked of such a small sample. The usual questions are in relation to comparing a suspect to a crime scene or the bigger question of trying to identify where a soil sample may have came from, with sample quantity barely visible to the naked eye.

Historical Perspective

Some of the earliest accounts of the uses of forensic geology describe what we would consider 'trace' evidence. Block (1958) describes the work of Oscar Heinrich using sand grains, such as those found on the knife used in the murder of Fr. Patrick Heslin. Subsequently, forensic geology publications have concentrated on more substantial amounts of evidence (Murray 2004). Murray describes cases such as lumps of mud fallen from car fenders in hit-and-run incidents. However, the first use of earth materials in an investigation can be seen as far back as 1856 by Professor Ehrenberg. Scientific American (1856) details the case involving the exchange of silver coins for sand somewhere along a length of railroad. Analysis of the sand from different stations along the track allowed the identification of the place where the exchange had been made. The use of geology in relation to criminal investigations can be seen in the fictional writings of Sir Arthur Conan Doyle in 1887. Conan Doyle's initial use of geology related to techniques which had never actually been tried and tested before in relation to criminal investigations but the basic principles and theories behind these ideas have proven to be an excellent starting point for forensic geologists to apply and follow from a scientific perspective. An example of this can be seen in the Sign of Four (Conan Doyle 1890) where a visual examination of a sample provides the famous fictional character Sherlock Holmes with information about where a person may have been: "observation shows me that you have been to the Wigmore Street Post-office this morning … *you have a little reddish mould adhering to your instep* … The earth is of this peculiar reddish tint which is found, as far as I know, nowhere else in the neighbourhood. So much is observation" (Conan Doyle 1890).

October 1904 saw the first criminal murder case to use earth materials as evidence with Georg Popp, a forensic scientist, being asked to examine evidence from a murder where a seamstress (Eva Disch) had been strangled in a field with her own scarf. Popp found at the scene, a filthy handkerchief containing not only nasal mucus but amongst it bits of *coal, particles of snuff and grains of minerals*, one of which in particular was the mineral hornblende. All had been left at the scene of the crime. A suspect was identified by the name of Karl Laubach who worked in a coal-burning gasworks, as well as part-time at a local gravel pit. Popp obtained samples from underneath the suspect's fingernails, and these samples were found to contain coal and mineral grains, particularly hornblende. Laubach also had two layers of dirt in his trouser cuffs, the lower layer of which matched soil from the crime scene and the upper layer, which characterised a particular type of mica particle, matched soil found on the path leading to the victim's home. The suspect confessed when confronted with this evidence against him (Murray 2004). Just like in Conan Doyle's fictional stories, using relatively simple trace geological observations, comparisons could be made between suspect and scene.

Locard's Exchange Principle

Locard's Exchange Principle states that: *Whenever two objects come into contact, there is always a transfer of material.* The methods of detection may not be sensitive enough to demonstrate this, or the decay rate may be so rapid that all evidence of transfer has vanished after a given time. Nonetheless, the transfer has taken place. Houck (2003) refers to the small amount of trace evidence which may be transferred during the contact of a criminal with a crime scene. Trace evidence can be defined as "microscopic material recovered as evidence that is used to help solve criminal cases. Because of their minute nature, trace materials can be easily cross-transferred from one surface or substrate to another without detection by a criminal" (Houck, 2003). Anything which has come into contact with the criminal at the scene of a crime will leave a trace on that person or clothing. Similarly, criminals will leave their mark behind them at the scene. "Wherever he steps, whatever he touches, whatever he leaves, even unconsciously, will serve as silent witness against him" (Hilderbrand 1999). Particles of any substance can be used to provide evidence that may prove that a transfer has occurred. Houck (2003) expands upon Locard's Exchange Principle stating that not finding traces on certain items may prove to be just as significant and that a negative finding does not automatically mean that no contact took place; this has been phrased informally as "absence of evidence is not evidence of absence" (Houck 2003).

Multi-proxy Techniques

Evidence found at a crime scene, whether positive or negative, may be used to help reconstruct the crime and counter fabricated alibis (Houck 2003). The importance of applying a number of different analytical techniques to a sample in order to ensure that it can be stated with confidence that as much information was obtained as possible can thus be seen. "Forensic geoscience requires techniques of exclusion rather than inclusion and an acknowledgement that analytical techniques may be diagnostic only in very specific situations" (Morgan and Bull 2007). Following forensic procedures and gaining as much information from a sample as possible may also become critical in that it could prove that, along with other evidence, a suspect is in fact innocent of the accused crime. This type of statement is important in relation to the criminal justice system today and is a key factor underlying this exhaustive analysis and investigation into crime. It is of paramount importance that the research is conducted in an objective manner to establish the possibility of the use of any of the analytical techniques particularly in relation to crime investigation.

The importance of ensuring that a sample has been accurately analysed and that it can be sworn on oath in a court of law that it significantly compares or does not compare to another sample taken from the scene of a crime is critical. In relation to crime investigation, Hellerstein's article (2005) states how, in America, "judges who do not value the truth are, at times, one of the obstacles to freeing the innocent" and at times the prosecutor in a case can become a serious obstacle to obtaining the truth. Hellerstein uses the case of Berger v. United States 295 US 78@ 88 (1935) as an example where the court said that the prosecutor "is the representative… whose obligation to govern impartially is as compelling as its obligation to govern at all; and whose interest, therefore, in criminal prosecution is not that it shall win a case, but that justice shall be done. As such, he is in a peculiar and very definite sense the servant of the law, the twofold aim of which is that guilt shall not escape nor innocence suffer". Similarly a forensic geologist, or geoscientist, analyses a sample which they are given to the best of their ability and ensures that any statement they make about their analysis is impartial to their personal views of the case and relates purely to the science behind the procedure that has been followed. Multi-proxy techniques will increase this confidence, which is critical to ensuring justice, as all areas of investigative procedures will have been exhausted. "Whilst 'every contact leaves a trace', it is imperative that we understand the nature of the traces that we are dealing with in forensic enquiries. Even under the most simplified transfers, soils and sediments undergo modification of various forms. It is essential that we understand the nature of the sample that we seek to analyze so as to apply the appropriate analytical techniques" (Bull et al. 2005). Bull et al. (2005) state a consideration which must be taken into account when analysing any soil sample seized for investigation, referring specifically to results obtained from SEM-EDS analysis used to discriminate soil, "The results from these experiments demonstrate the importance of understanding the nature of the forensic soil sample that is being dealt with before it is analyzed and subsequent results interpreted". This statement refers to the many different situations which may arise at a crime scene. For example, if a criminal steps directly onto the surface of the soil, the particles which are transferred to their shoes depends upon a number of variables, such as how wet the soil was at the time or the distribution of the particle grains at the surface of the soil. It is important to investigate how representative such a sample obtained can be of the area from which it came. It must be emphasised however that the samples obtained will not be considered in isolation but placed in context in relation to their surroundings. Conducting research into this will provide answers and advice regarding the best method to undertake for investigation.

When considering fingerprint evidence for use in forensic investigations, Olsen and Lee (2001) identify the two ultimate goals to which the efforts of latent print examiners are directed, viz. (i) the successful developing or enhancing of a latent print; (ii) the identification or elimination based upon the developed latent print. They go on to highlight that, although there have been many advances in fingerprint technology over the past number of years, these two basic principles have remained the same and the process itself can be related to any other type of forensic examination. The process is concerned with recognition, examination, identification, individualisation and evaluation (Olsen and Lee 2001), such as this study is attempting with soil, combined with the exclusionary principle as championed by Morgan and Bull (2007). Recognition in particular could be said to be perhaps the most important for soil evidence. If it is not recognised that a sample could possibly be taken from an area or if the sample is not taken in the correct way, information regarding the case could be permanently lost in the same way as if "crucial latent print evidence is not preserved, it will be lost, and the potential important links between a suspect and a crime may never be known or established" (Olsen and Lee 2001). P. Wiltshire and L. Donnelly (pers. comm.) highlight this particularly, as well as the need to make crime scene officers aware of the potential a crime scene holds in relation to all aspects of forensic investigation.

Analytical Techniques

The discriminatory power of sample analysis has been increased through the advancement of technology, allowing the rapid accumulation of data through analysis. It is the interpretation of this data, however, that is still key to ensuring that accurate and effective results are obtained. The fictional writings of Conan Doyle provided scientifically-unproven ideas on the ways soil, sediment or rocks could be analysed. Now advances in technology allow the analysis of samples in the way Conan Doyle depicted; however, extreme care is needed. Evett (1991) highlights how the expansion of a knowledge base does not necessarily get better as it gets bigger – "indeed, if not properly informed it could become an 'ignorance base' ". They discuss how it is at this stage then that calibration should be applied, carrying out experiments "to test the quality of the contents of the knowledge base under suitably controlled conditions: the result would serve to refine and improve the quality of the pool of expertise". This approach is being carried out in this research taking into account the current situation regarding the amount of material evidence obtained from crime scenes for forensic purposes. Mineralogical controls as well as fieldwork samples have been used to assess the potential for the use of different analytical methods in relation to criminal investigations where a sample is *in situ* on an item of clothing or fabric. The results which should be obtained are therefore known; thus the validity of the analytical techniques can be tested.

Some Examples of Non-destructive Analysis

Many analytical techniques exist which can be used to analyse a soil sample. Bull et al. (2006) discuss the importance of integrating different independent techniques in order to provide "forensic rigour" to an investigation. An important point is made in their paper where different independent analysis techniques were used during the analysis of soil samples obtained from a cast from a shoe print. They state that "given that there is enough material available for analysis…it should be possible to afford a meaningful analysis, comparison and interpretation of results" (Bull et al. 2006). This point is critical in relation to the research discussed in this chapter; only two of a number or analytical techniques which are currently being used for investigative purposes are presented here, namely colour and X-ray diffraction. However, the importance of non-destructive analysis can clearly be seen in that it allows a multi-proxy approach to the investigation as numerous techniques can be employed without destroying the sample. Independence of analysis techniques is particularly difficult with regards to non-destructive analysis as they are limited in number. Pollen analysis, for example, is an independent technique but results in the destruction of the sample.

Colour and Texture

In Conan Doyle's example of soil identification (1890), colour was used to discriminate where the suspect had been. Minerals contained within the soil contribute directly to the colour and it is therefore a key feature by which most geologic materials and soils can be identified. Murray (2004) states that "Colour is one of the most important identifying characteristics of minerals and soils". One of the main benefits in using colour as an analytical tool is the simplicity of use. There are, however, a number of factors that can affect the perception of colour (see Croft and Pye 2004).

The texture of a soil can also provide vital information with regards to the identification of where the soil may have originated. "Particle size is a fundamental property of any sediment, soil or dust deposit which can provide important clues to the nature and provenance" (Pye and Blott 2004). Use of microscopes can help differentiate those important distinguishing features which cannot be seen by the naked eye. In the 1920s Calvin Goddard, Charles Waite, Phillip O'Gravelle and John H. Fisher perfected the comparison microscope for use in bullet comparison (Inman and Rudin 2001). Use of the comparison microscope has greatly increased discrimination power between samples as it allows two items to be viewed concurrently, which can facilitate immediate examination of an item as being distinguishable from the relevant material upon visual analysis.

X- Ray Diffraction

Dr. Werner Kugler has already applied X-ray techniques to criminal cases. Research is required to test the limits of the technique in relation to *in situ* soil, rock dust and other crystalline materials adhered to fabric and other materials associated with crime suspects (Kugler 2003). Murray (2004) describes X-ray diffraction as "one of the most important and reliable methods of identifying the composition of crystalline substances". No two cases are ever the same. Therefore conducting research to highlight the limitations and benefits of the techniques available will aid the knowledge base and help in understanding the different sample preparations which may be needed in order to extract information from a sample. "The type, amount, and consistency of the suspected contact trace specimen, the involved contact trace carrier, and the forensic questions raised by the criminal offence determine the diffraction method used, the strategies of measurement applied, and the sample preparation technique selected" (Kugler 2003).

Discussion of Experimental Studies

Colour and Texture

Ongoing investigative work is discussed in which colour and texture analysis was undertaken alongside other analytical techniques of soil obtained from a wide variety of areas throughout Northern Ireland. Two field visits are presented relating to this area of work. During the first field visit, 16 soil samples (from puddles) in total were taken from four distinct locations, four samples per site. A field assistant also collected an additional nine blind samples from within these four locations, marked with coding unknown to the examiner. The blind samples were collected from areas within the same puddles that had been sampled in the original batch of 16. The aim of the exercise was to see if the blind samples could be compared to the 16 samples from known locations to identify where they may have originated. During the second field visit, 16 samples were taken by the examiner and a further 16 blind samples by the field assistant. During this visit, at each site one of the blind samples was taken from an area near the same puddle that the examiner had sampled and the other was taken from an area further away from the puddle sampled by the examiner but within the same location. Preliminary findings illustrate that visual comparison of samples for both colour and texture is easier when the sample is dry. During the first visit, seven out of nine could be seen to be comparable in terms of colour and texture when wet, nine out of nine when dry, and during the second visit, 13 out of 16 were comparable when wet and 16 out of 16 when dry.

X-Ray Diffraction

Initial research experimented with samples of known mineralogy, well crystalline kaolinite and sodium montmorillonite. Four substrates were chosen to try to replicate common types of material often procured at contemporary crime scenes; these were cotton, polyester, denim and rubber (Figure 14.1). Prior to impregnation each material had a 5 cm^2 of sample cut out from a grid system and analysed as a control

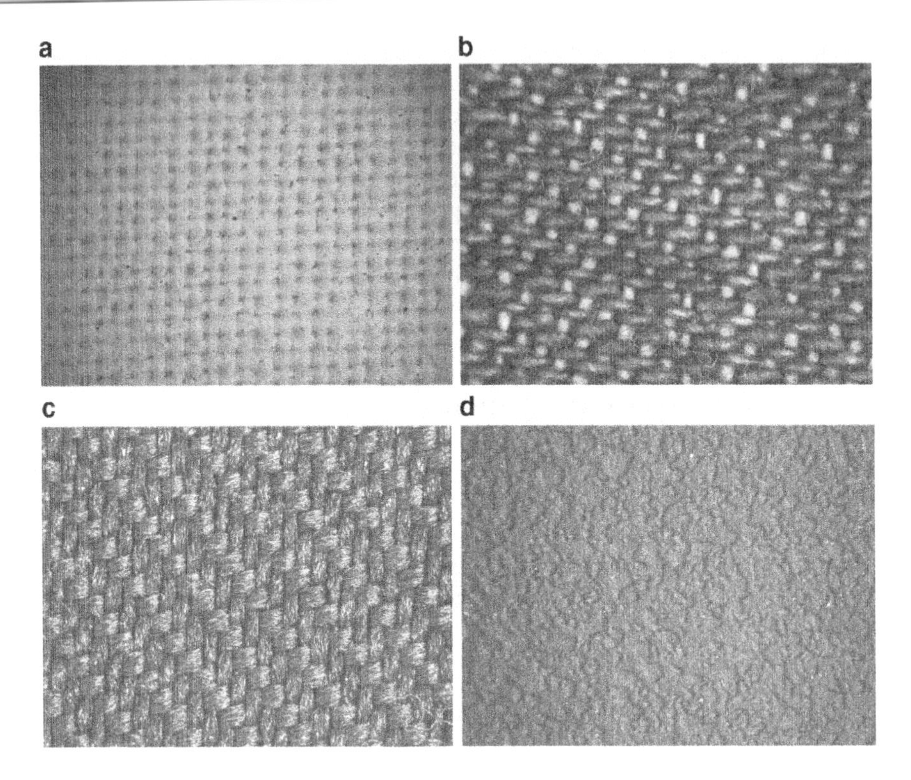

Fig. 14.1 Micrographs (1 cm wide) showing **(a)** cotton; **(b)** denim; **(c)** polyester; **(d)** rubber car mat. Note contrasting textures of each of the materials (*see colour plate section for colour version*)

of uncontaminated material, for the purposes of ascertaining whether they contain crystalline material introduced during material manufacture, item assembly or packaging, storage and sale (Figure 14.2).

The grid system was used in order to systematically analyse samples. For example, A1 was splashed with clay minerals, B1 was left blank, C1 was splashed with clay minerals, D1 was left blank, etc. A solution of pure kaolinite (Figure 14.3) and pure sodium montmorillonite was added to the four substrates as well as a mixture of the two minerals: 70% kaolinite 30% sodium montmorillonite; and 30% kaolinite 70% sodium montmorillonite (Figure 14.4). Figure 14.4 illustrates the XRD traces which should be visible on the different materials. Figure 14.5 shows the effect that the substrate material itself has on the XRD graphs. An experiment using kaolinite on rubber (Figure 14.5a) highlighted two issues, the first being that mud splashes do not adhere very well to this substrate for extended periods of time. This would be influential for the analysis of rubber mats obtained from a potential vehicle used in the execution of a crime or house mats from the scene of a crime.

As examples, their storage would be very important and they would need to be handled carefully to ensure that precious amounts of sample are not lost during transportation. The second issue was that kaolinite peaks could also be seen to be

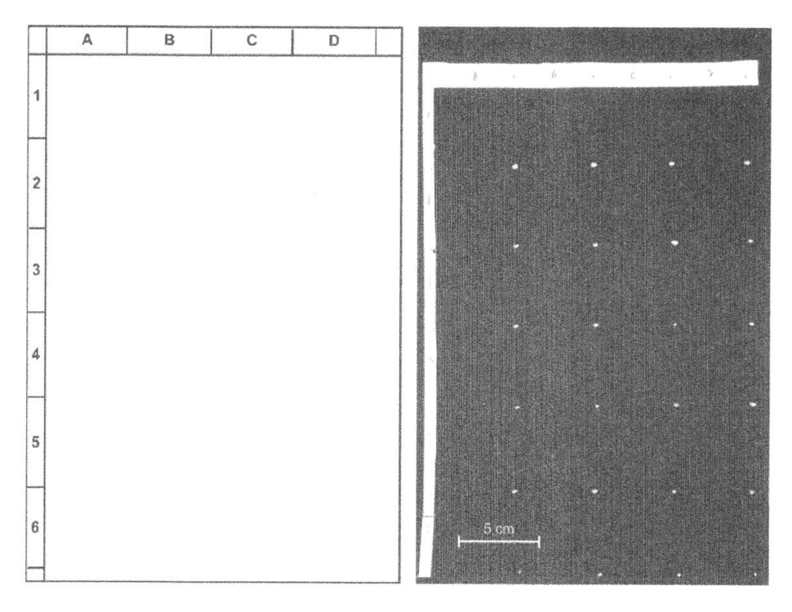

Fig. 14.2 Layout of grid system illustrating the systematic analysis of samples

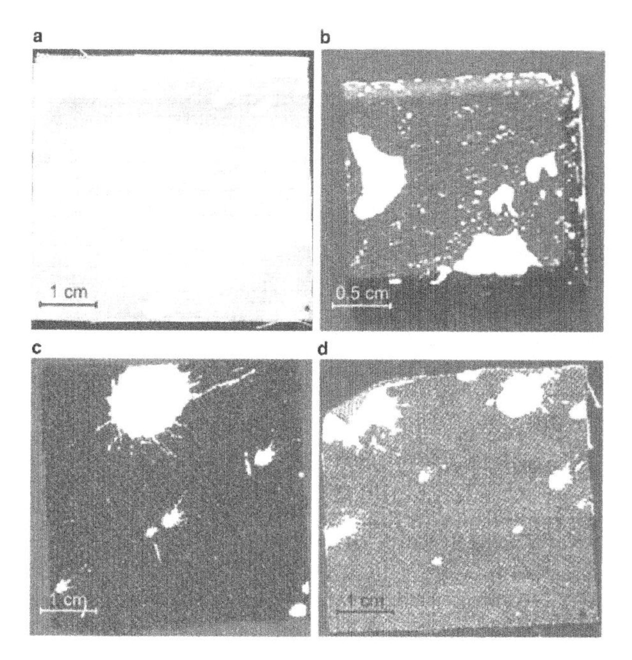

Fig. 14.3 Photographs of the cut portions of 5 × 5 cm materials with splashes of kaolinite to be irradiated by XRD. (a) Cotton; (b) rubber; (c) polyester; (d) denim (*see colour plate section for colour version*)

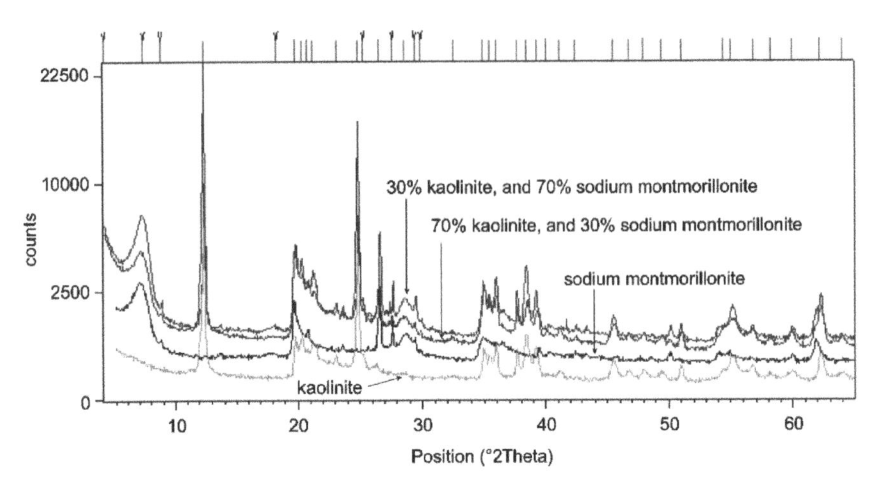

Fig. 14.4 X-ray diffractogram of standards kaolinite, sodium montmorillonite, 70% kaolinite – 30% sodium montmorillonite, 30% kaolinite – 70% sodium montmorillonite

apparent on the pure rubber substrate. The reason for this is because the clay mineral kaolinite is used in the production of rubber as a filler and therefore appeared on the control sample of rubber (British Geological Survey 2006). XRD analysis would therefore not be possible as a non-destructive method of analysis for rubber samples; the clay or soil adhering to the sample would need to be removed before analysis. Figure 14.5 shows the effect that the substrate can have on the results obtained. Figures 14.4 and 14.6 highlight this issue further. Figure 14.4 shows the graph that should be obtained from pure sodium montmorillonite, while Figure 14.6 illustrates the graph that was produced when sodium montmorillonite was applied to polyester. The fabric can clearly be seen to be masking the sodium montmorillonite which is known to be on the fabric. The peak at the beginning of the graph (characteristic of sodium montmorillonite) can still be seen. With further investigation it may be possible to highlight problematic mud splashes on particular substrates during analysis and provide information regarding the best way to proceed in relation to the next stage for the investigation.

Concluding Remarks

Investigative work into this type of criminal investigation relating to *in situ* sediment on fabric could also be applied to other aspects of criminal investigation. Suspect or suspect vehicle movement can be assessed using techniques such as those discussed above. A non-destructive, multi-proxy approach will ensure that as much information is obtained from the sample as is possible. In recent years it has also become increasingly important to apply analysis techniques to other areas of investigative work. An example of this can be seen in the shipping industry. As technology has improved, the amount of valuable goods being transported around the globe has increased dramatically as have

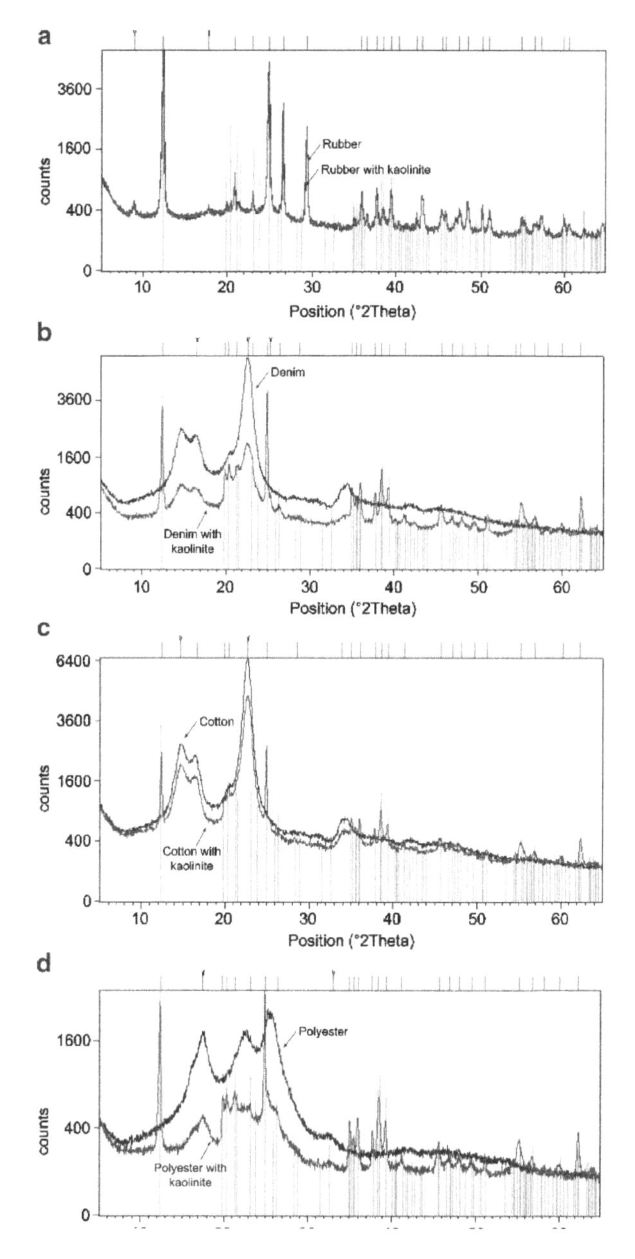

Fig. 14.5 X-ray diffractograms of cut sections of (a) rubber; (b) denim; (c) cotton; (d) polyester with splashes of kaolinite

the number of crimes associated with the shipment of packages. Commonly criminals use geological material as a replacement for new items which are stolen as they can easily replicate the weight of the parcels. Use of this geological evidence means that the route through which the packages were transported can be scrutinised and analysis

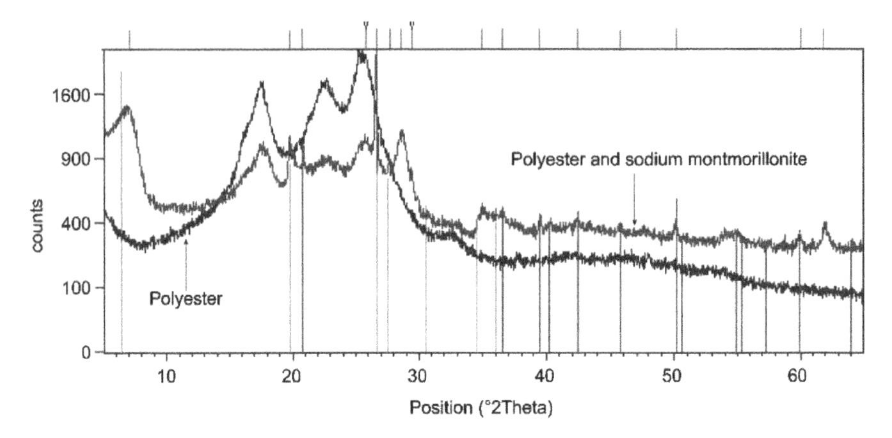

Fig. 14.6 X-ray diffractogram of 5 cm^2 cut section of polyester with splashes of sodium montmorillonite

of materials at each port or area of the country can be investigated to establish where the transfer may have taken place, similar to the work conducted by Professor Ehrenberg (Scientific American 1856). This type of work is invaluable not only to the companies from which the parcels have been stolen but also to insurance companies wishing to settle claims between the different firms employed to transport the packages. Kugler (2003) highlights another application of non-destructive analysis by analysing a suspected white anthrax specimen which was sent by mail after the attack on the World Trade Centre in New York. Kugler identified sucrose and gypsum to be present.

Cases such as these can be used to further highlight the importance for non-destructive methods of analysis. With expert evidence being called into question, destruction of a small amount of sample found at a crime scene would mean that that sample could never go on to be examined again. A multi proxy approach means that as much information as possible is obtained from a small amount of sample. It also means that other experts can come along independently of the case and analyse it in whatever way they wish to, for example, affording the defence a rare opportunity to instruct their own expert witness to corroborate or undermine previous results, thus providing an independent identification of where it is considered that the soil sample came from much in the same way that a piece of fingerprint evidence can be called into question by another examiner. If non-destructive techniques had been used when the case study presented at the beginning of the chapter had been examined, potentially more information could have been obtained from the minimum amount of sample which was available. Different independent techniques could have been used in a multi-proxy approach to the investigation before the sample was destroyed. It is hoped that with extensive research into this area, and given the non-destructive nature of the analysis techniques, it will in fact become the first port of call, providing a screening method enabling an educated decision to be made with regards to what other analytical techniques should be employed next.

Acknowledgements We thank Gary Galbraith for his assistance in the field, John Meneely and Mark Russell for their guidance with XRD analysis, Yoma Megarry for her assistance in the laboratory and Gill Alexander for help and guidance producing figures and illustrations. The work was considerably improved by the comments of the reviewers.

References

BBC (2007). Madeleine evidence 'may be lost.' www.news.bbc.co.uk/1/hi/uk/6761669.stm. Accessed 20th June 2007.

Bennett WW and Hess KM (2004). Criminal Investigation (7th Edition). Thompson Wadsworth, Belmont, CA.

Block EB (1958). The Wizard of Berkeley. Coward-McCann, New York.

British Geological Survey (2006). Mineral Planning Factsheet – Kaolin. www.mineralsuk.com/britmin/mpfkaolin.pdf. Accessed 20th June 2007.

Bull PA, Morgan RM and Dunkerley S (2005). Letter to the editor – 'SEM-EDS analysis and discrimination of forensic soil' by Cengiz et al. Forensic Science International 155:222–224.

Bull P, Parker A and Morgan R (2006). The forensic analysis of soils and sediment taken from the cast of a footprint. Forensic Science International 162:6–12.

Conan Doyle A (1890). Sherlock Holmes: The Sign of Four. www.sherlock-holmes.classic-literature.co.uk/the-sign-of-the-four/. Accessed 10th January 2007.

Croft DJ and Pye K (2004). Colour theory and evaluation of an instrumental method of measurement using geological samples for forensic applications. In: Forensic Geoscience Principles, Techniques and Applications (Eds. K Pye and D Croft), pp. 49–62. The Geological Society of London, London.

Evett IW (1991). Interpretation: A personal odyssey. In: The Use of Statistics in Forensic Science (Eds. CGG Aitken and DA Stoney), pp. 9–22. Ellis Horwood, London.

Hellerstein WE (2005). Freeing the innocent: Why so hard? New York Law Journal 233: No. 46.

Hilderbrand DS (1999). Footwear, the Missed Evidence – A Field Guide to the Collection and Preservation of Forensic Footwear Impression Evidence. Staggs, Wildomar, CA.

Houck M (2003). Trace Evidence Analysis. Elsevier Academic, London.

Inman K and Rudin N (2001). Principles and Practice of Criminalistics: The Profession of Forensic Science. CRC, New York.

Kugler W (2003). X-ray diffraction analysis in the forensic science: The last resort in many criminal cases in JCPDS – International Centre for Diffraction Data 2003. Advances in X-ray Analysis 46:1–16.

Morgan RM and Bull PA (2007). The philosophy, nature and practice of forensic sediment analysis. Progress in Physical Geography 31:1–16.

Morgan RM, Wiltshire P, Parker A and Bull PA (2006). The role of forensic geoscience in wildlife crime detection. Forensic Science International 162:152–162.

Murray RC (2004). Evidence from the Earth: Forensic Geology and Criminal Investigation. Mountain Press, Missoula, MT.

Olsen RD and Lee HC (2001). Identification of latent prints. In: Advances in Fingerprint Technology (Eds. HC Lee and RE Gaensslen), pp. 41–63. CRC, Boca Raton, FL.

Palenik SJ (2000). Microscopy. In: Encyclopaedia of Forensic Science (Eds. J Siegel, G Knupfer and P Saukko). pp 161–166 Academic, Press, New York.

Pye K and Blott SJ (2004). Particle size analysis of sediments, soils and related particulate materials for forensic purposes using laser granulometry. Forensic Science International 144:19–27.

Ruffell A and McKinley J (2008). Geoforensics. Wiley, Chichester.

Scientific American (1856). Science and Art: Curious Use of the Microscope. Scientific American, 11: 240. See: www.cdl.library.cornell.edu/cgibin/moa/pageviewer?coll + moa&root = %2Fmoa%2Fs. Accessed 21st November 2007.

Chapter 15
New Observations on the Interactions Between Evidence and the Upper Horizons of the Soil

Ian Hanson, Jessica Djohari, Jennifer Orr, Patricia Furphy, Claire Hodgson, Georgina Cox and Gemma Broadbridge

Abstract Since Darwin's work on the movement of objects in the soil due to earthworm action, interest has continued in determining how bioturbation affects the archaeological record. The work of Darwin is being continued with an additional focus on forensic implications of evidence moving over time. The actions of earthworms and the rates at which they cause small objects to sink into a given soil environment is predictable. Objects can accumulate over time on buried horizons, where once the horizon is identified they can be recovered. Small objects disappear from view in certain outdoor environments in a short timeframe with respect to forensic considerations. Experiments are being undertaken to test rates of sinking. Normal visual search techniques do not locate such evidence. As well as earthworm action, maggot masses feeding on a cadaver can rapidly cause small objects and bone to sink from view. The effects on vegetation growth and soil colour from decomposition of a body can indicate the primary deposition site of the body, even if it has been moved or dispersed. Case studies demonstrate how specific archaeological techniques have been used to maximise location and recovery of important evidence. Methods for consideration by Senior Investigating Officers suggest specialist professional support for some crime scene examinations will benefit forensic investigations.

Introduction

Charles Darwin first undertook observations and experiments to look at the effects of earthworm action on small objects in the soil (Darwin 1837), noting a tendency for objects to become covered and move downwards over time. Today, experiments and observation inspired by his work by the authors and others continue (e.g. Canti 2003). There are archaeological and forensic implications for the effects of earthworm action

---

I. Hanson(✉), J. Djohari, J. Orr, P. Furphy, C. Hodgson, G. Cox and G. Broadbridge
Centre for Forensic Sciences; Centre for Archaeology, Anthropology and Heritage, School of Conservation Sciences, Bournemouth University, Bournemouth BH12 5BB, UK
e-mail: ihanson@bournemouth.ac.uk

K. Ritz et al. (eds.), *Criminal and Environmental Soil Forensics*

(one part of the bioturbation process), especially those caused by invertebrates, on the position of evidence on the ground and in the soil. Forensic evidence, in the form of small objects, can be difficult to detect when deposited in an outdoor environment. For example, bullets, shell cases, small human bones, bone fragments, teeth, coins, clothing elements (such as buttons), murder weapons, personal effects and other trace evidence may be small in size and difficult to detect in the 'background noise' of a crime scene. These physical items provide intelligence and are often vital in identifying victims and perpetrators, and associating both of these to scenes, and timelines at scenes. Recovering and predicting where to find such evidence is obviously central to strategies for managing crime scenes. In outdoor scenes this evidence often does not remain in its original place of deposition but moves. From an archaeological and forensic perspective, developments in the prediction of the movement of such objects in the upper horizons of soil such as leaf litter, topsoil and upper subsoil (i.e. O, H and A horizons*) can inform how search, location and recovery strategies for artefacts and evidence can improve and become more efficient and accurate.

We can view bioturbation as the interaction between animals, plants and soil materials during which the soil fabric is altered (Grave 1999). This has been widely discussed over time in terms of earthworm activity and object movement (e.g. Darwin 1881; Hudson 1919; Keith 1942; Webster 1965; Yeates and Van der Meulen 1995; Canti 2003). The main focus has been on how surface objects and the archaeological record are affected by the processes of soil homogenisation and differential size sorting in soils (Grave 1999).

Although bioturbative activity may be caused by root action, drying, frost heave, burrowing etc. (Rolfsen 1980), much of the focus of the literature above is on earthworm action. It is the uniformity of their activity within a particular soil and environment that has created much interest; a soil in a particular environment with an effective earthworm population will generally undergo the same effects across its whole area. This is of interest archaeologically because it offers scope for predictive effects.

We also know that soils and their earthworm populations vary between environments and so rates of bioturbation due to earthworm activity differ and are affected by variables such as species, population, latitude, climate, temperature, season, geology, ecosystem, soil content, pollution, moisture, pH, food sources etc. (e.g. Lee 1985; Ligthart and Peek 1997; Shipitalo and Butt 1999). Taking this into account, depending on environment and climate, we know that a soil worked by earthworms may be moved at predictable rates in its upper horizons. Earthworm action is rapid in both an archaeological and forensic time frame.

To predict and determine where objects may come to rest, all that is needed is quantative verification of these effects in a given environment to assist with assessing probabilities of movement. It is also important to realise that, where earthworms are absent, movement of objects may be limited (Limbrey 1975). These points are

*See soil science textbooks such as Brady and Weil (2001).

important for archaeological and forensic investigation; knowing where to look and where not to look for these effects may be key to effective search.

Predicting the Movement of Small Objects in the Soil

Darwin maintained his interest in earthworms and their activity throughout his years of scientific research (Darwin 1837, 1840, 1881). This spurred a continued interest in earthworm action in upper soil and archaeological deposits by his son (Darwin 1901) and others (see above). This work is very important as it has changed perceptions that once artefacts are in the archaeological record they remain static. There is still a common presumption and misconception in archaeological fieldwork that this is the case, and the extent of earthworm action may often be underestimated on excavations. Stasis may be the case in some environments but, in many soils and archaeological deposits, earthworm action is great enough to cause objects to sink and accumulate on surfaces and horizons. It is also true that trace evidence such as plant material, fibres, hairs or small bones may to be dragged into burrows and migrate downwards through archaeological boundaries (Davis et al. 1992; Canti 2003; Figure 15.1). This may cause misinterpretation of the context of artefacts, and affect some fundamental archaeological principles such as Worsae's Law (Rowe 1962) and Cornwall's premise (1958) that the relative sequence of objects will not be changed by earthworm action. These principles are of course dated; we have more recent research for rates and sizes of objects sinking (Stein 1983; Armour-Chelu and Andrews 1994; Yeates and Van der Meulen 1995; Canti 2003) and these suggest there is potential for objects to move between upper soil horizons.

Fig. 15.1 Earthworm burrows through archaeological deposits. Section showing effect of vertical and horizontal burrows through agricultural topsoil into the numerous layers within an Aceramic Neolithic (10,000 BP) plaster mixing pit. Such earthworm action can disrupt attempts to define contexts, accurately carbon date layers and fills, differentiate and date deposits by artefact type and define palaeobotanical assemblages by context. Scale bar = 50 cm (source: Hanson) (*see colour plate section for colour version*)

Much research needs to be done to refine the testing of these principles. New observations at Down House in Kent on object movement by Keith (1942), Hanson (2006) and Butt et al. (2008) have allowed continuation of Darwin's work, relocating and re-examining his experimental areas and further testing their properties. New test pits have demonstrated the location and also the properties of artefacts placed by Darwin can still be evaluated beyond 160 years. This provides a rare opportunity to continue an experiment over a considerable time (Butt et al. 2008). New experimental plots have been set up to provide a comparison to Darwin's findings. This allows data to be looked at from a new perspective, comparing long term stasis and short term change at the same location; this assists in developing procedures for archaeological and forensic investigation.

Darwin noticed cinders and chalk he placed on fields (in the 1840s) had descended to a depth of 18 cm when he excavated test pits after 29 years (Darwin 1881). He, as have others, determined that the movement was due to two actions, firstly earthworms bringing soil in the form of casts up from their burrows, and horizontal and vertical burrows collapsing over time forming voids into which objects drop (Canti 2003). Keith (1942) recorded that the objects noted by Darwin were in a band at 15 to 18 cm depth. The objects were in a band at the same depth in 2003 when the authors dug test pits, the larger cinders resting on a flint layer appearing at 14 to 18 cm. The implication is that there is a general trend for objects to sink to a distinct horizon over time in certain conditions, and reach a stasis. This is determined by two things, firstly prevention of further sinking by a solid barrier (at Down House this is a layer of flints) or the lowest depth at which earthworm action is effective in bioturbative terms. Most earthworm species operate energetically in combination in the topsoil, often the top 15 to 20 cm (Atkinson 1957; Canti 2003), though some species such as *Lumbricus terrestris* are deep burrowers, and suitable soil conditions in archaeological deposits have seen earthworms observed living in very deep habitats (Hudson 1919).

Darwin and Keith both observed that objects other than those that were deliberately deposited ended up on this horizon. Although they did not consider this to be of particular significance, it is for those interested in archaeology and forensics. Objects in certain environments will collect over time on horizons which are, in effect, artefact/evidence traps. This allows predictable location and collection from an identified interface. Since 2003 at Down House, glass, ceramics, clay pipe and barbed wire have been collected from the 18 cm horizon, and some of their dates of manufacture were after the 1840s (Figure 15.2). A clay pipe stem found was from a type not in production until 1850; barbed wire was not patented until 1867. They provide a *terminus post quem* for the date they could have entered the ground, and must have been dropped and descended to the horizon long after Darwin began his experiment. Some artefacts were, no doubt, there before he started, such as 18th Century ceramic sherds and brick fragments. While the phenomenon aids in collecting evidence and artefacts, it limits the ability to date horizons using the objects accumulated or determine when objects came to rest there. Of course taphonomic effects will mean that artefact survival is selective; some materials (such as ceramics) last longer than others (such as hair) in certain buried environments and this needs to be considered.

Fig. 15.2 Contemporary observations at Down House. Test trench in plan revealing the cinders deposited on a horizon at 14–18 cm. A brick fragment can also been seen. The flint layer on which the cinders come to rest can be seen at 18 cm (source: Hanson) (*see colour plate section for colour version*)

Does Movement Happen Fast Enough to Have an Evidential Impact?

Although most studies have considered long term bioturbative effects, objects do sink fast enough to become buried within the time frame of investigative interest for cases such as homicide and missing person's searches.

Darwin's hypothesis, which he confirmed experimentally, was that small objects would after some years be found lying at the depth of some distance (of the order of inches, i.e. some centimetre) beneath the ground, but still forming a layer. He calculated the rate of descent on the test area – chalk downland pasture – as approximately 0.60 cm per year. However, he appears to have calculated this by dividing the depth of objects (18 cm) by the time of burial (29 years, at which time he excavated test trenches). There is no way to know, therefore, when the objects reached this depth during that time period. This means the rate of burial may be faster than the rate Darwin calculated for that environment. To test this, contemporary experiments are being undertaken to gain comparative data both at Down House and in the lab. In controlled containers of soil with mixed earthworm populations of *Lumbricus terrestris* and *Eisenia foetida* that match natural population levels (Svendsen 1955; Loh et al. 2005), objects of varying sizes were placed, and rates of sinking due to casting and collapsing burrows were observed. After a month experiment, results suggested that small objects (standardised metal disks) in these conditions may sink at rates of approximately 3 cm per annum when the diameter of the object is less than 3 cm. By contrast, in an experimental plot at Down House, after 9 months objects less than 3 cm in diameter had sunk 1 cm. In both cases the larger the object diameter area, the slower the rate of sinking. From these initial results it can be suggested the objects Darwin placed might have reached their horizon at a faster rate than he calculated; between 6 to 18 years. The implication is also that small objects disappear from surface view rapidly in some environments. Objects with larger surface areas may have earthworms casting underneath them and may be raised rather than sink, and are limited by effects of burrow collapse. Small objects sink at different rates depending on surface area, diameter and weight. It is also clear that small objects are, to a limited extent, moved horizontally as well as vertically by earthworm action.

Experiments show that in environments that include vegetation growth and death cycles, the timeframe in which objects are lost from view is rapid. For example, an experiment to determine how rapidly objects filtered down from view in beech woodland leaf litter found that objects such as keys, coins, small bones and teeth, when placed on the ground surface, were lost from view in 3 to 4 weeks due to *Lumbricus terrestris* earthworms moving the decaying vegetation. Experiments in a different environment, on a turf plot at Down House, show that filtering of cinders and small metal discs downwards due to plant/root growth and earthworm movement and casting, lead to a less rapid loss from view. Much of this was due to grass growth in the first 3 months and, after 6 to 9 months, objects had filtered to the base of plant stems and had started to be covered by earthworm casts and root growth (Figure 15.3). The same effect and timeframe was seen with personal jewellery on a turf plot in Dorset. On a plot of bare soil at Down House, it took 9 months for objects less than 3 cm in diameter to be mostly hidden and obscured by earthworm casts and start to drop into voids caused by collapsing earthworm burrows. There is a clear difference in the speed at which small objects are lost from view when placed on vegetation and leaf litter, compared to descent rates once they are in contact with the

Fig. 15.3 Lost from view: Filtering of cinders (originally deposited by Darwin, now white washed and re-used) into the turf root mat at Down House 9 months after deposition. Placed in early August, they were lost from view in 3 months due to grass growth. Earthworm casts are also starting to cover them (source: Hanson) (*see colour plate section for colour version*)

topsoil. This has implications for search strategies. If a crime scene is being searched concerning events that occurred 6 to 12 months previously, in some environments evidence of small size will not be seen by normal procedural techniques such as line searches; they will be under leaf litter, vegetation, casts and moving downwards. This barrier may also limit the effectiveness of search dogs. This impacts upon crime scene procedure, and specific strategies need to be used to find such evidence. At present, many searches rely on lines of officers using visual observation and moving at relative speed to locate evidence across wide areas. This approach will miss (and has missed) evidence that has moved out of view.

The Effect of Maggot Masses on Objects in the Soil

As well as earthworm action, other invertebrate activity rapidly moves small objects into leaf litter and topsoil. Observation of decomposition of deer carcasses during experiments to observe scavenger behaviour has noted rapid movement of small bones, personal effects and trace evidence such as hairs due to the action of maggot masses. On fresh corpses during warmer months of the year, blow flies will rapidly find and lay eggs on wounds and orifices (Haskell et al. 1997; Gennard 2007). Maggots develop rapidly and in favourable conditions form a large feeding mass in the corpse (Haskell and Williams 1990). The maggot mass is made up of thousands of individuals and forms what has been described as a 'feeding entity' weighing several kilograms (Fitzgerald B, pers. comm. 2007) that is capable of moving clothing and small objects. Experiments undertaken to observe scavenging activity have recorded the energetic effect of maggot masses (Komar and Beattie 1998; Morton and Lord 2006). Experiments in Dorset on scavenging effects on deer carcasses have shown the movement of maggots away from the centre of the mass to cool down and their subsequent return to the centre to re-warm and feed appear to create what might be called a 'convective' cyclical motion that causes movement of trace evidence and objects downwards (Figure 15.4).

The liquid 'soup' in which the maggot mass develops also encourages filtering of objects downwards as it envelops surrounding leaf litter, vegetation and topsoil. Maggot masses rapidly consume a body's tissues (Morton and Lord 2006) and when they disperse (Gennard 2007), it may take only 1 to 2 weeks from death to the greatest degree of soft tissue loss and filtering of small bones, hair and objects into the leaf litter and top soil. Unlike earthworm action, maggot mass activity tends to take place only in close proximity to corpses and during warmer weather.

Experiments in Dorset have also shown that, after dispersal of remaining tissue and skeletal elements by scavengers, staining of the soil and the dying-off of vegetation indicate where a corpse lay in its primary position of deposition. These can remain as an indicator for many months, depending on environment. Vegetation may die off, due to the nature and effects of decomposition fluids such as cadaverine (Figure 15.5), and clay soils may undergo a colour change, turning soil a distinctive dark grey/ blue colour, thought to be the result of reduction of iron in soil by bacteria. This has

Fig. 15.4 Maggot mass purging from rear of a roe deer 17 days after death: Note the transport of hair away from the body into the leaf litter. Scale bar = 20 cm (source: Hodgson) (*see colour plate section for colour version*)

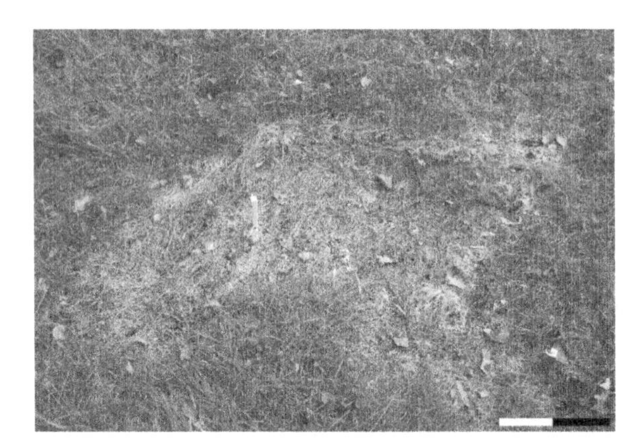

Fig. 15.5 Vegetation die-off caused by decomposition of a sika deer: the original position of the cadaver can clearly be seen, giving the body outline. Visible for 8 months after carcass is dispersed by scavenger activity, until obscured by new seasonal vegetation growth. Scale bar = 20 cm (source: Broadbridge) (*see colour plate section for colour version*)

been noted in graves investigated for the War Crimes Tribunals (c.f. Wright et al. 2005). These are excellent indicators of the primary deposition site of a body. The dragging of a body by scavengers to secondary locations can lead to misinterpretation of the original dump site and these phenomena can be of great assistance to crime scene investigators; something of evidential value is always left at the primary site. The half section excavation of a body deposition site of a sika deer in open mixed woodland in Dorset (Figure 15.6) found unfused epiphyses, teeth and hair had filtered to depths of up to 7 cm after 3 months. To search to these depths beneath the surface, especially when a body has been scavenged and dispersed, may not seem intuitive to crime scene investigators.

Fig. 15.6 Half section excavation underneath deer deposition site: This located small bones and hair brought down by maggot activity to the horizon between the leaf litter/humic layer and the more compact subsoil. The stain to the soil can clearly be seen (source: Orr and Furphy) (*see colour plate section for colour version*)

Do these experimental observations match what is observed in real forensic cases? Searches for skeletal remains and personal artefacts and evidence on police cases in the UK and during international missions for the UN in Africa have revealed the evidential horizons described. One case concerning a search for remains of an individual revealed a skull and major bones on heath land in the south of England. Only partial recovery of skeletal remains was managed using several line searches and subsequent cadaver dog search. The remains were known to have been scavenged, and the presumption was that they had been dispersed and no more skeletal elements would be recovered. Bournemouth University Centre for Forensic Sciences then undertook subsequent searches 4 and 5 years after the death of the individual. Gridding the area to ensure systematic ground coverage and a fingertip search were undertaken, with removal of leaf litter debris layers, and root mat layers to a horizon at the interface with the humic layer. Some bones such as ribs were found in these layers; it was extremely difficult to differentiate these from small bleached heather twigs. Removal of these layers revealed a horizon to which many small fragmented bones, teeth, small whole bones and personal items had descended. This horizon was not visible from the ground surface. Systematic uncovering of this horizon led to the location of a concentration of remains providing evidence of the original deposition site for the body. Some 100 additional bones and fragments were recovered that had not been detected by conventional search methods. However this intensive search took five working days with 25 student volunteers.

Similar results were found in a different environment searching for graves at a suspected execution site in a compound in tropical Africa in an area of dense undergrowth that required a stripping of vegetation, leaf litter and topsoil layers. Nothing was visible from the ground surface and a metal detector failed to detect many metal objects. A test trench revealed a rich organic humic topsoil some 20cm thick overlying a compact clay subsoil. The humic layer contained abundant fauna, including earthworms, beetle grubs, termites, ants, other invertebrates and fine roots, indicating significant bioturbative potential. This was an environment with a climate, soil type and biology where soil homogenises at faster rates than those encountered in temperate climates (Madge 1965). Stripping off this topsoil not only revealed the outline of several shallow graves cut into the subsoil (the cuts for which had been lost from view in the humic layer due to soil homogenisation), but also shell cases scattered across the area and sitting on the distinct horizon between the topsoil and subsoil. The head stamps from the shell cases indicated the artefacts dated to different periods, and were from different countries of origin. Other metal objects were found to have accumulated on this horizon, just out of range of the metal detector used at ground level. Subsequently it was found the graves were not related to the shell cases, which appear to have accumulated over 60 years during the compound's use as a military base in different conflicts, and pre-dated the events of investigative interest. The concentration of shell cases was misleading and might have been interpreted as evidence of mass execution, if the phenomenon of artefact accumulation had not been appreciated.

Procedures for Recovery of Buried Small Objects

What are the implications of these phenomena for search and recovery? There is an evidential paradox for small objects in that, while they filter down into leaf litter and topsoil over time and in doing so are potentially lost to searches, this burial provides a predictable collection zone and acts as potential protection for evidence (Hanson 2004; Cheetham et al. 2007), that may otherwise disperse beyond recovery in the environment.

Searches for buried horizons, even if a few centimetres beneath the surface of the ground, are more complex than surface searches. A visual scan of the ground surface by personnel in a shoulder-to-shoulder line can rapidly identify clusters of evidence and individual objects, and considerable areas can be searched relatively quickly. This can be undertaken by relatively unskilled personnel. Searches for buried horizons need personnel with archaeological skills, training and experience to find and follow horizons on which evidence may settle.

Buried horizons can be found by excavating test trenches to reveal a complex stratigraphy of the leaf litter; the H, O and A horizons. Geologically and archaeologically, these horizons are often not appreciated or are ignored or seen as peripheral. However, careful examination can reveal stratigraphy within these layers representing recent build up of deposits, on which and through which objects may filter. To maximise preservation and detection, stratigraphic principles of excavation should be employed to reveal the horizon, removing the uppermost deposits in sequence. Interpretation of test trenches dug within the search area allows an assessment of the potential for bioturbative effects in that particular environment.

Locating a horizon on which objects are accumulating is the most straightforward part of recovering such evidence. Following the horizon and extending its exposure to reveal relevant evidence to the widest extent means undertaking a sub-surface archaeological finger-tip search; this is a complex archaeological procedure. Layers 'feather edge' and fade away, others may sub-divide indicating additional stratigraphic events at that location. Horizons may transect or overly earlier stratigraphy or may be cut by later intrusive features or disturbed by natural phenomena such as animal burrowing or water erosion. Observing, identifying and interpreting the nature and potential discontinuity of such fragile stratigraphy is a complex, technical, ongoing interpretative archaeological exercise. 'Peeling' away each stratigraphic deposit that contains evidence with trowels should be done systematically by teams working in lines moving in one direction. All material removed needs to be sieved; perhaps only 40–60% of small objects or trace evidence within layers will be located during excavation, the rest will be recovered during sieving. The act of trowelling creates loose material in which evidence becomes obscured even as a deposit is revealed and removed. Sieving also functions as a confidence measure; it is a demonstration of having been systematic and thorough in a search.

If a horizon on which evidence rests is uncovered, the percentage recovered *in situ* will be higher, as the layer above the horizon can be 'peeled' off, exposing the accumulated objects. It should be born in mind than *in situ* for evidence that has filtered down through bioturbation is not *in situ* in terms of its original evidential crime scene position, but often retains a general horizontal spatial relationship with its original position, and other evidence. How removal of stratigraphic layers affected in this way is undertaken (e.g. whether by 10, 5 or 1 cm spits) therefore depends on investigational imperatives in terms of time, resources and staff available and the degree to which trace and other evidence is required to answer investigative questions. Of course, unless the value of these techniques is understood for certain crime scenes, Senior Investigating Officers may not fully appreciate the evidential potential at a scene; appreciation that should mould the procedural approaches at the outset to ensure investigative questions have the best chance of being answered.

If there is a requirement to maximise recovery of small objects or trace evidence *in situ* to fully understand their contextual value, then excavation should be in 1 cm spits; evidence will not be found in place using a less refined technique. The implication of time, excavating skills and staffing for this kind of detail are obvious. The methods described have value for both archaeological and forensic investigations; all excavation techniques are simply ways to achieve an ever more detailed and focused search for evidence (Cheetham and Hanson 2008).

Detailed Evidence Recovery that Requires Informed Specialist Search

To maximise evidence recovery from scenes, understanding the site formation processes before and after a crime will assist in evidence identification, recovery, scene reconstruction and interpretation. The potential for determining evidence locations depends on recognition of the variables that affect a particular environment and its innate potential for bioturbation. This requires a combination of archaeological, pedological, botanical, ecological and environmental appreciation and assessment; specialist inputs for scene search and examination. Recovering evidence from these scenes requires skilled archaeological excavation. The examples described show that evidence and remains that are not detected and may be undetectable using conventional methods of line search, cadaver dog and metal detector search can be located and recovered. Locating horizons to which evidence filters down and revealing their extent in the area of interest is an advanced and time consuming process. It recovers vital contextual evidence and timeline intelligence that answer investigative questions and allow maximum evidence recovery and effective crime scene interpretation to occur.

Acknowledgements Many thanks to Barry Fitzgerald, Dorset Police and Paul Cheetham, Bournemouth University for assistance in design and implementation of field experiments; Kevin Butt and Chris Lowe of UCLAN for data on earthworm populations from Down House; Toby Beasley, Head Gardener, Down House for permissions, advice and access; Renee Kosalka, Joe Partridge, Jo Laver, Ambika Flavel and Danny and Laura Webb in gathering data from Down House; Emeritus Professor Richard Wright, Sydney University and Graham C Wilson, Archaeological Consultant, Sydney for advice on artefact dating; and the many Bournemouth University students who have assisted with these experiments.

References

Armour-Chelu M and Andrews P (1994). Some effects of bioturbation by earthworms (Oligochaeta) on archaeological sites. Journal of Archaeological Sciences 21:433–443.
Atkinson RJC (1957). Worms and weathering. Antiquity 31:219–233.
Brady N and Weil R (2001). The Nature and Properties of Soil. 13th Edition. Prentice Hall, New Jersey.
Butt KR, Lowe CN, Beasley T, Hanson I and Keynes R (2008). Darwin's earthworms revisited. European Journal of Soil Biology 44(3):255–259.
Canti MG (2003). Earthworm activity and archaeological stratigraphy: a review of products and processes. Journal of Archaeological Sciences 30:135–148.
Cheetham P and Hanson I (2008). Excavation and recovery. In: World Archaeological Congress Handbook of Forensic Anthropology and Archaeology (Eds. D Ubelaker and S Blau). Left Coast, California.
Cheetham P, Cox M, Flavel A, Hanson I, Haynie T, Oxlee D and Wessling R (2007). Search, location, excavation and recovery. In: The Scientific Investigation of Mass Graves (Eds. M Cox, A Flavel, I Hanson, J Laver and R Wessling). Cambridge University Press, Cambridge.
Cornwall IW (1958). Soils for the Archaeologist. Phoenix House. London.
Darwin C (1837). On the formation of mould. Transactions of the Geological Society of London 2:574–576.
Darwin C (1840). On the formation of vegetable mould. Transactions of the Geological Society of London 5:505–509.
Darwin C (1881). The Formation of Vegetable Mould through the Action of Worms, with Observations on Their Habits. John Murray, London.
Darwin H (1901). On the small vertical movements of a stone laid on the surface of the ground. Proceedings of the Royal Society 68(446):253–261.
Davis B, Walker N, Ball D and Fitter A (1992). The Soil. Harper Collins, London.

Gennard DE (2007). Forensic Entomology. Wiley, England.
Grave P (1999). Assessing bioturbation in archaeological sediments using soil morphology and phytolith analysis. Journal of Archaeological Science 26:1239–1248.
Hanson I (2004). The importance of stratigraphy in forensic investigation. In: Forensic Geoscience: Principles, Techniques and Applications (Eds. K Pye and DJ Croft). The Geological Society of London Special Publication 232, Bath.
Hanson I (2006). A report on the continuation of Darwin's experiments on the formation of vegetable mould at Down House and further considerations on the movement of objects in soil due to worm action. A Report for English Heritage. Bournemouth University.
Haskell NH and Williams RE (1990). Collection of entomological evidence at the death scene. In: Entomology and Death: A Procedural Guide (Eds. EP Catts and NH Haskell). Joyce's Print Shop, Clemson, SC.
Haskell NH, Hall RD, Cervenka VJ and Clark MA (1997). On the body: insects' life stage presence, their post-mortem artefacts. In: Forensic Taphonomy: The Post-mortem Fate of Human Remains (Eds. WD Haglund and M Sorg). CRC, Boca Raton, FL.
Hudson WH (1919). The Book of a Naturalist. Hodder and Stoughton, London.
Jewell PA (1958). Natural history and experiment in archaeology. The Advancement of Science 59:165–172.
Keith A (1942). A postscript to Darwin's "formation of vegetable mould through the action of worms". Nature 149:716–720.
Komar D and Beattie O (1998). Post-mortem insect activity may mimic perimortem sexual assault clothing patterns. Journal of Forensic Sciences 43(4):792–796.
Lee KE (1985). Earthworms: Their Ecology and Relationships with Soil and Land Use. Academic, London.
Ligthart TN and Peek GJCW (1997). Evolution of earthworm burrow systems after inoculation of lumbricid earthworms in a pasture in the Netherlands. Soil Biology and Biochemistry 29:453–462.
Limbrey S (1975). Soil Science and Archaeology. Academic, London.
Loh TC, Lee YC, Liang JB and Tan D (2005). Vermicomposting of cattle and goat manures by *Eisenia foetida* and their growth and reproduction performance. Bioresource Technology 96:111–114.
Madge DS (1965). Leaf fall and litter disappearance in a tropical forest. Pedobiologia 5:273–288.
Morton RJ and Lord WD (2006). Taphonomy of child-sized remains: a study of scattering and scavenging in Virginia, USA. Journal of Forensic Sciences 51:475–479.
Rolfsen P (1980). Disturbance of archaeological layers by processes in the soil. Norwegian Archaeology Review 13:110–118.
Rowe JH (1962). Worsae's law and the use of grave lots for archaeological dating. American Antiquity 28:129–137.
Shipitalo MJ and Butt KR (1999). Occupancy and geometrical properties of *Lumbricus terrestris L* burrowing affecting infiltration. Pedobiologia 43:782–794.
Stein JK (1983). Earthworm activity: a source of potential disturbance of archaeological sediments. American Antiquity 48(2):277–288.
Svendsen JA (1955). Earthworm population studies: a comparison of sampling methods. Nature 175:804.
Webster R (1965). A horizon of pea grit in gravel soils. Nature 206:696–697.
Wright R, Hanson I and Sterenberg J (2005). The archaeology of mass graves. In: Forensic Archaeology: Advances in Theory and Practice (Eds. J Hunter and M Cox). Routledge, London.
Yeates GW and van der Meulen H (1995). Burial of soil-surface artefacts in the presence of lumbricid earthworms. Biological Fertility of Soils 19:73–74.

Chapter 16
The Forensic Analysis of Sediments Recovered from Footwear

Ruth M. Morgan, Jeanne Freudiger-Bonzon, Katharine H. Nichols, Thomas Jellis, Sarah Dunkerley, Przemyslaw Zelazowski and Peter A. Bull

Abstract The forensic analysis of sediments recovered from footwear has the potential to yield much useful information concerning the movements of a person before, during and after a crime has taken place. Three experimental studies and a number of examples of forensic casework provide insight into the complexity of the spatial distribution of geoforensic materials on the soles of footwear and the persistence of these materials over time on the soles and uppers. These findings have implications for both the geoforensic sampling protocols and procedures for footwear submitted for analysis in a criminal investigation and also for the analysis of any materials recovered. The preservation of sediment on a shoe sole will vary, with certain areas generally retaining more sediment than others. The sequential layering of sediments that have been transferred to the shoe will be preserved in some cases and in certain areas, but generally undergoes complex mixing. Such mixing of sediment from different sources occurs both across the shoe sole and also through time. It is therefore important to be aware of these variations when taking samples for analysis if representative samples are to be taken and meaningful interpretation of any analysis derived is to be effected. Furthermore, such mixing of pre-, syn- and post-forensic event sources has implications for the appropriateness of different analytical techniques. Visual identification techniques which are able to identify where such mixing has taken place are preferred to forms of analysis that require homogenisation of the sample prior to analysis, as this reduces the possibility of false negative or positive associations when undertaking comparison of samples in a forensic context.

R.M. Morgan(✉)
UCL Jill Dando Institute of Crime Science 2nd Floor Brook House,
Torrington Place London, WC1E 7HN, UK.
e-mail: ruth.morgan@ucl.ac.uk

J. Freudiger-Bonzon, K.H. Nichols, T. Jellis, S. Dunkerley, P. Zelazowski and P.A. Bull.
Oxford University Centre for the Environment University of Oxford,
South Parks Road Oxford, OX1 3QY, UK.

J. Freudiger-Bonzon.
Faculty of Geosciences, University of Fribourg, Switzerland.

J. Freudiger-Bonzon.
Faculty of Geosciences and the Environment University of Lausanne Switzerland.

K. Ritz et al. (eds.), *Criminal and Environmental Soil Forensics*

The context within which any sampling or analysis is undertaken is crucial for a meaningful and accurate interpretation of the geoforensic evidence.

Introduction

The forensic analysis of soils and sediments is a rapidly developing field that has its roots in the geosciences and which applies geoscience principles to the forensic arena (Morgan and Bull 2007a, b). The underlying premise in geoforensic study is that evidence will be transferred from sources to recipient mediums (such as clothing, vehicles, etc.). The concept that 'every contact leaves a trace' was first articulated by Locard (1928, 1930) and these ideas have been developed more recently by Inman and Rudin (2002). The analysis and interpretation of geoforensic evidence also draws upon the body of literature concerning other forms of trace physical evidence, particularly with regard to the nature of evidence transfer and persistence (e.g. Pounds and Smalldon 1975a, b, c; Hicks et al. 1996; Roux et al. 1999; Wiggins et al. 2002).

The deposits found on and in footwear have proved an attractive source of comparator samples in geoforensic studies. Since footwear is in contact with the ground, there appears to be a reasonable opportunity to compare materials from the footwear with pertinent scenes related to a forensic event. The reality is, however, far more complicated – the devil is in the detail: Firstly, the very transfer of materials onto the soles of footwear will vary in relation to a number of physical characteristics (grain type and size, organic content) (Chazottes et al. 2004; Virtanen et al. 2007) and secondly, the transfer of materials is rarely onto a surface that does not already hold materials which may have been deposited before the forensic event in question. Likewise, materials on, or in, a shoe may derive from sources encountered after the particular forensic event and thus there may be at least three phases of sediment transfer which may themselves not be evenly distributed across the sole or upper of the shoe. Once transferred to footwear, the absolute amount of trace material will start to decrease and the persistence of such materials becomes a very relevant consideration. Paradoxically, the longer material is able to survive on footwear, the more problematic the interpretation of such evidence may become. However, if the material does not persist long enough to be collected, it will not be of any evidential value and will, in turn, impact upon the results of the crime reconstruction.

Sampling procedures, whether for physical, chemical or biological analyses, must tation will depend heavily upon these previous constraints (Morgan and Bull 2007b). Whether it is possible to identify an exclusion or to differentiate a false-positive exclusion is a matter that is considered here. Further, we address a number of other pertinent issues relating to the collection, analyses and interpretation of geoforensic evidence on footwear which fall broadly within the themes of spatial distribution and persistence.

The spatial considerations are: whether sequentially deposited layers of material are preserved in the chronological order in which they were transferred to the footwear sole; whether mixing (from pre-, syn- and post-forensic event) takes place on the sole of footwear in an ordered or more random manner; and whether sampling vertically or horizontally through the sediment deposit on the sole is affected by sample mixing and layer distortion. The temporal (persistence) considerations are whether geoforensic evidence transferred onto footwear (both the upper and sole) survives for a sufficient period of time for subsequent collection and analyses.

Spatial Mixing of Soil and Sediment on Footwear

Experimental Studies

Contact with the ground almost inevitably results in the transfer of materials onto the sole (or uppers) of footwear (one-way transfer), but may also initiate the transfer of material on the footwear to the underlying surface thus initiating two-way transfer (Locard 1930). Such transfers have been reported and analysed in the published literature in both experimental and case work studies (Horrocks et al. 1999; Bull et al. 2004; Bull et al. 2006; Morgan et al. 2006; Morgan and Bull 2006). Recent experimental work has concentrated on the problem of discrete sampling of mud on footwear on the soles of footwear by studying the distribution and movement of three layers of Plasticine during experimental runs which involved the wearer of the shoes walking and running. Plasticine was chosen as a soil proxy particularly of clay materials which constitute an important component of many soils. Although lacking the coarse silt and sand component of many soils, the Plasticine appeared to mimic results observed in case work described below. Here, three different colours of Plasticine were chosen to represent pre-, syn- and post-forensic event sources of soil. All three layers of Plasticine (the same size and thickness, 5 mm) were applied sequentially to the soles of identical pairs of flat-soled training shoes. Following the application of each layer, the wearer walked 250 m on paved ground in dry conditions so that Layer 1 was eventually walked on for 750 m, and Layers 2 and 3 were walked on for 500 and 250 m respectively. The experiment was replicated under the same conditions. A further experiment was undertaken which followed the same procedure, the only difference being that the wearer ran on each layer of Plasticine. A template of each shoe was constructed (using a model derived by Hessert et al. 2005) and sampling points (on a grid system) were identified. Vertical plugs of the Plasticine remaining on the sole of the training shoes were taken at each sample point for subsequent analysis. These plugs were then photographed in cross-section and the resultant digital images were rasterised using MATLAB to provide numerical comparison of the proportion of each layer (by colour) of Plasticine preserved at each sampling point. The general results for the right shoe from each experiment (both walking and running) are presented in Figure 16.1 and more detailed presentation can be seen in Figure 16.2.

Visual inspection of both figures shows that the three layers on the right shoe (Layer 1 was yellow, Layer 2 was blue and Layer 3 was red), although originally of the same thickness, now comprise layers of different thicknesses for both the experiments involving walking and running (Figure 16.1) with red the predominant colour preserved in each plug. Spatial variation of the Plasticine on the right shoe (Figure 16.1) shows a predominance of the last layer applied (red) and this is shown in Figure 16.2 where the relative area proportions of each layer are presented for the medial arch area of the shoe (MA) and for the toes area of the shoe (T) (for location of MA and T see Figure 16.1).

The spot sampling of footwear, in order to compare a footwear sample with a sample from a forensic site, requires an assumption that the area of footwear sampled provides an accurate comparator for materials derived from a forensic site. If our forensic event is the middle layer of the three presented in Figures 16.1 and 16.2, then the blue layer (Layer 2) assumes great importance. Multivariate statistical analysis (by canonical discriminant analysis) of the 'toe' (H, T, MT1, 3 and 4) and 'middle' area (MA and LA) of the shoe show that these two areas cannot be discriminated (Wilks lambda = 0.996, $P > 0.05$). However, when the 'heel' area (MC and LC) was included in the statistical analysis, the three designated areas of the shoe sole (toe, middle and heel) could be discriminated from each other at the 99% significance level (Wilks lambda = 0.761, $P < 0.01$). There is therefore a statistically significant difference between the layers preserved in the different spatial areas of the footwear sole. These results have dramatic implications for the production of un-testable false-positive or false-negative associations between materials taken from the sole of a shoe and materials taken from a site of forensic interest, especially when general sole samples are taken or analysis of the materials requires homogenisation, such as when using chemical analysis e.g. inductively coupled plasma spectrometry (ICP) or certain physical analyses (colour, particle size, etc.). Such caution has been suggested in the

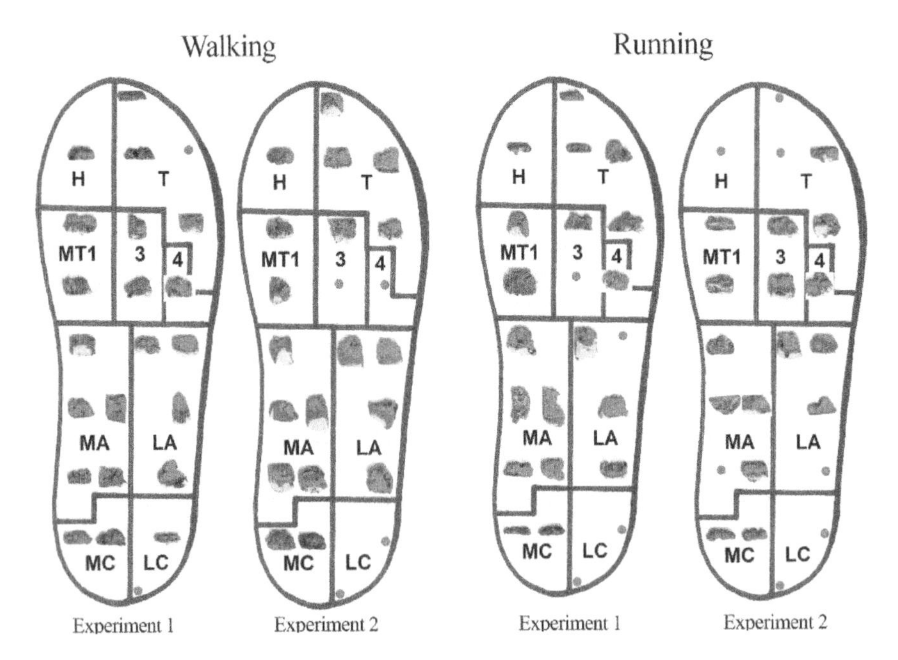

Fig. 16.1 Plasticine plugs recovered from each sample point on the right shoe sole for both walking and running experiments (MC = medial calcaneus, LC = lateral calcaneus, MA = medial arch, LA = lateral arch, MT1 = first metatarse, 3 = second and third metatarse, 4 = fourth and fifth metatarse, H = hallux and T = toes) (*see colour plate section for colour version*)

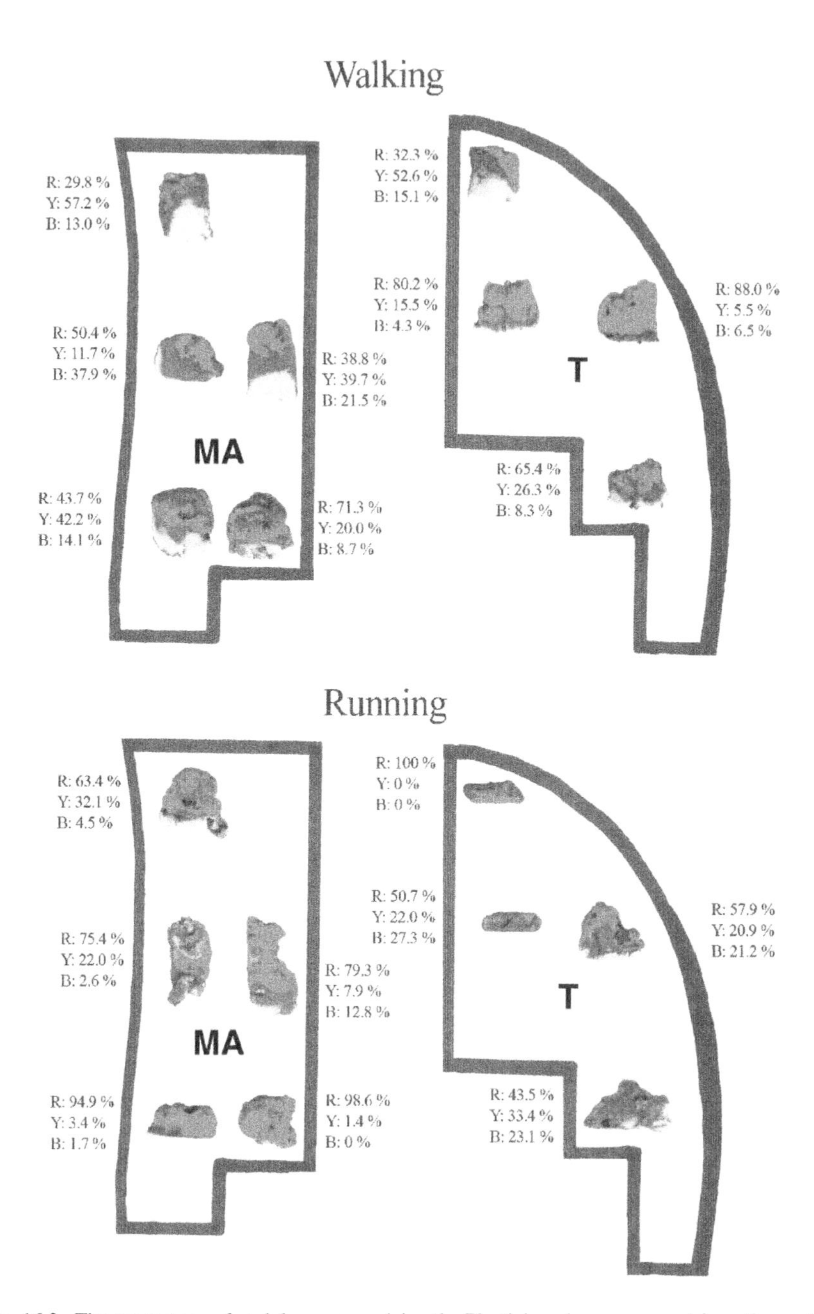

Fig. 16.2 The percentage of each layer comprising the Plasticine plugs recovered from the medial arch area and the toes area of the right shoe soles for both walking and running experiments (*see colour plate section for colour version*)

geoforensic literature in relation to geoforensic analysis more generally (Morgan and Bull 2007a, b). It becomes critical, therefore, that the establishment of sample procedures and protocols occurs and that they are adopted when investigations of sediments on the soles of footwear are used in forensic investigation (see below).

Specific rheological observations of the right shoe under walking or running conditions (Figure 16.2) show differences between the movements of the three layers depending upon the movement of the wearer. During the walking experiment, the sequential chronology of the three layers appears to be broadly preserved in both the MA and T sections of the sole. In contrast, after running, mixing appears to occur amongst the layers, disrupting their chronological sequence, indeed, in the medial arch area (MA) of the sole of the footwear, Layers 1 and 2 (yellow and blue) are broadly removed from the area leaving Layer 3 (red) as the only or predominant material present. This will also have severe implications for the sampling and subsequent analysis of sediment recovered from footwear. If mixing is occurring, it is suggested that analytical techniques are used which can identify the different layers even if they have undergone significant mixing with sediment from different sources.

The experiment above highlights the problems of spot sampling of a shoe, the necessity to avoid homogenisation of samples prior to analysis, and the importance of visual rather than automated forms of analyses (particularly when this analysis requires homogenisation of the sample). Interpretations of the results are further complicated by the possible false positive or negative association between the sample from the footwear or the sole of the footwear with that of the desired forensic event area. Some practitioners recognise some of these problems and choose to take various samples from the soles of footwear submitted for analysis. There appears to be no fixed protocol and only suggestion in the literature with, for example, Murray and Tedrow (1975) who advocate the methods of Georg Popp in the need for sequential layer analysis to reconstruct the pre-, syn- and post-forensic event history of mud deposition on the sole of footwear.

Casework Example

This example provides a case study of the prosecution of a man accused of digging up a badger sett in the Oswestry area of central England. Fundamental to the case was the comparison of soils taken from a pair of boots, two shovels and the much disturbed dug-out badger sett site. The soil exhibit taken from the badger sett site comprised hundreds of grams of sediment, whilst the sample recovered from the shovels comprised only tens of grams and the footwear only a few grams of material. In order to overcome the problem of comparison between exhibits, a number of spot samples was taken from each exhibit (see Figure 16.3 and for further details Morgan et al. 2006).

The physical (colour, particle size analysis, quartz grain surface texture analysis) and chemical analyses (conductivity, pH, atomic absorption spectrometry and Dionex (AAS/Dionex)) undertaken on these soil samples is documented in Morgan et al. (2006). None of these techniques were able to discriminate between the soil samples taken from the boots with the soil samples taken from the badger sett site. Further analysis of the quartz grain surface textures revealed that the quartz component

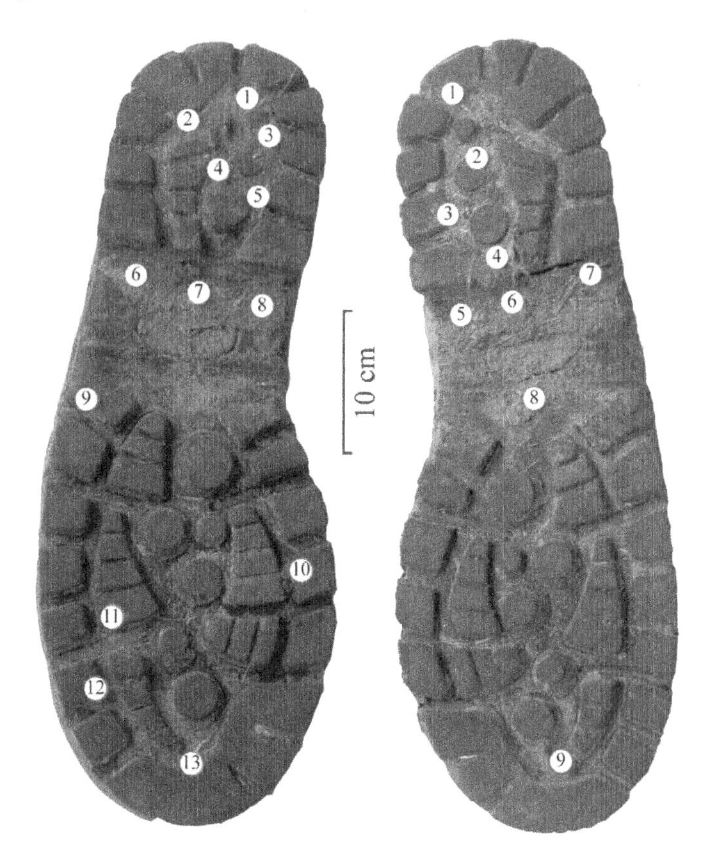

Fig. 16.3 The sampling points on the soles of the boots submitted for analysis

of the soil was made up of three distinct types of grain. Type I grains were characterised as deriving from a diagenetic sandstone exhibiting a suite of diagenetic features including both anhedral and euhedral crystal growth typically without subsequent edge abrasion (Figure 16.4a and 16.4b). Type II grains were well rounded grains with subaqueous impact features (such as found after river transportation) with later chemical smoothing (Figure 16.4c and 16.4d). Type III grains were characteristically high relief with angular/subangular grains with some subrounded additions with no edge abrasion but later chemical smoothing (Figure 16.4e and 16.4f). The grains were classified according to the system designated by Bull and Morgan (2006), from which it can be seen that only 0.5% of the quartz grains included in the database derived from English forensic soil samples (approximately 35,000 grains) were of the same form as Type I identified here. The palaeo-environmental assessment of these soils is that the Type I and III grains derive from the local sandstone, having not been transported by wind or water and thus exhibiting no grain edge abrasion, whilst Type II grains represent most likely a fluvial input into the area and mixing on site with the Type I and III grains. This very limited assemblage of quartz grain types makes the similarity of the materials found in samples taken from both boots and the badger sett site significant. Indeed, multivariate statistical analysis (canonical discriminant function

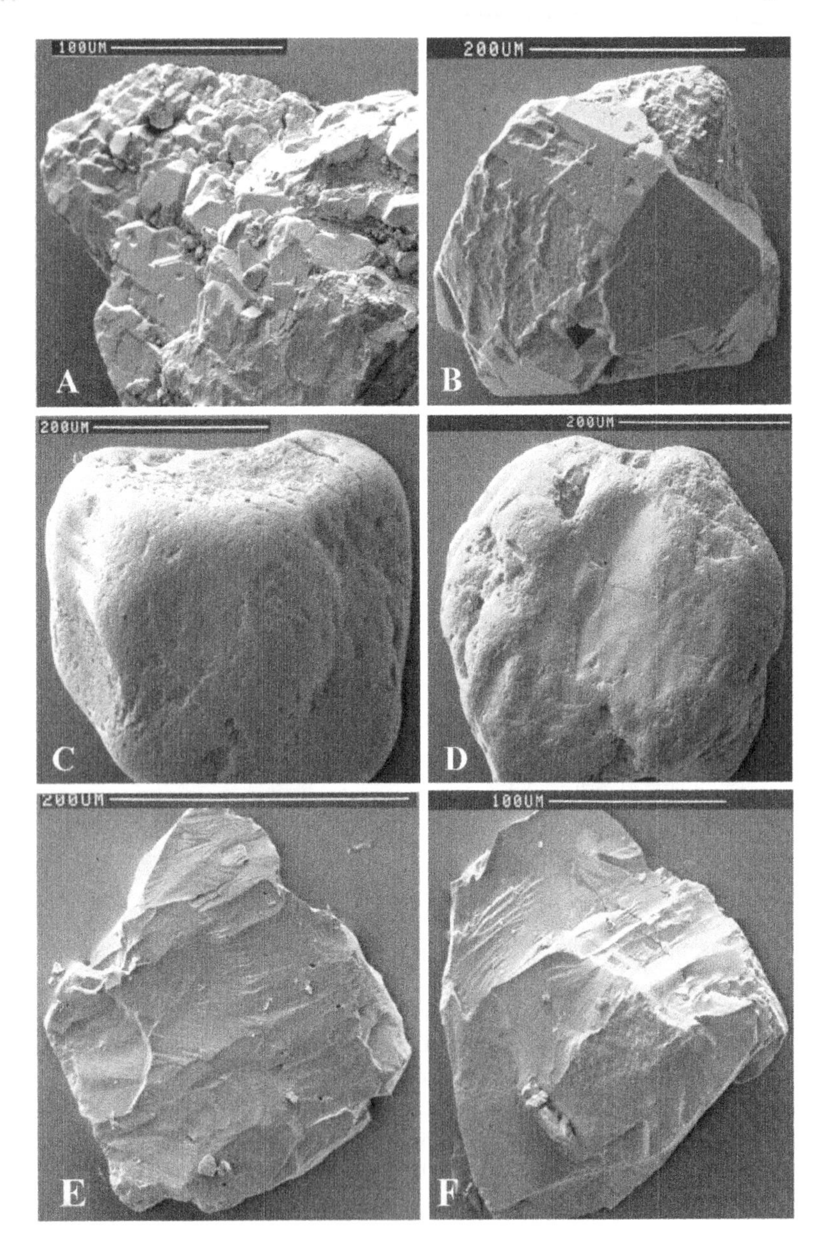

Fig. 16.4 Quartz grain types found in samples taken from the badger sett site and boots submitted for analysis; A and B Type I grains, C and D Type II grains, E and F Type III grains

analysis) demonstrated that it was not possible to discriminate between the quartz grain type assemblages from each location presented in Table 16.1 (Wilks lambda = 0.685, $P > 0.05$). In this case, it was not possible to exclude the soil samples taken from the footwear from the soil samples taken from the badger sett site.

Table 16.1 Quartz grain types identified in the soil samples taken from the crime scene and the boots

Case samples		Type I	Type II	Type III	Total
Badger sett site	Sub-sample 1	41	19	7	67
	Sub-sample 2	43	15	3	61
	Sub-sample 3	47	7	3	57
	Sub-sample 4	41	9	3	53
	Sub-sample 5	48	14	1	63
	Sub-sample 6	38	12	2	52
	Sub-sample 7	66	11	4	81
	Sub-sample 8	47	13	6	66
	Sub-sample 9	28	10	5	43
	Sub-sample 10	35	8	4	47
Right boot	Composite	34	16	6	56
	Composite	29	14	1	44
Left boot	Point 1	38	5	6	49
	Point 2	34	5	8	47

In order to come to a meaningful interpretation, it was necessary to take multiple samples (in this case 22 spot samples were able to be collected) from the sole of the footwear to compare with multiple samples (in this case 10 samples) taken from the bulk soil sample recovered from the badger sett site. It was not merely an analysis involving two comparator samples; over 30 samples were eventually analysed. It must also be stressed that this analysis represented only one strand of forensic investigation which itself utilised geoforensic results taken from independent techniques such as pollen analysis and this was undertaken by separate scientists who worked independently in this case.

Various aspects of the complexity of the spatial distribution and movement of pre-, syn- and post-forensic event soil on the soles of footwear have been identified in these experimental and casework studies. Footwear often provides the starting point for geoforensic enquiry, but this work highlights the need for caution when making comparison between soil samples recovered from the soles of footwear and forensic sites. There is no such thing as simple comparison nor is there any philosophical basis for attempting to 'match' samples (Morgan and Bull 2007b; Bull et al. 2008).

Persistence of Trace Evidence on Footwear

Experimental Studies

The long-standing view that trace materials persist on clothing stems from the experimental works undertaken by Robertson and Roux (2000), Hicks et al. (1996) and Pounds and Smalldon (1975a, b, c) where studies have generally provided decay curves of 4 to 8 h duration. More recent work has sought to extend the experimental decay curve timeline to hundreds of hours in an attempt to utilise the power of electron microscopy (Bull et al. 2006a). The experiments presented herein deal with the quantities of materials left on the soles of footwear over time, and the persistence of pollen particulates on the uppers of shoes over even longer periods of time.

The obvious advantages of finding trace particulates on clothing and footwear many hours after their transfer at the relevant forensic event could be argued to be outweighed by the very problems of persistence where the picture is complicated by pre-, syn- and post-event mixing. These problems, similar to those described above, are best overcome with resort to visual identification methodologies.

The Persistence of Trace Materials on the Soles of Footwear Through Time

An experimental study was undertaken to establish the nature of the persistence of silt-sized trace materials on the soles of footwear over time. In order to quantify the amount of trace sediment present, a UV powder (<15 μm) was mixed with soil and applied to the soles of training shoes. At a number of intervals (after 0, 100, 250, 350 and 450 m), the sole of each training shoe was photographed under an ultra-violet light. This digital image was then rasterised in IDRISI to provide an indication of the amount of silt-sized material remaining on the sole. This experiment was repeated three times and the results (Figures 16.5 and 16.6) show a general trend of an initial reduction in the amount of trace material adhering to the sole after 100 m of walking on a smooth concrete surface. However, after 250 m the amount of silt-sized material has increased with a subsequent decrease after 350 m. This pattern is due to the larger conglomerated materials decaying rapidly during the initial stages of walking. However, the remaining material is then spread out by the pressure applied through the foot during subsequent walking which in turn increases the distribution of silt-sized material. After this stage the trace material again rapidly decays; however, it is interesting to note that a sufficient amount of material remains on some parts of the soles for sampling to take place even after 450 m of walking.

The Persistence of Pollen Particulates on the Uppers of Footwear Over Time

In a second study, two pairs of shoes (a cotton plimsoll and a suede shoe) were used to assess the transfer and persistence of pollen particulates on the uppers of footwear. The participant brushed past a flowering shrub (*Jasminium nudiflorum*) wearing each pair of shoes and then wore the shoes for a total of 7 days undertaking general activities. Tapings were taken from pre-marked locations (to prevent repeat sampling) across the toe of each shoe at different time intervals (0, 2, 4, 6, 8, 10, 12, 24, 36

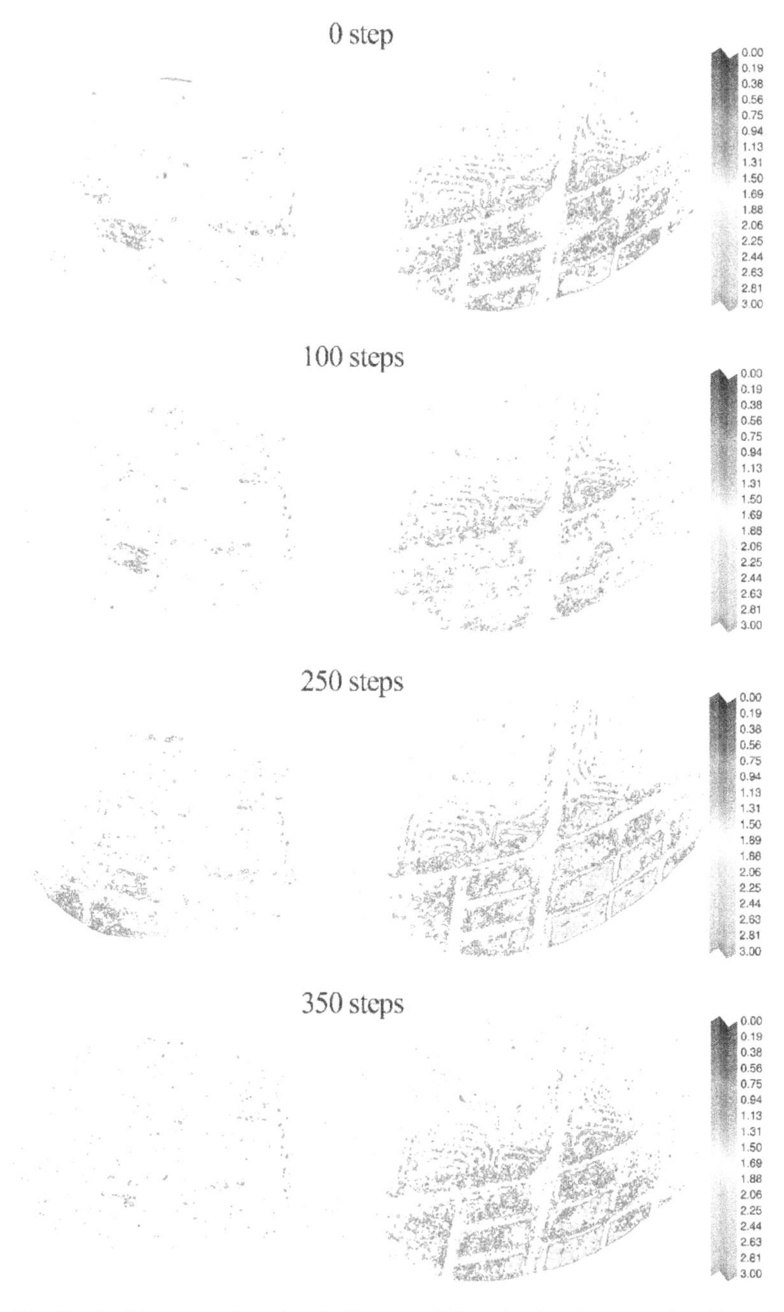

Fig. 16.5 Pixelated image to show the silt-sized material retained on the shoe soles after walking different distances (mean brightness indicates the amount of silt-sized material remaining on the sole) (*see colour plate section for colour version*)

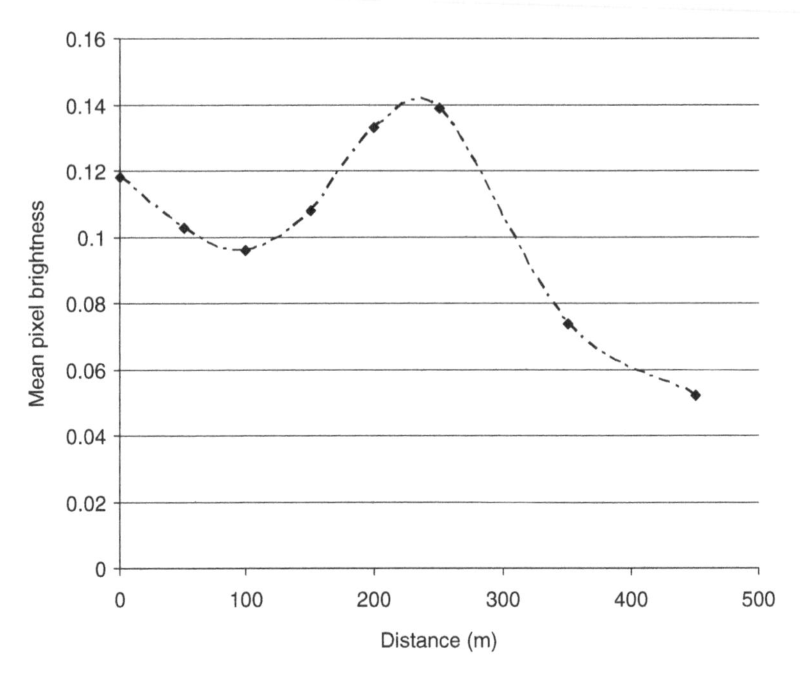

Fig. 16.6 General trend in mean pixel brightness (proxy for amount of silt-sized sediment) on the soles of footwear over distance (n = 6)

and 168 h) and observed under a scanning electron microscope at x330 magnification where the pollen grains present were counted. The results of each replicated experiment are presented in Figure 16.7a and 16.7b.

The persistence of pollen on these two different types of footwear appears to follow the previously identified trend of two/three stage decay (Pounds and Smalldon 1975a, b, c; Bull et al. 2006b). The loss of pollen is particularly rapid during the first 4 h, with subsequently less rapid loss between 4 and 10 h, followed by a period of much slower decay. After 168 h, 4.8% and 1.5% of the original pollen remained on the suede shoes whilst 1.8% and 1.0% remained on the cotton plimsolls. Whilst these appear to be only small percentages of remaining pollen, they are only taken from a very small area of the shoe upper (approximately 0.5 cm^2). These results therefore, have important implications for the forensic examinations of footwear; pollen particulates are highly likely to persist for many hours, days or even weeks. Such evidence has great potential to aid criminal investigations (Horrocks et al. 1999; Horrocks and Walsh 1999; Mildenhall 2006; Mildenhall et al. 2006) and this present study demonstrates that pollen is likely to remain on footwear for significant periods of time during normal wear thus enabling its recovery and analysis. Indeed this phenomenon makes the likelihood much greater of pre- and post-forensic event mixing with the relevant materials taken from the forensic scene. Thus the quality of persistence has both 'advantages' and 'disadvantages' for the interpretation of forensic evidence.

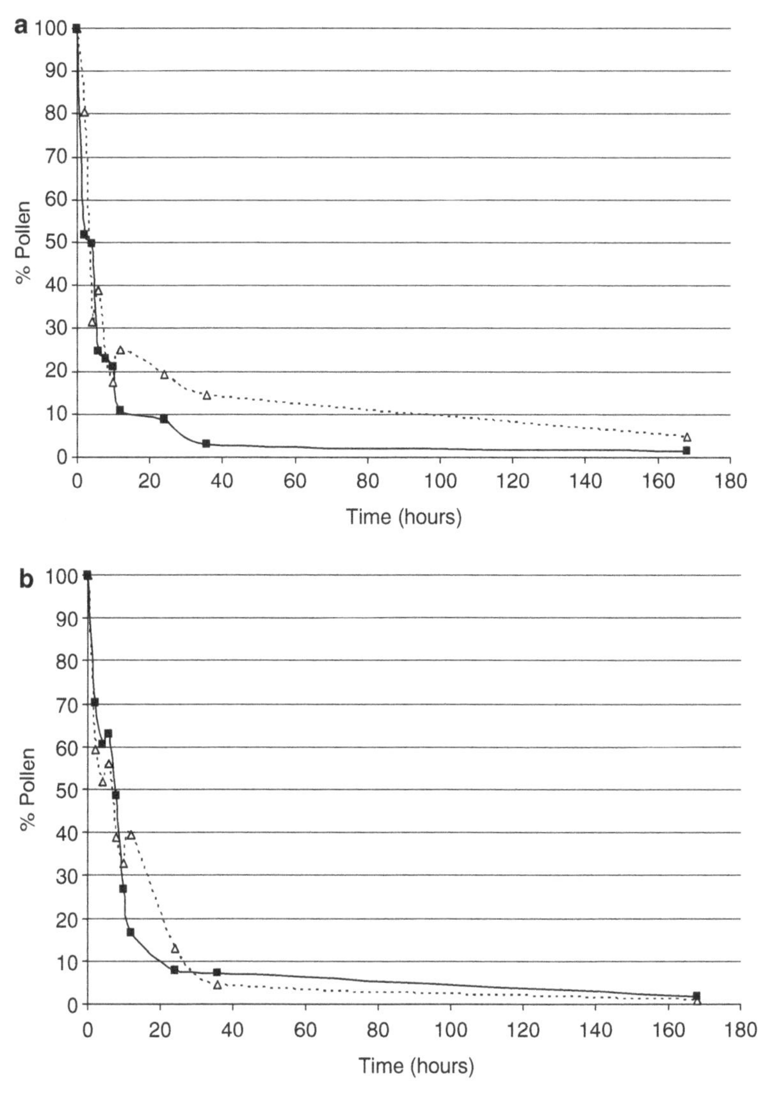

Fig. 16.7 Graphs showing persistence of pollen particulates. (**a**) The suede shoes over time. (**b**) The cotton shoes over time

Case Studies

The experimental studies outlined above have demonstrated that geoforensic evidence has the potential to persist on footwear (both soles and uppers) for significant periods of time. As mentioned previously, this not only has implications for the recovery and sampling of such evidence from footwear, but also has implications for the type of analytical technique employed to analyse such evidence. If trace geoforensic evidence persists for long periods of time, it will be important (as mentioned above) that materials from different sources can be identified during any subsequent analyses.

In our experience there have been a number of criminal investigations that have utilised the presence and recovery of geoforensic evidence from footwear. In the case of R v Wren (2002), distinctive gravel was recovered from the footwear sole of a suspect which could not be excluded from having derived from the crime scene. Similarly, in R v Hunt and Fawley (2002), a number of pairs of shoes were seized by the police and the comparison of trace geoforensic evidence recovered from the footwear and the crime scene enabled all but two pairs of footwear to be excluded from the investigation which aided the police in their crime reconstruction.

It is, however, not always evidence collected directly from footwear that can aid criminal enquiry. In the case of R v Flavious (2005), a car was used to transport a body to a grave site and then driven back to the home of the suspect. The car was seized and the driver footwell mat was found to be very muddy, with one distinct footprint present (Figure 16.8). The general debris was sampled and the mud from the footprint also collected and compared to the soil sampled from the body deposition site. Additionally, a pair of boots were seized which had mud present on the soles. A number of analytical techniques were employed (elemental chemistry, mineralogy, colour, pH, particle size analysis), in addition to the quartz grain surface texture analysis. A distinctive quartz grain 'type' was identified in the samples collected from the car footwell and the body deposition site but it was conspicuous in its absence in the boot samples. Another search at the home of the suspect yielded a pair of training shoes which had been washed in a washing machine. Whilst this footwear appeared to be very clean, inspection of the inners of the shoes underneath the in-sole yielded a small amount of clean debris. The quartz grains were analysed and the distinctive grain type was found to be present. In this instance, it was the geoforensic evidence deposited by the footwear that proved to be very important as well as the evidence contained within it. Furthermore, due to the mixing of sediment from different sources in the car footwell and in the training shoes, it was necessary to employ a visual technique (in this case, quartz grain surface texture analysis: Bull and Morgan (2006) and above) that was able to identify this mixing, to be more confident of the interpretation of the analysis and avoid false-exclusionary conclusions.

Thus, footwear has the potential to provide very useful information during the course of a forensic investigation. Not only the sediment transferred and preserved on the footwear (both inners and outers, uppers and sole) but also the sediment deposited as muddy footprints or collected by a plaster cast of footprint impression (see Bull et al. 2006) can be recovered and provide valuable contextual information.

Conclusions

The implications of these experimental studies and case work examples are twofold and have a bearing upon both sampling and analysis of geoforensic evidence recovered from footwear. In terms of sampling footwear for geoforensic evidence, we have

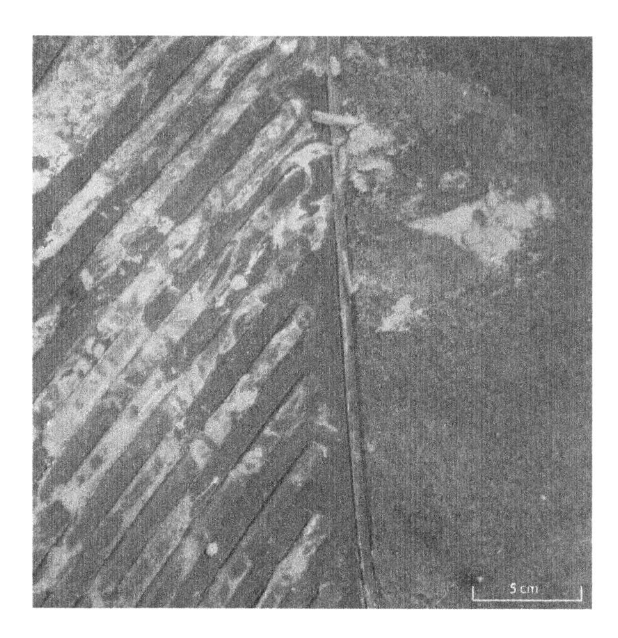

Fig. 16.8 Muddy footwell mat and the distinct footprint in case study (*see colour plate section for colour version*)

shown that certain areas of the sole are more likely to retain sediment than others. Of the soil/sediment evidence that is retained, it is unlikely that a sequential chronology of the different sources of sediment will be preserved; mixing of evidence from different sources does occur (in some areas more than others) on the soles of footwear.

Geoforensic evidence has been shown to persist for reasonable periods of time on the soles of footwear, and the uppers as well as after washing in the inners of shoes. The persistence of geoforensic evidence on footwear means that it is likely to be present and should be sampled to see if any such evidence can be recovered. However, it also means that the analytical technique employed must be able to identify when mixing of evidence from different sources which has been introduced at different stages has occurred, if meaningful interpretation of the analysis is to be made. Indeed, the spatial variation of geoforensic evidence preserved on the sole of a shoe also means that it is crucial to employ visual techniques that are able to identify materials derived from different sources. It is vital, however, that a proper preliminary optical examination initially under low magnification be conducted before more advanced analytical procedures are employed. Contemporary advances in digital photography which enable high magnification and resolution provide an excellent opportunity for preliminary visual analysis. Identification of materials derived from different provenances is of great importance because, if different sources of geoforensic material are not identified, there is a real possibility for un-testable false negative interpretations of the evidence to be reached. Such un-testable conclusions can have no place in a court of law. Indeed, this chapter also serves to highlight the importance of experimental studies for deepening our knowledge and increasing our ability to interpret forensic evidence accurately and meaningfully within the appropriate context. Such studies have a valuable role to play in the development of forensic science (Morgan et al. 2008).

References

Bull PA and Morgan RM (2006). Sediment fingerprints: a forensic technique using quartz sand grains. Science and Justice 46:64–81.

Bull PA, Morgan RM, Wilson HE and Dunkerley S (2004). Multi-technique comparison of source and primary transfer soil samples: an experimental investigation. A comment. Science and Justice 44:173–176.

Bull PA, Morgan RM, Sagovsky A and Hughes GJA (2006a). The transfer and persistence of trace particulates: experimental studies using clothing fabrics. Science and Justice 46:182–191.

Bull PA, Parker AJ and Morgan RM (2006b). The forensic analysis of soils and sediment taken from the cast of a footprint. Forensic Science International 162:6–12.

Bull PA, Morgan RM and Freudiger-Bonzon J (2008). A critique of the present use of some geochemical techniques in geoforensic analysis. Forensic Science International 178:e35–e40.

Chazottes V, Brocard C and Peyrot B (2004). Particle size analysis of soils under simulated scene of crime conditions: the interest of multivariate analyses. Forensic Science International 140:159–166.

Hessert MJ, Vyas M, Leach J, Hu K, Lipsitz LA and Novak V (2005). Foot pressure distribution during walking in young and old adults. BMC Geriatrics 5:8.

Hicks T, Vanina R and Margot P (1996). Transfer and persistence of glass fragments on garments. Science and Justice 36:101–107.

Horrocks M and Walsh KAJ (1999). Fine resolution of pollen patterns in limited space: differentiating a crime scene from an alibi scene seven metres apart. Journal of Forensic Sciences 44:417–420.

Horrocks M, Coulson SA and Walsh KAJ (1999). Forensic palynology: variation in the pollen content of soil on shoes and in shoeprints in soil. Journal of Forensic Sciences 44:119–122.

Inman K and Rudin N (2002). The origin of evidence. Forensic Science International 126:11–16.

Locard E (1928). Dust and its analysis. Police Journal 1:177.

Locard E (1930). Analyses of dust traces parts I, II and III. American Journal of Police Science 1:276–298, 401–418 and 496–514.

Mildenhall DC (2006). *Hypericum* pollen determines the presence of burglars at the scene of a crime: An example of forensic palynology. Forensic Science International 163:231–235.

Mildenhall DC, Wiltshire PEJ and Bryant VM (2006). Forensic palynology: why do it and how it works. Forensic Science International 163:163–172.

Morgan RM and Bull PA (2006). Data interpretation in forensic sediment geochemistry. Environmental Forensics 7:325–334.

Morgan RM and Bull PA (2007a). The philosophy, nature and practice of forensic sediment analysis. Progress in Physical Geography 31:43–58.

Morgan RM and Bull PA (2007b). Forensic geoscience and crime detection. Identification, interpretation and presentation in forensic geoscience. Minerva Medicolegale 127:73–90.

Morgan RM, Wiltshire P, Parker A and Bull PA (2006). The role of forensic geoscience in wildlife crime detection. Forensic Science International 162:152–162.

Morgan RM, Allen E, Lightowler Z, Freudiger-Bonzon J and Bull PA (2008). A forensic geoscience framework and practice. Policing: A Journal of Policy and Practice 2: 185–195.

Murray RC and Tedrow JCF (1975). Forensic Geology: Earth Science and Criminal Investigation. Rutgers University Press, Chapel Hill, NC.

Pounds CA and Smalldon KW (1975a). The transfer of fibres between clothing materials during simulated contacts and their persistence during wear part I – Fibre transference. Journal of the Forensic Science Society 15:17–27.
Pounds CA and Smalldon KW (1975b). The transfer of fibres between clothing materials during simulated contacts and their persistence during wear part II – Fibre persistence. Journal of the Forensic Science Society 15:29–37.
Pounds CA and Smalldon KW (1975c). The transfer of fibres between clothing materials during simulated contacts and their persistence during wear part III – a preliminary investigation of the mechanisms involved. Journal of the Forensic Science Society 15:197–207.
Robertson J and Roux C (2000). Transfer and persistence. In: Encyclopedia of Forensic Sciences (Eds Siegel JA, Saukko PJ and Knupfer GC), pp. 834–838. Academic, London.
Roux C, Langdon S, Waight D and Robertson J (1999). The transfer and persistence of automotive carpet fibres on shoe soles. Science and Justice 39:239–251.
Virtanen V, Korpelainen H and Kostamo K (2007). Forensic botany: usability of bryophyte material in forensic studies. Forensic Science International 172:161–163.
Wiggins KG, Emes A and Brackley LH (2002). The transfer and persistence of small fragments of polyurethane foam onto clothing. Science and Justice 42:105–110.

Chapter 17
Using Soil and Groundwater Data to Understand Resistivity Surveys over a Simulated Clandestine Grave

John R. Jervis, Jamie K. Pringle, John P. Cassella and George Tuckwell

Abstract Geophysical electrical resistivity surveys have been used in a number of attempts to locate clandestine 'shallow' graves, based on the valid assumption that a grave may represent a contrast in the electrical properties of the ground compared to 'background' values. However, the exact causes of measurable geophysical signals associated with graves are not well understood, particularly for electrical methods. In this study, soil and groundwater samples have been obtained from a simulated grave containing a domestic pig (*Sus domestica*) carcass, in order to better understand how the presence of a grave may influence the bulk electrical properties of the soil. This information is used to explain observations based on repeat resistivity surveys over a period of 6 months over a second simulated grave at the same site. An area of low resistivity values was observed at the grave location in the survey data obtained from 4 to 20 weeks post-burial, with the grave being difficult to identify in survey data collected outside of this interval. The low resistivity grave anomaly appeared to be caused by highly conductive fluids released by the actively decomposing carcass and this is consistent with the relatively short timescale during which the grave was detectable. It is then suggested that the most appropriate time to use resistivity surveys in the search for a grave is during the period in which the cadaver is most likely to be undergoing active decomposition. However, other authors have observed low resistivity anomalies over much older graves and it is possible that, for graves in different environments, other factors may contribute to a detectable change in the bulk electrical properties of the soil.

J.R. Jervis(✉) and J.K. Pringle
Applied and Environmental Geophysics Group, School of Physical Sciences and Geography, Keele University, Staffordshire ST5 5BG, UK
e-mail: j.jervis@epsam.keele.ac.uk

J.P. Cassella
Department of Forensic Science, Faculty of Sciences, Staffordshire University, Stoke-on-Trent, Staffordshire ST4 2DE, UK

G. Tuckwell
Stats Limited, Porterswood House, St. Albans, Hertfordshire AL3 6PO, UK

K. Ritz et al. (eds.), *Criminal and Environmental Soil Forensics*

Introduction

Forensic geophysical surveys have been used in a number of attempts to search for clandestine graves (e.g. Nobes 2000; Buck 2003; Scott and Hunter 2004; Cheetham 2005; Ruffell 2005). Ground penetrating radar (GPR) is perhaps the most commonly used geophysical method in the search for clandestine graves (Cheetham 2005) and its capability for this purpose has been well studied in a number of surveys over controlled animal (Schultz et al. 2002, 2006; Schultz 2008) and human burials (Freeland et al. 2002; Miller et al. 2002). Of the other geophysical techniques that have been used, electrical resistivity surveys have demonstrated success in locating simulated graves (Cheetham 2005; Pringle et al. 2008) and a number of published reports detail the use of such surveys in searches for murder victims (Buck 2003; Scott and Hunter 2004; Cheetham 2005). Graves commonly appear as areas of reduced resistivity compared to background values in resistivity survey data (Cheetham 2005) or, equivalently, areas of increased conductivity in data from electromagnetic surveys (France et al. 1992; Nobes 2000). The possible cause of the reduced resistivity and increased conductivity observed in survey data over both real and simulated shallow graves has been suggested to be the increased porosity of the backfilled soil (France et al. 1992; Scott and Hunter 2004) or moisture trapped within the grave (Nobes 2000). Additionally, decomposing bodies are known to release fluids with a greater ion concentration than normal groundwater (Vass et al. 1992), and this could also, theoretically, cause a low-resistivity anomaly in geophysical data. However, no previous study involving electrical resistivity or conductivity surveys over graves has been supported by direct measurement of porosity, moisture levels or fluid conductivity. Hence, the exact contributions of any of these factors to the anomalies associated with graves in resistivity and conductivity data remains largely unknown. A greater understanding of what causes graves to be identifiable in data from resistivity and conductivity surveys may give more insight into the relative strengths and weaknesses of these techniques for locating graves, giving potential future forensic search coordinators a better appreciation of when and where the use of these methods would be most appropriate.

Measurements made using resistivity survey equipment are equal to the apparent resistivity (which is an idealised version of the true resistivity) of the sub-surface multiplied by a geometrical factor that depends on the relative orientation and separation of the electrodes (Reynolds 1997). Hence, in order to gain an understanding of what may cause a grave to be detectable using electrical resistivity surveying, it is necessary to be aware of which soil physical properties influence the bulk resistivity of the ground. Archie (1942) developed an empirical law that relates the bulk resistivity (ρ_b) of both unconsolidated partially saturated sands and consolidated sandstones to their porosity (θ), degree of saturation (S) and the resistivity of the pore water (ρ_w) that can be written:

$$\rho_b = \theta^{-m} S^{-n} \sigma_w^{-1} \quad (1)$$

where σ_w is the conductivity of the pore water (i.e. ρ_w^{-1}), *m* appears to depend on the level of consolidation of the medium in question and is generally found to lie in the range 1.3 to 2 and *n* has a value of approximately 2 (Archie 1942). Although only originally tested on sands and sandstones, Archie's law may be used to describe the electrical properties of soils, although in clayey soils it may be necessary to add an extra 'surface conductivity' term to Equation 1 to account for the fact that clay minerals may exchange ions with the soil solution (Friedman 2005). It can be seen from Equation 1 that an increase in any one, or a combination, of soil porosity, saturation and water conductivity may explain the low resistivity and high conductivity anomalies associated with shallow graves observed in previous studies. Hence, by independently monitoring each of these three variables in a controlled experiment involving a simulated grave, it may be possible to determine the relative importance of each variable to any changes in the measurable bulk resistivity.

Aims

This study had three mains aims: (1) to conduct repeat electrical resistivity surveys over a simulated clandestine grave at regular intervals in order to monitor the time-varying bulk electrical response of the grave; (2) to collect soil and groundwater samples from a second simulated grave and 'undisturbed' ground in order to determine any variations in porosity, saturation or water conductivity and map these variations to the recorded resistivity data; (3) to draw from the analysis of the results any general conclusions regarding the use of electrical resistivity survey techniques in searching for shallow graves.

Methods

Study Site

The site chosen for the study was the back garden of Staffordshire University's 'Crime Scene House' in Stoke-on-Trent, Staffordshire, UK. The garden is grassed, surrounded by hedges and trees and approximately 40 m long by 10 m wide. British Geological Survey borehole data (borehole record SJ84NE2579) from a borehole located on a raised bank approximately 10 m from the study site show a 3 m thick 'made ground' layer above a 1 m thick layer of sandy gravel, beneath which is sandy, silty clay. The 'made ground' is described as a mix of clayey ash, sand and gravel. Digging of the graves revealed a significant amount of debris, including tree roots, whole bricks, concrete and coal fragments in the made ground layer, which extended to a depth of approximately 0.5 m in the Crime Scene House garden. The heterogeneous nature of the shallow subsurface at this site suggests that geophysical data might be expected to exhibit considerable variation across the study site. These localised variations have the potential to mask any subtle geophysical signal from the grave, meaning that the site offers a realistic test environment for geophysical search methods. A previous study at the same site found some difficulty in identifying buried matter using a number of geophysical techniques, including ground penetrating radar and magnetic gradiometry (Pringle et al. 2008).

Simulated Graves

In this study, buried domestic pig (*Sus domestica*) cadavers weighing approximately 31 kg each were used as a proxy for clandestine human graves, due to the ethical and legal issues surrounding the use of human cadavers in experiments in the UK. Two simulated shallow graves were created; one to be surveyed with the geophysical equipment and a second for the collection of soil and water samples, so that the removal of soil and water from the grave did not affect the geophysical data. The graves were approximately 1.1 m long, 0.6 m wide and 0.6 m deep. Carcasses were between 1.0 and 1.1 m in length and had had all internal organs, with the exception of brain, kidneys and bladder, removed via a long (approximately 0.4 m) incision in the abdomen. After placing each of the carcasses in a grave most of the excavated soil was backfilled, tamped down and the turf replaced. This left a mound, raised by a few centimetres relative to the surrounding grass at each grave. Excess soil was then disposed of at the edge of the garden.

Geophysical Data Collection

A survey grid measuring 8 × 4.5 m with one of the pig graves at the centre was marked out for geophysical survey (Figure 17.1). Plastic pegs were used to permanently mark both ends of each survey line in order to ensure that the surveyed area was consistent throughout the study. Electrical resistivity measurements were made using an RM4 resistance meter (Geoscan Research) mounted on a custom-built twin-probe array, which features two mobile probes 0.5 m apart on a mobile frame. The two reference probes were situated 0.75 m apart at a fixed location approximately 17 m from the survey area. Resistivity measurements were obtained every 0.25 m along survey lines 0.25 m apart, a measurement spacing that is recommended for surveys over relatively small graves (Cheetham 2005). Surveys were conducted every 2 weeks between March and October 2007, commencing 2 weeks after burial of the pig carcasses.

Geophysical Data Processing

Raw resistivity data in 'x, y, z' format were median filtered to remove small-scale data 'spikes'. This was achieved by using a rolling filter to take the median of

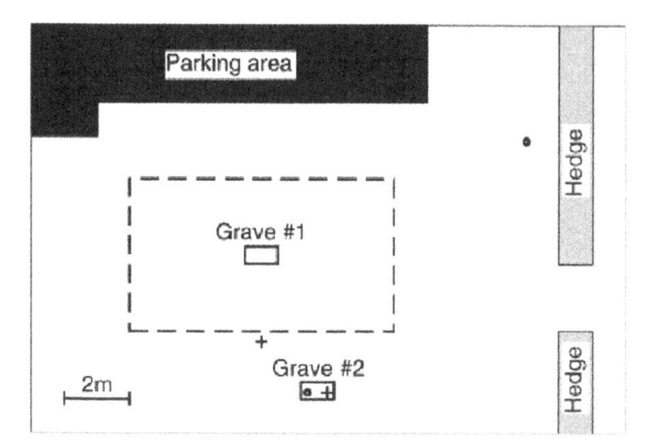

Fig. 17.1 Plan of study site showing: location of simulated graves, geophysically surveyed area (dashed line), lysimeters (circles) and soil sampling positions (crosses)

each triplet of adjacent points in the direction parallel to the 8 m long side of the survey grid. The data were subsequently interpolated to give data points every 2.5 cm using a continuous curvature surface gridding algorithm (Smith and Wessel 1990). Long wavelength trends were then removed by fitting a cubic surface to the gridded data and subsequently subtracting this surface from the data. The principal purpose of the median filtering and trend removal is to eliminate variations at length scales that are much shorter and much longer than that of any signal associated with the grave. Each dataset was then normalised by division by its standard deviation. As the trend removal process involved removal of the mean of each dataset, normalisation results in a set of values that represent the variance of each value in the original (post-interpolation) dataset from the mean in standard deviations, which allows datasets obtained at different times to be more easily compared. The processed resistivity datasets were then plotted using a common grey-scale palette, ranging from black at minus two standard deviations to white at plus two standard deviations.

Groundwater Sampling

Site groundwater and grave water samples were obtained using model 1900 soil water samplers (Soilmoisture Equipment Corp.), also known as lysimeters, which were installed in both the second simulated grave and at a control location at a point approximately 4 m outside the survey area (Figure 17.1). After the pig carcass had been placed in the second grave, a small amount of the excavated soil was mixed with water from the nearby River Trent to create a 'slurry', a small amount of which was then deposited onto the base of the grave between the hind quarters of the pig and the grave wall. The porous end cap of the lysimeter was then inserted vertically into the slurry and the grave was then back-filled. The presence of slurry was necessary to ensure good hydraulic conductivity between the soil and the lysimeter. The control lysimeter was installed by digging a narrow hole (approximately 0.3 × 0.3 m wide) to 0.6 m depth, depositing some slurry at the base of the hole, into which the tip of the lysimeter was inserted, and then back-filling the hole. Once installed, the open ends of the lysimeters at the surface were sealed with a rubber stopper and a vacuum pump was used to generate a suction of 65 kPa within the lysimeters, in order for the instruments to draw moisture from the soil. On each day that a resistivity survey was performed, the rubber stopper was removed from each lysimeter and any water present was extracted using a plastic syringe with a narrow tube attachment. The rubber cap was then replaced and the suction pressure restored. The conductivity of each sample was measured immediately after collection using a multiline P4 multi-parameter meter (WTW Inc.). Conductivity was not measured for the first two water samples obtained from each lysimeter (i.e. the samples obtained 2 and 4 weeks after burial), as these were likely to contain a significant amount of the water used to make the slurry. As such, these samples were considered not to be representative of the site groundwater or the grave water.

Soil Sampling

Narrow (1.5 cm diameter) steel augers were used to collect soil samples to a depth of 0.7 m below ground level. Control samples were obtained from a location near the edge of the surveyed area and grave samples from just inside the edge of the second grave at the opposite end to the grave lysimeter, so that the water removed by the grave lysimeter did not affect saturation levels of the soil samples. Samples were visually inspected upon extraction from the ground and those with sections of soil missing were discarded and a repeat sample was obtained. The augers containing the soil were then immediately returned to the laboratory. Sections from the part of the auger that were 0.1 to 0.3 m below ground level (henceforth referred to as 'shallow soil samples') and 0.4 to 0.6 m below ground level ('deep soil samples') were removed and placed into pre-weighed sample trays and oven dried at 105 °C for 24 h. Each sample along with its container was weighed before oven drying ('wet weight') and after drying ('dry weight'). The weight of water in each sample was calculated by subtracting the dry weight from the wet weight, and the soil weight was calculated as the dry sample weight minus the container weight. Soil volume was calculated using the soil weight and an assumed soil particle density of 2.65 g/cm^3, which is deemed appropriate for most soil types (Hillel 1980). The density of the water lost on drying was assumed to be 1.0 g/cm^3. The cross-sectional area of the auger samples was 0.72 cm^2, meaning that the volume sampled for each 20 cm segment removed from the auger (V_T) was 14.4 cm^3. Porosity and saturation were then calculated for all samples using standard formulae (e.g. Barnes 2000):

$$\theta = \frac{V_a + V_w}{V_T} \qquad (2)$$

$$S = \frac{V_w}{V_a + V_w} \quad (3)$$

where V_w is the volume of water in each sample, calculated from the water weight and the assumed water density and V_a is the volume of air in each sample, which was calculated by subtracting the soil volume and the water volume for each sample from the total sample volume.

Statistical Analysis of Soil Data

Statistical analysis was conducted using NCSS (Hintze 2001). Statistical hypothesis testing was used to determine whether differences in four measured parameters between the grave soil and the control soil were significant (i.e. due to an actual difference between the two soil types) or non-significant (i.e. due to random variation in measurements). The four tested parameters were shallow porosity, deep porosity, shallow saturation and deep saturation. A paired samples hypothesis test was used as this is most appropriate for experiments involving repeat measurements (Warner 2008): for a given parameter, each pair consisted of the values measured for the grave soil and the control soil each fortnight. Consequently, the test statistic was the mean difference between each pair (grave measurement – control measurement) and the null hypothesis that the mean difference (μ_d) was equal to zero was used. Tests were two -tailed with the level of significance set at 0.05.

Results

Resistivity Data

The processed geophysical data (Figures 17.2 and 17.3) showed a low resistivity anomaly associated with the grave that varied considerably in shape, extent and magnitude between individual datasets. The week 2 resistivity data did not show any obvious features associated with the grave. However, in the data from weeks 4 to 20, distinct areas of low-resistivity were visible at both the head and foot ends of the grave. In the data from weeks 22 and 24, only the low-resistivity feature at the head end of the grave was easily identifiable.

Soil Porosity and Saturation Data

Measured values of porosity and saturation data are shown in Figure 17.4a and b and results of the statistical analysis of these data are given in Table 17.1. For all

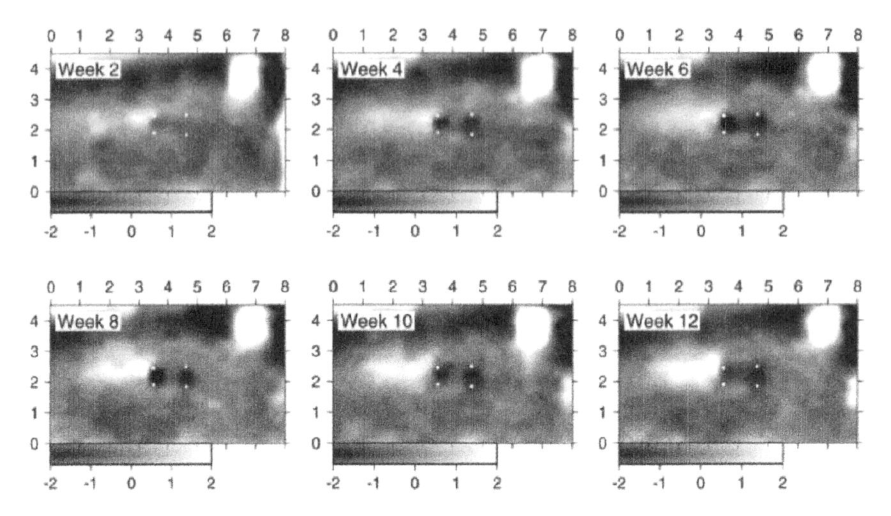

Fig. 17.2 Normalised resistivity survey data acquired over the simulated grave between 2 and 12 weeks post-burial. The grey scale shows the variation from the mean of each dataset in standard deviations. The corners of the grave are indicated by the white circles, with the head end of the grave at the left side of the marked area. Labels at the top left of each plot indicate the time of each survey relative to the time of burial.

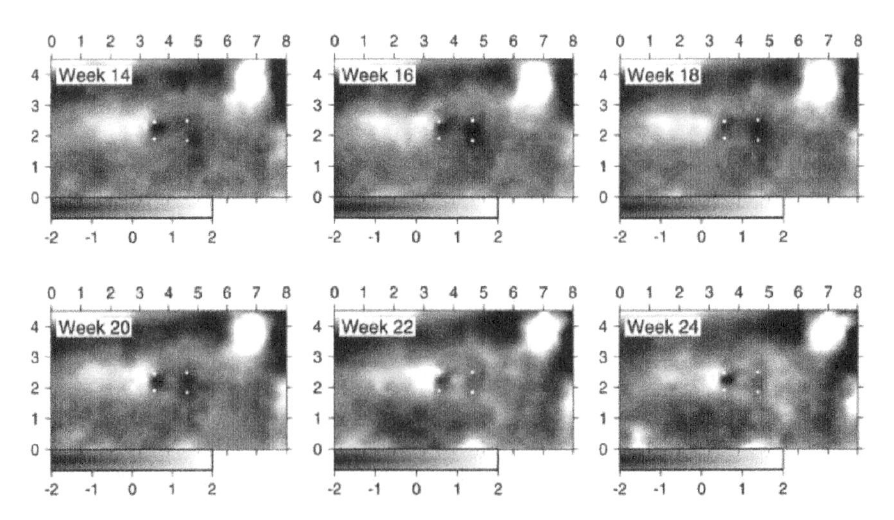

Fig. 17.3 Normalised resistivity survey data acquired over the simulated grave between 14 and 24 weeks post-burial

tests, the p-value is well above the significance level of 0.05, suggesting that differences in saturation or porosity between the grave soil and the control soil are simply due to random variation. When H_0 is accepted, it is necessary to consider the statistical power of the test, which defines the test's ability to reject H_0 when it is false and, ideally, should be greater than 0.8 (Warner 2008). In the tests used here, the

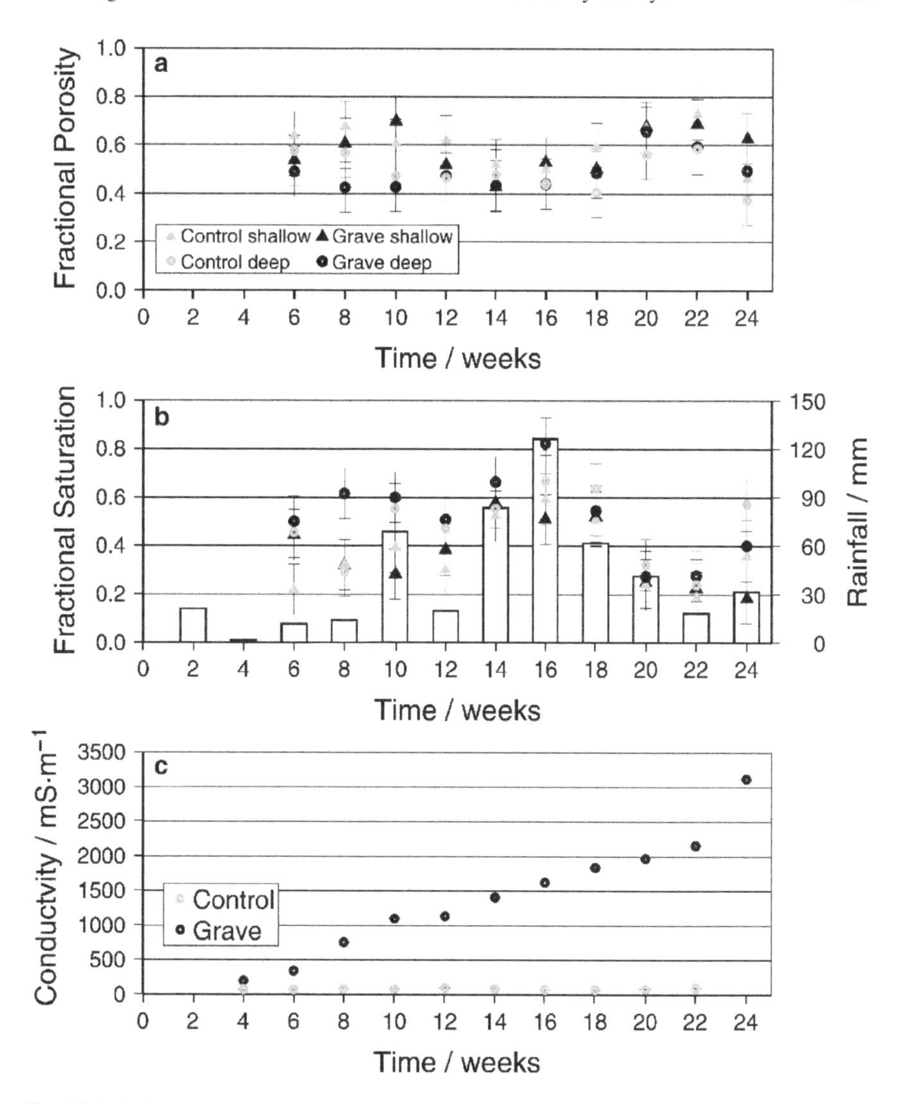

Fig. 17.4 Soil and water data collected during the project. (**a**) Soil porosity data for deep and shallow samples collected from the grave and control locations. (**b**) Soil saturation data plotted with the same symbol key as the porosity data. Bars show rainfall for each fortnight plotted against the right-hand scale. (**c**) Water conductivity for samples collected from the grave and control lysimeters. Error bars in (**a**) and (**b**) represent combined measurement errors from all values used in the calculation of porosity and saturation, respectively

statistical power is at best approximately 0.15, which suggests that the experiment is not well suited to detecting differences in porosity or saturation between the grave and control samples. Low statistical power often indicates a poor signal to noise ratio. Here, this may be a result of too few measurements being made, high measurement error or too much natural variability within the soil samples.

Table 17.1 Outcomes of statistical analysis of soil data

Test parameter	Null hypothesis	Mean difference	Standard error	p-Value	Decision	Power of the test
Shallow porosity	$H_0: \mu_d = 0$	−0.019	0.029	0.531	Accept H_0	0.090
Deep porosity	$H_0: \mu_d = 0$	0.000	0.026	0.998	Accept H_0	0.050
Shallow saturation	$H_0: \mu_d = 0$	0.005	0.035	0.883	Accept H_0	0.052
Deep saturation	$H_0: \mu_d = 0$	0.044	0.043	0.330	Accept H_0	0.152

Groundwater Data

The conductivity of the water samples (Figure 17.4c) obtained from the grave rose in near-linear fashion from 196.6 mS/m in weeks 4 to 2,150 mS/m in week 22, before rising suddenly to 3,110 mS/m in week 24. In contrast to this, the conductivity of the control groundwater was approximately constant, with a minimum of 71.5 mS/m (week 18) and a maximum of 94.5 mS/m week 22). No water was present in the control lysimeter in week 24.

Discussion

Despite the challenging nature of the study site, the simulated grave is clearly delineated in the geophysical survey data, albeit only from 4 to 20 weeks after burial. This is similar to the results of Bray (1996, as cited in Cheetham 2005), in which monthly resistivity surveys did not reveal a low resistivity anomaly over a pig grave until 2 months after burial. Furthermore, a reduction in the contrast between another pig grave and background readings was reported after one year of burial (Bray 1996, as cited in Cheetham 2005).

Analysis of the soil porosity and saturation data is inconclusive and because of the low statistical power of the hypothesis tests it must be concluded that the contribution of changes in soil porosity and saturation to the low resistivity anomaly of the grave cannot be determined in this study. However, a definite increase in fluid conductivity within the grave was observed and is likely to be at least partly responsible for the observed low resistivity anomaly. The increase in fluid conductivity in the grave may be a result of the decomposition of the pig cadaver. The ‘active decay’ stage of decomposition has been observed to be associated with a release of cadaveric fluids into the ground, causing soil beneath bodies deposited on the surface to take on a darker, stained appearance (Carter et al. 2007) and a similar staining of the soil has been noted in the case of buried pig cadavers (Wilson et al. 2007). Experimental work involving human cadavers deposited on the surface has shown the release of fluid during decomposition to be associated with elevated concentrations of a number of different ions in water extracted from soil samples taken from directly beneath the body (Vass et al. 1992). Fluid conductivity increases with the concentration of dissolved solids (e.g. Drever 1982). Therefore, the elevated ion concentrations in the soil water close to decomposing human remains described by Vass et al. (1992) would have resulted in a localised increase in water conductivity. It is then suggested that a similar phenomenon is responsible for the increase in grave water conductivity observed here. From week 8 onward, the water samples obtained from the pig grave had a brown discoloured appearance and a distinctive 'cheesy' odour which gave way to a more noxious fetid odour after approximately 16 weeks of burial. Similar odours have been associated with excavated pig cadavers (Turner and Wiltshire 1999), adding further support to the notion that alteration of the soil solution in the graves is a result of the decomposition of the pig carcasses.

The possibility that the fluid released by the decomposing pig cadaver is responsible for the low resistivity of the grave appears to be consistent with a number of features in the geophysical data. For example, the two-part nature of the low resistivity grave anomaly (e.g. the data from week 8) mirrors the distribution of organic matter within the grave; as organic matter is the likely source of decompositional fluid, this would be expected. The organic matter with the grave was concentrated at the two ends of the grave because the pig carcasses had been eviscerated; this left only ribs and a thin layer of skin present in the middle of the carcass, whilst the head (complete with brain) and forelegs (with all associated muscle) were situated at one end of the grave and the kidneys, bladder and hind legs (with all associated muscle) were situated at the other end of the grave. The timescale over which the low resistivity anomaly is visible may also be explained in terms of the decomposition of the pig carcasses. That no significant anomaly is visible in the survey data obtained in week 2 could be due to the fact that, after 2 weeks in the ground, the cadaver had either not reached the stage of active decay, or the active decay process had not produced a large enough volume of fluid for the grave to be detectable. The diminished area of low resistivity values at the grave location in the survey data from weeks 22 and 24 may be attributed to a reduction in the volume of highly conductive fluid once the main phase of active decay is over. Such a decrease in the volume of highly conductive fluid within the grave would be consistent with the decrease in saturation observed in the deep grave soil samples compared to the control samples from week 18 onward. A decrease in the volume of conductive fluid within the grave could also explain the fact that the low resistivity anomaly in the survey data reduces in size after week 20 despite the continued rise of the conductivity of the water samples extracted from the grave during this period.

Hence, it seems likely that the low resistivity of the simulated grave is at least partly due to the conductive fluids released by the decomposing cadaver, whilst the contribution of any changes in the soil porosity and saturation to the geophysical anomaly are undetermined in this case. This suggests that if a cadaver is undergoing active decomposition, it should be detectable using the electrical resistivity technique, even if the ground conditions are complex and challenging, such as those at the study site discussed here. However, it should be noted that if the low resistivity anomaly is solely due to the cadaveric fluids, it is possible that the results of an electrical resistivity survey over a cadaver that is wrapped (e.g. in a blanket or tarpaulin), or even one that is clothed, may differ significantly from those that are presented here.

Finally, the relatively short-lived nature of the low-resistivity anomaly presented here conflicts somewhat with other published reports of electrical resistivity and conductivity surveys used in murder enquiries. For example: a low resistivity anomaly associated with a murder victim buried 16 years before a resistivity survey was undertaken (Cheetham 2005) and a high conductivity anomaly in data collected over a murder victim buried approximately 12 years previously (Nobes 2000) demonstrate that detectable areas of altered soil resistivity associated with a grave can be present several years after burial. Comparison of the GPR response of several buried pig cadavers ranging from approximately 30 kg (Schultz 2008) to approximately 64 kg (Schultz et al. 2006) suggests that heavier bodies can be detected for longer post-burial intervals than lighter ones. If the same is true for resistivity surveys, this may explain the relatively short period over which the small (approximately 31 kg) pig cadavers used in this study were detectable using resistivity surveys. However, only one location is considered in this study and it is possible that burials in other environments, where the soil type or climate is different to that discussed here, may also produce a longer-lasting resistivity anomaly.

Conclusions

The electrical resistivity survey method used here over a buried pig cadaver shows considerable promise for the location of shallow graves in the first few months after burial, despite the difficult survey conditions provided by the study site and the relatively small size of the buried pig cadaver. The low-resistivity anomaly, which allows the grave to be easily identified, appears to be at least partly caused by an increase in fluid conductivity within the grave. This increase in fluid conductivity is suggested to be a result of an increased concentration of dissolved ions in the water within the grave as a result of fluids released by the buried cadaver. The short-lived nature of the anomaly associated with the grave in the geophysical data raises concerns over the suitability of resistivity surveys for the location of small cadavers that have been buried for longer than approximately 6 months. However, further work is necessary to understand how a grave may be detectable using resistivity surveys in other environments, soil types, burial conditions and over longer timescales.

Acknowledgements John Jervis' PhD is supported by a Co-operative Awards in Science (CASE) studentship, funded by the Engineering and Physical Sciences Research Council (EPSRC) and Stats Limited. We thank; Staffordshire University's Faculty of Sciences for allowing this experiment to be conducted on their campus; Nigel Cassidy at Keele University and Tim Grossey of Stats Limited for informative discussions on the geophysical data; Ian Wilshaw and Zoe Robinson of Keele University for assistance with the lysimeters and supplying local weather data; several Keele University undergraduate students for assistance in acquisition of the geophysical data. Figures were prepared with the GMT software. We would also like to thank two anonymous reviewers for their constructive comments, which helped to improve the manuscript.

References

Archie GE (1942). The electrical resistivity log as an aid in determining some reservoir characteristics. Transactions of the American Institute of Mining and Mineralogical Engineers 146:54–62.
Barnes GE (2000). Soil Mechanics: Principles and Practices, 2nd edition. Palgrave Macmillan, Basingstoke.
Buck SC (2003). Searching for graves using geophysical technology: field tests with ground penetrating radar, magnetometry and electrical resistivity. Journal of Forensic Sciences 48:5–11.
Carter DO, Yellowlees D and Tibbett M (2007). Cadaver decomposition in terrestrial ecosystems. Naturwissenschaften 94:12–24.
Cheetham P (2005). Forensic geophysical survey. In: Forensic Archaeology: Advances in Theory and Practice (Eds. J Hunter and M Cox), pp. 62–95. Routledge, London.
Drever JI (1982). The Geochemistry of Natural Waters. Prentice-Hall, New Jersey.
France DL, Griffin TJ, Swanburg JG, Lindemann JW, Davenport GC, Trammell V, Armbrust CT, Kondratieff B, Nelson A, Castellano K and Hopkins D (1992). A multidisciplinary approach to the detection of clandestine graves. Journal of Forensic Sciences 37:1445–1458.
Freeland RS, Yoder RE, Miller ME and Koppenjan SK (2002). Forensic application of sweep-frequency and impulse GPR. In: Proceedings of the Ninth International Conference on Ground Penetrating Radar (Eds. SK Koppenjan and H Lee), pp. 533–538. SPIE – The International Society for Optical Engineering, Bellingham.
Friedman SP (2005). Soil properties influencing electrical conductivity: a review. Computers and Electronics in Agriculture 46:45–70.
Hillel D (1980). Fundamentals of Soil Physics. Academic, London.
Hintze J (2001). NCSS and PASS. Number Cruncher Statistical Systems. NCSS Systems, Kaysville, UT.
Miller ML, Freeland RS and Koppenjan SK (2002). Searching for concealed human remains using GPR imaging of decomposition. In: Proceedings of the Ninth International Conference on Ground Penetrating Radar (Eds. SK Koppenjan and H Lee), pp. 539–544. SPIE – The International Society for Optical Engineering, Bellingham.
Nobes DC (2000). The search for "Yvonne": a case example of the delineation of a grave using near-surface geophysical methods. Journal of Forensic Sciences 45:715–721.
Pringle JK, Jervis JR, Cassella, JP and Cassidy NJ (2008). Time lapse geophysical investigations over a simulated urban clandestine grave. Journal of Forensic Sciences 53 DOI: 10.1111/j.1556-4029.2008.00884.x
Reynolds JM (1997). An Introduction to Applied and Environmental Geophysics. Wiley, Chichester.
Ruffell A (2005). Searching for the IRA "disappeared": ground penetrating radar investigation of a churchyard burial site, Northern Ireland. Journal of Forensic Sciences 50:1430–1435.
Schultz JJ (2008). Sequential monitoring of burials containing small pig cadavers using ground penetrating radar. Journal of Forensic Sciences 53:279–287.
Schultz JJ, Collins ME and Falsetti AB (2006). Sequential monitoring of burials containing large pig cadavers using ground-penetrating radar. Journal of Forensic Sciences 51:607–616.
Schultz JJ, Falsetti AB and Collins ME (2002). The detection of forensic burials in Florida using GPR. In: Proceedings of the Ninth International Conference on Ground Penetrating Radar (Eds. SK Koppenjan and H Lee), pp. 443–448. SPIE – The International Society for Optical Engineering, Bellingham.
Scott J and Hunter JR (2004). Environmental influences on resistivity mapping for the location of clandestine graves. In: Forensic Geoscience: Principles, Techniques and Applications (Eds. K Pye and DJ Croft), pp. 33–38. The Geological Society of London, London.
Smith WHF and Wessel P (1990). Gridding with continuous curvature splines in tension. Geophysics 55:293–305.
Turner B and Wiltshire P (1999). Experimental validation of forensic evidence: a study of the decomposition of buried pigs in a heavy clay soil. Forensic Science International 101:113–122.

Vass AA, Bass WM, Wolt, JD, Foss, JE and Ammons, JT (1992). Time since death determinations of human cadavers using soil solution. Journal of Forensic Sciences 37:1236–1253.

Warner RM (2008). Applied Statistics: From Bivariate Through Multivariate Techniques. Sage, Los Angeles.

Wilson AS, Janaway RC, Holland AD, Dodson HI, Baran E, Pollard AM and Tobin DJ (2007). Modelling the buried body environment in upland climes using three contrasting field sites. Forensic Science International 169:6–18.

Chapter 18
Spatial Thinking in Search Methodology: A Case Study of the 'No Body Murder Enquiry', West of Ireland

Jennifer McKinley, Alastair Ruffell, Mark Harrison, Wolfram Meier-Augenstein, Helen Kemp, Conor Graham and Lorraine Barry

Abstract Geographical information systems (GIS), recognised as instrumental in the documentation, mapping and analysis of spatial crime data, can provide a framework for the integration and analysis of spatial data whether remotely generated on a regional scale, ground surveyed or sampled on a local scale at a crime scene. In this paper a missing person homicide case study is discussed to illustrate the role of spatial thinking in search methodologies and the application of GIS and spatial analysis techniques. Differential global positioning systems were used to collect data from an area of mixed moorland, bog and agricultural ground in the west of Ireland where police intelligence suggested human remains may have been hidden by a murderer. These data allowed the creation of a digital terrain model (DTM) at a resolution not achieved by conventional terrain mapping. The resultant topographic maps and 3D visualisations allowed a sector, or topographic domain, approach to be used at a scale finer than usual in geomorphology. This in turn allowed small water catchments to be defined. These data informed the sampling of shallow groundwater for carbon content and isotope analysis. Two anomalies were indicated, in places consistent with known criminal behaviour. The locations were surveyed by a ground penetrating radar system, and by a cadaver dog. Ground penetrating radar (GPR) failed to indicate any subsurface disturbance or grave, yet the cadaver dog indicated a point of interest close to the location of one anomaly identified from groundwater sampling. Further searching near this location failed to discover human remains, yet the isotope, topography and dog indications showed

---

J. McKinley(✉), A. Ruffell, C. Graham, and L. Barry
School of Geography, Archaeology and Palaeoecology, Queen's University, Belfast
Belfast BT7 1NN, UK
e-mail: j.mckinley@qub.ac.uk

M. Harrison
National Policing Improvement Agency, Wyboston Lakes, Green North Road, Wyboston MK44 3AL, UK

W. Meier-Augenstein and H. Kemp
Stable Isotope Laboratory, Scottish Crop Research Institute,
Invergowrie, Dundee DD2 5DA, UK.

K. Ritz et al. (eds.), *Criminal and Environmental Soil Forensics*

that some anomaly existed. This finding may be a false-positive, the result of previous excavation activity or that the remains had 'returned to earth', in the light of the missing person never having been found. Regardless, the spatial search methodology described is an innovative combination of new technology, traditional landscape interpretation and hydrological chemical analysis. Adaptation, testing and use of this protocol for similar searches are recommended. The approach also has broader application to environmental, humanitarian and military investigations.

Introduction

An appreciation that criminal activity takes place over different geographic scales enables an investigator to utilise a spatial approach to analyse the criminal activity (Canter et al. 2000; Hirschfield and Bowers 2001). Behavioural profiling and/or intelligence-led information (e.g. the last known movements of victim and/or suspect) may define a search area for a deposition site (of a body or illegal environmental waste), the areal extent of which can cover a range of scales, from a few metres to tens or hundreds of kilometres. It becomes unfeasible as regards cost and time to attempt to intensively investigate a search area across large areas: a spatial approach may be used to address this issue.

Harrison (2006) describes a scenario-based search method to focus an investigation of criminal activity potentially covering tens or hundreds of kilometres to several key locations as indicated by behavioural profiling and intelligence led information. Depending on the spatial scale at which the search is conducted, the emphasis of the search method changes: from a scenario-based, at broad scales (kilometres), to a feature-based approach, at fine scales (metres). The feature-based approach uses landforms, vegetation, access and points of geographic note (trees, streams, hills, buildings) as markers used by possible perpetrators to navigate and return to scenes. The nature of the environment (e.g. urban and rural) will have implications for both types of search strategy and this must be addressed when considering spatial scale in the analysis of the criminal activity. In a feature-based search approach, human and physical anchor points such as a key building, windbreaks or a fork in a stream will determine the geographic extent of the search subsequently informing any spatial approach adopted.

Geographical Information Systems (GIS), widely recognised and utilised in the documentation, mapping and analysis of spatial crime data (e.g. Hirschfield and Bowers 2001; Chainey and Ratcliffe 2005), can provide a framework for the integration and analysis of spatial data whether remotely generated on a regional scale or ground surveyed or sampled on a local scale. The use of GIS and spatial analysis techniques have been used at an urban crime scene (McKinley and Ruffell 2008) where sampling at a fine scale (centimetres) indicated spatial variability that placed the suspect close to an identified location (within 0.50 m) at the scene of crime, confirming witness reports. This paper investigates the multi-scale aspect of search methodology and discusses a missing person homicide case to demonstrate the role of spatial thinking in search methodologies. Aspects of the missing person case are investigated in more detail in Ruffell and McKinley (2008). This paper describes the spatial approach that was deployed in the investigation and the importance of integrating spatial information generated at multiple scales. Broader applications of the approach to environmental, humanitarian and military investigations are also discussed.

Case Study: Missing Person (No Body Murder) West of Ireland

Background

In late September 1995, a teenage girl attending a club at a seaside town in the west of Ireland accepted a midnight lift home by a mature man who was known, informally, to her, although she did not know of his previous criminal records, including abduction and rape. Neither person can be named because the enquiry is ongoing and, as such, some details of the case have been changed. The man's estate car was noticed by an off-duty member of the police some three hours later, parked in a lay-by to a farm track, on a quiet country road. The officer, two colleagues and a dog team returned to this location some 30 minutes later and a police dog indicated a point of interest at a location along the farm track where fields passed into scrubland ('turning area', Figure 18.1). Concerned at their proximity to the Northern Irish – Irish Republic border, and thus both causing problems of jurisdiction and at the presence of republican terrorists, the police abandoned their

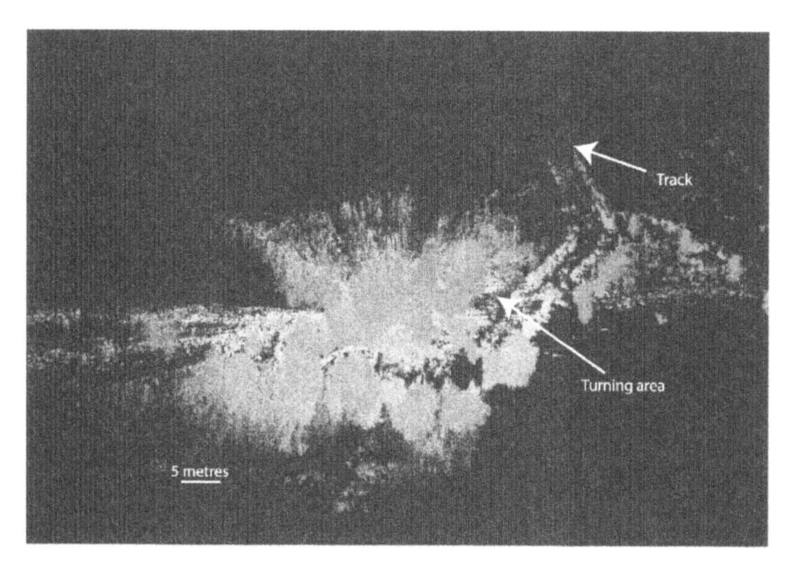

Fig. 18.1 Composite terrestrial Lidar image of the search site. Image created by a composite of laser scanned images of the area (*see colour plate section for colour version*)

search until daylight. A follow-up search was made, but heavy rain and tractor and cattle movement limited the use of tyre marks or footprints as meaningful clues or evidence. The teenage girl was reported missing some days later. The reason for the delay in reporting of her disappearance was stated as her habit of staying with friends and even the possibility that she had absconded to her separated father's house in Dublin.

Enquiries led to the mature man who was the last person to see the girl, but conflicting reports of her being seen elsewhere in Ireland and England were inconsistent with her disappearance. Only when the mature man was arrested, four years later, for the abduction and rape of a 15-year old girl was a possible connection made, and a search made, the focal point of which was his flat at the time of the disappearance of the teenager in 1995. As the search widened from this point, so the area of his last known position on the night in question was considered.

The Search Location (Scenario-Based Search Methodology)

A broad scale (covering tens of kilometres) scenario-based search method was employed to investigate the disappearance of the teenage girl. Intelligence, in the form of the last known actions of the teenager and the sighting of unusual activity by the last known person to have been in the company of the teenager, along with geographical profiling of the suspect, refined the search to an area of scrubland. The search location includes a range of vegetation and soil types. The vegetation around the area comprises planted pine adjacent to the main road and silver birch along the track and around the ridge that formed the core of the search area. Scrubby grassland with metre-high stands of gorse or reeds, with some bedrock exposure, occurs throughout the area. Low-lying ground is characterised by reeds growing on bog, some of which showed evidence of being drained in the past. The vegetation was initially used to classify the landscape, but proved unsuitable as a basis for focusing a search.

Search Limitations and Domains (Feature-Based Search Methodology)

No surface remains of clothing, persons or personal items were found, indicating that if the victim was in the location, she would be buried. Four years had passed since the area was last searched: the location has high rainfall (1400 mm/year) with extensive cattle movement, masking the expression of surface features. Some fundamental controls on the search were established. A main road forms a border to the east. A major river occurs to the south, with housing some 500 m distant, with open, arable farmland to the west. An extensive bog that cannot be walked upon even in summer borders the area to the north. Within the search area, individual locations were limited by views, inaccessible and were due to thick trees, bog, river, or rock (i.e. insufficient depth of soil for burial). To facilitate a

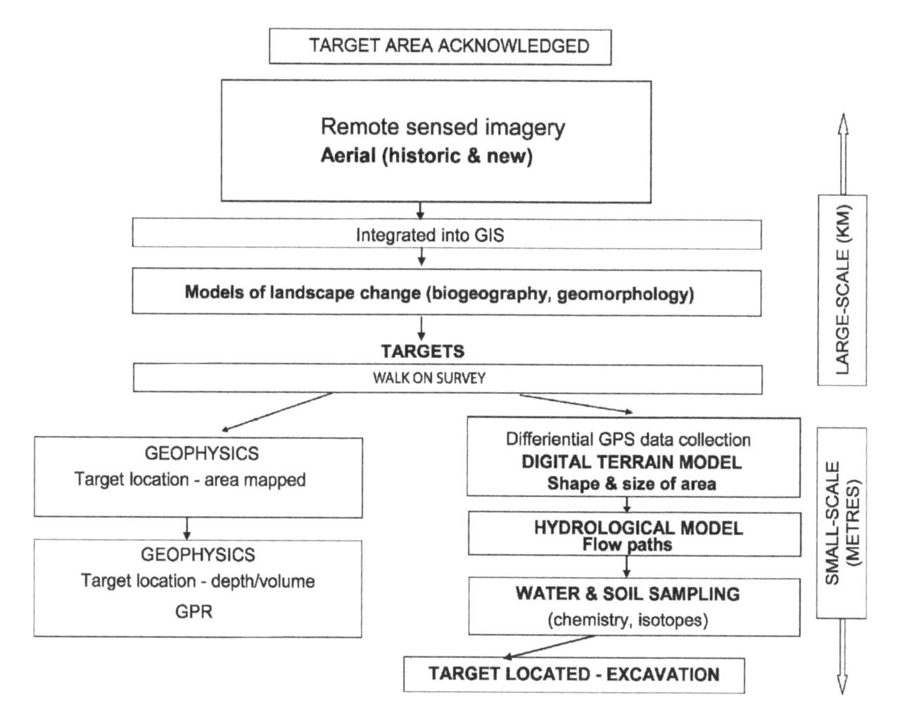

Fig. 18.2 Flow diagram illustrating the multi-scale search methodology and the range of geographical and spatial information used

feature-based search scenario for potential body deposition or burial, a landscape domain-based approach was deployed. The search strategy (Figure 18.2) entailed the acquisition of remotely-sensed imagery (aerial photography), a ground-based elevation survey using differential GPS (DPGS) to facilitate landform mapping and the generation of elevation derived maps (e.g. slope) and other derived analysis (e.g. line of sight mapping). This approach enabled the most appropriate location for water sampling to be defined. A complementary geophysical survey was undertaken to map target locations using ground penetrating radar (GPR).

Global Positioning Information

Aerial photography was used to provide an overview of the full geographic extent included in the search area and to provide an initial visualisation of the landscape. However, since topography was important to define landscape domains, elevation data were also collected by DPGS. Decisions on the most appropriate spatial resolution of generated elevation data were based on a balance between the accurate representation of the topography and the time available to collect the elevation data. In this case, DGPS data were collected over one day using a base station and rover system using a stop and record procedure to collect DGPS data.

An important consideration in undertaking the ground GPS survey was to combine the requirement of representing the topography accurately with an awareness of what the accused was suspected of doing. Hence, the need for a spatial resolution of generated elevation data to adequately record the presence of low ground. Such an area would be consistent with the covert activity of burial requiring isolation and low ground, perceived as not being visible from the farm track off the quiet country road. DGPS data were taken approximately every metre over the area but higher resolution data were collected (at 0.5 m) when required to adequately record changes in terrain (e.g. the presence of low ground). Approximately 100 DGPS points were recorded in total.

Elevation data were transferred to the Irish Grid co-ordinate system and stored and managed in a GIS. DGPS data enabled the creation of a Digital Elevation Model (DEM) and provided a georeference system for the integration of aerial photography, and the production of several derivative maps (slope map) and other surface analyses (viewshed analysis).

Landform (or Geomorphological) Mapping

Geomorphological mapping, based on the concept of dividing up the landscape using the boundaries between landforms, most especially changes in relief, remains one of the fundamental research methods of geomorphology (Waters 1958; Cooke and Doornkamp 1990; Minár 1992). Landforms can be summarised with regard to relief units with increasing complexity. In this way, the identification of elementary landform units forms the first stage in the study of the spatial interaction between landforms, soil, vegetation, local climate or hydrological regime. Such a study can be facilitated further by the integration of remotely-sensed data and GPS technologies in a GIS framework. Collection of DGPS elevation data and the creation of a terrain model (overlain with aerial photography) allows a view of topography at a greater resolution than is possible from paper maps. Most especially, this topography can be assessed with regard to external views into the site, and when linked to visual observations, ground-based photography, and ground-based Lidar, to views through trees and bushes. Topography can then be linked to soil depths likely offender behaviour. Most critical for this study is that the fine-resolution DEM can be used to define small fluvial catchments (often no more than ditches, streams and rivulets) that can be used to inform the location of water samples. Ten individual catchments were defined from topography alone by this means. Areas defined in Figure 18.3 include the boundary features and the landscape domain-based approach described previously. Clockwise, from the southeast of the area defined by the road, these are labelled on Figure 18.3 as 6, 9, 8, 8, 8, 5, 3, 2, 1, 9, 7, 4 and 10.

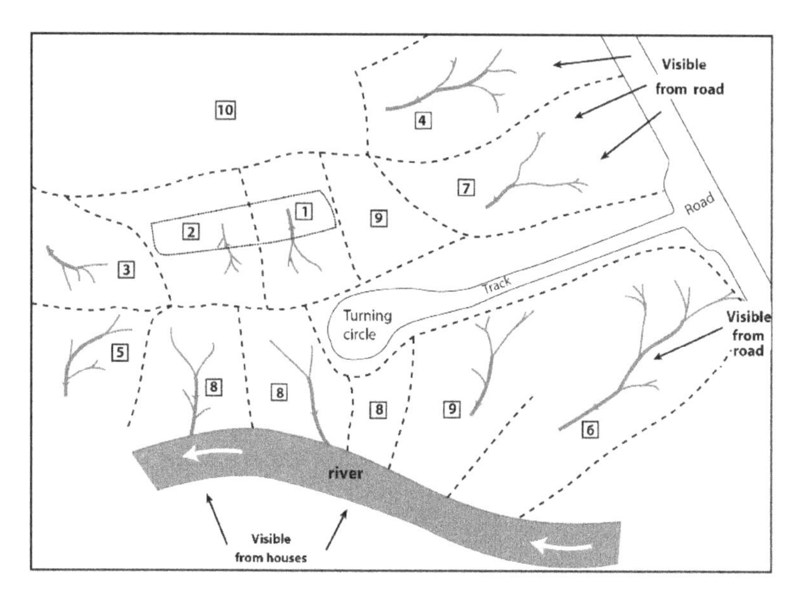

Fig. 18.3 Landscape mapping domain-based search approach (adapted from Ruffell and McKinley 2008). Numbers 1 to 10 refer to domains discussed in the text. Highlighted area refers to target search area with Domains 1 and 2

Catchments and Soils

The ten catchments were then considered with regard to soil thickness and type. In the same order as above (Figure 18.3), the southern catchments (6, 9, 8, 8, 8) were all located in a field of mixed rough pasture and rushes, with occasional bedrock exposures (9). Catchments 5 and 3 were located in areas of low rocky outcrop with silver birch and thin (1 m augered depth) peat. Catchments 2, 1 and 9 comprised peat (1–2 m depth) with silver birch. Catchments 7 and 4 comprised peat with pine forest. Catchment 10 is a flat peat bog of over 3 m depth.

Domains

A composite landform mapping approach (combining the above topographic analysis and soil type/depth information) was used to create a priority of ten search domains on which to focus a feature-based search approach for the missing person. Areas undefined in Figure 18.3 include the boundary features described previously. The domains are described in terms of the spatial interaction between landform, soil type/depth, vegetation, catchments and offender profiling.

Domain 10 (Figure 18.3) comprised the bog, in which a dug hole became quickly waterlogged. Domain 9 was the opposite – having bedrock exposed or within 20–30 cm of ground level. Nonetheless, pockets of deeper, softer material could occur, for instance if this was in a limestone area prone to deep weathering. The bedrock comprises hard sandstones with limited hollows and pockets. The domains were collectively grouped as eight were all on grazed land, which could be viewed from houses. Under cover of darkness, this ground could have been excavated but the resultant scar on the landscape would have been visible for sometime thereafter.

The potential burial locations in Domain 7 are limited by bedrock, bog and by views from the road, most especially where police vehicles parked on the night in question. Domain 6 has the same limitation as 7 and 8, being visible from the road and the houses. Domain 6 is also actively grazed by cattle and is an unlikely target area. Domain 5 has all the correct spatial attributes for a burial location in being covert and underlain by soft ground with no waterlogged soil. However, the distance to Domain 5 that the suspect would have been forced to walk, or would have to drag a victim, through rough terrain is considerable compared to other locations. As the *modus operandi* of the suspect shows him to be a lone-operator, this reduces Domain 5 as a likely body deposition site. Domain 4 has the correct mix of cover and proximity to vehicular access. The possible area within Domain 4 where a covert burial could occur is limited by views from the road and the waterlogged bog to the north and west.

Domain 3 is similar to Domain 5 in location and topography. The only factor which makes this domain a possible burial site is access. Domain 5 is inaccessible because of the thick trees and scrub vegetation, whilst Domain 3 is accessible across rough grassland on a central plateau area. Thus, distance to a vehicle is about the same as Domains 4 and 5, but Domain 4 would necessitate movement over rocks, ditches and bog, whereas Domain 3 is a relatively easy walk westwards over grass.

Domains 1 and 2 are much the same, with their northern limits, adjacent to the bog being below the 2 m elevation of the plateau, and with no dwellings to the north. A highlighted area in Figure 18.4 indicates where the search limits within these areas could be, with bog to the north, and visibility to the south. Domains 1, 2 and 4 all have the advantage of pre-existing ditches, which at their upslope end could have been dry enough to be re-excavated.

Domain 1 represented the priority search location in being proximal to the vehicle access point (Figure 18.3 'turning circle'). The combined use of geomorphological mapping, integrated with other feature-based search criteria, presents an example at one scale of how the human mind may divide up the landscape in order to target resources of time and cost. A similar methodology could be used at smaller or larger scales, depending on the area which has to be searched and on the complexity of the terrain.

Landform classification enabled the division of the area into manageable tracts of ground, each with individual characteristics that focused a feature-based search approach and refined the application of specific search techniques. Broad (metres to decimeters) areas of soft ground provided ideal locations for the application of geophysics whereas much smaller (metre or less) areas, more problematic for geophysics, were found to be more suitable for investigation by probing.

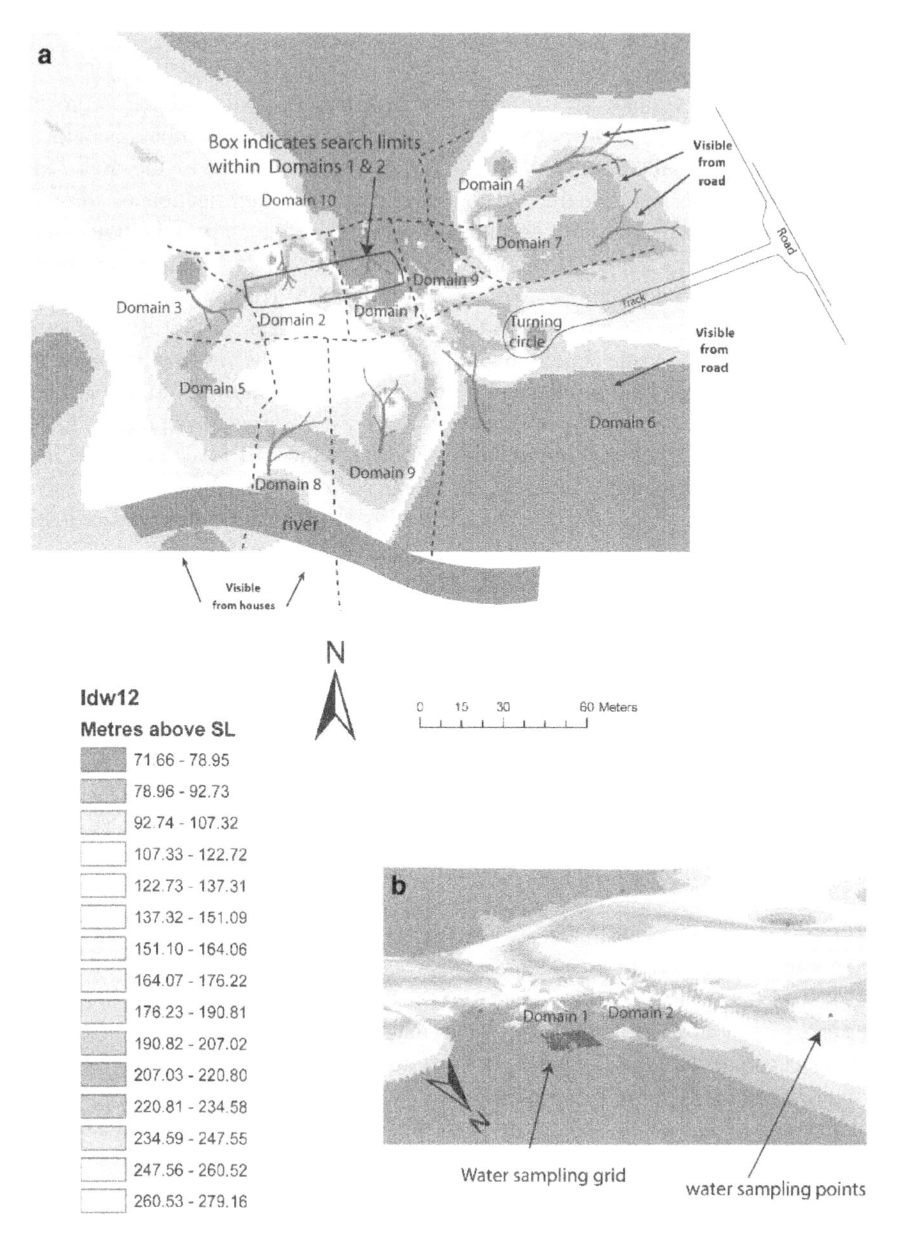

Fig. 18.4 Digital elevation maps (DEM) generated in GIS using inverse distance weighting (IDW) interpolation of DGPS elevation data (IDW 12 indicates the number (12) of neighbouring DGPS data used in the interpolation procedure). (a) Two-dimensional. (b) Three-dimensional. 3D DEM has been overlain with the domain-based search areas shown in Figure 3 described in the text. Highlighted area refers to target search area with Domains 1 and 2. Key locations for water sampling throughout the area along with a larger area in Domain 1 are shown on the 3D DEM. The viewing perspective has been changed to aid landscape interpretation (*see colour plate section for colour version*)

Integration of Spatial Search Criteria in GIS

GIS provided the framework to integrate the landscape domain-based approach with aerial photography, elevation data and derived maps. The outputs generated using the GIS provided a 3D visualisation, or representation, of the area that, although involving some degree of generalisation and simplification of the real world, provided the opportunity to interactively explore the entire search area and scrutinise different search domains. Secondary attributes, comprising aerial imagery and search domains, were draped over the DEM (Figure 18.4a).

Shaded relief maps (using information from slope and aspect) increased the level of realism of the 3D visualisation in which the angle of view, viewing azimuth and viewing distance could all be adjusted (Figure 18.4b). This allowed further verification of the mapped domains, and enabled feature-based search strategies to be worked through, to reduce time and cost on the ground and focus resources and provide optimum time to conduct the field search. Search domains could be ruled in or out at this stage with those such as Domains 1 and 2 warranting greater field-based investigation (e.g. water sampling Figure 18.4b).

Part of the landscape mapping approach to define areas to focus search efforts involved evaluating visibility of the different domain: defining areas of the topography that were not visible from the track, country road or nearby dwellings. Certain domains could be ruled out very quickly in this aspect; namely Domains 6 and 7. At the time of the disappearance of the teenager, the suspected body deposition site was accessible by vehicle down a narrow country track. Vehicle accessibility, however, would have diminished some 50 m along the track, suggesting that any further movement or transport of a body would have to be made on foot over the rough moorland and bog. It was particularly useful, therefore, to know which areas of the terrain would have been hidden from view from the furthest access point of the country track.

The DEM and derived slope map were used to produce a line-of-sight or viewshed map of the area (Figure 18.5), and in particular search Domains 1 and 2. The viewshed map enabled a feature-based assessment of the terrain, and in this case the important features were hollows or hidden depressions that would have concealed covert activities of burial or deposition. However, vegetation cover (bracken with interspersed saplings) would also have provided some concealment (Figure 18.6a).

Terrestrial Lidar generated using a Leica HDS3000 laser scanner recorded the vegetation of the area at the time of the current search (Figure 18.6b). Laser scanning is an increasingly popular and practical technique in the visualisation of landscapes. A laser scanner effectively 'blankets' a scene generating a rich dataset in the form of high density 3D point clouds (thousands of data points are recorded). Although the application of a technique such as laser scanning is obviously more useful when deployed at the time of an incident, and may be of limited value for cold cases, the use of the technique was justified in this case by the limited change in the vegetation involved (i.e. rough moorland and bog). Terrestrial Lidar combined with viewshed analysis provided a comprehensive picture of the degree of concealment afforded to covert activities in the area.

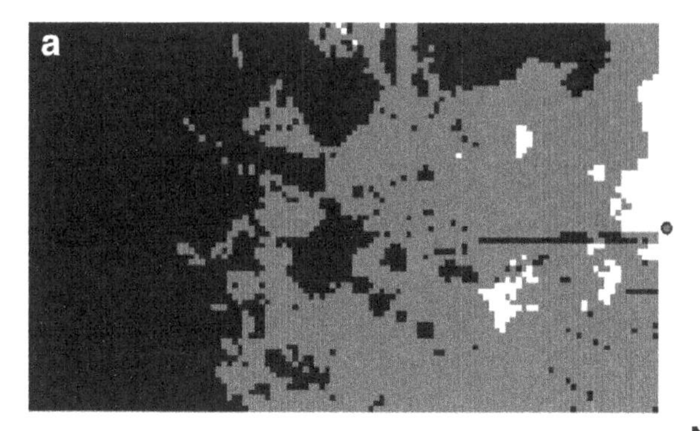

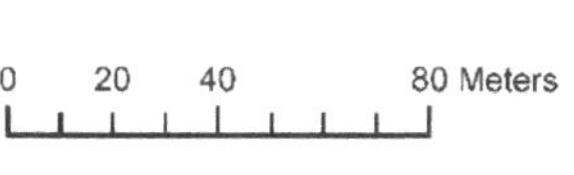

IDW 12 and IDW32 Viewshed overlaps

Visibility from view point

0 Not visible

1 Visible using IDW32

2 Visible using IDW12 and IDW32

• View point

0 15 30 60 Meters

Fig. 18.5 (a) 2D and (b) 3D Viewshed (line of sight) map generated in GIS from the DEM and slope map. IDW 12 and IDW 32 indicate the number of neighbouring DGPS data (12 and 32) used in the interpolation procedures

a

b

Fig. 18.6 (a) Photograph of vegetation in the search area and use of GPR. Figure is approximately 1.8 m. Field of view is 10 to 15 m. (b) Terrestrial Lidar image of the vegetation generated using a Leica HDS3000 laser scanner. Field of view is 15 to 20 m (*see colour plate section for colour version*)

Hydrological Chemical Analysis

The creation of the DEM, resultant topographic maps and 3D visualisations enabled a hydrological assessment of the area and allowed the derivation of small water catchments within the search domains. Combined with 'hits' from cadaver canine searches, hydrological mapping was used to inform a sampling strategy of shallow groundwaters for carbon content and isotope analysis. Stable isotope analysis of organic and inorganic carbon in groundwater and shallow sub-surface water may provide an indication of a clandestine burial, similar to use in environmental forensics to detect contaminant plumes. The average human body contains approximately 20% body weight carbon. When a body decays, proteins, fats and carbohydrates will start to break down to their constituent parts: amino acids, fatty acids and sugars. The collective carbon contained within the breakdown products termed 'dissolved organic carbon' (DOC) and will be released from human remains relatively shortly after death or deposition (dependant on burial conditions and pH of the environments). As the DOC starts to breakdown over time, dissolved inorganic carbon (DIC) levels will rise in 'aged' depositions. This human-remain derived carbon (DOC and DIC) can then leach into the surrounding environment (through soil and groundwater movement), thus altering the usual carbon signature of the local environment. The supposition is therefore that pure stable isotope profiles of DIC/DOCs from human origin should exhibit a distinct profile that is markedly different from that derived solely from environmental, plant material or animal origin. In the current missing person case, given the terrain (acidic, peaty conditions), a slow rate of decomposition (i.e. bog-body/mummification) seemed plausible and as such any sub-surface water coming into contact with a body might well still exhibit elevated levels in the composition of either DIC or DOC or both, providing a distinct 'human-carbon stable isotope profile'. Key locations throughout the area were sampled along with a larger area within Domain 1 where 50 sub-surface water samples were collected using a regular grid scheme at a site where some scoping excavation had taken place (Figure 18.4b). These samples were analysed for Total Organic Carbon (TOC) content only.

Spatial Analysis

Spatial variation in the distribution of isotope data, concentrations in ppm of TOC in water, collected from Domain 1 was characterised using spatial analysis techniques (i.e. GIS and geostatistics). The variogram, central to geostatistics and used for spatial prediction and simulation, characterises spatial dependence in the property of interest: in this case carbon isotope content of shallow groundwaters. A spherical model was used (Figure 18.7a), estimated using Gstat (Pebesma and Wesseling 1998) and fitted using the weighted least squares (WLS) functionality of Gstat with a nugget and two spherical components. The various components of the model fitted to the variogram can be related to the spatial structure of the carbon content of shallow groundwaters at the search site.

The nugget effect of the variogram represents unresolved variation and indicates a degree of randomness, which may be explained by a mixture of spatial variation at a finer scale than the sample spacing and measurement error (Journal and Huijbregts 1978). The ranges of the variogram model give information about the dominant scale of spatial variation or correlation distance. In this instance, the ranges of

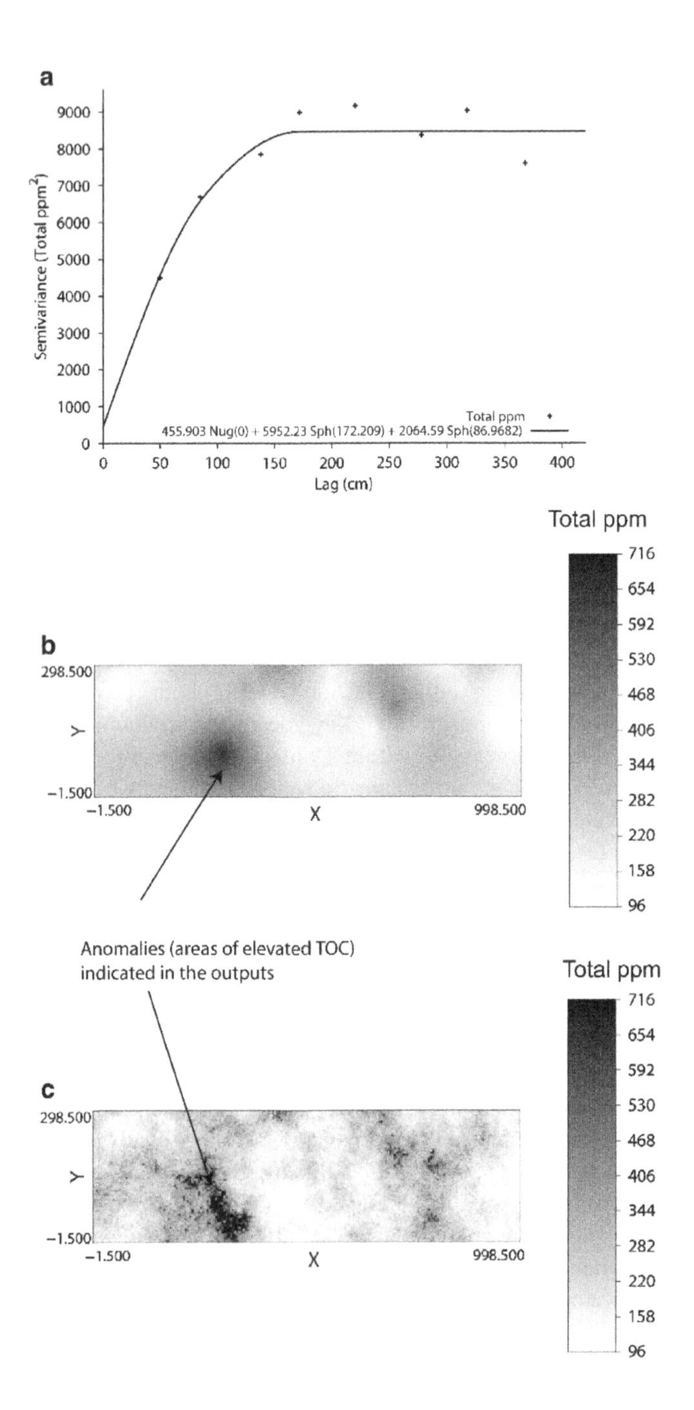

Fig. 18.7 (a) Omnidirectional variogram. (b) Kriged map. (c) Simulated realisation (adapted from Ruffell and McKinley 2008)

the variogram model indicate a restricted geographical distribution within which a similar carbon isotope signature is found (1.72 and 0.87 m). The use of geostatistics through variogram analysis maximises the information available on the spatial distribution of carbon isotope content from sampling shallow groundwaters. However, the unresolved variation and degree of randomness indicated by the nugget effect of the variogram needs to be acknowledged and accounted for in any conclusions drawn from variogram analysis. The modelled variogram is only one possible solution and is dependent on several factors including the model used (spherical in this case) and user experience. In this case, the unresolved variation is attributed to background variability in decaying vegetation. Indeed, the value of the nugget effect, comprising approximately 7% of the total sill value (structured and nugget component) indicates a limited amount of unresolved variation and a low degree of randomness. Readers are encouraged to read Ruffell and McKinley (2008) and other dedicated geostatistics texts (e.g. Deutsch and Journel 1998) for a full explanation of the methods used. The coefficients of the model fitted to the variogram were used to generate kriged predictions (Figure 18.7b) as a first step to investigate spatial patterns in isotope values across the search site. Kriged predictions are weighted moving averages of the sampled water and, as such, represent a smoothed interpolation or approximation of the carbon isotope signature across the sampled area.

With the use of GIS and geostatistics in forensic applications, any mapped output needs to be accompanied by information on the sampling strategy employed, any transformation methods applied to the data, the interpolation and geostatistical techniques used and the associated uncertainty. The kriging variance (not shown) provided a measure of confidence in the kriged predictions and was used to indicate places where a higher degree of certainty can be placed on the output. However, the kriging variance is not conditional on (does not honour) the sampled values of carbon isotope content and for this reason conditional simulation was used to estimate spatial uncertainty over the area.

Simulated values in conditional simulation are conditional on the original data (carbon isotope content) and previously simulated values (Deutsch and Journel 1998), and therefore conditional simulation is not subject to the smoothing associated with kriging. For this reason, a conditional simulation approach, sequential Gaussian simulation, was used to generate many different possible realisations (100 simulated realisations were produced) representing a 'possible reality' of the search site (Figure 18.7c). This allowed greater assessment of the uncertainty in the simulated outputs of the search site (Goovaerts 1997; Chilès and Delfiner 1999; Deutsch 2002).

Zones of high and low carbon isotope signature (TOC) are indicated on the output maps (Figure 18.7b and 18.7c). The kriged map (Figure 18.7b) has a smoothed appearance, whereas the sequential Gaussian simulation realisation (Figure 18.7c) appears visually noisy but does not seem to be entirely random. Areas of more or less constant TOC concentration in the outputs represent the 'background' and can be inferred as derived from decaying vegetation. However, two anomalies are indicated through this approach where 'ripples', 'peaks' or 'spikes' in the distribution may indicate concentrations comprising the sum of background and decaying soft tissue. This information was used to further inform survey locations for cadaver dogs and the two anomalous locations were surveyed by GPR.

GPR failed to indicate any subsurface disturbance or grave, yet the dog indicated close to one isotope anomaly. The advantage of this approach enabled a feature-based search scenario to be better defined and the subtlety of 'peaks' or 'spikes' against background TOC concentration values to be visually highlighted, which could have indicated a body deposition or burial location. Further search near this location failed to discover human remains, yet the TOC results, topography and dog indications showed that some anomaly existed. This may be a false-positive or artefacts caused by previous excavation activity. However, in the light of the missing person never having been found, the isotope analysis, topography and dog indications cannot be discounted but instead the anomaly may indicate that the remains had 'returned to earth.'

Concluding Comments

One of the main advantages of the search methodology described is the opportunity to incorporate a large range of spatially-located data sets, digital terrain mapping, landform classification and water sampling for isotope analysis, into a GIS framework, to enable an informed assessment of the search area. The use of GIS in the integration of spatial data in the Missing Person case allowed a sector, or topographic domain, approach to be used at a scale finer than usual in geomorphology. Small water catchments were used to define an intensive sampling scheme of shallow groundwater for carbon content and isotope analysis.

The interpolated surface (kriged) and simulated realisation of water chemistry (isotope data) of the sediment could then be overlain on the DEM (Figure 18.8) to provide a 3D visualisation of the search domains, and used to highlight anomalies of interest. In this case, the visualisations were used to interpret potential body deposition sites but they could be applied to environmental criminal activity (e.g. illegal dumping of waste). The integrated domain-based approach of partitioning the landscape into topographic or hydrological search units, with clearly defined boundaries, created an object-based priority scenario of search areas. This was most easily described within a GIS framework, which facilitated the spatial interaction between landform, soil type, vegetation, and hydrological aspects of the search area.

The priority scenario of search areas combined with 'hits' from cadaver canine searches was used to inform a sampling strategy of shallow groundwaters for carbon content and isotope analysis. Integral to this was sampling the spatial variability of soil and sediment water characteristics of the scene using a spatial analysis (geostatistics) model of prediction.

Although the missing person has never been found, the methodology used in this case describes an innovative combination of GIS and spatial analysis techniques,

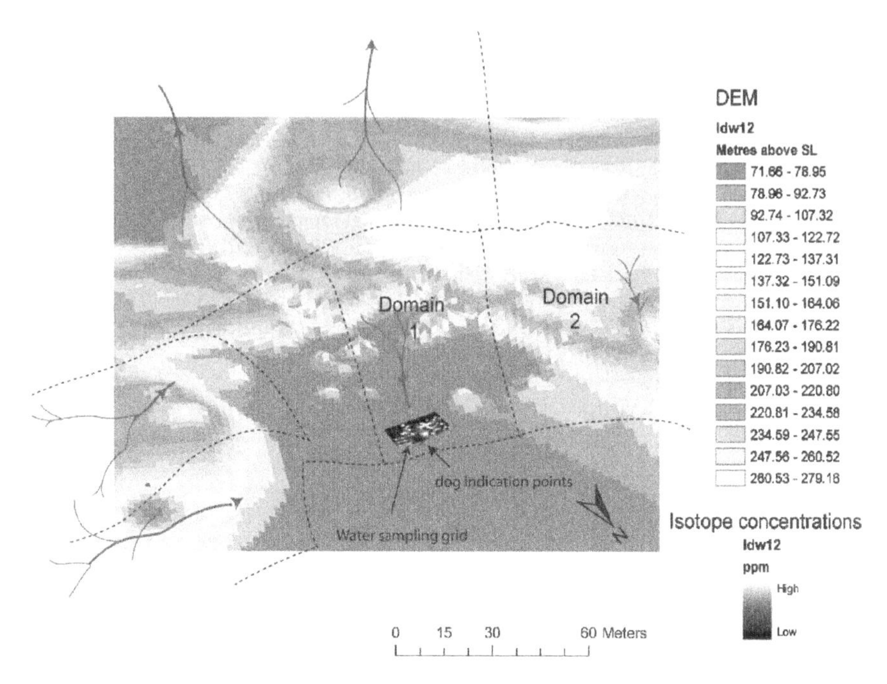

Fig. 18.8 3D DEM visualisation with overlain domain-based search areas shown in Figure 3 described in the text. Target search areas Domain 1 and 2 and the interpolated TOC distribution of the water sampling grid. The DEM and isotope distribution are generated in GIS using inverse distance weighting (IDW) interpolation (IDW 12 indicates the number (12) of neighbouring data used in the interpolation procedure) (*see colour plate section for colour version*)

comprising traditional landscape interpretation and hydrological chemical analysis, which offers a potential protocol for similar search scenarios.

Acknowledgements Thanks to John Gilmore, Raymond Murray and Steve McIlroy for their guidance and assistance in the field. The work was considerably improved by the comments of the anonymous reviewers.

References

Canter D, Coffey T, Huntley M and Missen C (2000). Predicting serial killers home base using a decision support system. Journal of Quantitative Criminology 16:4.

Chainey S and Ratcliffe J (2005). GIS and Crime Mapping. Wiley, Chichester.

Chilès JP and Delfiner P (1999). Geostatistics: Modeling Uncertainty.Wiley, New York.

Cooke RU and Doornkamp JC (1990). Geomorphology in Environmental Management. Clarendon, Oxford.

Deutsch CV (2002). Geostatistical Reservoir Modelling. Oxford University Press, New York.

Deutsch CV and Journel AG (1998). GSLIB: Geostatistical Software Library and User's Guide, Second Edition. Oxford University Press, New York.

Goovaerts P (1997).Geostatistics for Natural Resources Evaluation. Oxford University Press, New York.
Harrison M (2006). Search Methodologies. Geoscientists at Crime Scenes Conference Abstract. The Geological Society of London, 20th December 2006.
Hirschfield A and Bowers K (2001). Mapping and Analysing Crime Data: Lessons from Research and Practice. Taylor & Francis, New York.
Journel AG and Huijbregts CJ (1978). Mining Geostatistics. Academic, London.
McKinley J and Ruffell A (2008). Contemporaneous spatial sampling at scences of crime: advantages and disadvantages. Forensic Science International 172:196–202.
Minár J (1992). The principles of the elementary geomorphological regionalization. Acta Facultatis Rerum Naturalium Universitatis Comenianae. Geographica 33:185–198.
Pebesma EJ and Wesseling CG (1998). Gstat, a program for geostatistical modelling, prediction and simulation. Computers in Geoscience 24:17–31.
Ruffell A and McKinley J (2008). Geoforensics. Wiley, UK.
Waters RS (1958). Morphological mapping. Geography 43:10–17.

Chapter 19
Localisation of a Mass Grave from the Nazi Era: A Case Study

Sabine Fiedler, Jochen Berger, Karl Stahr and Matthias Graw

Abstract Between late 1944 and early 1945, 66 concentration camp prisoners who had died of starvation were buried in a municipal forest close to Stuttgart, Germany. When World War II ended 10 months later, the mass grave was opened in order to remove the corpses and give them a dignified burial in a Jewish cemetery. Little is known about the original location of the mass grave. The aim of our study was to identify the exact location of the original burial site. At first, historical evidence was gathered, which included a contemporary aerial photograph. Interpretation of the photograph suggests that the original mass grave was located somewhere in an area extending across approximately 5,000 m^2. The area of interest was overgrown with closely spaced maple trees, making it inaccessible to investigations involving modern techniques such as ground penetrating radar (GPR). We therefore used quantitative probe measurements to determine the exact position of the grave. Following a regular 5 × 5 m grid of 128 georeferenced sampling points we characterised the soil material, including texture, colour, number of soil horizons, penetration depth and resistance. Based upon interpretation of the soil characteristics, using a geographic information system (GIS), it was possible to specify three points at which there was a high degree of disturbance. On detailed observation of these points, three shallow pits were revealed in close proximity to each other and a ditch was cut through them. Here, numerous corroded iron objects were found which clearly indicated anthropogenic activities. The depth profile of organic carbon and total phosphorus content also suggested previous human interventions, as did the discovery of a 'Richmond crown', a special kind of dental prosthesis, which was commonly used in Germany up to around 1950. The combination of historical evidence with information from soil forensic and medical analyses eventually led to the

S. Fiedler(✉), J. Berger, and K. Stahr
Institute for Soil Science and Land Evaluation, University of Hohenheim,
Emil-Wolff-Strasse 27, 70599 Stuttgart, Germany
e-mail: fiedler@uni-hohenheim.de

M. Graw
Institute of Legal Medicine, University of Munich, Nußbaumstrasse 26, 80336
Munich, Germany

K. Ritz et al. (eds.), *Criminal and Environmental Soil Forensics*

determination of the original location of the grave with almost complete certainty. However, in contrast to graves containing human remains, the present study does not, despite a great deal of evidence, provide final proof that the area was actually the location of the former mass grave.

Introduction

Several methods are available for the identification of clandestine graves (France et al. 1992; Hunter et al. 1996; Killham 1990; Owsley 1995). These methods look at the detailed characteristics of the area that are associated with the digging of graves. Grave areas usually have patterns of vegetation and plant growth which differ from that of undisturbed areas (France et al. 1992). Moreover, the natural composition of the soil, air and water budget (Hunter et al. 1996), as well as microrelief, considerably differs from that of undisturbed areas (Killham 1990).

The degree of success in the discovery of clandestine graves depends on the methods used, which in turn depend on factors such as the size and accessibility of the area, as well as the time that has elapsed since inhumation (France et al. 1997). The identification of degradation products in soil air, or the use of thermal cameras, only prove successful if a relatively short amount of time has elapsed since inhumation (France et al. 1992; Killham 1990). In contrast, soil properties such as homogeneity and looseness are still discernable hundreds of years after inhumation (Reed et al. 1968; Stein 1986).

Soil sounding rods have frequently been used to locate disturbed areas along a specific grid (Owsley 1995; Ruffel 2005) and have proved useful in identifying clandestine graves as long as 150 years after they were dug (Owsley 1995). The results obtained using soil sounding rods require to be confirmed by digging holes in areas of disturbed soil (Killham 1990; Owsley 1995). The discovery of human remains always confirms the presence of a grave.

But what about if the graves were only occupied for a short period of time and no longer contain human remains? In the Nazi era, around the end of 1944 and beginning of 1945, the bodies of 66 starved concentration camp prisoners were secretly transported to a remote municipal forest (Bernhäuser Forst) close to Stuttgart, Germany (Silberzahn-Jandt 1994). According to eyewitnesses, the bodies were transported in a truck and buried in shallow rectangular pits (Back 2005). In October 1945, approximately 10 months after inhumation, the American Military Government ordered that the corpses be exhumed and given a dignified burial in a Jewish cemetery.

Knowledge of the exact location of the mass grave was lost over time and information from eyewitnesses regarding the original location of the grave proved to be contradictory. The people of the city of Filderstadt were interested in locating the grave as they wished to erect a monument in memory of those dark times in German history. This paper reports on a study which aimed to identify the exact location of the former mass grave by using soil surveys in combination with historical investigations.

Historical Investigation

The first stage of the project was to collect all available information with the aim of reducing the size of the geographical area that would need to be investigated in greater detail (Table 19.1). We found an aerial photograph that dated from the time close to inhumation (10.04.1945). Interpretation of this photograph suggests that the original mass grave was located somewhere within an area approximately 5,000 m^2 in extent.

The area of interest was visited in February 2006. What was previously a windfall area was now overgrown with 4 m high maple trees (*Acer platanoides, Acer pseudoplatanus*). The trees, which were closely spaced, made the area inaccessible to investigations involving modern technologies (e.g. ground penetrating radar, GPR) which would otherwise have been able to provide information on soil disturbance.

The following soil types, relating to local relief and prevailing source material, were found in the area under investigation: Luvisols (loess, even), Vertisols or Cambisols (Triassic marlstone (Norian) Lower Jurassic – claystone (Hettangian)) and Planosols (elongated flattened surfaces, Upper Triassic sandstone (Carnian)) (Kösel and Weis 2000). Soil development naturally led to three (Vertisol, Cambisol: A, B, C or R) and four (Luvisol, Planosol: A, E, B, C or R) horizons which can be clearly differentiated from each other.

Table 19.1 Historical evidence and interpretation

Information	Interpretation
Burial of dead concentration camp prisoners in the municipal forest of the city of Filderstadt, Germany	Areas that were privately owned in 1944, were excluded as potential sites of the mass graves
66 bodies were temporarily buried in 2–3 pits (not very deep)	Area must be large enough to hold 66 bodies
Inhumation in December 1944 – exhumation in October 1945	No human remains expected; post-mortal degradation during the exposition of 10 months (organic carbon as well as phosphorus contamination), soil must have been disturbed twice (disturbed profile)
Exhumation requires that water be removed from the burial pits	Very wet area
Bodies were washed at the site of exhumation	An open water source must be available close to the grave
Transportation to and from the grave in trucks	This requires the presence of a suitable road in close proximity to the grave
Aerial photo taken on 10.04.1945 shows a vegetation-free area of approximately 5,000 m^2 in the municipal forest area	Eyewitnesses report that the mass grave was dug in this area; however, statements on the exact location are contradictory
Area where mass grave is located was planted with spruce after 1945; the spruce were destroyed by the winter storm "Lothar" in 1999	Destruction of the natural soil structure through windfall

However, it can be assumed that the preparation of the mass grave led to the destruction of the natural soil structure. In spring 2006 (March, April), core samples were taken using a grid pattern of 5 × 5 m.

The aerial photographs revealed a clearing of approximately 5,000 m^2 which was surrounded by high trees. Investigations were extended to an area in shadow at the southern limit of the clearance, and to some of adjacent, privately owned, areas (Figure 19.1a). The private area was perfect to provide reference points for the subsequent mapping of the soil since it could be assumed that the soil composition in this area remained unaltered over time and that the mass grave was dug on municipal (Table 19.1) rather than on private land.

Grid Mapping: Results and Interpretation

Of the 234 points, georeferenced using a GPS system TRIMBLE 4700 (Trimble Navigation LTD, Sunnyvale, CA, USA), 128 were selected for boring. The core-sampling tool (length: 100 cm, diameter: 28 mm) was slowly driven into the earth using a hammer. The selection depended on: (1) whether or not the area was privately owned in 1944/1945 (it is known that the north-western and north-eastern sections were privately owned at the time when the mass grave was dug) (Figure 19.1a), and (2) whether the soil layer was deep enough for a grave to be dug (the southern section of the area under investigation had a soil depth of 30 cm above solid rock and was therefore considered to be unsuitable).

The mapping of the area suggested that the soil in the vicinity of the grave had been dug up twice, as evident from the homogenous soil pattern in the disturbed area (Table 19.2). The properties used to characterise the soil material were determined according to FAO (2006), with a location-specific mapping key using a parameter-specific evaluation scale from 0 to a maximum of 4 (Table 19.2).

No spatial pattern of soil texture was found. The soil texture of all samples was predominantly silty clay loam, and clayey loam. The presence of calcium carbonate ($CaCO_3$) is established by adding some drops of 10% HCl to the soil. The degree of effervescence of carbon dioxide gas is indicative of the amount of calcium carbonate present (Table 19.2). The carbonate concentration in the soils did not suggest previous human activities. Rather, the spatial distribution of secondary chalk precipitates (chalk concretions 1 to 6 mm in diameter; compaction areas of several centimetres) suggested the influence of a spring layer in the northeast of the area. The soil was very moist at the time of sampling, rendering it impossible to make reliable statements about the soil structure. Therefore, soil texture, structure and carbonate concentration proved to be unsuitable for identifying disturbed soil areas.

The colour of the soil ranged, for the large part, between yellow-brownish to brown (7.5 to 10 YR). Some areas had a pale greyish or dark grey colour (2.5 to 5 YR), which were mainly present in moist areas. Approximately 50% of all sample

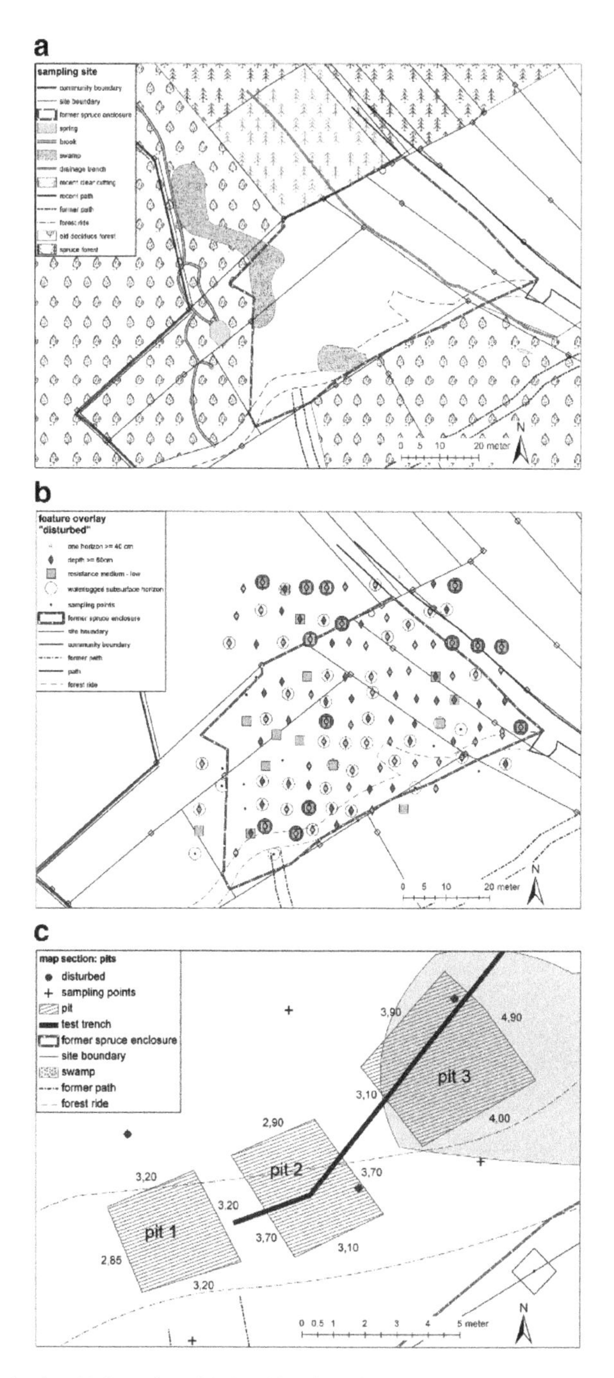

Fig. 19.1 Study site. (a) Location. (b) Combination of single parameters. (c) Pits in an area with the greatest degree of disturbance

Table 19.2 Parameters used to identify the disturbance

Parameter	Notes
Mass grave leads to soil disturbances = homogenisation of the soil material – in contrast to reference points (private areas)	Description of the parameters used to describe the disturbance according to FAO (2006)
No distinct horizons	Horizon borders: 0 = unclear, 1 = diffuse, 2 = clear, abrupt
Increase in horizon thickness	Number of horizons
Uniform texture	Rapid determination (tested with fingers)
Uniform colour	Munsell (soil colour charts)
Lower bulk density	Penetration resistance (penetration > 6 cm per hammer stroke)
Disturbed water and air budget	Redoximorphic properties: Fe-/Mn-concretions: 0 = none, 1 = very few, 2 = few, 3 = common, 4 = many Fe-/Mn-mottles: 0 = none, 1 = very few, 2 = few, 3 = common, 4 = many
Disturbed soil structure	Description of aggregate shape
Presence of foreign materials	Abundance of anthropogenic artefacts (e.g., charcoal, bricks): 0 = none, 1 = few, 2 = common, 3 = many, 4 = abundant
Irregular or invisible depth gradient:	
pH	Estimated using indicator papers
Carbonate concentration	Reaction of carbonates with HCl was defined as follows: 0 = No detectable visible or audible effervescence 1 = Slightly calcareous: audible effervescence but not visible 2 = Moderately calcareous, visible effervescence 3 = Strongly calcareous: strong visible effervescence. Bubbles form a low foam 4 = Extremely calcareous: extremely strong reaction. Thick foam forms quickly
Stone concentration	Abundance of stones: 0 = none, 1 = few, 2 = common, 3 = many, 4 = abundant

points revealed redoximorphic properties (concretions and mottles). Charcoal, burnt clay brick residue and organic substances (at a depth of at least 50 cm) were also discovered in a small part of the area.

The penetration depth, calculated from the soil depth, ranged between 35 and 100 cm, with a large part of the area exhibiting a medium to high degree of penetration resistance (estimated by the number of hammer strokes). Connected areas which had a lower degree of penetration resistance were found towards the north-eastern, eastern and south-western edges of the mapped area.

In order for further conclusions to be drawn on the extent and cause (natural or anthropogenic) of the disturbances observed, the data obtained were further processed in a Geographic Information System (GIS), ArcGIS 9.1 (ESRI Inc., Redlands, CA, USA, 2005). The selection of points with a high degree of disturbance was based on the combination of characteristics and distinctive features of the individual pore points. Of particular importance were: (1) penetration depth (the thickness of the solum), (2) penetration resistance, (3) occurrence of horizons with a thickness of >40 cm (i.e. <3 horizons), and (4) hydromorphic features (uniform grey colour of the bore material). The effect of windfalls was not apparent in the area investigated.

The combined information about the soils in the area directed attention to bores at 11 points at sites with the highest degree of disturbance. Seven points were located in the privately owned areas, which were excluded as a potential burial site (Table 19.1). Of the four points located in the municipal area (Figure 19.1b), three were in close proximity to each other and were regarded as the most promising because they covered an area that was large enough to hold a mass grave.

A second investigation of the area, in early May, revealed three shallow pits (maximum height difference of 20 cm) (Figure 19.1c). The pits were not discovered during the first field observations at the site as at the time they were filled with leaves and covered with snow. This experience emphasises the importance of visiting the areas under investigation several times, preferably in different seasons.

Digging: Results and Interpretation

During investigations, a ditch was cut through Pit 2 and 3 (Pit 1 was not approved for investigation) (depth: 90 to 100 cm, breadth: 60 cm, length: 9 m) (Figure 19.2a). Pit 2 revealed anthropogenic influences, in that several severely corroded iron objects were found (Figure 19.2a). It can be assumed that most of these objects were disposed of when the pits were filled with soil after exhumation of the bodies. No distinct horizons were present in the area of the two pits (Figure 19.2a). The soil material had an almost uniform colour up to a depth of approximately 60 to 70 cm (10YR 4/3). The areas between had soils with colour differing by depth (topsoil 10 YR 6/3, subsoil 10YR 3/1), and consisted of typical Gleysol and Planosol horizons. The estimated position of the grave was confirmed by the discovery of a pivot crown found in Pit 2 at a depth of approximately 60 cm (Figure 19.3).

In anatomical terms, according to Zuhrt et al. (1978), the crown could have been a replacement for tooth number 12 in the upper jaw or tooth number 31 of the lower jaw. In the first case, the tooth would have been a slender representative of number 12, and in the second case it would have been a very broad front tooth. There was no evidence to distinguish between the two possibilities.

The crown consisted of a blend of silver, gold and palladium, as well as small amounts of copper and zinc. The latter two are no longer used for the preparation of crowns. In contrast to current-day pivots, the tooth replacement found is a ready-made pivot mounted with a crown. The pivot was soldered to a ring-shaped lid. The porcelain tooth had a protective back plate (Figure 19.3)

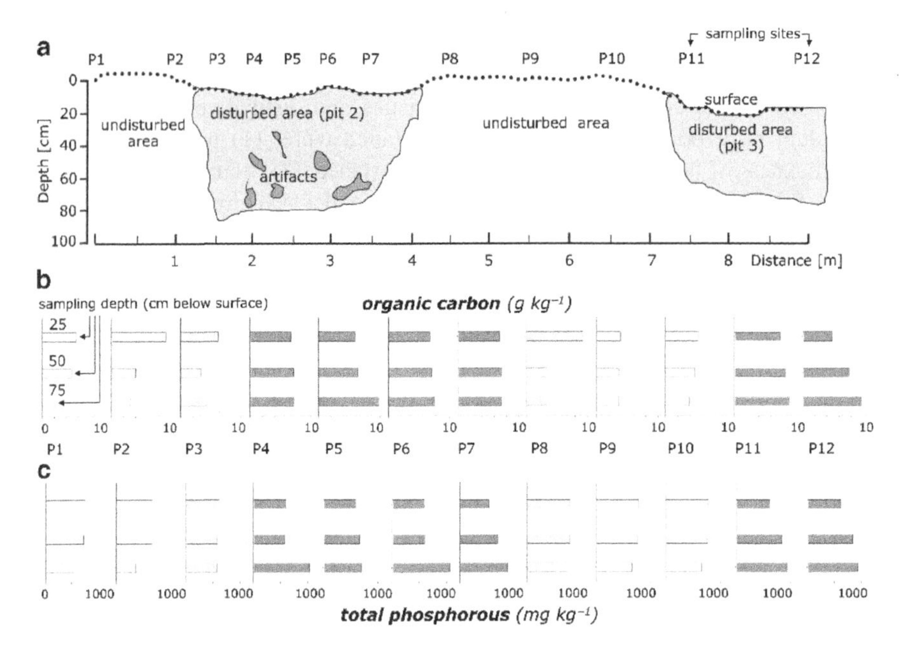

Fig. 19.2 Ditch across Pits 2 and 3. (a) Cross-section with sampling sites (P1 to P12). (b) Spatial distribution of organic carbon (g kg^{-1}). (c) Distribution of total phosphorus (mg kg^{-1}) depending on sampling depths (25, 50 and 75 cm below surface) and the sampling site (P1 to P12). White shading denotes undisturbed areas (P1, P2, P9, P10), dark grey shading denotes disturbed areas (P4 to P7, P11 and P12), light-grey shading denote areas regarded as transition zones (P3 and P8)

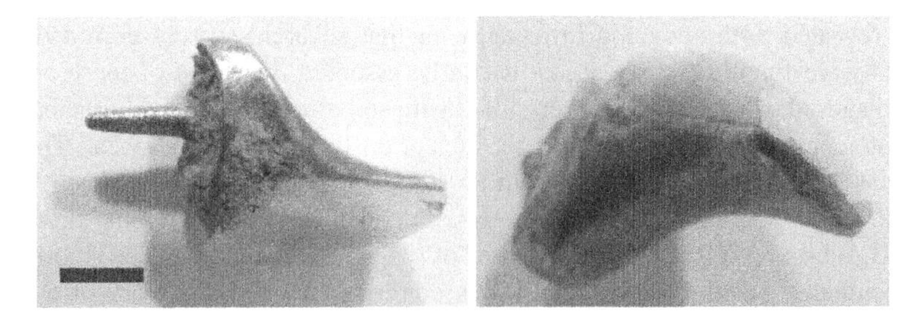

Fig. 19.3 A so-called 'Richmond crown' which was found in Pit 2 at depth of 60 cm below surface. Scale bar = 5 mm

which was also soldered to the pivot. All these signs point to a typical 1950s Central and east Europe tooth replacement. It can be assumed with certainty that the tooth is, what is commonly referred to as, a 'Richmond crown', which is a type of tooth replacement produced in Central Europe up until around 1950 (personal communication, G. Lindemann).

Chemical Analyses and Interpretation

Further soil samples (n = 36) were taken along the ditch at depths of 25, 50 and 75 cm and at 50 to 150 cm intervals (P1 to P12, Figure 19.2a) for analysis. Due to morphological particularities (soil colour, horizons), the sampling points P1, P2 and P10 were eventually taken to be undisturbed areas (white graphs in Figure 19.2b and 19.2c), whereas P4 to P7, P11 and P12 were regarded as disturbed areas (dark grey graphs in Figure 19.2b and 19.2c) and P3, P8 and P10 were regarded as transition areas (light grey graphs in Figure 19.2b and 19.2c).

Soil parameters were determined in accordance with Schlichting et al. (1995). Bulk density was measured by undisturbed cores (100 cm^3, five replicates). Standard soil analyses were performed in duplicate on the fine earth (≤2 mm) including particle size distribution, total nitrogen and carbon (dry combustion, measured by Leco Instruments GmbH, Krefeld, Germany). The pH of the solid phase was determined potentiometrically in a 1:2.5 mixture of soil and 0.01 M $CaCl_2$ solution. In all samples, organic carbon (C_{org}) was equal to total carbon. Total concentrations of phosphorus were extracted with aqua regia in a microwave oven (concentrated HCl and concentrated HNO_3 vol/vol:3/1) and quantified photometrically (John 1970). Results are expressed on the basis of the oven-dry (105 °C) soil weight.

An increase in organic carbon was observed in the disturbed cores with increasing depth. At the greatest depth 9.7 g C kg^{-1} was measured (Figure 19.2b). In contrast, decreasing concentrations from topsoil to subsoil were found in the core samples from undisturbed soils (Figure 19.2b, P1: 6 vs. 3 g C kg^{-1} P2: 9 vs. 4 g C kg^{-1}). This is typical for natural soils of the area of investigation.

The human body consists of approximately 1% phosphorus, which enters surrounding soil when degradation takes place. The same is true for carbon. In soil, phosphorus is more resistant to export than other elements (e.g. calcium). This is why phosphorus is often used in archaeology as evidence for the existence of (pre) historical graves or settlement areas (Dietz 1957).

According to Holliday and Gartner (2007), it is necessary to determine the total phosphorus content. Phosphorus can be used for determining the original position of corpses as well as that of corpses that have only been buried for a short time (maximum 10 months). The highest concentrations of total phosphorus (1,037 mg kg^{-1}) were found at the greatest depth (75 cm) in disturbed areas (Figure 19.2c). The sampling depths of 25 and 55 cm revealed considerably lower concentrations (600 mg P kg^{-1}). This suggests that phosphorus must have accumulated at a depth where the corpses were originally buried. The undisturbed areas had lower concentrations of phosphorus, decreasing from topsoil (750 mg kg^{-1}) to subsoil (400 mg kg^{-1}). The soils could not be clearly differentiated on the basis of texture, bulk density (largely due to the pressure of the backhoe that was used), pH value or content of total nitrogen between the pits and the areas connecting them.

Assessment of the Grave Area

The assessment as to whether the area of the pits was large enough for 66 corpses was based on the following assumptions: (1) grave depth = 80 cm; (2) size of the corpses: 35/40 cm width and 20 cm height (to thorax), 180 cm length; (3) arrangement of the corpses: tightly packed, facing opposite directions.

Assuming that two corpses were placed next to each other (a width of approximately 1 m) and three deep (approximately 60 cm), an area of 19.8 m^2 would have been required to contain all 66 bodies. If two corpses were arranged next to each other (occupying a width of 1 m) and two deep (40 cm), an extra 9.9 m^2 would have been needed. In either case, the entire area of the three pits (35 m^2) would have been large enough for 66 corpses.

The spring layer identified earlier is located at approximately 50 m from the mass grave. People living in the vicinity refer to the spring as the 'Jew spring', which in turn suggests that the assumption that the exhumed corpses were washed here is correct. The pits are located 1 to 2 m from a woodland path which may have been used to transport the corpses to and from the graves.

Conclusions

Locating the former mass grave was a particular challenge. The grave was only used for a short time (the soil matrix therefore revealed low postmortem C and P contaminations), the use of the grave dates back a very long time and the area was impassable. This limited the choice of methods from those which would be commonly used for the discovery of clandestine graves.

However, the current study clearly shows that it is possible to narrow down the potential location of the mass grave to a certain area, despite the many unfavourable conditions. This achievement is based on: (1) the hierarchical system of investigation that is generally used to clarify forensic issues (Morgan and Bull 2007) and (2) the interdisciplinary research approach employed. The order of events in our investigation (Figure 19.4) can be universally employed in the search for clandestine graves.

It is only possible to isolate the area of the ancient mass grave if significant historical information is available (Step 1). Although eyewitness statements differed enormously, some of the information was, nevertheless, found to be correct. However, the few correct statements made could only be verified by another step – pedological mapping. It became evident that the aerial photographs taken in 1945 could only be used to locate the mass grave approximately. The location of the mass grave was assumed to be within an illuminated area (i.e. treeless) interpreted from the aerial photographs. It was found in a bright area of the photography, bordered by shadows cast by adjacent trees.

Subsequent digging (Step 3), and the detailed investigation of the soil (Step 4) in the area from step 2, showed that the area under investigation was most likely the

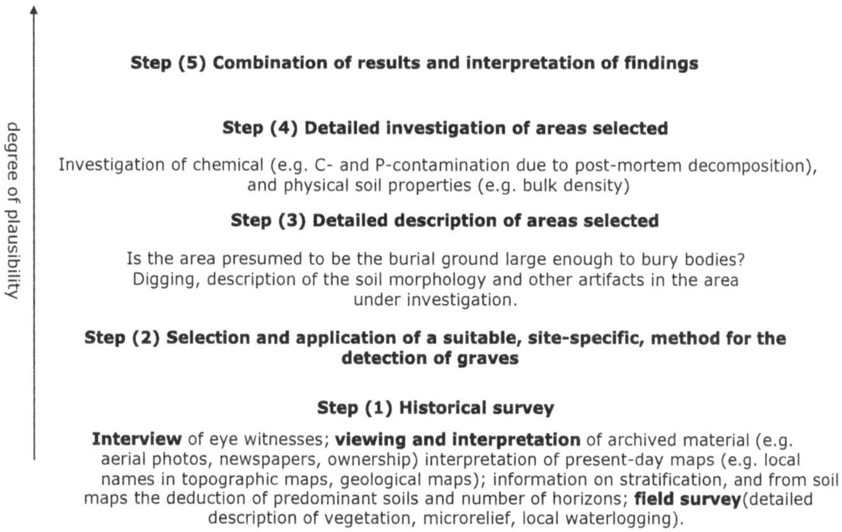

degree of plausibility

Step (5) Combination of results and interpretation of findings

Step (4) Detailed investigation of areas selected

Investigation of chemical (e.g. C- and P-contamination due to post-mortem decomposition), and physical soil properties (e.g. bulk density)

Step (3) Detailed description of areas selected

Is the area presumed to be the burial ground large enough to bury bodies? Digging, description of the soil morphology and other artifacts in the area under investigation.

Step (2) Selection and application of a suitable, site-specific, method for the detection of graves

Step (1) Historical survey

Interview of eye witnesses; **viewing and interpretation** of archived material (e.g. aerial photos, newspapers, ownership) interpretation of present-day maps (e.g. local names in topographic maps, geological maps); information on stratification, and from soil maps the deduction of predominant soils and number of horizons; **field survey**(detailed description of vegetation, microrelief, local waterlogging).

Fig. 19.4 Sequence of events in the search for a grave that was only occupied for a short time and no longer contains human remains

location of the former mass grave. This finding was substantiated by the discovery of dental prostheses. In contrast to graves with human remains, the considerable evidence used in the present study did not prove conclusively that the area under investigation was the actual site of the mass grave.

Acknowledgements The authors would like to thank Dr G. Lindemaier for the dental analyses and Dr. Martina van de Sand for critical discussions.

References

Back N (2005). Das ehemalige Massengrab im Waldstück Bernhäuser Forst. In: Filderstadt und sein Wald (Ed. City of Filderstadt), pp. 164–166. Reihe Filderstädter Schriftenreihe zur Geschichte und Landeskunde 18, Filderstadt.

Dietz EF (1957). Phosphorus accumulation in soil of an Indian habitation site. American Antiquity 22:405–409.

FAO (2006). Guidelines for Soil Description, 4th Edition (Ed. Food and Agriculture Organization of the United Nations), 97 pp. Rome.

France L, Griffin TJ, Swanburg JG, Lindemann JW, Davenport GC, Trammell V, Armbrust CT, Kondratieff B, Nelson A, Castellano K and Hopkins D (1992). A multidisciplinary approach to the detection of clandestine graves. Journal of Forensic Sciences 37:1445–1458.

France L, Griffin TJ, Swanburg JG, Lindemann JW, Davenport GC, Trammell V, Traavis CT, Kondrtieff B, Nelson A, Castellano K, Hopkins D and Adair T (1997). NecroSearch revisited: further multidisciplinary approaches to the detection of clandestine graves. In: Forensic Taphonomy (Eds. Haglung WD and Sorg MH), pp. 497–509. CRC, Boca Raton, FL.

Holliday V and Gartner WG (2007). Methods of soil P analysis in archaeology. Journal of Archaeological Science 34:301–333.
Hunter JR, Roberts CA and Martin A (1996). Locating buried remain. In: Studies in Crime: An Introduction to Forensic Archaeology (Ed. Hunter JR with contributions by AL Martin), pp. 86–100. Batsford, London.
John MK (1970). Colorimetric determination of phosphate in soil. Journal of Soil Science 21:72–77.
Killham EW (1990). The Detection of Human Remains. Charles C Thomas, Springfield.
Kösel M and Weis M (2000). Bodenkarte von Baden-Württemberg 1:25000. Blatt 7321 Filderstadt. Karte und Erläuterungen. Freiburg i. Br.: Landesamt für Geologie, Rohstoffe und Bergbau Baden-Württemberg.
Morgan RM and Bull PA (2007). Forensic geoscience and crime detection: Identification, interpretation and presentation in forensic science. Minerva Medicolegale 127:73–89.
Owsley DW (1995). Techniques for locating burial sites, with emphasis of the probe. Journal of Forensic Sciences 40:735–740.
Reed NA, Bennett JW and Porter JW (1968). Solid core drilling of Monks Mounds: Technique and findings. American Antiquity 33(2):137–148.
Ruffel A (2005). Burial location using cheap and reliable quantitative probe measurements. Forensic Science International 151:207–211.
Schlichting E, Blume H-P and Stahr K (1995). Bodenkundliches Praktikum. Blackwell Publishing, Berlin.
Silberzahn-Jandt B (1994). Vom Pfarrberg zum Hitlerplatz. Fünf Filderdörfer während der Zeit des Nationalsozialismus: Eine Topographie. Filderstadt. Reihe Filderstädter Schriftenreihe zur Geschichte und Landeskunde 9, pp. 175–179.
Stein JK (1986). Coring archaeological sites. American Antiquity 51(3):505–527.
Zuhrt R, Rottstock F and Winterfeld RI (1978). Möglichkeiten und Methoden der Stomatologie bei der Identifizierung. In: Identifikation (Eds. Hunger H and Leopold D), pp. 287–340. Springer, Berlin Heidelberg New York.

Part III
Taphonomy

Chapter 20
Research in Forensic Taphonomy: A Soil-Based Perspective

Mark Tibbett and David O. Carter

Abstract Forensic taphonomy is the use of processes associated with cadaver decomposition in the investigation of crime. For example, these processes have been used to estimate post-mortem interval, estimate post-burial interval and locate clandestine graves. In recent years, significant advances have provided a better understanding of cadaver decomposition and its effect on associated soil (gravesoil). These are reviewed in the context of soil-based information. In this chapter, we consider the effect of a cadaver on gravesoil and how these processes might be used in the legal system. In addition, we attempt to introduce the idea of contrived, experimental work to forensic taphonomy.

Introduction

Significant advances have been made in the decade since Haglund and Sorg (1997a) released their landmark text on forensic taphonomy. Estimates of post-mortem interval have improved through a better understanding of intrinsic cadaver decomposition processes (Vass et al. 2002) and the development of forensically important insects (Higley and Haskell 2001; Huntington et al. 2007). More effective methods to locate clandestine graves have resulted from a more detailed understanding of the effects that a cadaver has on the environment (Carter and Tibbett 2003; Lasseter et al. 2003; Vass et al. 2004; Carter et al. 2007; Carter et al. 2008a), while improved determinations of cause and manner of death have resulted from investigation into the taphonomic changes associated with trauma (Calce and Rogers 2007). Yet despite

M. Tibbett(✉)
Centre for Land Rehabilitation, School of Earth and Geographical Sciences,
The University of Western Australia, 35 Stirling Highway, Crawley WA 6009, Australia
e-mail: mark.tibbett@uwa.edu.au

D.O. Carter
Department of Entomology, College of Agricultural Sciences and Natural Resources,
202 Plant Industry Building, University of Nebraska-Lincoln, Lincoln, NE, USA

K. Ritz et al. (eds.), *Criminal and Environmental Soil Forensics*

these significant contributions to forensic taphonomy, an extensive gap in knowledge exists in the relationship between cadaver decomposition and soil, particularly soil biology and chemistry.

The poor understanding of decomposition processes in gravesoils is due to several factors. Most research in forensic taphonomy has focused on pathology (e.g. Clark et al. 1997), entomology (e.g. Nabity et al. 2006) and anthropology (e.g. Calce and Rogers 2007) rather than soil processes. This approach is arguably justified, as many death investigations occur in urban settings (e.g. within buildings) rather than in or on soil. However, when soil is used as physical evidence, it is typically used as associative evidence (see Fitzpatrick 2008) rather than as a medium with which to understand cadaver decomposition. Soil as associative evidence has assisted countless criminal investigations but it represents only a part of what soils can contribute, particularly in areas of low population density where a cadaver can be left to decompose in association with gravesoil for several weeks or years. Therefore, to maximise the forensic potential of soils, it is necessary to investigate the processes associated with cadaver decomposition in gravesoils. To contribute toward this goal, the purpose of this paper is to (1) discuss new ways that soils might contribute to forensic taphonomy and (2) attempt to introduce the concept of contrived, replicated, experimental work to forensic taphonomy rather than a reliance on case studies and anecdotal evidence. Thus, this chapter will emphasise the knowledge that soils might contribute to forensic taphonomy and the need for taphonomy to use properly designed experimental studies to address major questions in cadaver breakdown.

Gravesoil Processes

In reality, cadaver decomposition is a dynamic process that begins at the time of death and continues until all cadaver components have been cycled into the wider ecosystem. Although this is a continual process, many stages of decomposition have been proposed in an attempt to help understand what occurs during the breakdown of a cadaveric resource (Fuller 1934; Bornemissza 1957; Payne 1965; Payne and King 1968; Vass et al. 1992). Recent research has shown that a cadaver can have a significant effect on the biology and chemistry of associated soils and these effects can change as cadaver decomposition proceeds (Table 20.1).

Aboveground Decomposition

The first stage of decomposition, Fresh, is associated with little change in gravesoil biology and biochemistry other than that which can result from soil disturbance. Typically, soil disturbance tends to result in a brief increase in soil microbial activity (e.g. Carter et al. 2008b), as it exposes previously unavailable food sources and

Table 20.1 Stages of above ground cadaver decomposition (after Payne 1965) and their effect on associated soil (gravesoil). Volatile fatty acids include propionic, iso-butyric, n-butyric, iso-valeric and n-valeric acid (Vass et al. 1992)

Stage of decomposition	Effect on soil	References
Fresh	Initial disturbance	
Bloat	Initial introduction of cadaveric fluids from mouth, nose, anus, ears and increase in nutrient concentration and pH: Ammonium Calcium Chloride Magnesium Ninhydrin-reactive nitrogen potassium Sodium Sulphate Volatile fatty acids	Vass et al. (1992); Spicka et al. (2008)
Active decay	Increased concentration of nutrients and pH: See Bloat Ninhydrin-reactive nitrogen Volatile fatty acids	Vass et al. (1992); Spicka et al. (2008)
Advanced decay	Peak levels of gravesoil nutrient concentrations and soil pH: See Bloat Ninhydrin-reactive nitrogen Volatile fatty acids	Vass et al. (1992); Spicka et al. (2008)
Dry and remains	Gradual decrease in nutrient concentration levels and gravesoil pH with elevated levels of: Ammonium Calcium Carbon (total) Chloride Nitrate Nitrogen (total) Phosphorus (Bray) Phosphate Potassium Sodium Volatile fatty acids	Vass et al. (1992); Towne (2000); Danell et al. (2002); Melis et al. (2007)

results in the death of microbial cells, which are also used as food by living microbes. As the enteric micro-organisms break down the cadaver, evolved gases result in the bloating of the cadaver and the initial release of cadaveric fluids into gravesoil, which might represent the initial change in gravesoil chemistry and biology. This initial change, thus far observed as an increase in the concentration of ninhydrin reactive nitrogen, can occur as early as 48 h after death during the warm summer months (Spicka et al. 2008). During this initial release of cadaveric fluids, maggot activity will reach its peak, thus designating the onset of Active Decay. This stage is associated with the majority of cadaver mass loss, some of which is introduced into gravesoil. Although cadaveric materials are being introduced into gravesoil during Active Decay, peak nutrient concentrations are associated with Advanced Decay (Vass et al. 1992; Carter and Tibbett 2008) (Table 20.1). Advanced Decay begins with the migration of the blow fly larvae from the cadaver. The Dry and Remains stages are the final stages of decomposition and it is currently understood that nutrient concentrations remain elevated, but it is not known how long this effect can persist. Sagara et al. (2008) have reported that the post-putrefaction fungi can form fruiting structures for up to 10 years following soil nutrient amendment. Fungi have been observed in association with above ground cadaver decomposition as soon as one month post-mortem (Carter et al. 2007). Thus, this phenomenon might indicate an extended persistence of elevated nutrient concentration in gravesoil.

Belowground Decomposition

Decomposition processes in gravesoil following burial has received less experimental attention than above ground decomposition. Payne and King (1968) proposed an alternative set of decomposition terminology because the decomposition in these two settings is sufficiently different (Table 20.2). This is primarily due to the absence of insects and scavengers. Thus, below ground decomposition is primarily mediated by micro-organisms and proceeds less rapidly than above ground decomposition. It has been estimated that burial results in a rate of decomposition that is eight times slower than above ground decomposition (see Rodriguez 1997). However, there is no experimental evidence to support this estimation. (For a more detailed description of below ground cadaver decomposition see Payne et al. 1968; Fiedler and Graw 2003; Dent et al. 2004.)

The initial stages of belowground decomposition, Fresh and Inflated, proceed similarly to the Fresh and Bloat stages observed above ground. During the Inflated stage, fluids are first introduced to the soil. This introduction, combined with the initial disturbance of the soil, results in an increase in soil microbial activity (Carter et al. in 2008b) (Table 20.2). The third stage however, Deflation and Decomposition, represents the time when most of the fluids are released into the soil (Payne et al. 1968). These fluids are released from natural orifices including the mouth, nose, anus, and ears and these fluids can support the initial proliferation of bacterial and fungal communities (Payne et al. 1968). This growth in the soil microbial biomass has been associated with enhanced protease and phosphodiesterase activity (Carter et al. in press). These extracellular enzymes are released to decompose protein and nucleic acids, respectively. By the fourth stage, Disintegration, bacteria and fungi can cover the cadaver completely (Payne et al. 1968). In addition, Payne et al. (1968) observed soil mites and collembola first appear during this stage. If flies are able to colonise a buried cadaver,

Table 20.2 Stages of below ground cadaver decomposition (after Payne et al. 1968) and their effect on associated soil (gravesoil)

Stage of decomposition	Effect on soil	References
Fresh	Initial disturbance associated with increased soil microbial activity (carbon dioxide respiration)	Carter et al. (in press)
Inflated	Initial release of decomposition fluids into soil result in elevated: Carbon dioxide (CO_2) Soil pH	Payne et al. (1968); Wilson et al. (2007); Carter et al. (2008b)
Deflation and decomposition	Peak release of decomposition fluids into soil associated with elevated: Electrical conductivity CO_2 Microbial biomass Protease Phosphodiesterase Soil pH	Payne et al. (1968); Carter et al. (2008b); Wilson et al. (2007); Janaway et al. (this chapter 22)
Disintegration	Established bacterial and fungal colonies with gradual decline in microbial activity. Elevated levels of: CO_2 Microbial biomass Protease Phosphodiesterase Soil pH	Payne et al. (1968); Wilson et al. (2007); Carter et al. (2008b)
Skeletonization	Elevated levels of: Ammonium Amino acid N CO_2 Total C Total N Microbial biomass Protease Phosphodiesterase Soil pH	Hopkins et al. (2000); Rapp et al. (2006); Wilson et al. 2007; Carter et al. (2008b)

maggot migration will occur at the end of Disintegration. The final stage of below ground decomposition, Skeletonisation, represents the period when the primary cadaveric carbon sources are hair, skin and nails. These cadaver components, as well as bone, occupy an island of soil that has been stained by decomposition fluids containing carbon and nitrogen, which can result in an increased nutrient concentration and soil microbial biomass for over 400 days following burial (Hopkins et al. 2000).

Potential Contributions from a Soil-Based Approach

Gravesoil is a complex and dynamic system of interdependent chemical, physical and biological processes that can be significantly affected by cadaver decomposition. Tables 20.1 and 2 clearly show that several biological and chemical changes occur in gravesoil as a body decomposes. However, only some of these phenomena have been investigated for forensic use. A more detailed understanding of gravesoil processes will likely contribute to forensic science in three primary areas: improved estimates of post-mortem interval and post-burial interval and enhanced methods to locate clandestine graves and gravesoils.

Estimation of Post-mortem Interval and Post-burial Interval

An accurate estimation of post-mortem interval (PMI) is one central objective to any medico-legal investigation of death, equal to victim identification and cause of death. Estimation of the PMI can direct or re-orientate an investigation by serving to accept or reject an alibi or elucidate the peri-mortem activities of a victim. Pathology, anthropology and entomology, from oldest to most recent, have developed criteria to enhance the estimation of PMI (Forbes 2008). Traditionally, in early post-mortem time the pathologist best ascertained the PMI using the soft tissue indicators of *rigor mortis, livor mortis* and *algor mortis* (DiMaio and Dana 2006). As the interval lengthens to include the visual cues of numerous gross morphological attributes of decomposition (i.e. bloating, discoloration, etc.), anthropology has become increasingly contributory at PMI estimation by temperature correlation (Megyesi et al. 2005). Most successful at the estimation of the PMI, overlapping pathology and anthropology, is entomology, which uses the developmental biology of blowflies (Higley and Haskell 2001).

Gravesoil research holds promise as it may provide a rapid and reliable technique to estimate PMI and help control for the increasing time error that accompanies extended decomposition stages. At present, only two soil-based techniques are available for the estimation of early PMI. The technique developed by Vass et al. (1992) to analyse fatty acids and nutrients can be used to estimate PMI from immediately following death to several years post mortem. In addition, Spicka et al. (2008) demonstrated that the concentration of ninhydrin-reactive nitrogen in gravesoil associated with juvenile to adult sized cadavers (20–50 kg) remains at basal levels until two days post mortem. This phenomenon can be used estimate early PMI when a fresh cadaver has been discovered, i.e. if the concentration of ninhydrin reactive nitrogen is similar to control values then the cadaver has been dead for less than two days.

Although forensic entomology is arguably the most successful way to estimate PMI, blow fly larvae are at their greatest forensic value up until Advanced Decay (see Payne 1965), which can occur as soon as 10 to 14 days after death in warmer months. As a consequence, forensic taphonomy lacks a precise method to estimate PMI once fly larvae have begun to pupate. This is a particular problem in rural areas where bodies can go undetected for several months following death. The time period that follows Advanced Decay, the extended PMI, is where gravesoil processes will likely have their greatest forensic impact. At present, few techniques exist to estimate extended PMI using soils. As mentioned above, the Vass et al. (1992) method has been developed. Another potential area of emphasis is the ecology of the post-putrefaction fungi (Sagara 1995). These fungi form fruiting structures in response to the cadaver breakdown and have been observed to fruit in two successional phases: Phase I fruits from 1 to 10 months post mortem while Phase II fruits from one year to four years post mortem. Although the forensic use of the fungi requires more detailed research, it might find successful use in cases where bodies have been missing for several years. Thus, a great need exists to develop rapid, reliable, and inexpensive techniques that use the biology and chemistry of gravesoil as a basis to estimate postmortem interval of cadavers that decompose above ground.

Some of the cadavers that are disposed of in terrestrial ecosystems are buried in soil. As a consequence, there is a great need for cadaver decomposition studies to investigate the gravesoil processes associated with buried cadavers. Perpetrators of crimes rely on the decomposition of corpses to hinder identification and obscure estimates of PMI or post-burial interval (PBI). Burial can greatly confound current methods of estimating PMI, such as entomology (Turner and Wiltshire 1999), because it often prevents the ability of insects and scavengers to access a cadaver as a resource. Thus, decomposition rates on the soil surface do not represent decomposition that occurs belowground. To further complicate matters, it is not uncommon for a body to be dead for some length of time prior to burial. Thus, PMI and PBI can be quite different (Forbes 2008). At present, only plant growth (Haglund and Sorg 1997b), palynology (Szibor et al. 1998), and microbial activity (Tibbett et al. 2004; Sagara et al. 2008) have been investigated as potential means to estimate PBI. However, the approaches described for above ground decomposition will likely provide insight into the relationships between edaphic parameters and the estimation of PBI. They simply must be tested on gravesoils associated with buried bodies.

It has been stated above that forensic entomology currently provides the most accurate way to estimate PMI. This is due to two primary factors: (1) blow flies can arrive at, and oviposit on, a cadaver within seconds of death (Mann et al. 1990) and (2) the development of these insect larvae is positively correlated to temperature (Higley and Haskell 2001). Thus, the estimation of PMI requires the determination of the age of the blow fly larvae along with a record of temperatures at the scene. This relationship has resulted in the regular use of accumulated degree days (ADDs) by forensic entomology. Of the soil-based cadaver decomposition studies, only Vass et al. (1992) and Carter et al. (2008b) have considered the use of ADDs. However, they might play a significant role in the development of further soil-based forensic methods. Vass et al. (1992) have demonstrated a significant relationship between temperature and gravesoil chemistry and a similar relationship might exist between temperature and gravesoil biology. Like insects, soil microbes respond to cadaver introduction in a short period of time (<24h) (Carter et al. 2008b). Thus, if a relationship between temperature, cadaver decomposition and soil ecology is to be developed, it might make significant contributions to the estimations of extended PMI.

Location of Clandestine Graves

It is not uncommon for an investigative agency to be aware that a clandestine grave exists, yet be unable to find it. As a consequence, several methods have been developed to locate human remains, whether they are on or in soil (see Killam 1990). Ultimately, these techniques aim to detect the changes that occur once a body is placed in a terrestrial ecosystem. Typically, the search for a clandestine grave is conducted in two stages. The first stage uses as little intrusion into the soil as possible. The most common methods include geophysical techniques (e.g. ground penetrating radar) (Schultz 2008) and the use of cadaver dogs (Lasseter et al. 2003) that detect changes in soil physics and chemistry, respectively. In addition, Vass et al. (2004, 2008) have recently developed an instrument to analyse the decomposition gases released from a cadaver during decomposition. Less common is the identification of the post-putrefaction fungi, although it represents a low-cost method for the detection of buried mammalian remains.

Following the detection of putative clandestine graves, soil samples are collected and tested to determine if intrusive exploration will occur. Due to the wide range of chemical and biological effects that a cadaver has on gravesoil following burial (Table 20.2), there is great potential for the development of a soil-based method to locate clandestine graves. Potential methods include each of those discussed for the estimation of PMI. If cadaver decomposition results in a significant change in gravesoil ecology, then fatty acids, nutrients, and carbon can be used to detect gravesoil. However, the measurement of ninhydrin reactive nitrogen (Carter et al. 2008a; Carter et al., Chapter 21) is currently the most rapid, inexpensive and simple method to presumptively test for gravesoil.

Considering Environmental and Edaphic Parameters: The Need for Experimental Research

While soil has been much studied as a decomposition environment for materials of relatively little forensic value such as leaf litter or dead roots (Cadisch and Giller 1997), there is clearly a need for experimental forensic taphonomy to provide rigorously tested information to practitioners and the courts to better understand gravesoils. However, forensic taphonomy must deal with the problem that it is difficult to acquire human cadavers for experimental use. Also, it is impossible to replicate human cadavers. This results in statistical deficiencies and a tendency to disturb cadavers during sampling, which can have a significant effect on the rate of decomposition (Adlam and Simmons 2007). Thus, it is necessary to conduct field- and laboratory-based research using human cadaver analogues, while continuing to use information from human cadaver decomposition studies and case studies.

However, experimental studies of the decomposition of human cadavers under controlled conditions have rarely been published. Field studies, occasionally using human bodies (Rodriguez and Bass 1983; Rodriguez and Bass 1985) but, more commonly, animal surrogates have been undertaken (Payne 1965; Payne et al. 1968; Micozzi 1986; Turner and Wiltshire 1999; Forbes et al. 2005c; Carter et al. 2008a). However, knowledge of the decomposition processes and the influence of the environment and edaphic parameters are limited because the primary sources of information are case studies and empirical evidence (Motter 1898; Mant 1950; Morovic-Budak 1965; Spennemann and Franke 1995). As a consequence, edaphic parameters were recognised as having little influence (Mant 1950; Morovic-Budak 1965; Mant 1987) on cadaver decomposition until the early 21st century (Fiedler and Graw 2003; Forbes et al. 2005a; Carter et al. 2008a).

It is now becoming increasingly apparent that the effect of the type of soil and prevailing environmental conditions can have a profound effect in the rate of cadaver decomposition and hence estimates of PMI, PBI and gravesoil detection (Forbes et al. 2005a,b; Wilson et al. 2007; Carter et al. 2008b). Examples of some basic soil characteristics that might affect the rate of cadaver decomposition include: physical texture (whether the soil is sandy, silty or clayey can profoundly affect the rate of decomposition by limiting the movement of gases and water to and from the cadaver); chemistry (the acidity or alkalinity of a soil may affect decomposition); and biological activity (a soil with an active faunal population may have the capacity to decompose cadaveric tissue more quickly) (Fiedler and Graw 2003). The key environmental parameters that need consideration are temperature and moisture (the main determinants of climate). The key edaphic parameters are less clear but are likely to include soil pH, salinity, redox potential and nutrient status.

Environmental Effects

Environmental determinants can have a critical effect on cadaver decomposition. For example, if the environment is permanently frozen or waterlogged, there can be close to zero decomposition and, by contrast, optimised conditions for temperature and moisture can lead to very rapid decomposition. In addition, recent work has shown that specific microenvironments can promote or delay the rate of cadaver decomposition in soils c.f. Janaway et al., Chapter 22. Currently better estimates can be made of the effect of environmental parameters compared with edaphic parameters on the rate of cadaver decomposition; however, there remains a paucity of experimental evidence to support these estimates, at least in the peer-reviewed literature.

Few published experiments investigate the effect of soil temperature on the rate of decomposition of cadaveric material (Carter and Tibbett 2006; Carter et al.2008b.

Table 20.3 Temperature coefficients (Q_{10} values ± SE) of carbon dioxide respiration in a sandy loam soil (100 g dry weight calibrated to 60% water holding capacity) of the Fyfield series, Lindens Farm, East Lulworth, Dorset, England following the burial of 1.5 g skeletal muscle tissue (Ovis aries). After Carter and Tibbett (2006)

Q_{10}	Day 21	Day 42
2 °C–12 °C	2.9 ± 0.1	2.5 ± 0.1
12 °C–22 °C	1.8 ± 0.1	1.5 ± 0.10

In one of these studies (Carter and Tibbett 2006) the effect of three temperature regimes (2 °C, 12 °C, 22 °C) was examined. The results provided the first definitive data of the effect of temperature on the rate of mammalian tissue decomposition in soil (see Table 20.3). The data show quite clearly that (for this soil type) decomposition rate can vary greatly with temperature (there was ca. 60% difference in the rate of mass loss between 2 °C and 22 °C after 14 days), yet that even at a very low temperature (2 °C), decomposition can proceed at a significant rate. This type of study is laboratory based, as it is difficult (and expensive) to control environmental parameters in a field setting. However, this study also highlights the potential for the use of ADDs in forensic soil science. As stated previously, further detailed experimental work should demonstrate whether ADDs can be applied to soil processes and used for the accurate estimation of PMI and PBI.

Edaphic Effects

Few examples exist where replicated experimental work has been carried out to quantify the effects of different edaphic characteristics on decomposition. One such study considered the effect of different soils of contrasting pH on the decomposition of skeletal muscle tissue (Haslam and Tibbett, unpublished data). In this study two types of soil were compared. One soil type, rendzina, had alkaline pH (7.8) the other type, podsol, had an acid pH (4.6) (Figure 20.1). The rate of decomposition of skeletal mammalian muscle tissue (1.5 g – cuboid) was measured along with any changes to the soil pH over the course of a six-week incubation. The methods used followed those described elsewhere (Tibbett et al. 2004) and organic lamb (Ovis aries) was used as an analogue for human tissue.

The results of this experiment have led to three important findings with some interesting implications for forensic taphonomy (Figures 20.1 and 20.2). Firstly, the study confirmed what had previously been described; that soil pH increases in the presence of a decomposing cadaver (Rodriguez and Bass 1985). This is thought to be due to the release of ammonium ions (Hopkins et al. 2000), a suggestion for which we have recently acquired supporting evidence (Stokes, Forbes and Tibbett, unpublished data). Secondly, that the autochthonous soil pH has a profound effect on the subsequent change caused by muscle tissue decomposition. In an alkaline

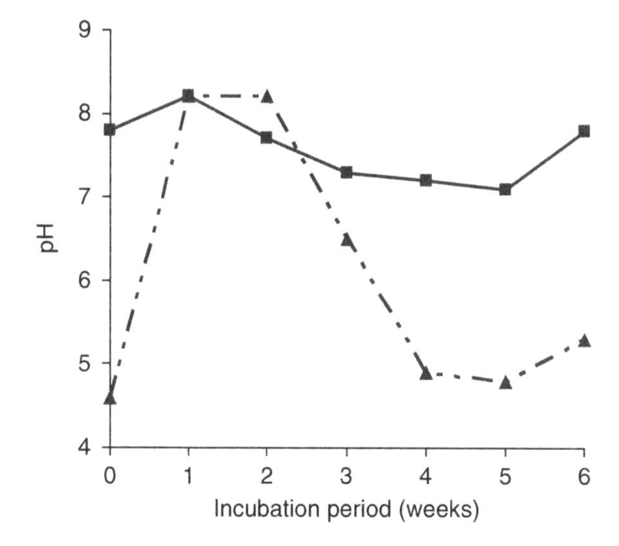

Fig. 20.1 The effect of burial of mammalian muscle tissue (*Ovis aries*) on soil pH in an acid soil (podsol, pH 4.6 – triangles and dashed line) and an alkaline soil (rendzina, pH 7.8 – squares with solid line) (Haslam and Tibbett, unpublished data)

soil, pH did not change by much, whereas in an acidic soil, pH rose by over three units. Thirdly, the dynamics of decomposition (the rate of mass loss) were different in the contrasting soils. Between two and three weeks the muscle tissue in the acidic podsol had decomposed twice as fast as in the alkaline rendzina. By the end of the experiment (six weeks), the muscle tissue in the podsol had completely decomposed whereas there was still a residual muscle tissue in the rendzina soil.

This type of experimental evidence begins to develop some predictive power to soil-based data, so that for a given soil type we may anticipate a particular decomposition dynamic and timeframe. However, the data may also be used retrospectively and will allow more scientifically sound estimates of PMI and PBI, especially for buried cadavers.

The experiment described above is not of the type that can directly be used in court tomorrow, however, it provides a framework for more predictive 'real-world' experiments with cadavers and in the field. These type of experiments are clearly more expensive and time consuming and it is up to the research funding agencies (including the law and order agencies) to step up the level of funding to an appropriate scale to allow real progress to be made to provide high quality experimental evidence that is admissible in court.

Admissibility of Soil Evidence

Ultimately, forensic taphonomy aims to contribute to criminal proceedings. Thus, the science must be admissible in a court of law, regardless of whether it is presented

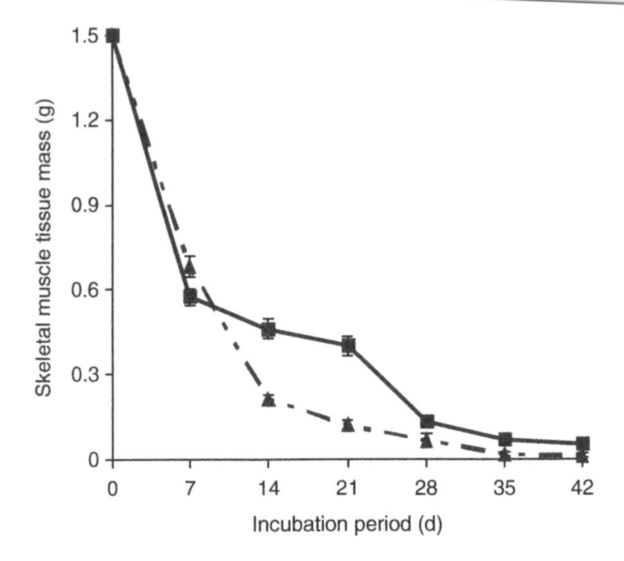

Fig. 20.2 The rate of mass loss (decomposition) of mammalian muscle tissue (*Ovis aries*) when buried in two soils on contrasting pH. The two soils were an acidic soil (podsol, pH 4.6 – triangles and dashed line) and an alkaline soil (rendzina, pH 7.8 – squares with solid line) (Haslam and Tibbett, unpublished data)

at trial or not. The admissibility of physical evidence has been a subject of great interest in the recent past (Kiely 2006), particularly in the USA. The federal ruling Daubert v Merrell Dow Pharmaceuticals, 1993, which has been adopted by several US states, established judges as the arbiters of scientific rigour and legal admissibility. Briefly, judges determine whether or not physical evidence is admissible by using the following guidelines: (i) can the science be replicated and tested? (ii) has the science been published in a peer-reviewed journal? (iii) does the science have known error rates and established standards? (iv) is the science generally accepted by the relevant scientific community and taught at university?

These guidelines for admissibility clearly show that, although case studies and anecdotal evidence can be published and their content can be taught at university, they typically cannot provide data regarding error rates or represent an established standard. Quite simply, forensic taphonomy must move toward the implementation of a contrived, replicated experimental approach if it is to garner future use in the legal system.

Conclusions

Forensic taphonomy holds great potential to contribute to the estimation of post-mortem interval, estimation of postburial interval, and location of clandestine graves (Carter and Tibbett 2008). Current research is starting to fill in the gaps in knowledge that inevitably exist in developing areas of science such as this. As a multidisciplinary science, forensic taphonomy requires contributions from anthropologists, entomologists, soil scientists, microbiologists, biochemists and chemists to work together in the exciting and expanding frontier of forensics. Currently, too little is known in forensic taphonomy from experimental research, and the science is, to date, dependant on the experience of practitioners and the logical inferences and estimates from carefully examined case studies. Although this dependence is understandable, the time has now come for forensic taphonomy to rely primarily on contrived experimental work, and an increasing number of studies are now based on carefully designed experimental protocols in the laboratory and field that should provide the forensic practitioner with more a robust science on which to base research that is admitted into the courtroom.

References

Adlam RE and Simmons T (2007). The effect of repeated physical disturbance on soft tissue decomposition – are taphonomic studies an accurate reflection of decomposition? Journal of Forensic Sciences 52:1007–1014.

Bornemissza GF (1957). An analysis of arthropod succession in carrion and the effect of its decomposition on the soil fauna. Australian Journal of Zoology 5:1–12.

Cadisch G and Giller KE (1997). Driven by Nature. Plant Residue Quality and Decomposition. 409 pp. CAB International, Wallingford, UK.

Calce SE and Rogers TL (2007). Taphonomic changes to blunt force trauma: a preliminary study. Journal of Forensic Sciences 52:519–527.

Carter DO and Tibbett M (2003). Taphonomic mycota: fungi with forensic potential. Journal of Forensic Sciences 48:168–171.

Carter DO and Tibbett M (2006). Microbial decomposition of skeletal muscle tissue (Ovis aries) in a sandy loam soil at different temperatures. Soil Biology and Biochemistry 38:1139–1145.

Carter DO and Tibbett M (2008). Cadaver decomposition and soil: processes. In: Soil Analysis in Forensic Taphonomy: Chemical and Biological Effects of Buried Human Remains (Eds. M Tibbett and DO Carter), pp. 29–51. CRC, Boca Raton, FL.

Carter DO, Yellowlees D and Tibbett M (2007). Cadaver decomposition in terrestrial ecosystems. Naturwissenschaften 94:12–24.

Carter DO, Yellowlees D and Tibbett M (2008a). Using ninhydrin to detect gravesoil. Journal of Forensic Sciences 53:397–400.

Carter DO, Yellowlees D and Tibbett M (2008b). Temperature affects microbial decomposition of cadavers (Rattus rattus) in contrasting soils. Applied Soil Ecology 40:129–137.

Clark MA, Worrell MB and Pless JE (1997). Post-mortem changes in soft tissue. In: Forensic Taphonomy: The Post-mortem Fate of Human Remains (Eds. WD Haglund and MH Sorg), pp. 151–164. CRC, Boca Raton, FL.

Danell K, Berteaux D and Braathen KA (2002). Effect of muskox carcasses on nitrogen concentration in tundra vegetation. Arctic 55:389–392.

Dent BB, Forbes SL and Stuart BH (2004). Review of human decomposition processes in soil. Environmental Geology 45:576–585.

DiMaio VJM and Dana SE (2006). Handbook of Forensic Pathology. CRC, Boca Raton, FL.

Fiedler S and Graw M (2003). Decomposition of buried corpses, with special reference to the formation of adipocere. Naturwissenschaften 90:291–300.

Fitzpatrick RW (2008). Nature, distribution and origin of soil materials in the forensic comparison of soils. In: Soil Analysis in Forensic Taphonomy: Chemical and Biological Effects of Buried Human Remains (Eds. M Tibbett and DO Carter), pp. 1–28. CRC, Boca Raton, FL.

Forbes SL, Dent BB and Stuart BH (2005a). The effect of soil type on adipocere formation. Forensic Science International 154:35–43.
Forbes SL, Stuart BH and Dent BB (2005b). The effect of burial environment of adipocere formation. Forensic Science International 154:24–34.
Forbes SL, Stuart BH and Dent BB (2005c). The effect of the burial method on adipocere formation. Forensic Science International 154:44–52.
Forbes SL (2008). Potential determinants of post-mortem and postburial interval. In: Soil Analysis in Forensic Taphonomy: Chemical and Biological Effects of Buried Human Remains (Eds. M Tibbett and DO Carter), pp. 225–246. CRC, Boca Raton, FL.
Fuller ME (1934). The insect inhabitants of carrion: a study in animal ecology. Council for Scientific and Industrial Research Bulletin 82:1–62.
Haglund WD and Sorg MH (1997a). Forensic taphonomy: the postmortem Fate of Human Remains. CRC, Boca Raton, FL.
Haglund WD and Sorg MH (1997b). Introduction of forensic taphonomy. In: Forensic taphonomy: the postmortem Fate of Human Remains. (Eds. WD Haglund and MH Sorg), pp. 1–9. CRC, Boca Raton, FL.
Higley LG and Haskell NH (2001). Insect development and forensic entomology. In: Forensic Entomology: The Utility of Arthropods in Legal Investigations (Eds. JJ Byrd and JL Castner), pp. 287–302. CRC, Boca Raton, FL.
Hopkins DW, Wiltshire PEJ and Turner BD (2000). Microbial characteristics of soils from graves: an investigation at the interface of soil microbiology and forensic science. Applied Soil Ecology 14:283–288.
Huntington TE, Higley LG and Baxendale FP (2007). Maggot development during morgue storage and its effect on estimating the post-mortem interval. Journal of Forensic Sciences 52:453–458.
Kiely TF (2006). Forensic evidence: science and the criminal law. CRC, Boca Raton, FL.
Killam EW (1990). The detection of human remains. Charles C Thomas, Springfield, IL.
Lasseter AE, Jacobi KP, Farley R and Hensel L (2003). Cadaver dog and handler team capabilities in the recovery of buried human remains in the Southeastern United States. Journal of Forensic Sciences 48:617–621.
Mann RW, Bass MA and Meadows L (1990). Time since death and decomposition of the human body: variables and observations in case and experimental field studies. Journal of Forensic Sciences 35:103–111.
Mant AK (1950). A study in exhumation data. London University, unpublished MD thesis.
Mant AK (1987). Knowledge acquired from post-war exhumations. In: Death, Decay and Reconstruction: Approaches to Archaeology and Forensic Science (Eds. A Boddington, AN Garland and RC Janaway), pp. 65–78. Manchester University Press, Manchester.
Megyesi MS, Nawrocki SP and Haskell NH (2005). Using accumulated degree-days to estimate the post-mortem interval from decomposed human remains. Journal of Forensic Sciences 50:618–626.
Melis C, Selva N, Teurlings I, Skarpe C, Linnell JDC and Andersen R (2007). Soil and vegetation nutrient response to bison carcasses in Białowieża Primeval Forest. Poland Ecological Research 22:807–813.
Micozzi MS (1986). Experimental study of post-mortem change under field conditions: effects of freezing, thawing and mechanical injury. Journal of Forensic Sciences 31:953–961.
Morovic-Budak A (1965). Experiences in the process of putrefaction in corpses buried in earth. Medicine Science and Law 5:40–43.
Motter MG (1898). A contribution to the study of the fauna of the grave. A study of one hundred and fifty disinterments, with some additional experimental observations. Journal of the New York Entomological Society 6:201–231.
Nabity PD, Higley LG and Heng-Moss TM (2006). Effects of temperature on development of Phormia regina (Diptera: Calliphoridae) and use of developmental data in determining time intervals in forensic entomology. Journal of Medical Entomology 43:1276–1286.
Payne JA (1965). A summer carrion study of the baby pig Sus scrofa Linnaeus. Ecology 46:592–602.
Payne JA and King EW (1968). Coleoptera associated with pig carrion. Entomologist's Monthly Magazine 105:224–232.

Payne JA, King EW and Beinhart G (1968). Arthropod succession and decomposition of buried pigs. Nature 219:1180–1181.

Rapp D, Potier P, Jocteur-Monrozier L and Richaume A (2006). Prion degradation in soil: possible role of microbial enzymes stimulated by the decomposition of buried carcasses. Environmental Science and Technology 40:6324–6329.

Rodriguez WC (1997). Decomposition of buried and submerged bodies. In: Forensic Taphonomy: The Post-mortem Fate of Human Remains (Eds. WD Haglund and MH Sorg), pp. 459–468. CRC, Boca Raton, FL.

Rodriguez WC and Bass WM (1983). Insect activity and its relationship to decay rates of human cadavers in east Tennessee. Journal of Forensic Sciences 28:423–432.

Rodriguez WC and Bass WM (1985). Decomposition of buried bodies and methods that may aid in their location. Journal of Forensic Sciences 30:836–852.

Sagara N (1995). Association of ectomycorrhizal fungi with decomposed animal wastes in forest habitats: a cleaning symbiosis? Canadian Journal of Botany 73(Suppl. 1):S1423–S1433.

Sagara N, Yamanaka T and Tibbett M (2008). Soil fungi associated with graves and latrines: toward a forensic mycology. In: Soil analysis in forensic taphonomy: Chemical and biological effects of buried human remains (Eds. M Tibbett and DO Carter), pp. 67–108. CRC, Boca Raton, FL.

Schultz JJ (2008). Sequential monitoring of burials containing small pig cadavers using ground penetrating radar. Journal of Forensic Sciences 53:279–287.

Spennemann DHR and Franke B (1995). Decomposition of buried human bodies and associated death scene materials on coral atolls in the tropical Pacific. Journal of Forensic Sciences 40:356–367.

Spicka A, Bushing J, Johnson R, Higley LG and Carter DO (2008). Cadaver mass and decomposition: how long does it take for a cadaver to increase the concentration of ninhydrin-reactive nitrogen in soil? Proceedings of the 60th Annual Meeting of the American Academy of Forensic Sciences 14:178.

Szibor R, Schubert C, Schoning R, Krause D and Wendt U (1998). Pollen analysis reveals murder season. Nature 395:450–451.

Tibbett M, Carter DO, Haslam T, Major R and Haslam R (2004). A laboratory incubation method for determining the rate of microbiological degradation of skeletal muscle tissue in soil. Journal of Forensic Sciences 49:560–565.

Towne EG (2000). Prairie vegetation and soil nutrient responses to ungulate carcasses. Oecologia 122:232–239.

Turner BD and Wiltshire PEJ (1999). Experimental validation of forensic evidence: a study of the decomposition of buried pigs in a heavy clay soil. Forensic Science International 101:113–122.

Vass AA, Bass WM, Wolt JD, Foss JE and Ammons JT (1992). Time since death determinations of human cadavers using soil solution. Journal of Forensic Sciences 37:1236–1253.

Vass AA, Barshick S-A, Sega G, Caton J, Skeen JT, Love JC and Synstelien JA (2002). Decomposition chemistry of human remains: a new methodology for determining the post-mortem interval. Journal of Forensic Sciences 47:542–553.

Vass AA, Smith RR, Thompson CV, Burnett MN, Wolf DA, Synstelien JA, Dulgerian N and Eckenrode BA (2004). Decompositional odor analysis database. Journal of Forensic Sciences 49:760–769.

Vass AA, Smith RR, Thompson CV, Burnett MN, Dulgerian N and Eckenrode BA (2008). Odor analysis of decomposing buried human remains. Journal of Forensic Sciences 53:384–391.

Wilson AS, Janaway RC, Holland AD, Dodson HI, Baran E, Pollard AM and Tobin DJ (2007). Modelling the buried human body environment in upland climes using three contrasting field sites. Forensic Science International 169:6–18.

Chapter 21
Can Temperature Affect the Release of Ninhydrin-Reactive Nitrogen in Gravesoil Following the Burial of a Mammalian (*Rattus rattus*) Cadaver?

David O. Carter, David Yellowlees and Mark Tibbett

Abstract Although temperature and soil type are well known to influence the decomposition of organic resources, the effect of these variables on the release of ninhydrin-reactive nitrogen (NRN) of cadavers in soil has received little experimental investigation. To address this gap in knowledge, juvenile rat *(Rattus rattus)* cadavers were buried in one of three contrasting soils from tropical savanna ecosystems in Queensland, Australia and incubated at 29 °C, 22 °C or 15 °C in a laboratory setting. Cadaver burial resulted in a significant increase in NRN in all gravesoils to a concentration of approximately 15 µg/g soil greater than basal concentration of NRN. Peak levels were observed between 105 and 154 accumulated degree days. This effect was significantly affected by temperature, as gravesoils incubated at 15 °C were associated with a slower accumulation of NRN. No difference between soil types was observed. These findings have important implications for forensic taphonomy because they show the time at which NRN becomes an effective means to identify gravesoils and estimate early (1 to 2 days after death; ≤105 accumulated degree days) post-mortem interval.

Introduction

Temperature is of fundamental importance to the decomposition of cadavers. In conditions that are favourable for decomposition (i.e. readily available moisture, accessible decomposer community), temperature typically has a positive relationship with the breakdown of a cadaver (Mann et al. 1990; Fiedler and Graw 2003;

D.O. Carter(✉) and D. Yellowlees.
School of Pharmacy and Molecular Sciences, James Cook University, Townsville, QLD 4811, Australia.
e-mail: dcarter2@unl.edu

M. Tibbett
Centre for Land Rehabilitation, School of Earth and Geographical Sciences, University of Western Australia, Crawley, WA 6009, Australia.

K. Ritz et al. (eds.), *Criminal and Environmental Soil Forensics*

Dent et al. 2004; Carter and Tibbett 2008). Ultimately, this relationship is due to the positive correlation that exists between temperature and abiotic chemical reaction rates (van't Hoff 1898), enzyme activity (Margesin et al. 2007) and the development of forensically important organisms (primarily insects; Higley and Haskell 2001). However, the large majority of cadaver decomposition studies are initiated in warmer months, when insects are active and conditions are ideal for breakdown (Carter et al. 2007a). As a consequence, very little is known about cadaver decomposition that begins at cooler temperatures. Since cadavers are deposited in terrestrial ecosystems during all seasons of the year, it is necessary to investigate the dynamics of cadaver decomposition at contrasting temperatures.

In addition to being exposed to a range of temperatures, a cadaver can be subject to several types of environments. One relatively common method of disposal is burial in soil (Manhein 1997). Because most of the cadavers that are recovered are found on the soil surface, most cadaver decomposition studies do not incorporate burial as an experimental variable (e.g. Hewadikaram and Goff 1991; Haglund 1997; Tomberlin and Adler 1998). Consequently, little is known about the decomposition of cadavers following burial. However, an understanding of cadaver decomposition and soils can contribute to the investigation of death in two primary ways. Processes of cadaver decomposition can be used to estimate post-mortem interval (PMI; Vass et al. 1992) and to locate clandestine graves (e.g. Carter and Tibbett 2003). The location of clandestine graves is of particular importance in areas of low population density where a body can remain undiscovered for an extended period of time (i.e. months or even years). In these situations, investigators might be aware that a cadaver is present within a terrestrial ecosystem but are unable to locate it.

Recently, Carter et al. (2008) discussed the use for ninhydrin to locate clandestine graves. Ninhydrin, a compound that is regularly used by investigative agencies to detect fingerprints on paper (Odén and von Hofsten 1954), reacts with the organic-nitrogen and ammonium–nitrogen (collectively referred to here as ninhydrin-reactive nitrogen, NRN) that is released from a cadaver during decomposition. Although this method has the potential to assist in the location of clandestine graves, little is known about the relationship between the release of ninhydrin and edaphic parameters such as temperature.

The current experiment tested the hypothesis that higher temperature results in a more rapid release of NRN into gravesoil. We also compared the NRN release in a range of soil types (three contrasting soils from tropical savanna ecosystems in Queensland, Australia).

Materials and Methods

Cadavers

Juvenile rat (*Rattus rattus*) cadavers (~18 g wet weight) aged 8 to 10 days were used as organic resource patches.

Soils

Three contrasting soil types from three sites in tropical savanna ecosystems of Queensland, Australia were used in the current study. The sites were: Yabulu (19°12'S, 146°36'E), Pallarenda (19°11'S, 146°46'E) and Wambiana (20°33'S, 146°08'E). Soil physicochemical characteristics are presented in Carter et al. (2007b).

Experimental Design

A sequential harvesting regime was implemented based on Tibbett et al. (2004) where cadaver and soil samples were destructively harvested following 7, 14, 21 or 28 days of incubation. Soil (500 g dry weight) was weighed into incubation chambers (2 l, high density polyethylene tubs: Crown Scientific, Newstead, Queensland, Australia; Product no. A80WTE + 9530C) and amended with distilled water to a matric potential of −0.03 MPa. Soils were equilibrated for 7 days at 29 °C, 22 °C or 15 °C, following the calibration of moisture content. Following equilibration, rats were killed with carbon dioxide (CO_2), weighed and buried in soil on their right side at a depth of 2.5 cm. Soils with a cadaver will be referred to as gravesoils. Control samples (soil without a cadaver) were disturbed to simulate cadaver burial and to account for any effect of soil disturbance and will be referred to as control soils. Gravesoils and control soils were incubated at 29 °C, 22 °C or 15 °C and the experiment was replicated three times.

At each harvest event designated cadavers were exhumed along with gravesoil directly surrounding the cadaver (detritisphere; Tilston et al. 2004) (approximately 50 g). The detritisphere represented the soil that adhered to the cadaver and was collected manually. In control samples the disturbed soil was collected at each harvest event. Gravesoils and control soils were then analysed for ninhydrin-reactive nitrogen (NRN) following the method described by Carter et al. (2008). Accumulated degree days (ADDs) were calculated after Vass et al. (1992) using 0 °C as the minimum developmental threshold.

Statistical Analyses

NRN data were analysed for normality and homogeneity of variance using the Kolmogorov–Smirnov test and Levene's test, respectively. Differences between means were analysed using a univariate analysis of variance. All statistics were generated using SPSS Version 15.

Results

All gravesoils contained greater ($P < 0.05$) levels of NRN than control soils throughout the incubation (data shown as gravesoil NRN concentration less control NRN concentration in Figure 21.1). A significantly ($P < 0.05$) greater concentration of NRN was observed in all soils by day 7 (<105 ADDs). Peak levels of NRN in all gravesoils were approximately $15\,\mu g\,g^{-1}$ soil greater than in control soils (Figure 21.1). Peak levels were reached within 7 days (154 ADDs) of burial at 29 °C and 22 °C. A slower release of NRN was observed at 15 °C, which demonstrated that peak levels of NRN in gravesoils were reached between 105 ADDs and 154 ADDs.

Discussion

The current results support previous findings that an elevated level of NRN is associated with gravesoils (Carter et al. 2008) and have furthered this knowledge by demonstrating that temperature can significantly influence the release of NRN into gravesoils. Thus, we accept our hypothesis. This finding is not new to soil science, as it is well established that higher temperatures can enhance the decomposition of organic resources (Swift et al. 1979) and, thus, the release of N into associated soil. However, the current findings are important for the development of gravesoil NRN as a potential forensic tool, as it is necessary to determine the minimum time required to result in a significant change in NRN concentration. Although the current results do not necessarily fully achieve this, they show that a significant increase occurred after 105 ADDs. This knowledge provides forensic investigators with a putative starting point from which to potentially use ninhydrin to detect gravesoil, while other methods should be used if a body has been missing for less than 105 ADDs under these circumstances. The current results also show that the measurement of NRN might be used to estimate post-mortem interval. Our data suggest that, when soil associated with a recently killed body does not contain an elevated level of NRN, then the victim might have been deceased for less than 105 ADDs. These findings, however, need confirmation under field conditions and with human cadavers.

It is unlikely that the NRN will ever be used as a confirmatory test for gravesoil, such as the use of prostate specific antigen to confirm the presence of semen (Greenfield and Sloan 2005). Several organic resources release NRN into soil during decomposition, so the presence of an elevated level of NRN does not necessarily indicate the presence of a cadaver. However, it appears that a cadaver can release a level of NRN that is much greater than any other organic resource. Thus, NRN will likely remain as a presumptive test for gravesoil that, when positive, should lead to more detailed analyses such as through excavation (Dupras et al. 2006).

Temperature has long been recognised as one of the primary regulators of cadaver decomposition (Gill-King 1997). As a consequence, temperature has received a lot

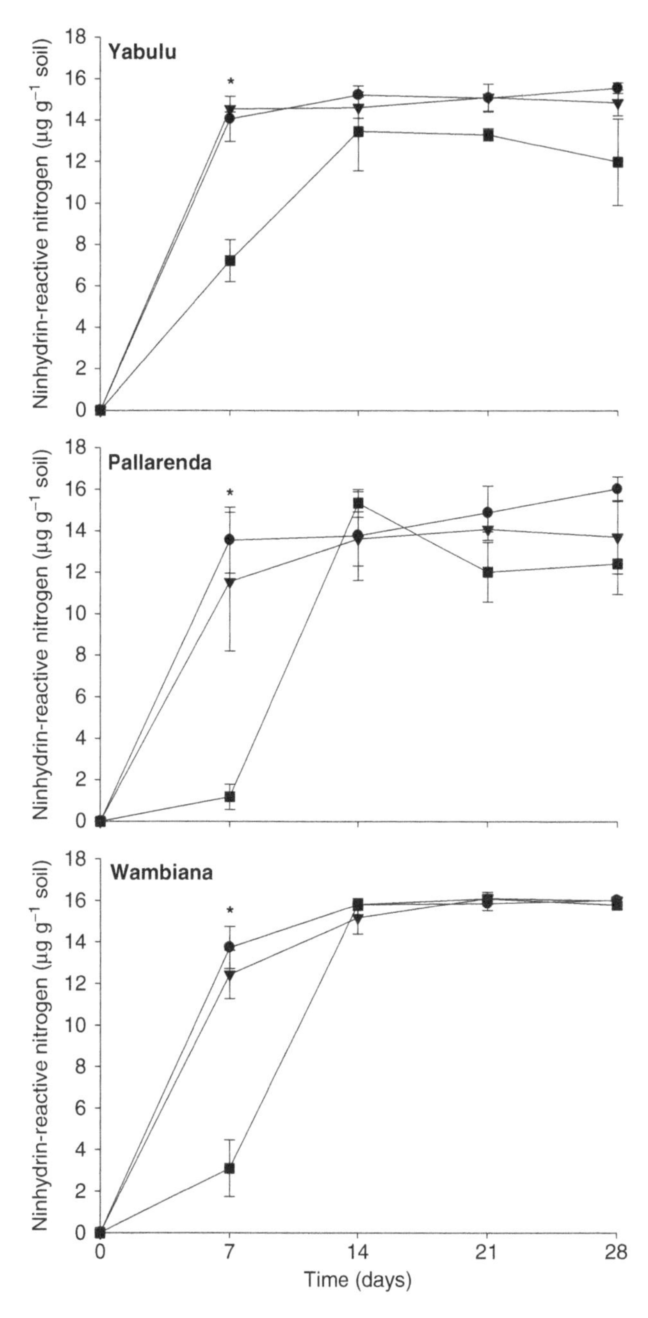

Fig. 21.1 Gravesoil ninhydrin-reactive nitrogen concentration (µg/g soil) following the burial (2.5 cm) of a juvenile rat (*Rattus rattus*) cadaver in 500 g (dry weight) sieved (2 mm) soil from Yabulu, Pallarenda, or Wambiana, Queensland, Australia and incubated at 29 °C (•), 22 °C (▾) or 15 °C (■). These data are presented as gravesoil NRN concentration less control NRN concentration. * represents a significant ($P < 0.05$) difference between treatments within time. Bars represent standard errors where n = 3

of attention in empirical and experimental cadaver decomposition studies (Mann et al. 1990; Vass et al. 1992; Janaway 1996; Vass et al. 2002; Nabity et al. 2006; Huntington et al. 2007; Carter et al. 2007a). Research in forensic entomology, in particular, has focused greatly on the use of temperature, and the calculation of ADDs to estimate post-mortem interval based on the development of blow fly (Diptera: Calliphoridae) larvae (Higley and Haskell 2001). Relatively untested, except the research conducted by Vass et al. (1992), is the use of ADDs in conjunction with gravesoil processes. The current research demonstrates the potential for using ADDs and gravesoil processes to estimate PMI. We suggest that, following further research, ADDs associated with gravesoil processes be investigated for their forensic use in conjunction with estimates of PMI based on entomological and other evidence. This might reduce the error associated with estimates of PMI in the future. Further, an understanding of the relationships between temperature and gravesoil processes might result in an additional forensic tool, when, for example, entomological evidence is not available (e.g. following burial) or there is a lack of precision (e.g. following maggot migration).

There is a great need for the development of methods to estimate the PMI of buried bodies (Forbes 2008). At present, the majority of cadaver decomposition studies are conducted on the soil surface, which will provide only limited insight into belowground decomposition processes. It is understood that burial can retard the most reliable of the PMI markers (Turner and Wiltshire 1999), but little has been done to address this need. It is unlikely that the measurement of NRN alone will be able to provide more than an estimate of minimum and maximum PMI, but the measurement of the individual compounds that comprise NRN (i.e. protein, peptide, amino acids, amines, ammonium) might be used to estimate PMI, as demonstrated by Vass et al. (1992) using volatile fatty acids and a range of nutrients. However, several more cadaver decomposition studies will need to be carried out before this technique will be admissible as evidence.

One potential concern associated with the use of gravesoil processes as forensic tools are their utility across different soil types. Physical, chemical and biological characteristics can vary greatly between soils (Fitzpatrick 2008) and, thus, it is logical to conclude that a soil process-based forensic tool might not work in all soils. In the current study, the effect of cadaver burial on gravesoil NRN was consistent across the three contrasting soils investigated. Therefore, potential exists for the current technique to be used universally. However, much more research should investigate the efficacy of gravesoil NRN in a wide range of ecosystems across the world.

Conclusions

The analysis of gravesoil NRN is a rapid and effective means to presumptively identify gravesoil. However, as discussed in Carter et al. (2008), a large amount of work still needs to be done before an elevated level of NRN is considered a reliable method to identify gravesoil. Variables such as soil moisture content, burial depth, cadaver mass, and cadaver species represent a few of the parameters that must be investigated further before the measurement of NRN can be considered for presentation in a courtroom. These investigations, along with the analysis of NRN components (e.g. amino acids) might lead to a robust method to estimate PMI.

References

Carter DO and Tibbett M (2003). Taphonomic mycota: fungi with forensic potential. Journal of Forensic Sciences 48:168–171.

Carter DO and Tibbett M (2008). Cadaver decomposition and soil: processes. In: Soil Analysis in Forensic Taphonomy: Chemical and Biological Effects of Buried Human Remains (Eds. M Tibbett and DO Carter), pp. 29–51. CRC, Boca Raton, FL.

Carter DO, Yellowlees D and Tibbett M (2007a). Cadaver decomposition in terrestrial ecosystems. Naturwissenschaften 94:12–24.

Carter DO, Yellowlees D and Tibbett M (2007b). Autoclaving can kill soil microbes yet enzymes can remain active. Pedobiologia 51:295–299.

Carter DO, Yellowlees D and Tibbett M (2008). Using ninhydrin to detect gravesoil. Journal of Forensic Sciences 53:397–400.

Dent BB, Forbes SL and Stuart BH (2004). Review of human decomposition processes in soil. Environmental Geology 45:576–585.

Dupras TL, Schultz JJ, Wheeler SM and Williams LJ (2006). Forensic Recovery of Human Remains. CRC, Boca Raton, FL.

Fiedler S and Graw M (2003). Decomposition of buried corpses, with special reference to the formation of adipocere. Naturwissenschaften 90:291–300.

Fitzpatrick RW (2008). Nature, distribution, and origin of soil materials in the forensic comparison of soils. In: Soil Analysis in Forensic Taphonomy: Chemical and Biological Effects of Buried Human Remains (Eds. M Tibbett and DO Carter), pp. 1–28. CRC, Boca Raton, FL.

Forbes SL (2008). Potential determinants of postmortem and postburial interval. In: Soil Analysis in Forensic Taphonomy: Chemical and Biological Effects of Buried Human Remains (Eds. M Tibbett and DO Carter), pp. 225–246. CRC, Boca Raton, FL.

Gill-King H (1997). Chemical and ultrastructural aspects of decomposition. In: Forensic Taphonomy: The Postmortem Fate of Human Remains (Eds. WD Haglund and MH Sorg), pp. 93–108. CRC, Boca Raton, FL.

Greenfield A and Sloan MM (2005). Identification of biological fluids and stains. In: Forensic Science: An Introduction to Scientific and Investigative Techniques (Eds. SH Jaes and JJ Nordby), pp. 261–278. CRC, Boca Raton, FL.

Haglund WD (1997). Dogs and coyotes: postmortem involvement with human remains. In: Forensic Taphonomy: The Postmortem Fate of Human Remains (Eds. WD Haglund and MH Sorg), pp. 367–382. CRC, Boca Raton, FL.

Hewadikaram KA and Goff ML (1991). Effect of carcass size on rate of decomposition and arthropod succession patterns. American Journal of Forensic Medicine and Pathology 12:235–240.

Higley LG and Haskell NH (2001). Insect development and forensic entomology. In: Forensic Entomology: The Utility of Arthropods in Legal Investigations (Eds. JJ Byrd and JL Castner), pp. 287–302. CRC, Boca Raton, FL.

Huntington TE, Higley LG and Baxendale FP (2007). Maggot development during morgue storage and its effect on estimating the post-mortem interval. Journal of Forensic Sciences 52:453–458.

Janaway RC (1996). The decay of buried remains and their associated materials. In: Studies in Crime: An Introduction to Forensic Archaeology (Eds. J Hunter, C Roberts and A Martin), pp. 58–85. Routledge, London.
Manhein M (1997). Decomposition rates of deliberate burials: a case of preservation. In: Forensic Taphonomy: The Postmortem Fate of Human Remains (Eds. WD Haglund and MH Sorg), pp. 469–482. CRC, Boca Raton, FL.
Mann RW, Bass MA and Meadows L (1990). Time since death and decomposition of the human body: variables and observations in case and experimental field studies. Journal of Forensic Sciences 35:103–111.
Margesin R, Neuner G and Storey KB (2007). Cold-loving microbes, plants and animals-fundamental and applied aspects. Naturwissenschaften 94:77–99.
Nabity PD, Higley LG and Heng-Moss TM (2006). Effects of temperature on development of Phormia regina (Diptera: Calliphoridae) and use of developmental data in determining time intervals in forensic entomology. Journal of Medical Entomology 43:1276–1286.
Odén S and von Hofsten B (1954). Detection of fingerprints by the ninhydrin reaction. Nature 173:449–450.
Swift MJ, Heal OW and Anderson JM (1979). Decomposition in Terrestrial Ecosystems. Blackwell Scientific, Oxford.
Tibbett M, Carter DO, Haslam T, Major R and Haslam R (2004). A laboratory incubation method for determining the rate of microbiological degradation of skeletal muscle tissue in soil. Journal of Forensic Sciences 49:560–565.
Tilston EL, Halpin C and Hopkins DW (2004). Genetic modifications to lignin biosynthesis in field-grown poplar trees have inconsistent effects on the rate of woody trunk decomposition. Soil Biology and Biochemistry 36:1903–1906.
Tomberlin JK and Adler PH (1998). Seasonal colonization and decomposition of rat carrion in water and on land in an open field in South Carolina. Journal of Medical Entomology 35:704–709.
Turner BD and Wiltshire PEJ (1999). Experimental validation of forensic evidence: a study of the decomposition of buried pigs in a heavy clay soil. Forensic Science International 101:113–122.
van't Hoff JH (1898). Lectures on Theoretical and Physical Chemistry. Part 1. Chemical Dynamics. Edward Arnold, London.
Vass AA, Bass WM, Wolt JD, Foss JE and Ammons JT (1992). Time since death determinations of human cadavers using soil solution. Journal of Forensic Sciences 37:1236–1253.
Vass AA, Barshick S-A, Sega G, Caton J, Skeen JT, Love JC and Synstelien JA (2002). Decomposition chemistry of human remains: a new methodology for determining the postmortem interval. Journal of Forensic Sciences 47:542–553.

Chapter 22
Taphonomic Changes to the Buried Body in Arid Environments: An Experimental Case Study in Peru

Robert C. Janaway, Andrew S. Wilson, Gerardo Carpio Díaz and Sonia Guillen

Abstract Despite an increasing literature on the decomposition of buried and exposed human remains it is important to recognise that specific microenvironments will either trigger, or delay the rate of decomposition. Recent casework in arid regions of the world has indicated a need for a more detailed understanding of the effects of burial over relatively short timescales. The decomposition of buried human remains in the coastal desert of Peru was investigated using pig cadavers (*Sus scrofa*) as body analogues. The project aims were to specifically examine the early phases of natural mummification and contrast the effects of direct burial in ground with burial in a tomb structure (i.e. with an air void). Temperature was logged at hourly intervals from both the surface, grave fill and core body throughout the experiment. In addition, air temperature and humidity were measured within the air void of the tomb. After two years all three pig graves were excavated, the temperature and humidity data downloaded and the pig carcasses dissected on site to evaluate condition. The results demonstrate that: (1) there were distinct differences in the nature/rate of decomposition according to burial mode; (2) after two years burial the carcasses had been subject to considerable desiccation of the outer tissues while remaining moist in the core; (3) the body had undergone putrefactive change and collapsed leading to slumping of soil within the grave fill following the curvature of the pig's back, although this was not evident from the surface; (4) there was a specific plume of body decomposition products that wicked both horizontally and also vertically from the head wounds in the sandy desert soil. These observations have widespread application for prospection techniques, investigation of clandestine burial, time since deposition and in understanding changes within the burial microenvironment under arid conditions.

A.S.Wilson(✉) and R.C. Janaway
Archaeological Sciences, School of Life Sciences, University of Bradford, Bradford, West Yorkshire, BD7 1DP, UK
e-mail: a.s.wilson2@bradford.ac.uk

G. Carpio Díaz and S. Guillen
Centro Mallqui, Ilo, Peru

K. Ritz et al. (eds.) *Criminal and Environmental Soil Forensics*

Introduction

The detection and recovery of human remains by forensic archaeologists relies on an understanding of the effects of disturbance of the ground in digging the grave coupled with the effects of body decomposition itself. Such changes are highly environmentally dependent (Hunter and Cox 2005). Most publications concerning search and location of single or mass graves have concentrated on temperate soil conditions within Europe and continental USA (Hunter and Cox 2005; Dupras et al. 2006). It has been demonstrated that after burial most bodies undergo rapid putrefactive change (Wilson et al. 2007), and this active decomposition phase has important implications for the detection of the grave through use of fieldcraft, cadaver dogs or geophysical techniques (Lynam 1970; France et al. 1992; Bray 1996; Threader 1997; Sorg et al. 1998).

There has been recent interest in arid environments for a range of disciplines including forensic archaeology and forensic taphonomy, for both current and historic criminal and military applications. Currently, pigs remain the most popular human cadaver analogues used in forensic experiments since they have a similar body mass, skin structure and fat-to-muscle ratio, making them ideal human cadaver analogues for soft-tissue studies (Schoenly et al. 1991). Most taphonomic experiments have used whole cadavers due to the importance of gut microflora in putrefactive change (Garland and Janaway 1989; Hopkins et al. 2000). Although some work has been done using isolated tissue, incubated under laboratory conditions (Aturaliya and Lukasewycz 1999), the opportunity for tighter environmental control and large replicated studies is offset by the difficulty in relating the results and timescales to whole bodies buried under real conditions. The dual utility of both laboratory and field experiment has been demonstrated in Chapters 20 and 23 by Tibbett et al. and Stokes et al. respectively in this volume.

Pig cadavers have widely been used in entomological studies (Payne 1965; Goff and Odom 1987; Goff et al. 1988; Lopes de Carvalho and Linhares 2001) and have been utilised in various arid environment studies in Western Australia focused on forensic entomology and an improved understanding of fat decomposition (Dadour et al. 2001; Forbes et al. 2005a, b).

This paper presents data from a two year taphonomic experiment in the coastal desert of southern Peru using pig cadavers. The experiment aim was to specifically examine changes to the body e.g. the onset of natural mummification and contrast the effects of direct burial in soil with burial in a structure with an air void. While the latter has direct relevance in forensic contexts, for example where a body may have been concealed in a culvert or pipe, it also has archaeological relevance where bodies were buried in small rock built tombs.

Coastal Peru was selected for this research for a variety of reasons including the arid climatic conditions of the region which favour natural mummification and the preservation of soft tissue (skin and muscle), hair, nail and textiles over archaeological timescales (Aufderheide 2003). Coastal Peru also suited us for logistics/security of the experiment based on current collaborations in the region as well as cultural/religious sensitivities which might otherwise preclude the use of pig carcasses furthermore, southern Peru currently offers better security for researcher personnel than many arid regions of the world.

Natural Mummification

While the phenomenon of natural mummification is well reported in both the forensic (Knight and Simpson 1997; Saukko and Knight 2004) and archaeological literature (Janaway 1996; Aufderheide 2003), detailed terms for tissue survival and condition are not always consistently applied. Normally soft tissue decomposition is characterised by bacterial putrefaction which leads to progressive liquefaction until only the skeleton remains. This process will be accelerated where insects can gain access with larvae feeding in large numbers. Natural mummification has been defined as "the drying of tissues in place of liquefying putrefaction" (Saukko and Knight 2004). The process requires the tissue water content to drop below a critical threshold whereby bacterial putrefaction is inhibited. The tissues desiccate due to environmental conditions and the body shrivels to a dry leathery mass of skin and tendons surrounding the bone (Pounder 2000). The phenomenon is perhaps best known in terms of Pre-Dynastic Egyptian burials of c.4400 to 3100 BC. In these burials the body was buried directly into the ground which had the effect of wicking moisture out of, and away from, the tissues (Dzierzykray-Rogalski 1986). This should not be confused with later Egyptian practices where the body was subjected to artificial treatment, usually with the mineral natron after evisceration and prior to wrapping and interment in a structure rather than directly in the ground (Buckley and Evershed 2001). Similar natural mummies and practices of artificial mummification are also well documented from various South American cultures (Guillen 2004). In forensic contexts from temperate regions natural mummification is usually associated with structures where temperature and airflow promotes tissue desiccation (Aturaliya and Lukasewycz 1999). The natural mummification of newborn infants, that are virtually sterile on emerging from the uterus, is well documented (Knight and Simpson 1997) especially if left in cool dry locations.

The forensic literature suggests that both mummification of the entire body, as well as localised mummification of specific body regions such as the extremities (e.g. fingers and toes) can occur. Despite these suggestions, even in the case of whole body mummification the outer layers of desiccated tissue tend to contain a 'bag of bones', often devoid of organs (Dix and Graham 2000). However, all these descriptions tend to oversimplify the processes of tissue survival. Firstly, under a wide range of environmental conditions soft tissue will be subject to bacterially-driven putrefaction. In the case of desiccating environments it is critical to establish at what stage during decomposition tissue and degradation products dry to below a critical threshold whereby putrefaction will effectively cease. If desiccation takes place rapidly then it is possible that re-hydration of tissue samples will reveal remnant tissue morphology (Lewin 1967; Aufderheide 2003). At the other extreme, however, it is also quite likely that the tissue may reach an advanced stage of putrefactive change and the resulting liquid will dry to form a 'skin' that retains no morphological characteristics, covering the skeletal elements. So it is not surprising that bodies recovered from desert soils range from mummies with extensive desiccated soft tissue through skeletons with some surviving tissue and often hair, to entirely skeletal material (Stover et al. 2003).

Cadaveric Decay in Arid Environments

While there have been extensive studies of cadaveric decay in temperate or tropical climates (Rodriguez and Bass 1985; Dent et al. 2004), work in arid regions has been more restricted. A notable exception is the work of Galloway and co-workers (Galloway et al. 1989; Galloway 1997) who conducted a retrospective study of casework from southern Arizona. The Arizona-Sonoran desert region has low population densities between the major towns that are connected by both highways and minor roads. This results in bodies that have been deposited in out of the way locations remaining undiscovered for long periods of time (Galloway 1997). The climate of southern Arizona is characterised by hot, arid summers and mild winters.

Data from the Airport at Tucson (lat N32° 07′; long W110° 56′; 789 m above sea level) show an average yearly high temperature of 27.6 °C, and average rainfall of 23.6 mm; however the heaviest periods of rainfall in July (64.5 mm) and August (51.6 mm) with the summer monsoon period (Galloway 1997; Hanson and Hanson, 1999). Using reports and photographs from case files, each body was classified according to five major stages of decomposition (A to E), further refined using secondary categories (C1, C2, etc.), as outlined in Table 22.1.

Bodies used in this study had been deposited in a range of locations with the majority 56% (n = 162) being deposited in the open air, while 27% (n = 79 were in closed structures (houses, trailers and other buildings) and only 10% (n = 30) were actually buried. Galloway et al. (1989) report that bodies buried directly in the soil exhibit very moist decomposition, with skin slippage and fungal development being common. The presence of fungi testifies to the body remaining sufficiently moist as fungal activity is severely inhibited by desiccation of the substrate (Hudson 1980). In addition different stages of adipocere formation are reported even in shallow graves (Galloway 1997) and the bodies eventually are skeletalised, although the data from these 30 cases are not presented separately.

The bulk of the material that Galloway (1997) looked at was either surface disposal, or referred to bodies within substantial closed structures in the Arizona-Sonoran desert region in which the onset of mummification and accelerated decomposition was markedly different, and therefore differed significantly from the closed cist structure as part of the experiments reported here. Although it should be noted that variables such as seasonality, air exchange, volume of closed structures, etc. are not fully discussed within the different versions of Galloway's paper (Galloway 1997; Galloway et al. 1989), the classification stages that she proposed (Galloway 1997) are more applicable to surface exposed bodies with some categories (e.g. skeletalisation with

Table 22.1 Categories of body decomposition

Category	Definition
A: Fresh	No visible trace of maggot activity, no discoloration of body
B: Early decomposition	Includes where discoloration has begun, bloating and post-bloating stages
	1. Skin slippage and some hair loss
	2. Grey to green discoloration
	3. Bloating with green discoloration
	4. Post bloating following rupture of abdomen, discoloration going from green to brown
	5. Brown to black discoloration of arms and legs, skin having leathery appearance
C: Advanced decomposition	Sagging of tissue, extensive maggot activity, mummification and desiccation
	1. Decomposition of tissues producing sagging of the flesh, caving in of the abdominal cavity. Often accompanied by extensive maggot activity
	2. Moist decomposition in which there is bone exposure
	3. Mummification, with some retention of internal structures
	4. Mummification of outer tissues only with internal organs lost through autolysis or insect activity
	5. Mummification with bone exposure of less than half the skeleton
	6. Adipocere development
D: Skeletonised	Majority of bones exposed, with decomposition fluids still present to dry bone
	1. Bones with decomposed tissue, sometimes body fluids present
	2. Bones with desiccated tissue or mummified tissue covering less than one half of the skeleton
	3. Bones largely dry retaining some grease
	4. Dry bone
E: Decomposition of skeletal remains	1. Skeletalisation with bleaching
	2. Skeletalisation with exfoliation
	3. Skeletalisation with cortical breakdown exposing cancellous bone

After Galloway et al. (1989), Galloway (1997).

bleaching; skeletalisation with exfoliation) that are redundant for buried remains. There would also be some benefit in breaking down the active decomposition category to distinguish between largely wet and largely mummified tissue.

Material and Methods

Field Site

The field site was selected adjacent to and up slope from the research centre for the study of archaeological mummies at Centro Mallqui[1] situated in the Lower Osmore

[1] Mallqui means 'mummy' in the Quechua language.

Fig. 22.1 Overview of site. (a) General view of experimental site with pig graves open while environmental monitoring is being set up; the nearest is Grave 1, the middle is Grave 2 (tomb), the furthest is Grave 3. (b) Soil section from control test pit (0.5 m scale/10 cm divisions). The profile consists of an upper layer of white/grey sand with rock fragments, overlying a red/brown sand horizon with rock fragments and a grey/white sand horizon with dense rock inclusions is at the base *(see colour plate section for colour version)*

valley, southern Peru (lat S17° 37′; long W71° 16′). The lower Osmore valley has yielded large quantities of naturally mummified pre-Columbian human remains from various cultures. The experimental site was at an elevation of 165 m above sea level roughly 8 km inland from the sea (Figure 22.1a). The surface consisted of a loose sand approximately 15 cm deep which overlays a red oxidised sand layer (Figure 22.1b; Table 22.2). The sands were compact but were easily dug to over 1 m depth across the experimental site. There were visible salt encrustations within the soil profile, also evidenced by high conductivity meter readings. Generic mineralogy for the recovered mixed fill of the two soil graves are shown in Table 22.3.

Use of Pig Carcasses

Pigs are well established as human body analogues, especially in relation to soft tissue decomposition. They are similar to humans due to a similar, not heavily haired, skin structure. They have a similar fat-to-muscle ratio, although their fat composition is similar but not identical to humans (Wilson et al. 2007). The thoracic cavity size of pigs is similar to humans and, being omnivores, their gut chemistry and flora is a closer match to humans than other models, e.g. sheep.

The pigs were sourced from a local farm less than 5 km from the experiment. Because of the local environmental conditions (in particular air temperature), it was

Table 22.2 Profile of the soil horizons in control test pit A (total depth 85 cm)

Horizon	Thickness	Compaction	Description	Conductivity (mS)	Total dissolved solids (g/l)	pH
1 (Top)	8 to 15 cm	Loose	White/grey sand with rock fragments	14.8	9.0	7.3
2	23 to 30 cm	Loose to medium compact	Red/brown sand with rock fragments	9.0	5.4	7.4
3 (Base)	40 cm to base of pit	Compact	Grey/white sand with dense rock inclusions	17.1	10.6	7.7

Table 22.3 Mineralogy of the grave fill (Graves 1 and 3) showing the presence of evaporite minerals (halite)

Mineral	Formula	Grave 1 (%)	Grave 3 (%)
Quartz	SiO_2	50.0	51.3
Albite	$(Na, Ca)(Si, Al)_4O_8$	38.7	39.7
Halite	$NaCl$	2.16	1.69
Cordierite	$Mg_2Al_4Si_5O_{18}$	2.01	3.13
Muscovite	$(K, Na)(Al, Mg, Fe)_2$	1.05	1.52
Magnetite	Fe_3O_4	0.98	
Chlorite	$Mg_5Al_2Si_3O_{10}(OH)_8$	0.59	0.59
Hematite	Fe_2O_3	0.44	0.38
Augite	$Ca(Mg, Fe, Al)(Si, Al)_2O_2$	0.34	1.27
Chloromagnesite	$MgCl_2$	0.29	
Niningerite	$(Mg, Fe, Mn)S$		0.42
Amorphous		3.43	

essential that they were buried rapidly after death. The pigs were killed with a single shot to the head by a local police officer using a police issue handgun and buried within 4 h of death. All three pigs used in this experiment were of similar carcass size (~130 to 140 cm in length), weight and condition.

Layout of Burial Pits

The experiment was sited on the southern slope of the Osmore valley. Three graves were dug 2 × 2 × 0.60 m deep. These were set out cross slope with a 1 m baulk between each grave. Graves 1 and 3 contained a pig buried directly in the soil. Grave 2 consisted of a stone-lined tomb containing a pig within an enclosed air void. The tomb was constructed out of local stone. Large stones were laid across the top of the side walls and then sealed with local mud cement made by adding water to the sand from the site. When dry this forms a hard natural cement called 'Arcilla'. The pig carcasses were orientated with their axes running cross slope.

When the graves were backfilled a durable square-mesh extruded plastic protection net (Gladiator™ predator net, Tildenet, Bristol) was pegged out at double thickness across each pit just below the surface and covered with sand. This was to deter village dogs and other potential scavengers from interfering with the graves.

Environmental Monitoring

The principal requirements of the environmental logging equipment needed for this experiment were that they were robust, capable of programmed recording of hourly temperature and humidity values with sufficient capacity to run independently for up to two years. Two types of logger (Gemini Dataloggers, UK) were used in this experiment: a Tinytag dual channel temperature logger (TGP1520) which has two external temperature probes and the TGP1500 (used in the tomb) with a single internal temperature plus relative humidity channel. The external temperature probes were custom made according to our previous design for use in temperate climates (Wilson et al. 2007). Each grave had a designated dual channel temperature logger (TGP1520) with one probe used to log core body temperature (via the anus) and the other probe buried in the grave fill. The exception was the tomb burial for which a probe was placed above and below the pig and a further dual channel temperature/humidity logger (TGP1500) was located in special recess in the wall of the tomb. Further TGP1500 loggers were located in the adjacent site building to monitor air temperature and humidity over the duration of the experiment.

Carcass Exhumation, Sample Collection and Analyses

After two years the graves were excavated stratigraphically using standard archaeological methods and the data loggers were downloaded. Each pig body was exposed and recorded *in situ* in the grave and then removed for dissection by participants from the Peruvian forensic team to evaluate the general condition of internal tissues.

Results and Discussion

Soil Burials (Graves 1 and 3)

Excavation of Soil Burials

Due to the loose nature of the sand, the location of the graves was not apparent on the surface due to any visible depressions or vegetational indicators. However, once the sand that had been covering the netting had been removed, it was clear that in

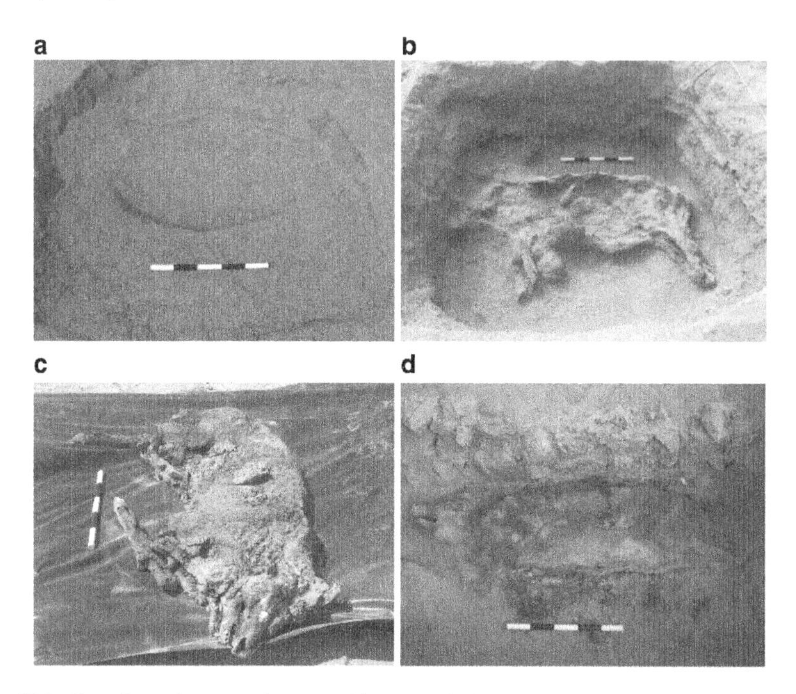

Fig. 22.2 Overview of carcass decomposition. (a) Netting buried subsurface above Grave 1 to deter scavengers, which shows soil movement associated with the collapse of the body cavity not apparent at the surface. (b) Carcass in grave after 760 days showing collapse of the body cavity and concretions associated with moisture derived from the decomposing carcass, especially evident adjacent to the cranial trauma. (c) Underside of Pig 3 after excavation, showing trauma and more advanced skeletalisation to the head. (d) Pig in tomb following removal of capping and part of the tomb construction. Dark pupal masses are visible adjacent to belly. All scales are 0.5 m with 10 cm divisions *(see colour plate section for colour version)*

Graves 1 and 3 this had documented soil movement associated with the collapse of the body cavity (Figure 22.2a). The sand below the depression had more moisture than the surrounding matrix and there was a distinct smell. As overburden on top of the pig was removed with hand trowels, there was a distinct odour and the sand 20 cm above the pig was more moist than the surrounding matrix. This indicated that after two years the pigs were still an active source of moisture. Immediately around the pig cadavers were patches of darker sand cemented together by a combination of salts and body decomposition products. Both soil burials contained a characteristic exudate from the head wound preserved by forming a solid matrix in the salt-laden soil. This exudate plume had wicked both laterally and vertically from the point of origin within the grave fill.

Carcass Condition

The excavated pigs revealed a characteristic collapse of the body cavity that has been seen in the archaeological excavation of quadrupeds buried on their side (Figure 22.2b).

It was this collapse that had caused the characteristic slump in the fill above. Body hair was clearly visible on the upper body surface. This detached easily from the skin. Once the body had been examined within the grave it was removed for a rapid on-site dissection. This revealed in both pigs that the partially desiccated exterior layer remained soft and pliable in places and contained a moist interior. Partial desiccation was more advanced in the superior body surfaces. While there were areas of skeletalisation, these were associated with areas of trauma, especially evident with the head of Pig 3 (Figure 22.2c). This is consistent with well-documented casework going back to the work Mant (1987) who recognised that trauma from cranial gunshot wounds can lead to increased decomposition or loss of tissue around the wound.

Soils

The soils within the grave fill of the plain burials had been subject to the effects of the active decomposition of the pig cadavers. Approximately 20 cm above and around pig was more moist than the surrounding matrix. Table 22.4 compares measured values for conductivity, total dissolved solids and pH from different soil depth locations within the fill of Grave 1; these were taken in a half section across Pig 1. The sample adjacent to Pig 1 was targeted from patches of darker sand cemented by decomposition products. The particular characteristics of the salt-laden soil and the effect of moisture derived from the decomposing carcass helped to form concretions of exudates from body decomposition – especially the characteristic plume associated with the head wounds.

These data show that the decomposition products from the pig are concentrated from the point of origin and dissipate with vertical distance in the grave. The sample adjacent to the pig, which is heavily contaminated with decomposition products, is more acidic and shows depressed conductivity/TDS values. As expected the soil profile from the grave fill is not the same as undisturbed soil from the control pits (Table 22.2).

Environmental Monitoring

The environmental conditions of Tucson, Arizona and Ilo Peru, while both arid environments, are markedly different. The principle differences relate to rainfall,

Table 22.4 Soil properties with respect to depth within the fill of Grave 1 (taken in a half section across Pig 1)

Location of sample (cm above carcass)	Conductivity (mS)	Total dissolved solids (g/l)	pH
30	16.5	10.0	7.9
20	17.4	10.4	8.0
10	15.6	9.3	8.2
0	7.9	4.7	7.7
Adjacent (rump of pig)	5.0	3.0	6.8

whereas Arizona is subject to a seasonal periods of wetting (with average monthly rainfall of ~24 mm), Ilo has scarce rainfall (typically less than 150 mm per annum) and is subject to El Niño events. In general, with the experimental burials we see subsurface soil temperature rise (from a minimum during August), associated with increase in seasonal ambient temperature. The tomb and the soil burials show different responses to this change (see below).

Tomb (Grave 2)

Excavation of Tomb

Removal of sand overburden revealed that the capping and mud cement seal of the top of the tomb was intact. Once the capping had been removed the collapsed pig carcass was revealed. The hair was a ginger, brown colour and no loose sand was covering the pig except for that which had been disturbed in removing the stones (Figure 22.2d). Unlike the plain burials there was no evident smell at this stage.

Carcass Condition

Examination of Pig 2 revealed much more advanced decomposition compared to Pigs 1 and 3. The exposed upper surface of the pig cadaver was desiccated and covered by small holes due to insect activity and a number of fly pupae cases were evident (Figure 22.3a). The skin on the lower fore limbs had split exposing the bone. There was a mass of fly pupae cases around snout and eye. The body in the tomb shows some characteristics of both massive bloat and insect attack that is to be expected due to decomposition in an air void in contrast to burial directly in the soil. Examination of the interior surfaces of the tomb revealed waxy material adhering to the rock along with insect remains. This constituted a mass of pupal cases, decomposition products and pig hair adhering to the tomb wall (Figure 22.3b). This is a clear indication that the body had bloated to fill the void prior to collapsing. Dissection of the pig from the tomb revealed that internal decomposition was more advanced than for the plain burials. There was considerably more insect activity compared to the plain burials. The Peruvian experiments lasted for 24 months, and using Galloway's classification, they exhibit sagging of the abdominal cavity, less than 50% skeletalisation, with mummification of the outer tissues, while retaining moist tissues within the body interior. These changes placed them firmly within the advanced decomposition categories (i.e. C1, C4, C5). When these experimental results are compared to those of Galloway (1997, Figure 22.3), the bulk of bodies from the open air are either desiccated or skeletonised (Decomposition Category D). This is perhaps not surprising since the Arizona data set includes all bodies dumped on the surface and it is to be expected that the decomposition rate would be much quicker. By contrast the Arizona data from closed structures at 24 months were largely skeletalised with

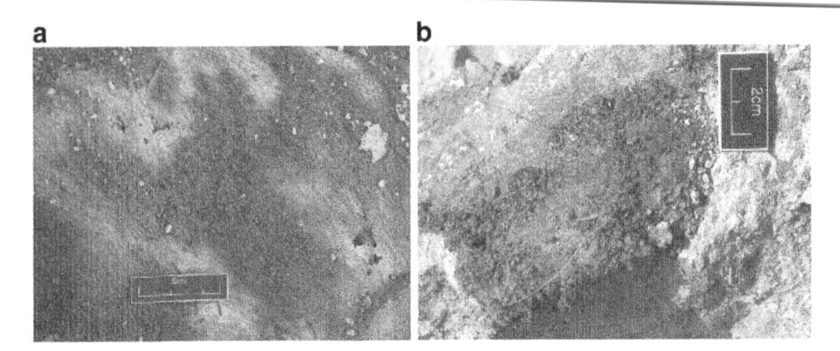

Fig. 22.3 Details of pig surfaces after decomposition. (a) Upper surface of Pig 2 flank showing pupal cases, loss of bristle and holes in the skin characteristic of insect damage. (b) Pig decomposition products deposited on the under-surface of the capping within the tomb, including adhering bristle and empty pupal cases *(see colour plate section for colour version)*

some still showing mummification. Again this data set is not strictly comparable with the Peruvian experimental data since the Arizona dataset includes all enclosed structure, including houses, trailer homes and other built structures.

Our interpretation of the process that has led to partial mummification of the carcass in the tomb is that entomological activity proceeded rapidly beneath an outer pliable but drier layer. On cessation of the main phase of insect activity, namely the maggot mass in the body core, the outer layer collapsed and hardened post-bloat. After 24 months the remains are characterised by a moist interior beneath a hard leathery shell. This is consistent with observations by Galloway (1997). The presence of pupal cases in the tomb that were not found within the two earth burials is consistent with the pattern of bodies exhumed from coffins, where, provided there has been access to gravid females in the time-interval between death and interment, in some circumstances the maggots can go through a full life cycle within the coffin (Reeve and Adams 1993). All carcasses were handled in the same manner and so had equal availability to oviposition prior to interment. What is significant is that we get a different pattern of insect activity based on depositional circumstances.

Environmental Monitoring

The air in the tomb reached a higher temperature than the soil fill of Graves 1 and 3 and elevated temperatures occurred more rapidly (Figure 22.4). As the body decomposed and released moisture, the relative humidity within the tomb steadily rose until day 8 (Figure 22.5). The humidity then held a plateau until day 84 suggesting that an equilibrium had been established in the immediate post-burial phase. The humidity then rose steadily again between days 84 and 110 suggestive of the pig being subject to the active phase of decomposition. On day 110 the sensor failed in a recognised fashion characteristic of the sensor being wetted and our interpretation is that this would have likely been the result of a purging episode. The temperature data shows distinct increased step changes at 120, 146 and 153 days.

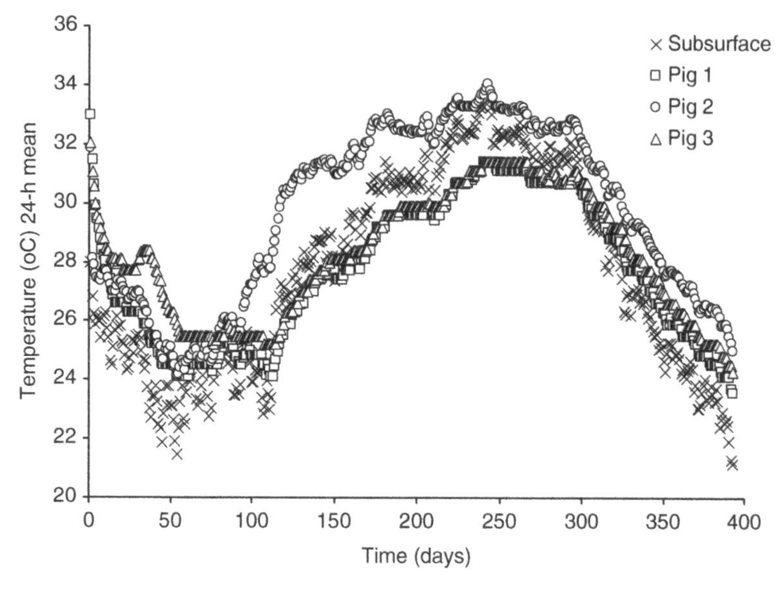

Fig. 22.4 Mean 24-h temperature data for all three pig burials for the first 400 days representing core body temperature alongside temperature immediately sub-surface

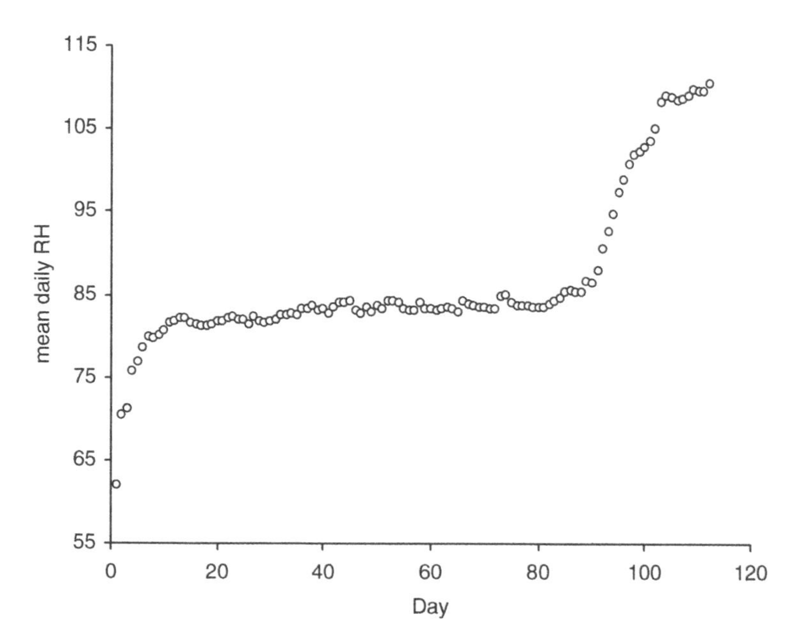

Fig. 22.5 Mean 24-h relative humidity in the tomb showing distinct step-wise changes up until failure of the sensor on day 113

Conclusions

While this experiment was limited in scope, in that it had minimal replication and was only run for two years at a single location, it does provide useful data applicable to the search and recovery of bodies from arid regions. Bodies buried directly in soil may remain a moisture source, via body decomposition products, for years after burial. These features provide a positive target for the use of cadaver dogs and some geophysical techniques. Although slumping was obscured by loose sand and not visible on the surface, the use of predator net (to deter scavenging) evidenced marked slumping within the grave fill, caused by collapse of the thoracic and abdominal cavities. In other regions of the world, in nutrient/water depleted desert soils, the location of bodies may be visible due to differential plant colonisation. This was not the case at the experimental site over the two year burial period.

In all three pig burials, the bodies had been subject to desiccation of the outer tissues and collapse of the body cavity. There was differential desiccation between the upper (dryer), and lower (more moist) body surfaces. Dissection of the bodies revealed moist tissue in the body core, and the extremities were more skeletalised (e.g. the limbs, and region associated with gunshot trauma). The specific nature of the salt-laden soil documented the plume of body fluids originating from the head trauma which had wicked both horizontally and vertically through the grave fill. The spatial relationship between the uppermost part of the plume and the geographic centre of the body/grave are at some distance. This phenomenon mirrors a trend found in cadaver dog searches, where the dog indicates some distance (horizontally) from the buried body. Bodies that are buried in structures with an air void may have markedly different patterns of decomposition compared to bodies buried directly in soil. In the tomb burial from this experiment, there was a marked difference in the level and nature of insect attack to the body, as evidenced by characteristic holes in the desiccated skin that were absent from the bodies buried directly in soil. The body in the stone cist exhibited a considerable bloating event that is witnessed by both the deposits on the stone lining of the grave and can also be correlated with the environmental data from within the tomb.

Our observations also parallel the variable condition of archaeological bodies excavated from different types of burial at archaeological cemetery sites within the lower Osmore valley, Peru (the subject of further work), and also mirror documented patterns of soft tissue preservation found in conflict cemeteries in locations such as Iraq.

Acknowledgements The authors would like to thank participants from the Peruvian Forensic Team (Marta and Roberto) and staff at Centro Mallqui, Peru. We wish to acknowledge the Wellcome Trust Bioarchaeology Programme for funding this work through a Fellowship to ASW (Grant 024661).

References

Aturaliya S and Lukasewycz A (1999). Experimental forensic and bioanthropological aspects of soft tissue taphonomy: 1. Factors influencing postmortem tissue desiccation rate. Journal of Forensic Sciences 44:893–896.

Aufderheide AC (2003). The Scientific Study of Mummies. Cambridge University Press, Cambridge.
Bray EJ (1996). The Use of Geophysics for the Detection of Clandestine Burials: Some Research and Experimentation. MSc Dissertation, University of Bradford, Bradford.
Buckley SA and Evershed RP (2001). Organic chemistry of embalming agents in Pharaonic and Graeco-Roman mummies. Nature 413:837–841.
Dadour IR, Cook DF, Fissioli JN and Bailey WJ (2001). Forensic entomology: application, education and research in Western Australia. Forensic Science International 120:48–52.
Dent BB, Forbes SL and Stuart BH (2004). Review of human decomposition processes in soil. Environmental Geology 45:576–585.
Dix J and Graham M (2000). Time of Death, Decomposition and Identification: An Atlas. CRC, Boca Raton, FL.
Dupras TL, Schultz JJ, Wheeler SM and Williams LJ (2006). Forensic Recovery of Human Remains: Archaeological Approaches. CRC, Boca Raton, FL.
Dzierzykray-Rogalski T (1986). Natural mummification. In: Science in Egyptology. (Ed. AR David), pp. 101–112. Manchester University Press, Manchester.
Forbes SL, Dent BB and Stuart BH (2005a). The effect of soil type on adipocere formation. Forensic Science International 154:35–43.
Forbes SL, Stuart BH and Dent BB (2005b). The effect of the burial environment on adipocere formation. Forensic Science International 154:24–34.
France DL, Griffin TJ, Swanburg JG, Lindemann JW, Davenport GC, Trammell V, Armbrust CT, Kondratieff B, Nelson A, Castellano K and Hopkins D (1992). A multidisciplinary approach to the detection of clandestine graves. Journal of Forensic Sciences 37:1445–1458.
Galloway A (1997). The process of decomposition: a model from the Arizona-Sonoran desert. In: Forensic Taphonomy: The Postmortem Fate of Human Remains (Eds. WD Haglund and MH Sorg), pp. 139–150. CRC, Boca Raton, FL.
Galloway A, Birkby WH, Jones AM, Henry TE and Parks BO (1989). Decay rates of human remains in an arid environment. Journal of Forensic Sciences 34:607–616.
Garland AN and Janaway RC (1989). The taphonomy of inhumation burials. In: Burial Archaeology: Current Research, Methods and Developments (Eds. C Roberts, F Lee and J Bintliff), pp. 15–37. B.A.R. British Series 211, Oxford.
Goff ML and Odom CB (1987). Forensic entomology in the Hawaiian Islands: three case studies. American Journal of Forensic Medicine and Pathology 8:45–50.
Goff ML, Omori AI and Gunatilake K (1988). Estimation of postmortem interval by arthropod succession: Three case studies from the Hawaiian Islands. American Journal of Forensic Medicine and Pathology 9:220–225.
Guillen SE (2004). Artificial mummies from the Andes. Collegium Antropologicum 28:141–157.
Hanson RB and Hanson J (1999). Sonoran desert natural events calendar. In: A Natural History of the Sonoran Desert Tucson (Eds. SJ Phillips and P Wentworth), pp.19–28. Arizona-Sonora Desert Museum Press, Tuscon, AZ.
Hopkins DW, Wiltshire PEJ and Turner BD (2000). Microbial characteristics of soils from graves: an investigation at the interface of soil microbiology and forensic science. Applied Soil Ecology 14:283–288.
Hudson HE (1980). Fungal Saprophytism. Edward Arnold, London.
Hunter J and Cox M, Eds. (2005). Forensic Archaeology: Advances in Theory and Practice. Routledge, Abingdon.
Janaway RC (1996). The decay of buried human remains and their associated materials. In: Studies in Crime: An Introduction to Forensic Archaeology (Eds. J Hunter, C Roberts and A Martin), pp. 58–85. Batsford, London.
Knight B and Simpson K (1997). Simpson's Forensic Medicine. Arnold, London.
Lewin PK (1967). Palaeo-electron microscopy of mummified tissue. Nature 213:416–417.
Lopes de Carvalho LM and Linhares AX (2001). Seasonality of insect succession and pig carcass decomposition in a natural forest area in southeastern. Brazilian Journal of Forensic Sciences 46:604–608.
Lynam JT (1970). Techniques of Geophysical Prospection as Applied to Near Surface Structure Determination. PhD Thesis, University of Bradford, Bradford.

Mant AK (1987). Knowledge acquired from post-war exhumations. In: Death, Decay and Reconstruction: Approaches to Archaeology and Forensic Science (Eds. A Boddington, AN Garland and RC Janaway), pp. 65–78. Manchester University Press, Manchester.

Payne JA (1965). A summer carrion study of the baby pig *Sus scrofa* Linnaeus. Ecology 46:592–602.

Pounder DJ (2000). Postmortem interval. In: Encyclopedia of Forensic Sciences (Eds. JA Siegel, PJ Saukko and GC Knupfer), pp. 1167–1172. Academic, San Diego, CA.

Reeve J and Adams M (1993). The Spitalfields Project: Across the Styx, volume1 – the Archaeology. Council for British Archaeology, York.

Rodriguez WC and Bass WM (1985). Decomposition of buried bodies and methods that may aid in their location. Journal of Forensic Sciences 30:836–852.

Saukko P and Knight B (2004). Knight's Forensic Pathology. Hodder Arnold, London.

Schoenly K, Griest K and Rhine S (1991). An experimental field protocol for investigating the postmortem interval using multidisciplinary indicators. Journal of Forensic Sciences 36:1395–1415.

Sorg MH, David E and Rebmann AJ (1998). Cadaver dogs, taphonomy and postmortem interval in the Northeast. In: Forensic Osteology: Advances in the Identification of Human Remains (Ed. KJ Reichs), pp. 120–143. Charles C. Thomas, Springfield, IL.

Stover E, Haglund WD and Samuels M (2003). Exhumation of mass graves in Iraq – Considerations for forensic investigations, humanitarian needs and the demands of justice. Jama-Journal of the American Medical Association 290:663–666.

Threader D (1997). An Assessment of Resistivity Surveying in the Investigation of Archaeological Burials. MSc Dissertation, University of Bradford, Bradford.

Wilson AS, Janaway RC, Holland AD, Dodson HI, Baran E, Pollard AM and Tobin DJ (2007). Modelling the buried human body environment in upland climes using three contrasting field sites. Forensic Science International 169:6–18.

Chapter 23
Decomposition Studies Using Animal Models in Contrasting Environments: Evidence from Temporal Changes in Soil Chemistry and Microbial Activity

Kathryn L. Stokes, Shari L. Forbes, Laura A. Benninger, David O. Carter and Mark Tibbett

Abstract Traditionally, soil evidence in forensic science has focused predominantly on the transference of soil particles from a victim or suspect and a crime scene. However, a recent increase in forensic taphonomy research has highlighted the potential of soil to provide key information to an investigation involving decomposed remains. A decomposing carcass can release a significant pulse of nutrients into the surrounding soil (gravesoil) resulting in the retention of decomposition products in the soil for a considerable period of time. In order to understand the complex associations between a decomposing carcass and the soil system, research must be conducted in both controlled laboratory environments and outdoor field environments. This chapter discusses two contrasting decomposition studies which aimed to investigate the cadaver/soil interaction. The first study investigated the decomposition of small mouse carcasses buried in soil and was conducted within a controlled laboratory environment in Western Australia. The second study investigated the decomposition of large pig carcasses placed on the soil surface and was conducted in an outdoor field environment in southern Ontario. Both studies investigated a range of decomposition products particularly focusing on carbon-based, nitrogen-based and phosphorus-based compounds as these were considered to offer the most valuable information to address the research questions. The results of both studies

---

K.L. Stokes
Centre for Land Rehabilitation, School of Earth and Geographical Sciences, Australia, Centre for Forensic Science, University of Western Australia, 35 Stirling Highway, Crawley 6009, Australia

S.L. Forbes(✉) and L.A. Benninger
Faculty of Science, University of Ontario Institute of Technology, 2000 Simcoe St N, Oshawa, ON, L1H 7K4 Canada
e-mail: shari.forbes@uoit.ca

D.O. Carter
Department of Entomology, College of Agricultural Sciences and Natural resources, University of Nebraska – Lincoln, 202 Plant Industry Building, Lincoln, NE, USA

M. Tibbett
Centre for Land Rehabilitation, School of Earth and Geographical Sciences, University of Western Australia, 35 Stirling Highway, Crawley 6009, Australia

K. Ritz et al. (eds.), *Criminal and Environmental Soil Forensics*

provide the opportunity to comment on the effect of carcass size, soil type and decomposition environment on the influx of decomposition products into the soil.

Introduction

The use of soils in forensic science traditionally focuses on the comparison of soil particles recovered from footwear, vehicles or dwellings and from a crime scene (Bull et al. 2006; Rawlins et al. 2006). However, recent studies have highlighted the influence of soils on a decomposing cadaver and their importance in a forensic taphonomy context (Tibbett and Carter 2008). The decomposition of a cadaver results in the release of the chemical components of the body through autolysis and putrefaction (Dent et al. 2004). Cadavers that are not readily consumed by vertebrate scavengers are subject to microbial and invertebrate decomposition (Putman 1978a). During decomposition, materials from a cadaver will enter associated soil (gravesoil) providing a localised pulse of nutrients which results in the formation of a concentrated island of fertility, also known as a cadaver decomposition island (CDI) (Carter et al. 2007). This island is associated with increased soil microbial biomass and microbial activity. In particular, the degradation of proteins, lipids and carbohydrates will yield carbon-based, nitrogen-based and phosphorus-based products which may be retained in the surrounding soil.

Although the decomposition of cadavers has been a neglected area of research in forensic science, ecological studies have long been investigating the effects of cadaver decomposition in terrestrial ecosystems (Towne 2000; Brathen et al. 2002; Melis et al. 2007). The impact of large herbivore carcasses on soil and vegetation changes has been shown to represent a potentially significant source of nutrient enrichment. Studies conducted in prairie (Towne 2000) and tundra (Brathen et al. 2002) environments have demonstrated that the effect of large herbivore carcasses on the surrounding soil and vegetation can be dramatic and still detectable after several years. A significant increase in inorganic N concentrations was detected in both soil and vegetation surrounding the carcasses. Fertile areas around the carcass favoured different components of the vegetation, stimulated biomass production and increased species richness and spatial heterogeneity (Towne 2000). Studies in a temperate forest (Melis et al. 2007) also demonstrated significant increases in nutrient concentrations. Calcium concentrations and pH were found to be higher directly underneath the carcass with a gradient decrease towards the periphery of the decomposition site. This effect was detectable for up to seven years after the death of the animal. Concentrations of nitrate (NO_3^-) in the soil also differed suggesting a fast turnover of nitrate in the forest ecosystem.

While the influx of nutrients into soil as a result of large herbivore decomposition is not particularly surprising, it has been shown that small carcasses can also infuse the underlying soil with nutrients that are qualitatively and quantitatively detectable (Putman 1978a, b; Carter and Tibbett 2006). It has been reported that 20 to 30 mg of organic matter can be released from rodent carcasses (weighing approximately 18 to 25 g) by metabolic processes alone and a further 4 mg of organic matter can leach into the soil and be utilised in the respiration of soil organisms during winter and spring months (Putman 1978a). Furthermore, this effect is significantly increased during the autumn and summer months when decomposition is more extensive and only 30% of the initial carcass remains after the degradation of soft tissue and loss of organic matter (in contrast to 85% for the winter months).

Controlled laboratory studies have demonstrated that skeletal muscle tissue (*Ovis aries*) can be used as a source of nutrients by the soil microbial biomass in a sandy loam soil (Carter and Tibbett 2006). In their study, Carter and Tibbett (2006) measured several parameters of decomposition, including skeletal muscle tissue loss, CO_2 evolution, microbial biomass, soil pH and total C and N content. Both temperature and skeletal muscle tissue were found to have a significant effect on all of the measured process rates.

Similar effects have been observed in shallow graves containing pig (*Sus domestica*) carcasses (Hopkins et al. 2000). Grave soils have been found to have higher levels of total C, microbial biomass C and total N, 430 days after burial. Increased rates of respiration and N mineralisation have also been detected when compared to control soils. Amino acid and ammonium (NH_4^+) concentrations were found to be consistent with the increases in both N mineralisation and pH. An increased concentration of S^{2-} indicated that substantial reduction of sulphur had occurred and reducing conditions were present in the grave sites.

Review of these ecological and taphonomic studies has highlighted the potential for the techniques used to be applied to forensic taphonomy research. It is apparent that decomposing cadavers produce a significant influx of decomposition products into the soil, regardless of the size of the carcass and whether the studies are conducted in a controlled laboratory environment or a field context. As a result, taphonomic research is now focusing on understanding the influence of the soil characteristics on cadaver decomposition and conversely the effect of the cadaver on the soil system. In order to do so, we must first have a better understanding of the fundamental processes of decomposition.

The aim of this chapter is to present the results of two contrasting decomposition studies conducted in different soil environments. It is not an attempt to compare the studies, as this would not be feasible, but rather to highlight important biomarkers that result from decomposition and have the potential to be used in forensic investigations. The first study investigates the decomposition of small mouse carcasses buried in controlled soil microcosms and carried out within the laboratory. The second study investigates the decomposition of larger pig carcasses on the soil surface and carried out in an uncontrolled outdoor environment. Both studies investigated a range of decomposition products particularly focusing on carbon-based, nitrogen-based and phosphorus-based compounds, as these were considered to offer the most valuable information. The results of both studies provide the opportunity to comment on the effect of carcass size, soil type and decomposition environment on the influx of decomposition products into the soil.

Laboratory Studies Conducted in Western Australia

Experimental Design

The experimental protocol was based on an established method (Tibbett et al. 2004). Prior to the decomposition events, the soil was sieved (2 mm) to remove any invertebrates, large soil clods and plant material. The water holding capacity for each soil was adjusted with sterilised deionised water to 50%, typically within the optimal range to facilitate microbial activity. Three experimental treatments were applied to each soil; fresh soil containing a cadaver, gamma irradiated (50 kGy) soil containing a cadaver and fresh soil with no cadaver (control). Juvenile (aged 1 to 2 days, 2 to 4 g) mouse cadavers (*Mus musculus*) were buried at a depth of 1.5 cm in soil microcosms and allowed to decompose in a constant temperature room set at 25 °C for the duration of the experiment. Juvenile cadavers were chosen due to their cartilaginous skeletal structure, which was favoured for later histological analysis. Control microcosms contained no cadaver but the soil was disturbed to account for any effect this may have on the soil microbial activity and chemistry. Destructive sequential harvests were conducted at 4, 8, 12, 16, 20 and 24 days after death.

The experiment was replicated six times for each treatment and at each harvest, giving a total of 216 soil microcosms. At each destructive harvest the cadaver was removed and soil surrounding the cadaver (detritosphere) was sampled (~40 g) for chemical analyses. The disturbed microcosms were then discarded. The microcosms were opened (~5 min, with fan) as required (indicated by CO_2 production) during the incubation to facilitate the exchange of oxygen with waste gases from the head space and help maintain an aerobic environment. The microcosms for the last harvest (day 24) contained sodium hydroxide (20 mL, 0.3 M) traps to monitor carbon dioxide (CO_2) respiration for the duration of the incubation (Rowell 1994). Phenolphthalein indicator was added and the NaOH was back titrated with hydrochloric acid (0.1 M) to ascertain CO_2 produced. Statistical analysis (ANOVA) was conducted using SAS 9.1; if data was not normally distributed it was transformed (log) before ANOVA analysis. Correlations (R^2) between ammonium and pH were calculated using MSExcel.

Soil Type

Three soils of contrasting texture were used in the laboratory-based decomposition trials in Western Australia. These comprised: sand, sandy clay loam and loamy sand. Each soil had contrasting chemical and physical properties (Table 23.1).

Table 23.1 Physical and chemical properties of three soil types used in these studies, located in Perth, Western Australia

	Soil		
Property	Sand	Sandy clay loam	Loamy sand
pH	5.5	5.5 to 6.0	8.5 to 9.0
EC (μS)	15 to 40	60 to 100	55 to 95
Ammonium (mg kg^{-1})	10.4	10.6	7.3
Nitrate (mg kg^{-1})	7.3	23.2	7.5
Phosphate (mg kg^{-1})	4.9	6.2	3.3
Potassium (mg kg^{-1})	5.8	86.2	5.9
Physical properties (air dry)	Grey (5Y 5/1), high organic content	Dark yellowish brown (10YR 3/6), gravelly	Reddish yellow (7.5YR 6/8), fine texture

Soil pH

One of the chemical changes which has been regularly noted within soil is a localised increase in the pH surrounding a site of decomposition (Vass et al. 1992; Hopkins et al. 2000; Carter and Tibbett 2006; Wilson et al. 2007). While it is rare for comparisons to be made between pH changes from different soil types and systems, some variations have been observed in the magnitude of pH increase when compared between different soil types with varying initial pH values (Wilson et al. 2007).

The testing of the pH of the soils evolved from the laboratory experiment using a 1:5 (soil:water) ratio (Rayment and Higginson 1992). The probe (Cyberscan20) was washed between the measurements of each sample. The soil types used in this study had differing initial pH levels, ranging from acidic (5.5 to 6.0) to alkaline (8.5 to 9.0). The detritosphere soil from each decomposition event was analysed and significant ($P < 0.001$) increases in pH were observed in soils with acidic basal pH values (see Figure 23.1). The pH increased to 7.5 to 8.0 in both the sand and sandy clay loam soils. However, no significant increase in pH was observed for the sandy loam soil which had an alkaline basal pH (8.5 to 9.0). In all soil types there was an observed decrease in pH towards the end of the study, which may correspond to decreased levels of ammonium as well as the production of organic acids due to the degradation of macromolecules in the carcass. The sandy clay loam demonstrated a rapid decline in alkalinity compared to both the sand and loamy sand systems which showed a more gradual pH decline at the completion of the study.

Microbial Activity

Carbon Dioxide Respiration

Soil microbial activity is commonly monitored through the measurement of carbon mineralisation (e.g. Carter and Tibbett 2006). It has been shown that increased microbial activity can begin within 24 h after the inhumation of a cadaver or skeletal muscle tissue (Putman 1978a; Tibbett et al. 2004; Carter and Tibbett 2006). Over the course of the current laboratory experiment, carbon mineralisation was measured through the use of sodium hydroxide traps (Rowell 1994). The introduction of an animal cadaver to the soil microcosms led to a significant ($P < 0.001$) increase in microbial activity (see Figure 23.2). Maximum microbial activity was observed within the first 10 days of burial and similar patterns were observed in each soil type. This result correlated with the findings of other, related, research (Tibbett et al. 2004; Carter and Tibbett 2006). Both the sand and sandy clay loam soils demonstrated a higher peak level of microbial activity in the initial stages of decomposition before returning to a stable level of activity. The loamy sand system had a smaller secondary peak of microbial

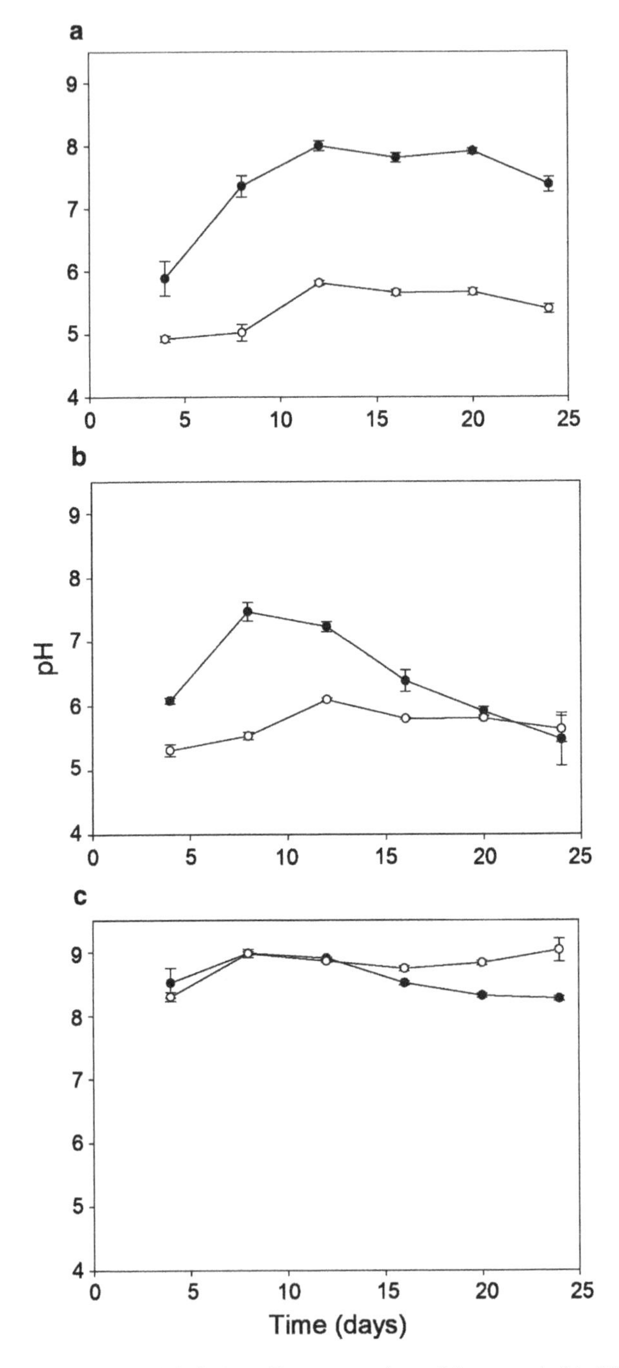

Fig. 23.1 Soil pH after the burial of a juvenile mouse cadaver (*Mus musculus*) in 350 g dry weight (**a**) sand, (**b**) sandy clay loam and (**c**) loamy sand. Cadaver (•) and control soil (o). Bars ± 1 standard error, n = 6

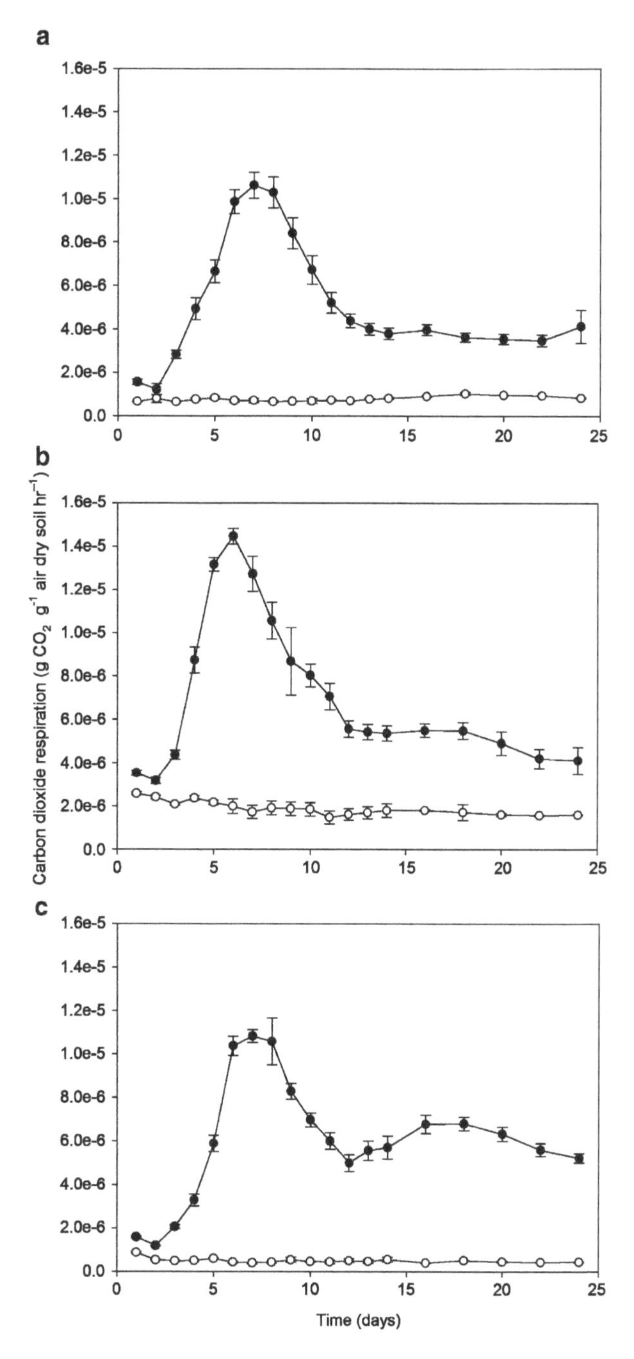

Fig. 23.2 Microbial CO_2 respiration from soil over a period of 24 days after the burial of a juvenile mouse cadaver (*Mus musculus*) in 350 g dry weight **(a)** sand, **(b)** sandy clay loam and **(c)** loamy sand. Cadaver (•) and control (o). Bars ± standard error, n = 6

activity that occurred from day 16 to 18. Taking basal respiration rates into account, the magnitude of microbial activity observed after the cadaver inhumation appears to be similar for sand and loamy sand, with sandy clay loam exhibiting a slightly higher peak of activity. The variation observed in the peak levels of microbial activity is thought to result from the different communities present within each soil. The sandy clay loam soil originates from an area of Western Australia which is used extensively for agriculture. It is considered to be more fertile and to contain a higher clay proportion than the other two soils used in the experiment.

Nitrogen Activity

A decomposing cadaver is a rich source of nitrogen. Concentrations are reported to be as high as 32 g kg^{-1} for an adult cadaver (Tortora and Grabowski 2000) or 19 g kg^{-1} for a neonate (Widdowson 1950). The nitrogen concentration for several animals such as pig, rat and rabbit has been reported as 26, 32 and 29 g kg^{-1} respectively (Spray and Widdowson 1950). These concentrations demonstrate that a significant influx of nitrogen into the surrounding environment can occur during decomposition of an animal or human cadaver (Hopkins et al. 2000).

Ammonium

Ammonium was analysed using a potassium chloride extraction (Rayment and Higginson 1992) and colorimetric analysis (Skalar autoanalyser, Breda, NL). All soils that contained a cadaver showed a significant ($P < 0.001$) increase in the concentration of ammonium extracted from the soil (see Figure 23.3). However, the concentrations of ammonium declined rapidly over time in the sandy clay loam and loamy sand soils. The levels of ammonium in the sandy soil remained relatively constant with time. The sandy loam soil showed lower concentrations of extractable ammonium in comparison to the other soils. At high pH, ammonium is released more readily as volatile gaseous compounds such as ammonia, which was a possible source of an odour detected in the loamy sand soil.

Published literature has suggested a correlation between the increase in pH and ammonium concentration in gravesoil (Hopkins et al. 2000). However, this has never been demonstrated in gravesoil. In the Australian study we note there was a strong correlation between pH and ammonium concentration of the sand soil type ($R^2 = 0.98$). As the initial pH of the soil system increased, the correlation decreased. Sandy clay loam showed a weaker correlation ($R^2 = 0.71$) whereas loamy sand showed no relationship between pH and ammonium concentration. Thus, the correlation of pH and ammonium concentration can only be applied to acidic soils and an alternative model is required for more alkaline systems (see Tibbett and Carter, Chapter 20).

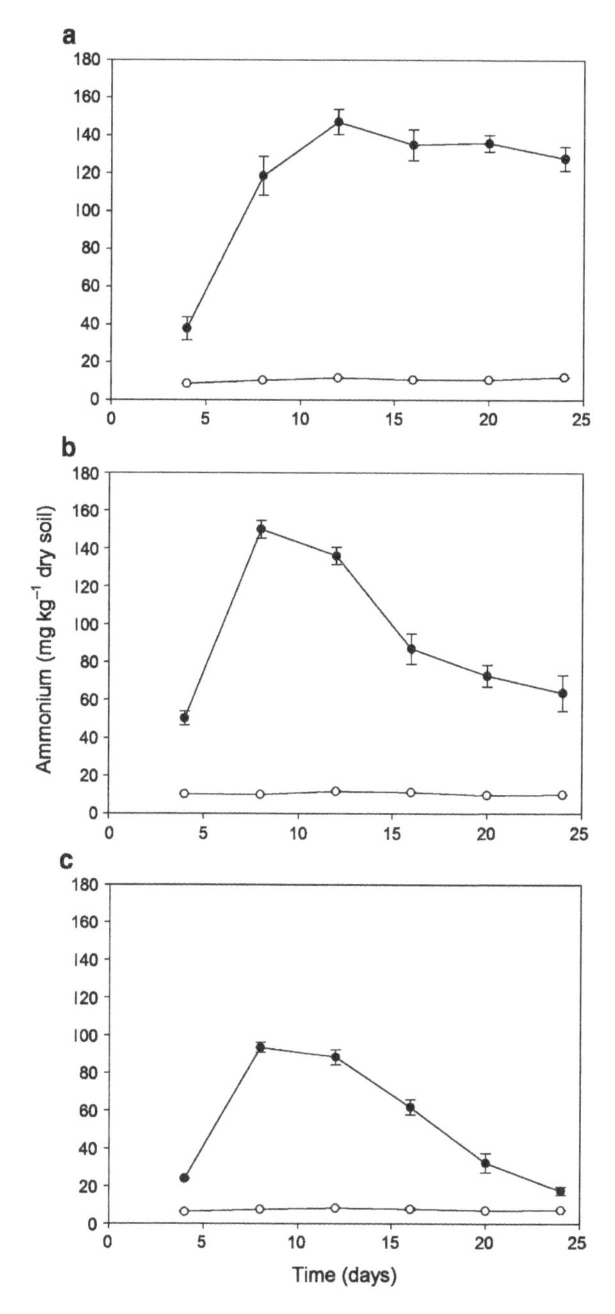

Fig. 23.3 Ammonium concentration (mg kg^{-1} dry soil) over a period of 24 days after the burial of a juvenile mouse cadaver (*Mus musculus*) in 350 g dry weight **(a)** sand, **(b)** sandy clay loam and **(c)** loamy sand. Cadaver (•) and control soil (o). Bars ± standard error, n = 6

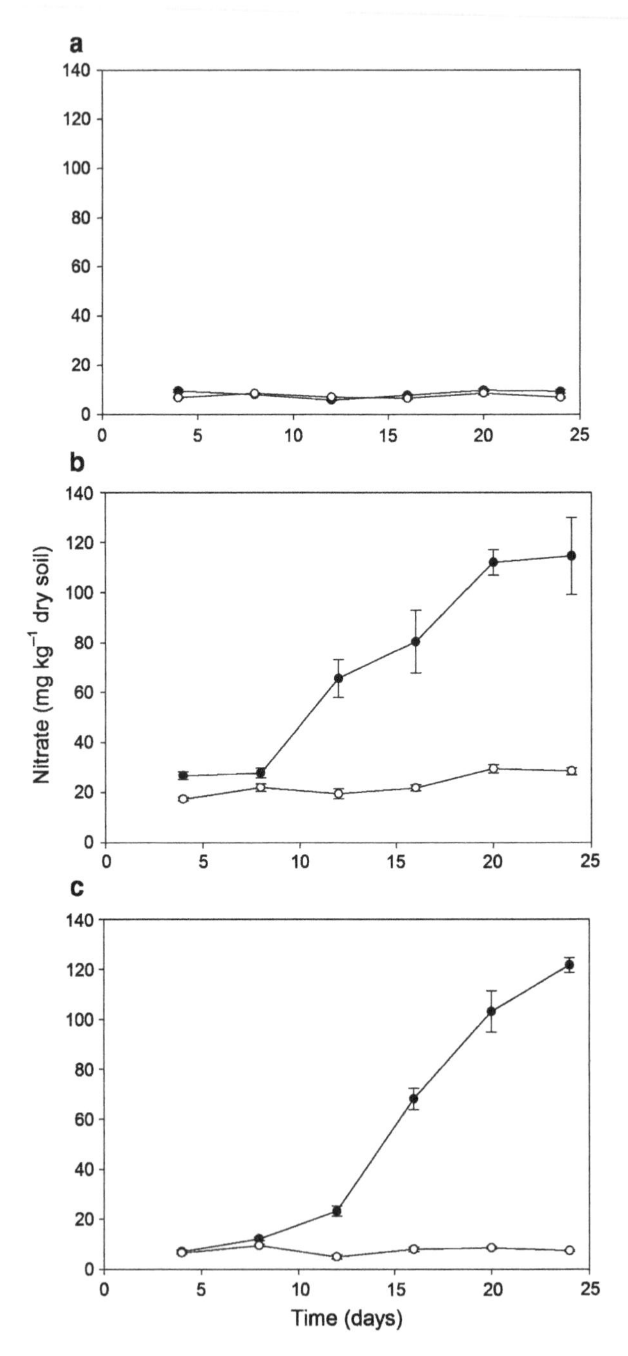

Fig. 23.4 Nitrate concentration (mg kg^{-1} dry soil) over a period of 24 days after the burial of a juvenile mouse cadaver (*Mus musculus*) in 350 g dry weight (**a**) sand, (**b**) sandy clay loam and (**c**) loamy sand. Cadaver (•) and control soil (o). Bars ± standard error, n = 6

Nitrate

Nitrate was measured with potassium chloride extraction (Rayment and Higginson 1992) and colorimetric analysis (Skalar autoanalyser). The soils demonstrated contrasting patterns of nitrification in response to the introduction of an animal cadaver (see Figure 23.4). The sandy soil showed no nitrification for the duration of the experiment. However, in a previous laboratory experiment using small quantities of skeletal muscle tissue decomposing in sand, minimal nitrification was observed (unpublished data). Both sandy clay loam and sandy loam soils showed significant ($P < 0.001$) nitrification over the course of the experiment. Nitrification appeared to start almost immediately in the sandy clay loam, whereas a lag in the loamy sand soil was observed prior to a rapid increase in nitrate concentration.

Nitrate is produced by the microbial oxidation of ammonium by a specialised group, chemoautotrophic archebacteria, and is part of the normal soil nitrogen cycle (Stevenson and Cole 1999). Previous research by one of the authors has also demonstrated highly variable levels of nitrification occurring at a decomposition site, even with the flush of ammonium which is observed (unpublished data). Hopkins et al. (2000) reported no nitrification for a pig carcass burial in heavy clay soil. In contrast, Melis et al. (2007) reported rapid nitrification in temperate forest soils below a decomposing bison carcass in Poland.

Nitrification is thought to be adversely affected by several factors including: anaerobic conditions, slow replication of the nitrifying community in soil systems, surrounding environmental pH (Kyveryga et al. 2004; Nugroho et al. 2007) and high concentrations of free ammonia which inhibit the nitrite conversion step (Balmelle et al. 1992; Philips et al. 2002; Vadivelu et al. 2007). Due to the complexity of a burial environment and the numerous chemical changes which occur in soil as a result of a decomposing carcass, it is difficult to clearly attribute nitrifying variability to a single factor in decomposition experiments.

Phosphorus Activity

Bicarbonate Extractable Phosphate

A cadaver is a recognised source of phosphorus which comprises approximately 1% of the body mass of an adult human (Tortora and Grabowski 2000). Increased concentrations of phosphorus have been reported at decomposition sites of animal carcasses (Towne 2000) as well as in ground water systems located near cemeteries (van Haaren 1951). The phosphorus concentrations in this study were tested using a bicarbonate extraction and colorimetric analysis (Rayment and Higginson 1992) to measure plant available phosphorus (phosphate) in soil. A significant ($P < 0.001$) increase in the levels of bicarbonate extractable phosphate was determined for all three soils (see Figure 23.5).

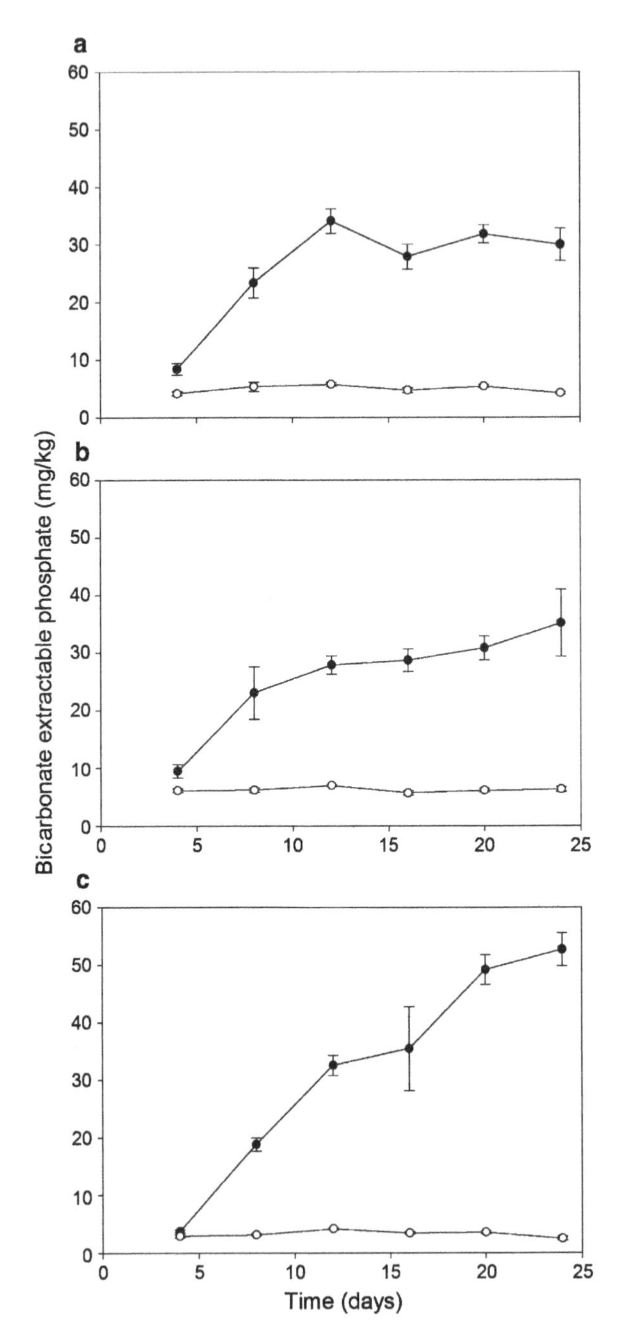

Fig. 23.5 Bicarbonate extractable phosphate concentration (mg kg^{-1} dry soil) over a period of 24 days after the burial of a juvenile mouse cadaver (*Mus musculus*) in 350 g dry weight **(a)** sand, **(b)** sandy clay loam and **(c)** loamy sand. Cadaver (•) and control soil (o). Bars ± standard error, n = 6

The results observed in the laboratory-based experiment correlated with experiments conducted in the field (Towne 2000). However, the concentrations varied between locations. It is well documented that sand systems experience rapid leaching of nutrients, particularly after fertilisation in agricultural fields (Meissner et al. 1995; Ulen 1999). This phenomenon would partially explain the lower levels of phosphate observed in both the sand and loamy sand systems. Phosphorus is also known to adsorb to mineral particles (Freeman and Rowell 1981) which may explain the increase in phosphate levels in the sandy clay loam system. It is possible that the phosphorus adsorbed to the mineral component of the system and, as a result, did not leach from the immediate decomposition zone as rapidly as the sand-based systems.

Field Studies Conducted in Southern Ontario

Experimental Design

The southern Ontario field study utilised adult pig (*Sus scrofa*) carcasses, each weighing approximately 55 lb and reared on the same diet (Schoenly et al. 2006). All animals were reared for the pork industry and were certified free of disease. Five carcasses were euthanised by a veterinarian with an overdose of anesthetic at a local pig farm prior to experimental treatment. Each carcass was enclosed in plastic and transported to the experimental site within 1 h of death. All carcasses were wrapped in mesh wire and placed on the soil surface to decompose for a period of 100 days. The study was conducted during the summer and autumn months of 2006 and the weather was typical of a southern Ontario climate (see Figure 23.6). The mean temperature for the study period was 17 °C and was 23 °C for the period when decomposition was most active (days 1 to 14). Rainfall during this time was minimal.

Reference soil samples were collected from the decomposition sites prior to the carcasses being placed on the surface. Soil samples (~50 g) were collected from directly underneath the torso of the carcasses (Carter et al. 2007) and control samples were collected approximately 2 m from each carcass. The samples were collected from a depth of 0 to 5 cm below the carcasses. Gravesoil and control soil samples were collected weekly for the first 6 weeks and then monthly for the duration of the decomposition trial.

Experimental Site

The experiment was conducted in a temperate woodland in Oshawa, southern Ontario, Canada. The study area was located approximately 51 km northeast of

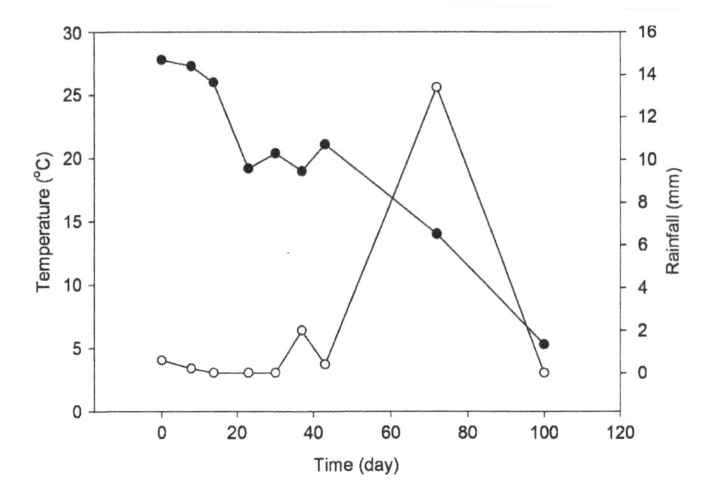

Fig. 23.6 Average temperature (•) and rainfall (o) per sampling day in Oshawa, southern Ontario, during the summer/autumn months of 2006

Toronto (43°56′N, 78°54′W) across a slight (5°) gradient, with some drainage/ depression areas. Dominant vegetation at the site included eastern white cedar (*Thuja occidentalis*), maples (*Acer* spp.) and trembling aspen (*Populus tremuloides*). The substrate at the site was a grey-brown podsolic soil of the Darlington Series (Hoffman et al. 1964), which had a loam texture with high organic content and pH 7.3, with EC 1.33 cmol, total C of 5.3%, total N of 0.3%, and total P of 0.0012 μg g^{-1}.

Soil pH

The loam soil type used in this study had an initial pH of 7.3. The pH of the control soil and gravesoil was monitored for the entire decomposition trial. Although small fluctuations in the pH of the gravesoil were observed, there was no significant difference in soil pH between the control or gravesoil samples (see Figure 23.7). This result is in common with the alkaline soil used in the Western Australia study.

Previous decomposition studies have provided conflicting results on the variation of soil pH over time (Rodriguez and Bass 1985; Vass et al. 1992; Towne 2000). A long-held belief is that cadaver decomposition results in increased pH values because of the accumulation of NH_4^+ in the soil (Hopkins et al. 2000; Carter et al. 2007; Melis et al. 2007). The studies in Western Australia and southern Ontario suggest that this is only true for soils starting with a pH value below neutral (see Tibbett and Carter, Chapter 20).

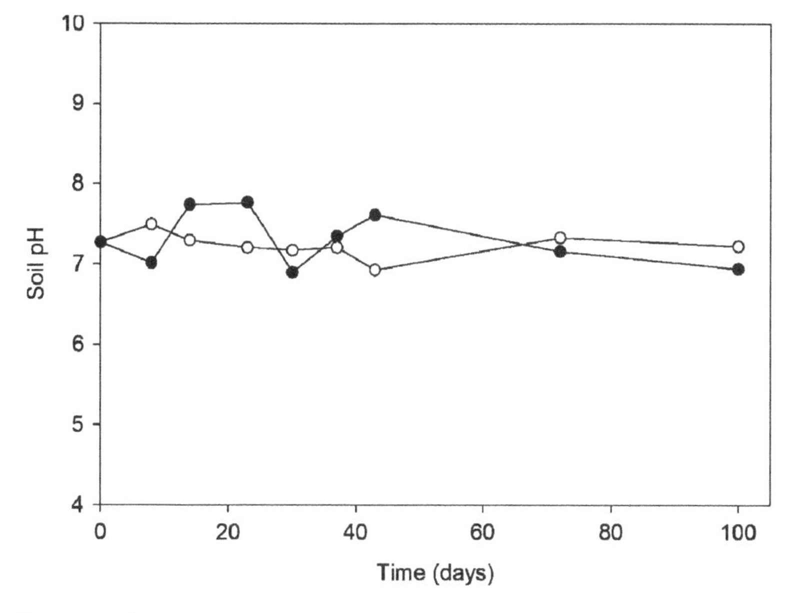

Fig. 23.7 pH variations observed during the decomposition of an adult pig cadaver (*Sus domesticus*) in loam soil. Gravesoil (•) and control soil (o), n = 5

Carbon Activity

Total Carbon

Naturally-occurring carbon forms are derived from the decomposition of both plants and animals. A wide variety of carbon forms are present in soils including leaf and branch litter, as well as highly decomposed forms such as humus (Schumacher 2002). Total carbon content in the soils was determined by combustion in a pure oxygen environment and analysis by an elemental analyser (Model LECO CNS-1000). It was theorised prior to the decomposition trials that total carbon would increase with decomposition. The results demonstrated a fluctuation in both the control and the carbon content of gravesoil; however, no significant difference between the two sets of samples was determined (Figure 23.8a).

Previous reports have indicated an increase in total C values in gravesoils sampled beneath pig carcasses (Hopkins et al. 2000). It is assumed in this study that most of the carbon released from the carcass was lost to the atmospheric environment as volatile gases and mainly as carbon dioxide. The pattern of carbon dioxide release in summer and autumnal months has been directly attributed to the activity of fly larvae on a carcass (Putman 1978a, b). It follows that high levels of carbon dioxide and other carbon-containing gases would have been released from the carcasses as a result of the insect activity and rapid tissue loss observed in this study. Although the Western Australia study did not involve entomological activity,

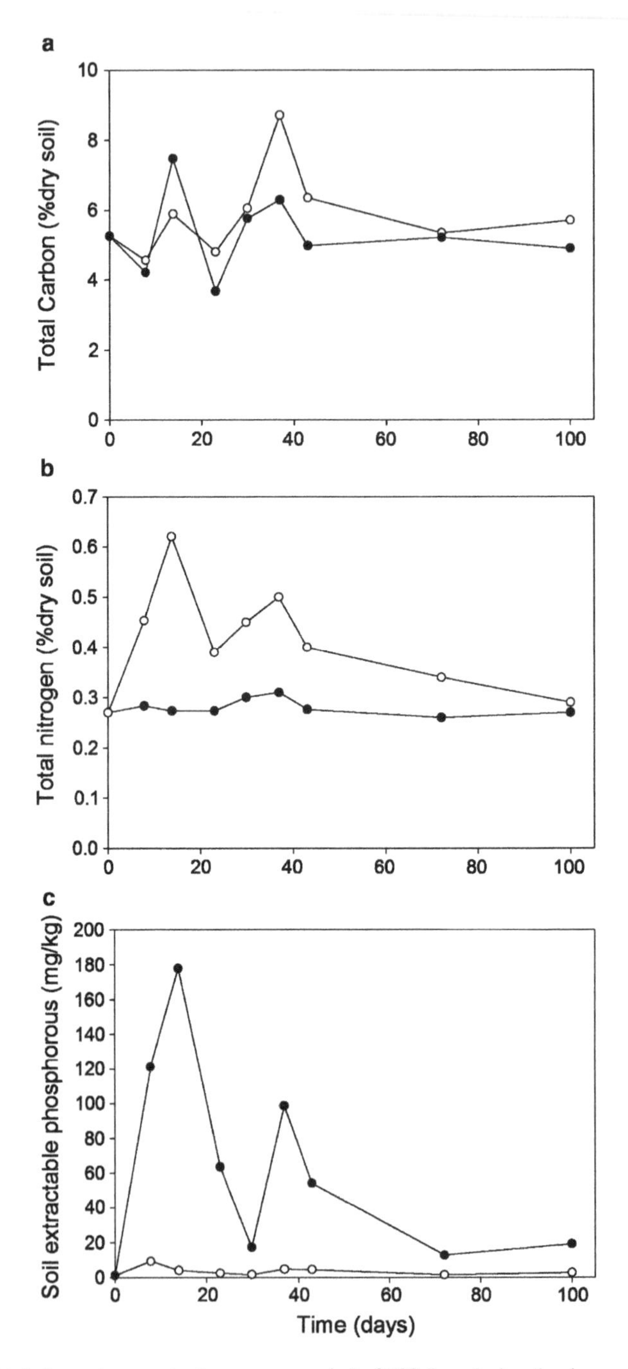

Fig. 23.8 Total element concentrations over a period of 100 days during the decomposition of an adult pig cadaver (*Sus domesticus*) in loam soil. **(a)** Carbon; **(b)** nitrogen (%); **(c)** phosphorus (mg kg^{-1} dry soil). Gravesoil (•) and control soil (○), n = 5

an increase in carbon dioxide was also measured as a result of the inhumation of a cadaver into the soil system.

Nitrogen Activity

Total Nitrogen

Total nitrogen content is the sum of the organic nitrogen and inorganic nitrogen (NO_3^- and NH_4^+) in soil. Total nitrogen was measured using the dry combustion technique and elemental analyser employed for total organic carbon measurements. There was a significant ($P < 0.05$) increase in the total nitrogen content of the gravesoils when compared to the control samples in the first 14 days of the decomposition trial. A second, smaller peak was also observed between days 21 and 42 of the decomposition period. However, as the decomposition period extended, the total nitrogen levels returned to basal levels (Figure 23.8b).

The significant ($P < 0.05$) increase in total N concentration is consistent with previous reports for larger carcasses (Hopkins et al. 2000; Towne 2000; Melis et al. 2007). The increase in N levels could be a result of either organic nitrogen or those inorganic forms, such as NO_3–N and NH_4–N, identified in the Western Australia studies. Further studies are recommended in this area to determine which nitrogen forms are being released during decomposition. It is important to note that the total N concentrations in the control samples were consistent, indicating that the total N in soil does not change significantly with temperature or rainfall. Thus, the large increase in the gravesoil can only be equated to the influx of decomposition fluids into the soil.

Phosphorus Activity

Soil Extractable Phosphorus

Several methods exist for measuring soil extractable phosphorus and include the Bray-P method (Bray and Kurtz 1945), lactate-P and sodium bicarbonate-P ($NaHCO_3$–P) method. Studies have shown that all three methods determine a similar concentration of soil extractable phosphorus, although the Bray-P and $NaHCO_3$–P methods are applicable to a wider range of soils (Zalba and Galantini 2007). This study utilised the Bray-P method. A remarkably large and significant ($P < 0.05$) increase in soil extractable phosphorus was observed over the 100 day cycle and, unlike the other nutrients measured, it did not return to basal levels during the experimental period (Figure 23.8c).

Towne (2000) also detected a significant ($P < 0.01$) increase in P concentrations beneath a decomposing ungulate (*Bos bison*) carcass and confirmed that available

P concentrations remained significantly higher ($P < 0.05$) at the carcass sites when compared to the control sites, up to 3 years post-mortem. In the body, the phosphorus store is found in a number of components such as proteins comprising nucleic acids and coenzymes, sugar phosphates and phospholipids (Dent et al. 2004). In addition to these compounds, animals can acquire phosphorus through the consumption of plants and other animals. Therefore, a decomposing carcass will release a large pulse of phosphorus into the surrounding soil. The residual P signature in the Ontario soils may be due to a combination of two factors related to the relatively short experimental period of 100 days: firstly; incomplete mineralisation of cadaver-derived P from the organic to the inorganic forms that continues to supply the extractable fraction; and, secondly, an incomplete uptake of the enhanced soil mineral P pool by plants and microorganisms.

Lipid-Phosphates

Lipid-phosphate content in soil was determined by the method of Kates (1986). Lipid-P is defined as those phosphorus-containing materials which are components of lipids and consequently represent phospholipids (Carter 1993). The lipid-P content of the control soils remained relatively constant throughout the entire 100 days. There was a statistically significant ($P < 0.05$) increase in the amount of lipid-P present in the gravesoil samples, demonstrating similar peaks to the total

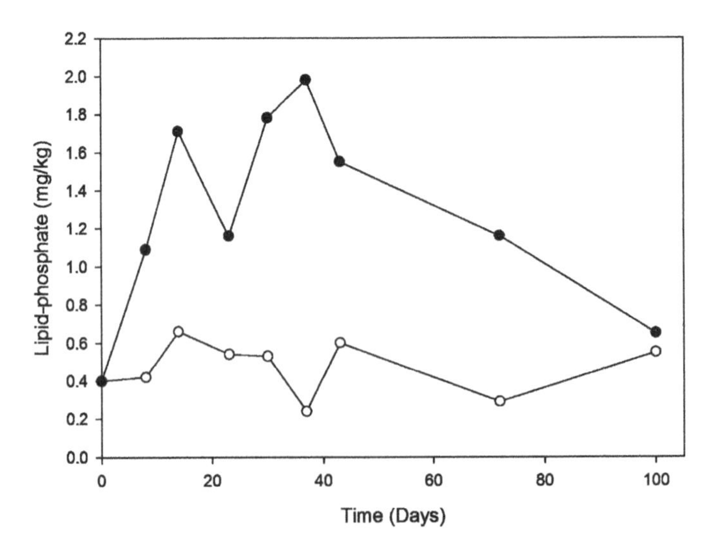

Fig. 23.9 Lipid-phosphate concentration (mg kg^{-1} dry soil) over a period of 100 days during the decomposition of an adult pig cadaver (*Sus scrofa*) in loam soil. Gravesoil (•) and control soil (o), n = 5

nitrogen and soil extractable phosphorus measurements. The change in the lipid-P concentration over the 100-day period followed a slight Gaussian curve, with greater concentrations being in the early decomposition period. As the decomposition period continued, the change in concentration decreased towards basal levels and was no longer statistically significant (Figure 23.9).

The concentrations of lipid-P in the gravesoils fluctuated considerably over the decomposition period but a significant variation was observed when compared to the control soil samples. While lipid-P is generally used to measure microbial biomass in soils and can be correlated to total bacterial fatty acids (Kerek et al. 2002), the measurement of lipid-P in this study cannot discriminate between phosphorus derived from bacterial communities or from the carcass itself. However, the significant increase in lipid-P does suggest that the influx of cadaveric material into the soil contributed towards an increased microbial biomass as a result of the decaying animal tissue.

Conclusions

The aim of this chapter was to provide results from two contrasting decomposition studies in order to highlight the effect of a decomposing cadaver on a range of soil systems. The results show that regardless of the type of animal model or size of the carcass, decomposition by-products are detectable in soil for a period of time after decomposition occurs. However, the concentrations of these by-products will vary accordingly based on the carcass size. Of the numerous soil characteristics studied, nitrogen and phosphorus-based products appeared to offer the greatest potential as forensic tools. Significant increases were observed in both the laboratory and field studies when compared to the control samples and, with further research, may demonstrate a direct correlation with the post-mortem interval.

The studies also demonstrated the integral role of the soil system on the formation and retention of decomposition by-products. Increases in nitrogen products were detected in all systems: loamy sand, loam, sand and sandy clay loam. Increases in soil pH were observed in the sand and sandy clay loam systems, but not in the loam or loamy sand system. These variations in decomposition products are clearly dependent on the original soil characteristics and highlight the need for caution when providing generalised statements about the impact of a decomposing carcass on a soil system.

Additionally, the decomposition environment must be considered when investigating variations in soil parameters beneath a decomposing carcass. It is well established that cadavers decompose at different rates aboveground compared to belowground. In the laboratory-based burial study, the small mouse carcasses were still in active decay on day 24 when the final harvest was conducted. In contrast, the larger pig carcasses were completely skeletonised by day 14 of the surface decomposition trial. The ambient temperature of both studies was comparable for the period of decomposition. Hence, the rate of production and influx of the decomposition products into the soil will vary considerably with environment and may dictate the period during which soil samples will provide useful information to a forensic investigation.

The purpose of the laboratory and field studies was not to investigate a large range of soil parameters but to focus on specific decomposition products which have shown potential in two contrasting environments. The studies discussed here are preliminary in nature and not designed to provide conclusive results. Rather they demonstrate the value of measuring decomposition by-products in soil and their potential application to the field of forensic taphonomy. Further research in this field is clearly necessary and strongly recommended with the intention of better understanding the interaction between a decomposing carcass and a soil system and ultimately providing accurate information regarding the post-mortem period.

References

Balmelle B, Nguyen KM, Capdeville B, Cornier JC and Deguin A (1992). Study of factors controlling nitrite buildup in biological processes for water nitrification. Water Science and Technology 26:1017–1025.

Brathen KA, Danell K and Berteaux D (2002). Effect of muskox carcasses on nitrogen concentration in tundra. Arctic 55:389–394.

Bray RH and Kurtz LT (1945). Determination of total, organic and available forms of phosphorus in soils. Soil Science 59:39–45.

Bull PA, Parker A and Morgan RM (2006). The forensic analysis of soils and sediment taken from the cast of a footprint. Forensic Science International 162:6–12.

Carter DO and Tibbett M (2006). Microbial decomposition of skeletal muscle tissue (*Ovis aries*) in a sandy loam soil at different temperatures. Soil Biology and Biochemistry 38:1139–1145.

Carter DO, Yellowlees D and Tibbett M (2007). Cadaver decomposition in terrestrial ecosystems. Naturwissenschaften 94:12–24.

Carter MR (1993). Soil Sampling and Methods of Analysis. CRC, Boca Raton, FL.

Dent BB, Forbes SL and Stuart BH (2004). Review of human decomposition processes in soil. Environmental Geology 45:576–585.

Freeman JS and Rowell DL (1981). The adsorption and precipitation of phosphate onto calcite. Journal of Soil Science 32:75–84.

Hoffman DW, Mathews BC and Wicklund RA (1964). Soil associations of southern Ontario, Report No. 30 of the Ontario Soil Survey. Canada Department of Agriculture, Ottawa, Ontario Department of Agriculture, Toronto.

Hopkins DW, Wiltshire PEJ and Turner BD (2000). Microbial characteristics of soils from graves: an investigation at the interface of soil microbiology and forensic science. Applied Soil Ecology 14:283–288.

Kates M (1986). Techniques of Lipidology: Isolation, Analysis and Identification of Lipids. Elsevier, St Louis, MO.

Kerek M, Drijber RA, Powers WL, Shearman RC, Gaussoin RE and Streich AM (2002). Accumulation of microbial biomass within particulate organic matter of aging golf greens. Agronomy Journal 94:455–461.

Kyveryga PM, Blackmer AM, Ellsworth JW and Isla R (2004). Soil pH effects on nitrification of fall-applied anhydrous ammonia. Soil Science Society America Journal 68:545–551.

Melis C, Selva N, Teurlings I, Skarpe C, Linnell JDC and Andersen R (2007). Soil and vegetation nutrient response to bison carcasses in Bialowieza Primeval Forest. Poland Journal of Ecological Research 22:807–813.

Meissner R, Rupp H, Seeger J and Sconert P (1995). Influence of mineral fertilizers and different soil types on nutrient leaching: results of lysimeter studies in east Germany. Land Degradation and Rehabilitation 6:163–170.

Nugroho RA, Roling WFM, Laverman A and MVerhoef HA (2007). Low nitrification rates in acid Scots pine forest soils are due to pH-related factors. Microbial Ecology 53:89–97.

Philips S, Laanbroek HJ and VerstraeteW (2002). Origin, causes and effects of increased nitrite concentrations in aquatic environments. Environmental Science and Biotechnology 1:115–141.

Putman RJ (1978a). Flow of energy and organic matter from a carcass during decomposition: decomposition of small mammal carrion in temperate systems 2. Oikos 31:58–68.

Putman RJ (1978b). Patterns of carbon dioxide evolution from decaying carrion: decomposition of small mammal carrion in temperate systems. Oikos 31:47–57.

Rawlins BG, Kemp SJ, Hodgkinson EH, Riding JB, Vane CH, Poulton C and Freeborough K (2006). Potential and pitfalls in establishing the provenance of earth related samples in forensic investigations. Journal of Forensic Sciences 51:832–845.

Rayment GE and Higginson FR (1992). Australian Laboratory Handbook of Soil and Water Chemical Methods. Inkata Press, Melbourne.

Rodriguez WC and Bass WM (1985). Decomposition of buried bodies and methods that may aid in their location. Journal of Forensic Sciences 30:836–852.

Rowell DL (1994). Soil Science Methods and Applications. Addison Wesley Longman, England.

Schoenly KG, Haskell NH, Mills DK, Bieme-Ndi C, Larsen K and Lee Y (2006). Recreating death's acre in the school yard: using pig carcasses as model corpses to teach concepts of forensic entomology and ecological succession. American Biology Teacher 68:402–410.

Schumacher BA (2002). Methods for the determination of total organic carbon (TOC) in soils and sediments. Report NCEA-C-1282 of the Ecological Risk Assessment Support Centre. United States Environmental Protection Agency, Las Vegas, NV.

Spray CM and Widdowson EM (1950). The effect of growth and development on the composition of mammals. British Journal of Nutrition 4:332–353.

Stevenson FJ and Cole MA (1999). Cycles in the Soil (2nd Edition). Wiley, New York.

Tibbett M and Carter DO (Eds.) (2008). Soil Analysis in Forensic Taphonomy: Chemical and Biological Effects of Buried Human Remains. CRC, Boca Raton, FL.

Tibbett M, Carter DO, Haslam T, Major R and Haslam R (2004). A laboratory incubation method for determining the rate of microbial degradation of skeletal muscle tissue in soil. Journal of Forensic Sciences 49:60–565.

Tortora GJ and Grabowski SR (2000). Principles of Anatomy and Physiology (9th Edition). Wiley, New York.

Towne EG (2000). Prairie vegetation and soil nutrient responses to ungulate carcasses. Oecologica 122:232–239.

Ulen B (1999). Leaching and balances of phosphorus and other nutrients in lysimeters after application of organic manures or fertilizers. Soil Use and Management 15:56–61.

Vadivelu VM, Keller J and Yuan Z (2007). Effect of free ammonia on the respiration and growth processes of an enriched Nitrobacter culture. Water Research 4:826–834.

van Haaren FWJ (1951). Churchyards as sources for water pollution. Moormans Periodieke Pers 35:167–172 (in Dutch).

Vass AA, Bass WM, Wolt JD, Foss JE and Ammons JT (1992). Time since death determinations of human cadavers using soil solution. Journal of Forensic Sciences 37:1236–1253.

Widdowson EM (1950). Chemical composition of newly born mammals. Nature 166:626–628.

Wilson AS, Janaway RC, Holland AD, Dodson HI, Baran E, Pollard AM and Tobin DJ (2007). Modelling the buried human body environment in upland climes using three contrasting field sites. Forensic Science International 169:6–18.

Zalba P and Galantini A (2007). Modified soil-test methods for extractable phosphorus in acidic, neutral and alkaline soils. Communication in Soil Science and Plant Analysis 38:1579–1587.

Chapter 24
Microbial Community Analysis of Human Decomposition on Soil

Rachel A. Parkinson, Kerith-Rae Dias, Jacqui Horswell, Paul Greenwood, Natasha Banning, Mark Tibbett and Arpad A. Vass

Abstract Human decomposition is a complex process, involving a multitude of microbial species. Currently, little is known about the microbial species and processes that occur as cadavers decompose, particularly in outdoor environments. With the development of molecular ecology tools that allow the study of complex microbial communities, the 'black box' of the microbiology associated with decomposition is being opened. A preliminary study was performed to evaluate the changes in the soil bacterial and fungal communities that occur in response to human cadaver decomposition, and to assess the potential for using such changes in either of these two microbial communities for forensic time-since-death estimation. Terminal restriction fragment length polymorphism (T-RFLP) and phospholipid fatty acid (PLFA) analyses were performed on soil samples collected from beneath decomposing human cadavers at the University of Tennessee's Forensic Anthropology Centre. The soil fungal and bacterial communities showed significant changes in response to decomposition events, with some succession patterns emerging as decomposition progressed. Evidence of sequential changes occurring in the soil microbial community provides support for the idea that specific microbes could be used as time-since-death biomarkers, although major differences in the way replicate cadavers decomposed in this study suggested that both the bacterial and fungal communities may need to be considered to ensure such a method is forensically useful. Some difficulties were also encountered with the polymerase chain reaction (PCR) step in the T-RFLP method being inhibited

R.A. Parkinson(✉) and J. Horswell.
The Institute of Environmental Science and Research Limited (ESR),
Kenepuru Science Centre, Porirua, New Zealand.
e-mail: rachel.parkinson@esr.cri.nz

K.-R. Dias.
The Centre for Forensic Science, University of Western Australia, Perth, Australia.

P. Greenwood, N. Banning, and M. Tibbett.
The Centre for Land Rehabilitation, University of Western Australia, Perth, Australia.

A.A. Vass.
Oak Ridge National Laboratory, P.O. Box 2008, Oak Ridge, TN 37831, USA.

K. Ritz et al. (eds.), *Criminal and Environmental Soil Forensics*

by soil or decomposition compounds, which suggest that while DNA-based 'profiling' methods may be useful for studying the effects of decomposition on soil, other more specific techniques are likely to be most useful in a forensic post-mortem interval context.

Introduction

The decomposition process of a human cadaver is a complex sequence of events that begins shortly after death. It starts with autolysis, which is initiated by an increase in cellular carbon dioxide causing increased acidity within cells. This pH decrease causes lysosomes to rupture releasing enzymes leading to the destruction of cellular components by enzymatic digestion. The subsequent release of nutrient-rich cellular fluids during cell lysis promotes a process known as putrefaction (Vass et al. 2002). The body's natural microflora, particularly the intestinal and lung flora, begin to break down soft tissues and may eventually migrate throughout the body via the blood and lymph systems (Janaway 1996). Gases are released as by-products of microbial metabolism (Vass et al. 2002) and collect within body cavities, causing bloating. The accumulated gas and fluids purge from natural body openings or wounds and can cause the abdominal cavity to rupture, releasing decomposition compounds and associated microflora into the external environment (Janaway 1996). Active decay follows and is characterised by intense microbial and insect activity that causes liquefaction and disintegration of the remaining soft tissues. Following the period of active decay, the skeleton and any remaining low-nutritive connective tissue begin to dry (mummification), and bone degrades slowly through a process known as diagenesis (Gill-King 1997). The progression of decomposition can therefore be divided into four generalised stages: fresh, bloat, decay, and dry (skeletonisation) (Reed 1958).

Decomposition is heavily dependant on micro-organisms and is thought to involve significant numbers of bacterial and fungal species, including both the body's own microflora and those species found in the surrounding environment (Janaway 1996; Vass 2001). Fungi are one of the primary decomposers in almost all terrestrial environments and are likely to be heavily involved in cadaver decomposition on soil (Bruns et al. 1991). The populations of soil-dwelling organisms in the area immediately surrounding the body are likely to be strongly affected by decomposition, as both their abiotic and biotic environments change during the process. Substantial quantities of organic compounds are released into the local environment, and the mix of compounds varies in composition depending on the stage of decomposition (Vass et al. 2004). These substances can be used by some microbial species as substrates but they may inhibit growth of others, either directly, or indirectly by making the environment unfavourable, for example by altering the pH. This change in substrate release by the body creates the potential for sequential changes in the composition of the local soil microbial community. The body microflora released during decomposition, and species introduced by visiting insects and scavengers, are also likely to impact the soil microbial community. Previous attempts to study the microbes actively involved in human decomposition have encountered difficulties due to the many species involved (Vass 2001) and the limitations of culturing and identification techniques. However, the rapid development of molecular ecology has produced techniques that can circumvent the requirement to isolate and culture micro-organisms, as a prerequisite to identification. Human microflora and environmental communities can now be routinely characterised using simple molecular techniques based on amplification of DNA targets from ribosomal genes (e.g. Heuer et al. 1997; Liu et al. 1997; Fisher and Triplett 1999; Hayashi et al. 2002; Sakamoto et al. 2003).

Bacterial community DNA profiling has subsequently been used by forensic scientists to compare forensic soil samples based on their bacterial communities (Horswell et al. 2002). Ammonia fungi (Sagara 1975) and post-putrefaction fungi (Sagara 1995) may act as above-ground grave markers of a buried and decomposing cadaver and the succession of fruiting behaviour, physiology and nitrogen utilisation can provide the basis for the estimation of post-burial interval (Carter and Tibbett 2003). Fungal growth identified on a human cadaver has also been used to estimate the minimum interval since death, thus displaying the potential of utilising microbes as a forensic tool (Hitosugi et al. 2006; Ishii et al. 2006).

Phospholipid fatty acid (PLFA) analysis, another culture-independent approach, was also used to investigate the biochemical diversity of the microbial communities present (Frostegard et al. 1993; Zelles 1997). PLFAs are essential membrane components of all living cells. Their chemical diversity, cellular abundance and rapid degradation upon cellular death make them good candidates to investigate the living and active soil microbial community (Drenovsky 2004). The PLFA profile of a sample is derived from the whole viable microbial community and each species contributes to the profile in proportion to its biomass (White 1979; Zelles 1997; Leckie 2005). Microbially diagnostic PLFAs provide a unique capacity to directly link microbial identity with specific processes or activity (e.g. cadaver decomposition).

This chapter describes a preliminary investigation into changes associated with human decomposition in soil bacterial and fungal communities to determine whether either of these microbial groups could be forensically useful. Understanding changes in these communities could ultimately enable estimation of the time elapsed since death in a way similar to that used by forensic entomologists. The timing and succession of insect colonisation of a body can be used to estimate the stage of decomposition, based on knowledge of the lifecycles of these species (Haskell et al. 1997). As with insect colonisation of a body, the microbial community associated with corpses is likely to be dynamic and heterogeneous, and may also display characteristics specific to decompositional stages, or have microbial species that could be used as biomarkers indicative of a certain stage. Forensic entomology is widely used to estimate the post-mortem interval (PMI) and, although numerous other methods have been developed (Vass et al. 1992; Babapulle and Jayasundera 1993; Swift et al. 2001; Bocaz-Beneventi et al. 2002; Vass et al. 2002, 2004; Ishii et al. 2006), PMI estimation becomes more difficult as the degree of decomposition increases (Vass et al. 1992). Thus, the more tools that can be made available for PMI estimation, the more likely it will be that accurate estimation of time since death will be possible.

Methods

Sample Collection

Two human decomposition experiments were performed, both at the University of Tennessee's Forensic Anthropology Centre. This research facility is located in a secluded, open-wooded area in Knoxville, Tennessee, USA, and is dedicated to the study of the decomposition of human remains in a natural environment.

Preliminary Experiment

The soil samples used in this initial study were collected as part of a wider decomposition study investigating volatile fatty acids and various anions and cations as markers for time since death estimation (Vass et al. 1992). An un-embalmed, un-autopsied cadaver (Cadaver A) was placed within the facility in early December 1988 (Table 24.1). The cadaver was placed naked and face down on bare soil that had not been used for any previous experimental work. During the spring and summer months, soil samples were collected every three days but during the autumn and winter months, when the rate of decomposition slowed, sampling was performed weekly. The soil surface under the cadaver, covered by the region from the pelvis to the shoulders, was divided into a grid with squares approximately 2 cm^2. At a depth of 3 to 5 cm, each area yielded 10 g of soil. Soil from a randomly selected grid-square was collected at each sampling. No area was sampled more than once to prevent collection of different soil layers. It is recognised that soil microbiology alters considerably with depth (Zhou et al. 2004), and this method was used to prevent sampling of different soil strata.

When taking samples, the cadaver was briefly tilted sideways by hand to enable soil collection from underneath. Soil sampling was performed using a

Table 24.1 Cadaver-related information for experimental decomposition studies.

ID	Experiment	Age	Sex	Weight (kg)	Height (cm)	Date of death	Cause of death	Control name	Date placed
A	Preliminary	59	F	44	160	28 Nov 88	Cancer	–	04 Dec 8
P	Secondary	69	M	159	184	20 Aug 06	Natural	O	22 Aug 06
R	Secondary	62	F	40	162	09 Sep 06	Renal failure	Q	11 Sep 06

spatula and soil corer, and the cadaver then returned to its original position. All samples to be used for bacterial community DNA profiling were placed in glass scintillation vials, frozen and then lyophilised for 24 h. The freeze-dried samples were stored at room temperature. A control soil sample was collected at the beginning of the experiment.

In accordance with established protocol, accumulated degree days (ADD) (Edwards et al. 1987) were used to describe the degree of human decomposition, rather than actual days since death. Temperature greatly affects the rate of decomposition, so using ADD allows for more reliable decomposition progression measurements as well as allowing direct comparison of rates of decomposition occurring at different times of the year (Vass et al. 2002). ADD were determined by summing the average daily temperatures (°C) for the period of time (days) over which the corpse had been decomposing. Daily temperatures were recorded at the site itself, allowing accurate ADD to be calculated. Samples were collected for a period of over one year (to ADD 5370) and a selection of 30 samples from throughout this time-frame were chosen for analysis.

Secondary Experiment

For the secondary experiment, two cadavers were placed at separate, but nearby, sites within the facility in 2006 (Table 24.1). Both were in the fresh stage of decomposition and did not display any visual signs of decay. The two cadavers were very different in size, with Cadaver P weighing 159 kg and Cadaver R 40 kg. This size difference was due to few cadavers being available (via donation) at the time of the experiment, so any that became available were accepted. Soil was prepared by raking to remove plant material and leaf debris, and an area of soil approximately 1 × 0.6 m was loosened and homogenised by raking and mixing. Sites were chosen for similarity based on soil content, similar local vegetation and lack of previous use. Root material and any small stones were removed by hand. Each cadaver was placed on a sheet of plastic mesh that allowed it to be rolled sideways for soil sample collection. The plastic mesh ensured minimal damage to the cadaver or disruption to the decomposition process.

Soil sampling was performed by scooping about 50 g from the surface of the decomposition area, by dragging the lip of a plastic centrifuge tube in an S-shaped pattern. This method was chosen over the grid-style sampling due to potential soil heterogeneity, even across small distances. Because the soil was loosened and homogenised to a depth of about 10 cm prior to cadaver placement, sampling of different soil strata over time was avoided. A control plot was prepared in exactly the same manner at a site approximately one metre, and slightly uphill, from each cadaver. Control samples were collected every time a decomposition sample was taken. Sampling was performed every second to third day for approximately three months. An ADD = 0 control sample was also collected from each decomposition plot immediately prior to cadaver placement. All samples had any insect larvae removed before storing at 4 °C. Bulk DNA was extracted from each sample on the day of collection and stored at −20 °C until required for analysis. As with the initial samples, ADDs were calculated.

Microbial T-RFLP Community Profiling

The profiling method used was adapted from that described by Osborn et al. (2000) and uses T-RFLP analysis (Liu et al. 1997; Osborn et al. 2000). The T-RFLP method generates a community profile based on DNA sequence variations within genes common to the group of microbes being studied. For bacterial species, the 16S ribosomal RNA (rRNA) gene was used, as this is common to all known bacterial species. For fungal analysis, the highly variable internal transcribed spacer (ITS) region of the fungal rRNA gene was used. Total community DNA was extracted and the gene targets obtained from microbes in the soil community amplified using the polymerase chain reaction (PCR), using fluorescently labelled primers, resulting in gene copies of similar length from all species, but each with quite different internal DNA sequences. Restriction enzyme digestion of the mix of PCR products generated DNA fragments of varying lengths, the terminal or end fragments of which are quantified using fluorescent detection methods. For both sets of samples, total DNA was extracted from 300 mg of each soil sample using the FastDNA Spin Kit for Soil (QBiogene, CA). The samples were homogenised by thorough mixing before sub-sampling. A DNAzol (Invitrogen) wash step was incorporated into the DNA extraction procedure to remove any remaining contaminants.

Bacterial Community T-RFLP Generation

Oligonucleotide primers were chosen that were specific to bacteria, by targeting the 16S rRNA gene. The bacterial primers used for the preliminary samples were: fluorescently-labelled FAM-F63 (5′ CAGGCCTAACACATGCAAGTC 3′) and unlabelled R1389 (5′ ACGGGCGGTGTGTACAAG 3′) (Marchesi et al. 1998). For the secondary experiment, a different reverse primer, HEX-R1087 (5′CTCGTTGCGGGACTTAACCC 3′) (Hauben et al. 1997) was used. For the initial samples, amplification was performed with 20 ng of extracted DNA, using 25 µl HotStarTaq DNA Polymerase Mastermix (Qiagen), 10 µm of each primer, and molecular biology grade bovine serum albumin (Invitrogen), added to a final concentration of 0.1 µg/µl. The thermocycling protocol included a 15 min hot-start step at 95 °C, followed by 30 cycles of 94 °C for 1 min, 55 °C for 1 min and 72 °C for 1 min. A final extension step was conducted at 72 °C for 10 min. For the secondary experiment, Qiagen Taq Core PCR Kits were used. Amplification was again performed with 20 ng of extracted DNA, 0.125 µl Taq Polymerase, 5 µl 10x buffer, 200 µM of each dNTP, 2 mM MgCl, 10 µm of each primer, and molecular biology grade bovine serum albumin (Invitrogen), added to a final concentration of 0.1 μg/μl. The thermocycling protocol for the secondary experiment samples included a 3 min denaturation step at 94 °C, followed by 30 cycles of 94 °C for 30 s, 55 °C for 30 s and 72 °C for 1 min. A final extension step was conducted at 72 °C for 20 min. For both experiments, the DNA was then purified using a QIAquick cleanup column (Qiagen), a silica-gel membrane spin column, and then 150 ng of cleaned DNA digested with 20 U of a restriction enzyme. The enzyme AluI (Promega) was used for the initial experiment, and MspI (Promega) for the second. Profiles were not able to be generated for all samples collected due to failure of the PCR step. This is most likely caused by inhibitory substances co-extracted from the samples.

Fungal Community ITS-T-RFLP Generation

The internal transcribed spacer (ITS) region, a polycistronic rRNA precursor transcript, is the most widely sequenced DNA region in fungi (Horton and Bruns 2001). Standard primers were used to amplify the highly variable ITS region of fungal ribosomal DNA. Amplification was facilitated using a fluorescently-labelled forward primer FAM ITS-1F (5′-CTTGGTCATTTAGAGGAAGTAA-3′) (Gardes and Bruns 1993) and an unlabelled reverse primer ITS4R (5′-TCCTCCGCTTATTGATATGC-3′) (White et al. 1990). A 50 μl reaction volume was used containing 5 to 20 ng of extracted DNA, 5 μl of 10x buffer (Bioline), 3 mM MgCl2 (Bioline), 1 μl of dNTPs with 25 mM of each dNTP (Bioline), 1.25 U of Taq and 20 pmol of each primer. The thermocycling protocol included a 3 min denaturation step at 95 °C and was followed by 30 cycles of 94 °C for 45 s, 55 °C for 45 s and 72 °C for 1 min. A final extension step was conducted at 72 °C for 20 min. The DNA was cleaned using the QIAquick PCR Purification kit. The cleaned DNA (80 to 300 ng) was digested with 2 μl of 10 U/μl of the restriction enzyme HhaI (Promega).

For both the bacterial and fungal community studies, the fluorescently labelled terminal restriction fragments were separated using an ABI PRISM 3,100 Genetic Analyzer (Applied Biosystems, Australia). The data produced were analysed using Genemapper v.4.0 software (Applied Biosystems, Australia). Terminal restriction fragments of the digested DNA in the 100 to 500 bp size range and above the threshold limit of 1% of the total fluorescence signal were included in subsequent analyses. Relative abundance of the peaks present in each profile was determined and multi-dimensional scaling (MDS) analysis was performed using PRIMER 6 software (Clarke 1993).

Phospholipid Fatty Acid Analysis

The PLFA analyses followed the protocol of White (1979) and Zelles et al. (1992) with a few modifications. Lipid extraction was performed using 2 g (±0.1 g) of thawed moist soil. A phosphate buffer (8.7 g K_2HPO_4/l, neutralised with 1N HCl to pH 7.4), 99% methanol and 99% chloroform mixture (v/v/v 2, 5, 2.5 ml) was added to the sample. The sample solution was then sonicated for 15 min in an ultrasonic bath, centrifuged for 5 min at 3,500 rpm and the supernatant decanted. Chloroform and deionised (Milli Q) water were added (3 ml each), the mixture kept on ice for 30 min and centrifuged. The chloroform phase (bottom phase) was isolated and reduced to low volume. The total lipid extract was remobilised in 2 ml chloroform and was separated into polarity-based fractions by successive elutions through silica-bonded columns (SPE-Si, Supelco, Poole, UK) with chloroform (2 ml) and acetone (2 ml) to remove the neutral fatty acids and free glycol fatty acids, respectively. Subsequent elution with methanol (2 ml) was removed the phospholipid fatty acids. Phospholipids were then subjected to fatty acid methyl ester (FAME) derivitisation. This involved resuspension in a mixture of methanol and toluene (1:1, v/v, 0.2 ml), addition of potassium hydroxide in methanol (0.2 M, 0.5 ml), heating and neutralising with acetic acid (0.2 M, 0.5 ml). Chloroform and deionised water were added (1 ml each) and the chloroform phase was separated and reduced to concentrate the derivitised fraction. A C19:0 fatty acid (Sigma) dissolved in hexane was added (20 μl, 10 ng/μl) as an internal standard. The PLFA extracts were characterised and quantified by gas chromatography-mass spectrometry (GC-MS). The preliminary identification of PLFAs was aided by corresponding GC-MS analysis of a standard mixture of authentic PLFA products. Peak assignments were based on retention times and mass spectral correlation. The peak area of selected PLFAs, in either the total or m/z 74 ion chromatograms were calculated by peak integration using the ChemStation software (Agilent Technologies, Paolo Alto, CA). The relative abundance of the PLFA peaks was quantified by comparison to the measured area of the C19:0 internal standard and was calculated as:

$$\mu\text{g individual fatty acid/g soil} = (\text{PFAME} \times \mu\text{g Std})/(\text{PISTD} \times \text{W})$$

where PFAME and PISTD were the GC-MS peak areas of fatty acid samples and internal standards respectively; μg Std is the concentration of the internal standard (μg/ul solvent); and W is the dry weight/cm^3. PLFA profiles were successfully generated for 32 of 35 control soil samples, and 33 of the 35 cadaver samples tested. MDS analysis was performed on the fatty acid concentration data using PRIMER 6 software (Clarke 1993).

Results

Preliminary Experiment

Discernible shifts occurred in the bacterial community in the underlying soil as decomposition progressed (Figure 24.1). The sample profiles tend to group loosely

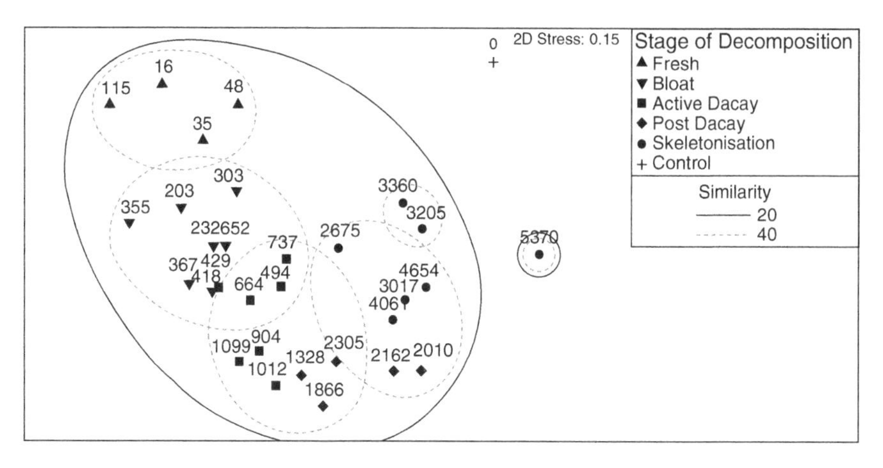

Fig. 24.1 Multi-dimensional scaling plot of bacterial T-RFLP data for Cadaver A. Accumulated degree days (adds) are used to denote the stage of decomposition when the sample was collected.

according to the visual stage of decomposition (fresh, bloat, active decay, post decay and skeletonisation) with evidence of a succession occurring. The samples from each stage of decomposition tend to merge slightly with those from the following stage, suggesting a reasonably continuous, gradual change in the soil bacterial communities as decomposition progressed, rather than punctuated community shifts. Control soil samples collected throughout the decomposition period were not available, so the degree of change occurring temporally in the soil community without the influence of decomposition could not be determined.

Secondary Experiment

Because of the size difference between the two cadavers used in this experiment, some physical differences in the decomposition process were noted. Cadaver P underwent a very rapid decay phase, being largely skeletonised by ADD 376 (Figure 24.2a). This was considered to be unusually rapid and is likely to be related to the obesity of the cadaver. Cadaver R, however, exhibited much slower decomposition, exhibiting very little abdominal bloating, but showing a high level of fungal growth on the skin. By ADD 684, when sampling ceased, Cadaver R was still in the active decay stage. The cause of death for this cadaver was listed as renal failure. A lengthy period of ill-health prior to death is likely to have contributed to this individual being under-weight, and may also have negatively impacted on the natural microflora of the body. It is unknown what antimicrobial medications, if any, were administered prior to death. These vastly different decay rates led to the two cadavers being unsuitable as experimental replicates, but did afford observational study of decomposition in two weight extremes.

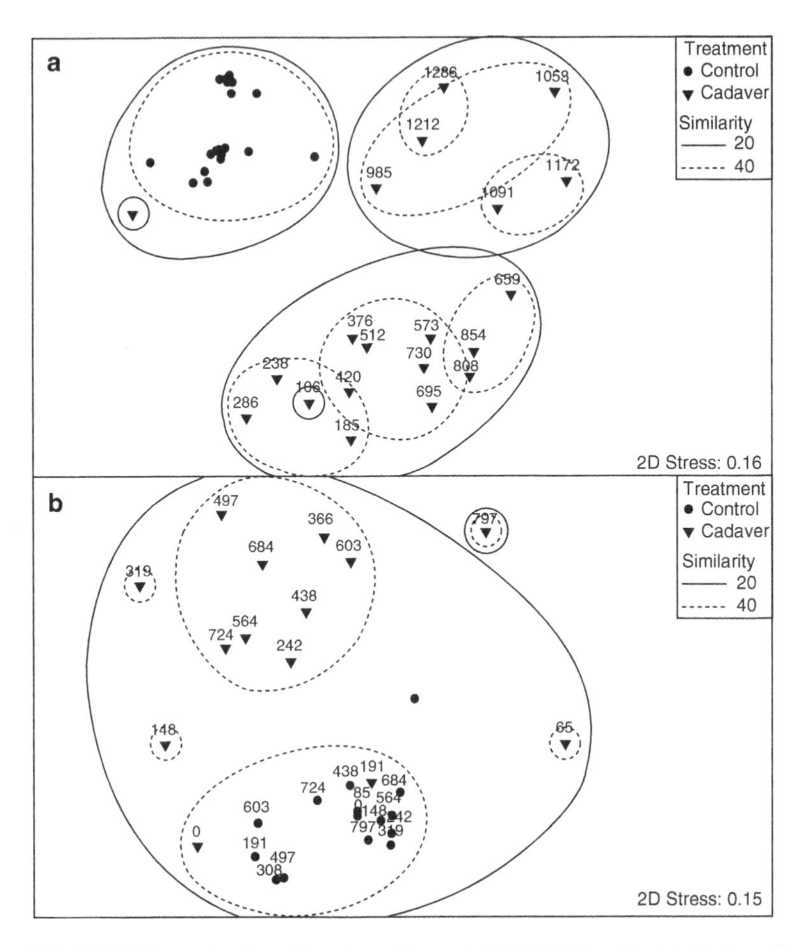

Fig. 24.2 Multi-dimensional scaling plot of bacterial T-RFLP data for (a) Cadaver P, and (b) Cadaver R, and associated controls. Accumulated degree days (adds) are used to denote the stage of decomposition when the sample was collected. Cadaver ADD = 0 sample was the control sample collected immediately prior to cadaver placement

Bacterial Community T-RFLP

The bacterial T-RFLP results from Cadaver P, when plotted using MDS, appear to cluster in three distinct groups. One cluster contains the control soil samples, plus the control collected from the decomposition site immediately prior to cadaver placement. The other two clusters consist of samples collected up to ADD 854, and samples collected after this time. Within these groups, however, there are more closely grouped samples, which, like the cadaver in the initial experiment (Figure 24.1), are loosely related to the stage of decomposition. Cadaver R showed less clear successional change in the bacterial soil community (Figure 24.2b), with the control soils and most of the decomposition samples forming two distinct clusters. This lack of succession is likely to be due to the very slow decomposition rate of this cadaver. The earliest samples from under the cadaver plot either within the control soils group or distinctly from either main group suggest early shifts in the bacterial community, followed by a period of less change when the later samples cluster more tightly. One-way ANOSIM analyses were performed to test the null hypothesis of no significant difference between the control and cadaver soil microbial populations. The control- and decomposition-derived bacterial communities were shown to be significantly different ($P < 0.001$) for both the Cadaver P and R samples, and controls.

Fungal Community ITS-T-RFLP

The MDS plot of the fungal T-RFLP data from Cadaver P (Figure 24.3a) shows some separation between the cadaver and control samples. A natural variation in the control soil fungal communities over time is seen, although the control samples seem to cluster a little more closely than the cadaver samples. No clear trend or succession is evident over time for these samples. The fungal T-RFLP profiles for the Cadaver R samples cluster separately from the control soil samples (Figure 24.3b). There was a potential temporal trend from early (ADD 0, 85,171 and 207) to mid (ADD 366, 438, 497, 603) to late phase fungi (ADD 564, 684, 724, 797), denoted by the arrow (Figure 24.3b).

Phospholipid Fatty Acid

The PLFA data for Cadaver P (Figure 24.4a) show, like the T-RFLP results, a clear separation of microbial communities between the cadaver and control soils, with the controls clustering tightly compared with the more spaced decomposition samples. The first cadaver sample, collected immediately prior to placement of the cadaver on the soil (ADD 0) falls within the control soil sample cluster. A one-way ANOSIM indicates the two microbial populations were significantly different ($P < 0.001$).

The PLFA profiles from Cadaver R (Figure 24.4b) show less separation from the control samples than was evident for the Cadaver P data. Some of the early decomposition samples fall within the control sample cluster, which is consistent with the bacterial T-RFLP results for this cadaver (Figure 24.2b). The control and decomposition derived microbial populations were significantly different ($P < 0.001$). The contribution of known fungal versus bacterial PLFAs to the cadaver-induced changes (Figures 24.3b and 4a) were investigated, but no clear trends were evident in the data from this study.

Discussion

The results of the preliminary experiment showed clear separation between samples collected at the different stages of decomposition. Unfortunately, control soils were not collected at each sampling for this study, so existing temporal changes in the

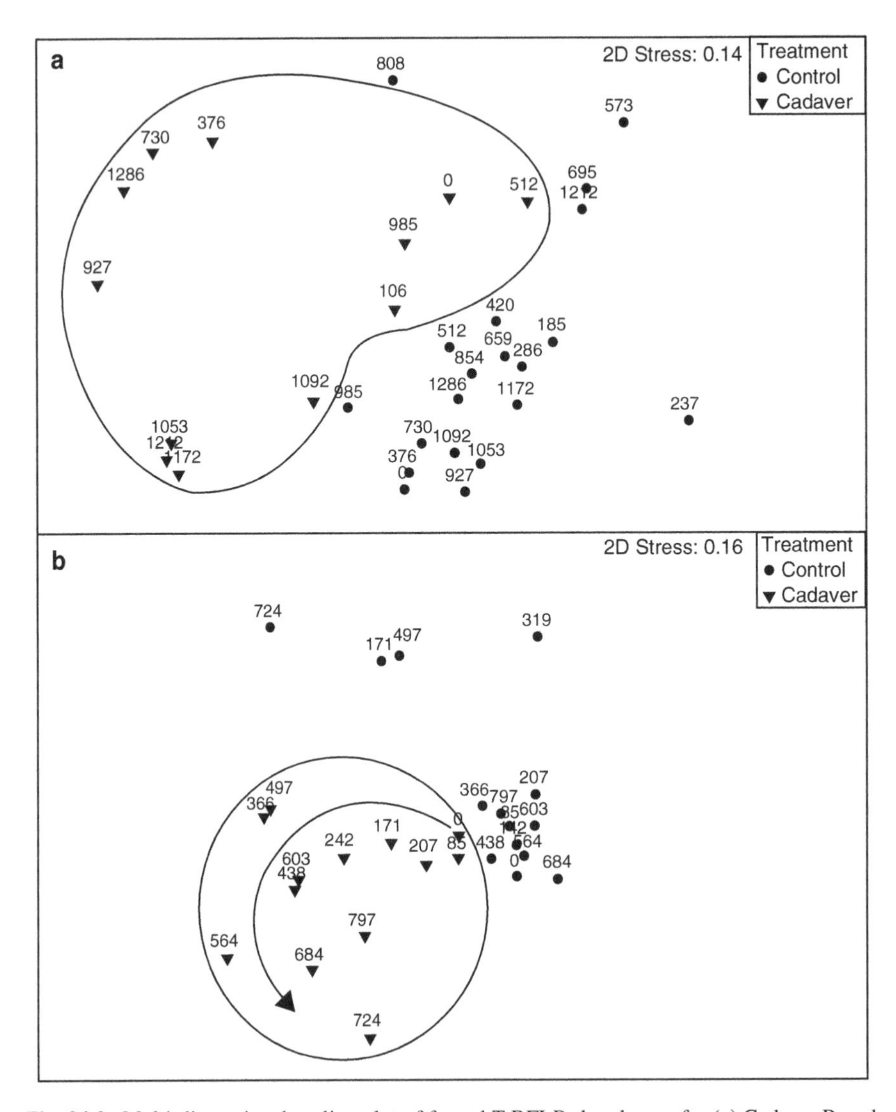

Fig. 24.3 Multi-dimensional scaling plot of fungal T-RFLP abundances for (a) Cadaver P, and (b) Cadaver R, and associated controls. The Cadaver P samples are enclosed by the ring. Accumulated degree days (adds) are used to denote the stage of decomposition when the sample was collected. The arrow in (b) shows possible fungal pattern of succession

bacterial community could not be compared. Control samples were collected each time the cadavers were sampled in the secondary experiment. The considerable differences in body morphology between the two cadavers, and the rates of decomposition exhibited meant that they could not reasonably be treated as replicates, but they afforded an opportunity to observe microbial community change in cadavers at the extremes of the adult human weight range, during the same season.

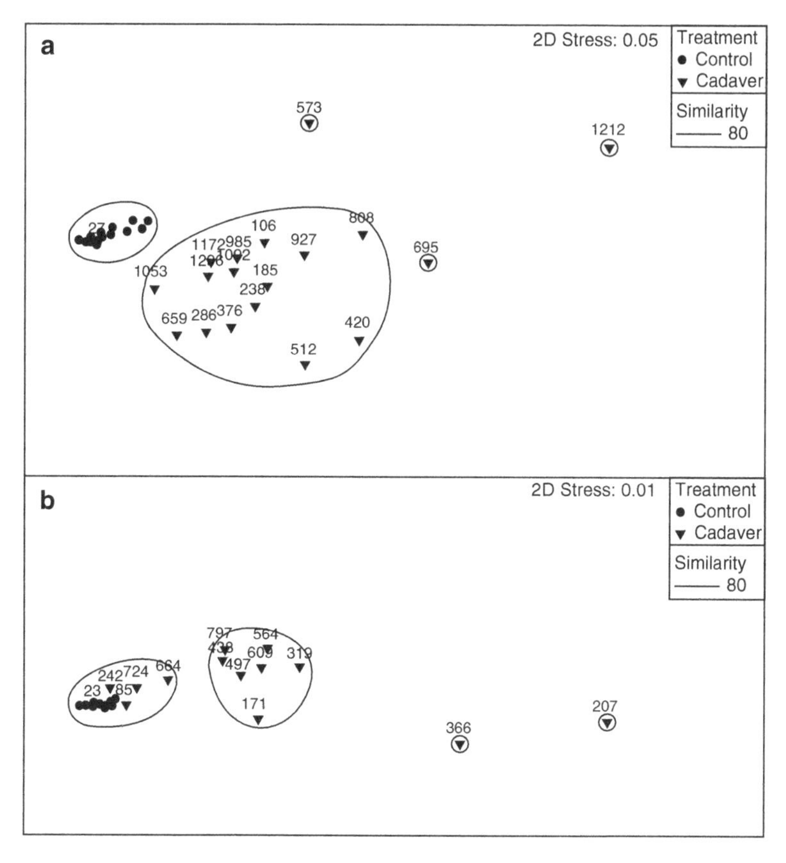

Fig. 24.4 Multi-dimensional scaling plot of microbial PLFA abundances for (a) Cadaver P, and (b) Cadaver R, and associated controls. Accumulated degree days (adds) denote the stage of decomposition when the sample was collected

The T-RFLP results for Cadavers A and P showed identifiable progressive change in the bacterial soil communities, and in the fungal community for Cadaver R. The bacterial T-RFLP data for Cadaver R showed some initial changes in the soil bacterial community, followed by a period where no obvious pattern could be identified. Cadaver P also showed little evidence of sequential change in the fungal community. This could in part be explained by the decomposition rates of the two cadavers. Cadaver R exhibited visible fungal growth on the skin and in the underlying soil. This could be in response to a disturbed body microflora caused by medical treatment such as antibiotic administration, which could retard normal bacterial putrefaction and allow proliferation of fungal species. Medical records, however, were not available to confirm this hypothesis. The PLFA data showed distinct clustering of the control and decomposition samples for Cadaver P, but less clear separation for Cadaver R, again likely to be caused by the slow decomposition rate of this cadaver. Further investigation into individual fungal and bacterial PLFAs did not show any clear trends in the data.

In light of these results, it appears that neither the bacterial nor fungal communities would be particularly useful in a forensic context on their own. Human decomposition is a complex process that is affected by many variables that can influence the way in which a body decomposes, and the microbial processes that occur. Much more research is required to understand the effects of these variables and to more closely examine the changes which are occurring within the different soil microbial communities. The microbial community profiling methods used in this study were useful to examine the overall changes in the microbial communities as decomposition progressed, but are unlikely to be sensitive or robust enough for applied forensic use. A more likely scenario is to use sequence-based technologies to identify microbial biomarker species that are indicative of specific decomposition time-points. The development of a post-mortem interval estimation tool would most likely rely on identification of a suite of both bacterial and fungal indicators for various time-points whose presence/absence or relative quantity could be detected using a rapid diagnostic technique such as microarray. Studies are currently underway to examine specific microbial groups within the greater community, and identify any such biomarkers. Further research into the effects of environmental factors such as soil type and climate is also required, as is a wider study of variation between individual cadavers. The opportunity for research using human cadavers is limited and many research groups are taking the lead from forensic entomology and using pig (*Sus scrofa*) carcasses as models for human decomposition (Hopkins et al. 2000; Forbes et al. 2004; Wilson et al. 2007). The data from this study supports the notion that model organisms are preferential to human cadavers for replicated studies because of the vast differences between individual human cadavers, and the ability to be able to reasonably control many of the variables influencing decomposition rates with a larger number of reared animals. A study is currently being performed to evaluate the similarity of the decomposition microbiology of pigs and humans, to determine their suitability as models for microbiological studies.

Acknowledgements The authors would like to thank the University of Tennessee's Forensic Anthropology Center for allowing access to the facility for experimental work, and the British Soil Science Society for conference attendance support.

References

Babapulle CJ and Jayasundera NP (1993). Cellular changes and time since death. Medicine, Science and the Law 33:213–222.

Bocaz-Beneventi G, Tagliaro F, Bortolotti F, Manetto G and Havel J (2002). Capillary zone electrophoresis and artificial neural networks for estimation of the post-mortem interval (PMI) using electrolytes measurements in human vitreous humour. International Journal of Legal Medicine 116:5–11.

Bruns TD, White TJ and Taylor JW (1991). Fungal molecular systematics. Annual Review of Ecology and Systematics 22:525–564.

Carter DO and Tibbett M (2003). Taphonomic mycota: fungi with forensic potential. Journal of Forensic Sciences 48:168–171.

Clarke KR (1993). Non-parametric multivariate analysis of changes in community structure. Australian Journal of Ecology 18:117–143.

Drenovsky RE, Elliot GN, Graham KJ and Scow KM (2004). Comparison of phospholipid fatty acid (PLFA) and total soil fatty acid methyl esters (TSFAME) for characterizing soil microbial communities. Soil Biology & Biochemistry 36:1793–1800.

Edwards R, Chaney B and Bergman M (1987). Pest & Crop Newsletter 2:5–6.

Fisher MM and Triplett EW (1999). Automated approach for ribosomal intergenic spacer analysis of microbial diversity and its application to freshwater bacterial communities. Applied and Environmental Microbiology 65:4630–4636.

Forbes SL, Stuart BH, Dadour IR and Dent BB (2004). A preliminary investigation of the stages of adipocere formation. Journal of Forensic Sciences 49:566–574.

Frostegard A, Baath E and Tunlid A (1993). Shifts in the structure of soil microbial communities in limed forests as revealed by phospholipid fatty acid analysis. Soil Biology & Biochemistry 25:723–730.

Gardes M and Bruns TD (1993). ITS primers with enhanced specificity of basidiomycetes: application to the identification of mycorrhizae and rusts. Molecular Ecology 2:113–118.

Gill-King H (1997). Chemical and ultrastructural aspects of decomposition. In: Forensic Taphonomy: The Postmortem Fate of Human Remains (Eds. WD Haglund and MH Sorg), pp. 93–104. CRC, Boca Raton, FL.

Haskell NH, Hall RD, Cervenka VJ and Clark MA (1997). On the body: Insects' life stage presence and their postmortem artifacts. In: Forensic Taphonomy: the Postmortem Fate of Human Remains (Eds. WD Haglund and MH Sorg), pp. 415–441. CRC, Boca Raton, FL.

Hauben L, Vauterin L, Swings J and Moore ERB (1997). Comparison of 16S ribosomal DNA sequences of all Xanthomonas species. International Journal of Systematic Bacteriology 47:328–335.

Hayashi H, Sakamoto M and Benno Y (2002). Phylogenetic analysis of the human gut microbiota using 16S rDNA clone libraries and strictly anaerobic culture-based methods. Microbiology and Immunology 46:535–548.

Heuer H, Krsek M, Baker P, Smalla K and Wellington EM (1997). Analysis of actinomycete communities by specific amplification of genes encoding 16S rRNA and gel-electrophoretic separation in denaturing gradients. Applied and Environmental Microbiology 63:3233–3241.

Hitosugi M, Ishii K, Yaguchi T, Chigusa Y, Kurosu A, Kido M, Nagai T and Tokudome S (2006). Fungi can be a useful forensic tool. Legal Medicine 8:240–242.

Hopkins DW, Wiltshire PEJ and Turner BD (2000). Microbial characteristics of soil from graves: an investigation at the interface of soil microbiology and forensic science. Applied Soil Ecology 14:283–288.

Horswell J, Cordiner SJ, Maas EW, Martin TM, Sutherland KB, Speir TW, Nogales B and Osborn AM (2002). Forensic comparison of soils by bacterial community DNA profiling. Journal of Forensic Sciences 47:350–353.

Horton TR and Bruns TD (2001). The molecular revolution in ectomycorrhizal ecology: peeking into the blackbox. Molecular Ecology 10:1855–1871.

Ishii K, Hitosugi M, Kido M, Yaguchi T, Nishimura K, Hosoya T and Tokudome S (2006). Analysis of fungi detected in human cadavers. Legal Medicine 8:188–190.

Janaway RC (1996). The decay of buried human remains and their associated materials. In: Studies in Crime: An Introduction to Forensic Archaeology (Eds. J Hunter, C Roberts and A Martin), pp. 58–85. Batesford, London.

Leckie SE (2005). Methods of microbial community profiling and their application to forest soils. Forest Ecology and Management 220:88–106.

Liu WT, Marsh TL, Cheng H and Forney LJ (1997). Characterization of microbial diversity by determining terminal restriction fragment length polymorphisms of genes encoding 16S rRNA. Applied and Environmental Microbiology 63:4516–4522.

Marchesi JR, Sato T, Weightman AJ, Martin TA, Fry JC, Hiom SJ, Dymock D and Wade WG (1998). Design and evaluation of useful bacterium-specific PCR primers that amplify genes coding for bacterial 16S rRNA. Applied and Environmental Microbiology 64:795–799.

Osborn AM, Moore ER and Timmis KN (2000). An evaluation of terminal-restriction fragment length polymorphism (T-RFLP) analysis for the study of microbial community structure and dynamics. Environmental Microbiology 2:39–50.

Reed HB (1958). A study of dog carcass communities in Tennessee with special references to the insects. American Midland Naturalist 59:213–245.

Sagara N (1975). Ammonia fungi: a chemoecological grouping of terrestrial fungi. Contributions of the Biological Laboratory of Kyoto University 24:205–290.

Sagara N (1995). Association of ectomycorrhizal fungi with decomposed animal wastes in forest habitats: a cleaning symbiosis? Canadian Journal of Botany 73:S1423–1433.

Sakamoto M, Hayashi H and Benno Y (2003). Terminal restriction fragment length polymorphism analysis for human fecal microbiota and its application for analysis of complex bifidobacterial communities. Microbiology and Immunology 47:133–142.

Swift B, Lauder I, Black S and Norris J (2001). An estimation of the post-mortem interval in human skeletal remains: a radionuclide and trace element approach. Forensic Science International 117:73–87.

Vass AA (2001). Beyond the grave – understanding human decomposition. Microbiology Today 28:190–192.

Vass AA, Bass WM, Wolt JD, Foss JE and Ammons JT (1992). Time since death determinations of human cadavers using soil solution. Journal of Forensic Sciences 37:1236–1253.

Vass AA, Barshick SA, Sega G, Caton J, Skeen JT, Love JC and Synstelien JA (2002). Decomposition chemistry of human remains: a new methodology for determining the postmortem interval. Journal of Forensic Sciences 47:542–553.

Vass AA, Smith RR, Thompson CV, Burnett MN, Wolf DA, Synstelien JA, Dulgerian N and Eckenrode BA (2004). Decompositional odor analysis database. Journal of Forensic Sciences 49:760–769.

White DC (1979). Lipid analysis of sediments for microbial biomass and community structure. In: Methodology for Biomass Determinations and Microbial Activities in Sediments (Eds. CD Litchfield and PL Seyfreid), pp. 87–103. American Society for Testing and Materials, Philadelphia, PA.

White TM, Bruns T, Lee S and Taylor J (1990). Amplification and direct sequencing of fungal ribosomal RNA for phylogenetics. In: PCR protocols: a guide to methods and applications (Eds. MA Innis, DH Gelfand, JJ Sninsky and TJ White), pp. 315–321. Academic, San Diego, CA.

Wilson AS, Janaway RC, Holland AD, Dodson HI, Baran E, Pollard AM and Tobin DJ (2007). Modelling the buried human environment in upland climes using three contrasting field sites. Forensic Science International 169:6–18.

Zelles L (1997). Phospholipid fatty acid profiles in selected members of soil microbial communities. Chemosphere 35:275–294.

Zelles L, Bai QY, Beck T and Beese F (1992). Signature fatty acids in phospholipids and lipopolysaccharides as indicators of microbial biomass and community structure in agricultural soils. Soil Biology & Biochemistry 24:317–323.

Zhou J, Xia B, Huang H, Palumbo AV and Tiedje J (2004). Microbial diversity and heterogeneity in sandy subsurface soils. Applied and Environmental Microbiology 70:1723–1734.

Part IV
Technology

Chapter 25
Analysis of Soils in a Forensic Context: Comparison of Some Current and Future Options

G. Stewart Walker

Abstract The increasing number of sophisticated techniques available for application to forensic, environmental, and forensic environmental investigations raises questions about their relative discriminating power, and the relative cost and return. A number of different variables including ease of use, availability, relative cost and admissibility in court should be considered to determine a hierarchy of techniques for aiding selection by investigating authorities. Of recently developed techniques, confocal RAMAN, laser induced breakdown spectrometry (LIBS) and laser ablation inductively coupled plasma mass spectrometry (LA-ICPMS) are considered for their potential for cost effective discrimination of soils. Outcomes of comparisons between techniques suggest that a hierarchy of techniques can be developed, which may change with each case depending on the resources available and the needs of the investigation.

Introduction

An increasingly large array of analytical techniques are being applied to forensic and/or environmental soil investigations. These range from visual interpretation of the tester (comparison with Munsell Soil Colour Charts), through gross scientific measures (pH), and non-destructive quantitative spectrometric methods (micro-spectrophotometry) to complex quantitative multi-elemental and isotopic analysis of each component of the soil (laser ablation inductively coupled plasma mass spectrometry, LA-ICPMS). This chapter draws on the experience of the author, and recent reports in the scientific and technical literature, to provide a critical comparison of how some of these methods could be applied in addressing issues which arise in soil forensics.

G.S. Walker
Forensic and Analytical Chemistry, School of Chemistry, Physics and Earth Sciences, Flinders University, GPO Box 2100, Adelaide, SA 5001, Australia
Stewart.Walker@flinders.edu.au

K. Ritz et al. (eds.), *Criminal and Environmental Soil Forensics*

Hierarchy

Forensic soil investigations, like other analyses (e.g. glass or fibres), require a hierarchy of techniques to determine the extent of discrimination achievable between two samples. An example of a hierarchy of analysis is divided into Initial Screen and Sample Preparation, Tier 1 for all samples, Tier 2 for selected samples, and Tier 3 for very important samples or research investigations (Figure 25.1).

The actual application of each technique in soil investigations will depend upon a number of factors, not only a choice of analytical technique based on cost, availability or discriminating power. Special attention should be given to comparison of the following factors: ease of operation; training required; mobility; sample size required; destructive/non-destructive; sample preparation; time for analysis; potential for personal bias; potential for automation; opportunity to quantify; opportunity to develop databases of analytical results; cost (initial set up and per analysis costs); ease or difficulty of presentation of results in court; and ease or difficulty of getting results accepted in court.

Cost may ultimately be the crucial and deciding factor in determining which analytical technique will be applied to an investigation. However, attempts to quantify

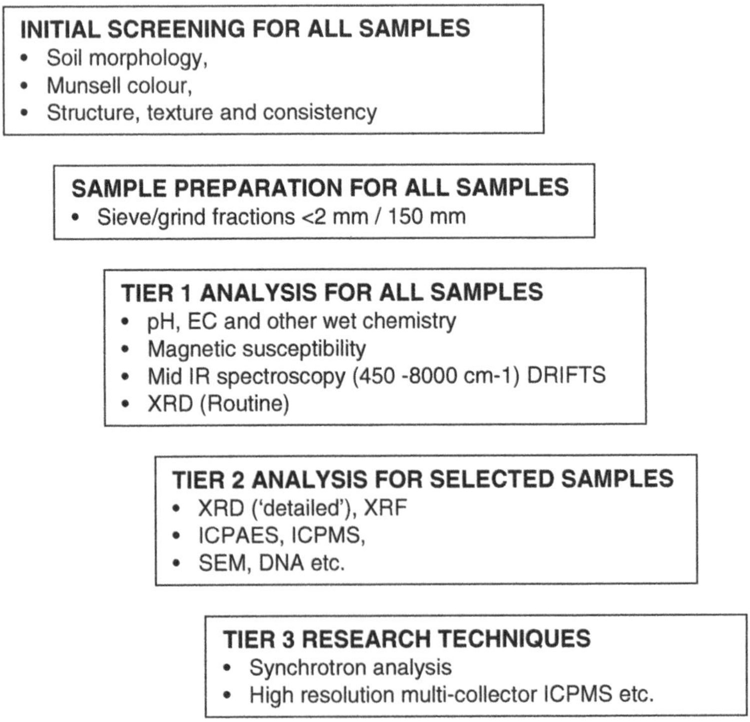

INITIAL SCREENING FOR ALL SAMPLES
- Soil morphology,
- Munsell colour,
- Structure, texture and consistency

SAMPLE PREPARATION FOR ALL SAMPLES
- Sieve/grind fractions <2 mm / 150 mm

TIER 1 ANALYSIS FOR ALL SAMPLES
- pH, EC and other wet chemistry
- Magnetic susceptibility
- Mid IR spectroscopy (450 -8000 cm-1) DRIFTS
- XRD (Routine)

TIER 2 ANALYSIS FOR SELECTED SAMPLES
- XRD ('detailed'), XRF
- ICPAES, ICPMS,
- SEM, DNA etc.

TIER 3 RESEARCH TECHNIQUES
- Synchrotron analysis
- High resolution multi-collector ICPMS etc.

Fig. 25.1 Hierarchy of analysis of soils in a forensics context, showing initial screening, and sample preparation.

and compare costs are largely futile because the relative costs of sample preparation and analysis vary depending upon the organisation and costing structures (e.g. free use, user pays or per sample costing), and the relative expense such as labour and electricity costs and time. Therefore, when costs are discussed in this chapter, they are indicative and are only one factor to be considered as, for example, a full spectrum LIBS will cost about twice that of a narrow spectrum LIBS but provide information from a wider range of wavelengths.

For the interpretation of soil morphology, a visual inspection by suitably trained experts, and a comparison with Munsell Colour Charts, can be a fast, effective and cheap starting point, which may give sufficient differentiation powers for excluding any similarity of samples. A recent example illustrates this point, where the author was requested to analyse two flints – one suspected of being of local/South Australian origin used by aboriginals to produce flint artifacts and one suspected of being ballast from the River Thames in England. When the client provided the two specimens (one from each trouser pocket) for analysis using a LA-ICPMS, one sample was black and flaky and the other a mousey brown, thus rendering LA-ICPMS unnecessary for discrimination as they were evidently not sourced from the same location.

Research is being undertaken (Dawson et al. 2007; Creeper 2008) to automate computer-based spectrometric and analytical techniques to enable searchable databases to be developed. Such databases would support more advanced discrimination techniques (e.g. MIR), being moved higher up the hierarchy, and visual comparison and spectral techniques being made more quantitative and searchable (Creeper 2008).

One technique where this is in operation is that of MIR Spectroscopy, which has been developed into a quick and relatively cheap diagnostic tool, with CSIRO Land and Water, Australia providing a full grinding and predictive analysis for around twenty AUS dollars a sample (2008). CSIRO claim to be able to determine from chemometric interpretation of MIR spectra of soil the following properties: total organic carbon (TOC), charcoal ('inert carbon'), particulate organic carbon (POC), total nitrogen (N), calcium carbonate, cation exchange capacity (CEC), exchangeable Ca, Mg, K and Na, New Zealand P Retention Index (NZPRI), P buffering capacity (O&S), (R&H) and P buffering index, pH_{water}, pH_{CaCl}, lime requirement, electrical conductivity (EC), exchangeable Na percentage (ESP), bulk density (BD), clay, silt, sand, volumetric water content % at 0, 1, 3, 5, 10 and 50 kPa, 5 and 15 bar, quartz, kaolin and smectite and Si, Al, Fe, Ca, Mg (CSIRO 2008).

Other researchers have applied chemometrics to the interpretation of data which can be related to soil analysis, including a new three-step chemometric process for determining similarity index (Guild 2005) and a method of determining numerical differences (Thomas 2008). One example of chemometrics applied to multiple data from time of flight secondary ion mass spectrometry of 'soils' comes from Engrand et al. (2006), who are preparing a study of the comet P67/Churyumov-Gerasimenko. To calibrate the instrument, a variety of terrestrial mineral samples were analysed. Initially, 36,000 inorganic ions and 61,000 organic ions were identified from mineral samples. Of these, 539 peaks were subjected to CORICO correlation plots to discriminate between talc, $Mg_3Si_4O_{10}(OH)_2Mg_{2.6}Si_{2.7-2.9}O_{4.8-4.1}(OH)_{7.2-7.9}$, enstatite $(Mg_{0.9}Fe_{0.1})Si_{1.0}O_3$, olivine, $(Mg,Fe)_2SiO_4(Mg_{1.8}Fe_{0.2})Si_{1.0}O_4$, and serpentine $Mg_{2.6-2.8}Al_{0.2-0.5}[Si_{1.7-2.0}Fe_{0.10}O_{5.8-4.5}](OH)_{3.2-4.5}$. The number of ions was reduced to 32 to achieve sufficient differentiation. Finally, a suitable differentiation screen was developed using only the proportion of normalised relative intensity of three peaks due to four ions, $^{24}Mg^{+,}$ $^{28}Si^{+}$ and $^{56}Fe^{+}$ and $^{28}Si_2^{+}$.

The sum of [$^{56}Fe^{+}$ and $^{28}Si_2^{+}$] was used as the MS was not able to separate the ions, requiring as it would a resolution of greater than 2000. With the sum of intensities set at 100%, if [$^{56}Fe^{+}$ and $^{28}Si_2^{+}$] < 2% it is predicted to be talc, if [$^{56}Fe^{+}$ and $^{28}Si_2^{+}$] > 8 then it is assigned as olivine. If [$^{56}Fe^{+}$ and $^{28}Si_2^{+}$] is greater than 2% but less than 8% and [$^{24}Mg^{+}$] < 65, then it is serpentine and if [$^{24}Mg^{+}$] > 65 it is assigned as enstatite. This is an example where differentiation can be achieved with a minimal number of ions and would therefore make a quick approach to the screening of soils in forensic examinations.

Some other techniques which have been applied to soil investigations will not be discussed here as they are either covered elsewhere in this publication, or in recent reviews and publications, for example Isotope ratio MS (Benson et al. 2006), NIR (Janik et al. 2007) and X-ray (Smith and Cresser 2004), which cover the material in more depth. In this chapter, attention will focus on outlining the recent developments of three techniques: Raman, LIBS and LA-ICPMS.

Spectrometric Techniques

Raman

Spectrometric techniques have the advantages that they can be non-destructive, are relatively cheap and can lend themselves to chemometric interrogation and computer processing. Raman, as a reflection rather than transmission spectrometric technique, is ideally suited to analysis of solid samples. However, especially in the presence of organic matter, it can suffer from fluorescence interference. Surface-enhanced Raman scattering (SERS), with enhancement by factors of 10^{15}, has been reported by Nie and Emory (1997) and Kneipp et al. (1997). Techniques, such as selection of a narrow laser wavelength from a tunable laser, can reduce the fluorescence, as can reducing the collection volume to minimise background fluorescence. The extreme example of this comes from a diamond shaped tip with an orifice at the apex of the tip in the nanometre range. If held within a few nanometres of the surface, this prevents non-directly reflected light from getting into the orifice, and therefore this background/interference light does not reach the detector. The theory of this method, called near-field scanning Raman, was described by Wessel (1985) and has been developed for surface scientists and increasingly used by nanotechnologists.

Stöckle et al. (2000) calculated that, assuming a tip diameter of 20 nm and a laser spot diameter of 300 nm, the tip-induced enhancement was larger than 40,000-fold.

This produces full Raman spectra of the surface with no interference and enhanced signal and with nanometre resolution (Hartschuh et al. 2003). Although currently restricted to surface scientists and nanotechnologists, there may be potential to utilise these new techniques in forensic soil analysis.

However, one limiting factor which has become apparent in the author's initial application of Scanning Confocal Raman, where an image can be created based on Raman spectra of individual spectra from a grid, is that the focus range of the instrument is very narrow. It is designed for imaging flat surfaces, so any natural variation on height of the surface causes the Raman laser to be out of focus at certain depths. Subsequently, fluctuations in the image may be due to variations in the surface topography rather than changes in the chemistry.

Laser Induced Breakdown Spectrometry

Laser induced breakdown spectrometry (LIBS) has advantages over ICPMS as it is mobile, has low power requirements and does not require gas supplies (however, some LIBS spectra are enhanced by an inert atmosphere). The basic components of a LIBS system are a laser, a means of focusing the laser on a sample, and a means of detecting and spectrally analysing the light emitted. Other features can be enclosed sample holders, gas flow controls, and a video camera for observation of the sample. The laser in LIBS is typically 50 or 200 mJ but can be lower for a hand-held device, where power supply is an issue, or higher for specific applications. However, the power of the laser is restricted to ensure that there is no over-loading of fibre optics. Therefore, the costs of the LIBS can vary considerably.

As with all emission spectrometric techniques, the narrower the wavelength interval processed the higher the resolution and finer the detail obtained. Higher resolution and wider spectral ranges come at a cost. For example, a simple LIBS can be ordered with a single bundle of fibres for a narrow spectral range at cost of US$5000 for the LIBS and US$250 for a single channel of fibre optics. When sample chamber, 50 mJ laser etc. is added, the total cost is around US$45,000.

Up to seven channels are available covering wavelength ranges 200 to 305, 295 to 400, 390 to 525, 520 to 635, 625 to 735, 725 to 820 and 800 to 980 nm. For the full wavelength range, the LIBS costs US$35,000 and the seven fibre bundles cost $1200, with a total cost of approximately US$77,000. An additional US$6,000 will get a 200 mJ laser. Therefore, irrespective of the currency, the initial cost of a LIBS system can vary by a factor of two. Compared to other analytical equipment (c.f. ICPMS which is in the hundreds of thousands of dollars) the LIBS system is relatively cheap but offers different capabilities.

However, theoretical studies (Mueller et al. 2007) of the application of Intensified Detectors (ICCD), and the cheaper and more robust non-intensified CCD to LIBS systems, showed a simulated LOD for silicon of 0.001% and 0.0007% respectively, and an actual LOD of 0.025% for short-gate ICCD, 0.022% for CCD and 0.018% for long-gate ICCD. Detection limits for other elements also showed variation with

ICCD and CCD selection, with best LODS for Cu and Cr at 0.008%V, 0.022%, Mo 0.028% and Ni at 0.025%. However, the overall findings were that signal-to-noise ratios and limits of detection for the cheaper CCD were 'comparable or even better' to those with ICCD.

LIBS has been used in some specialised applications such as: QA in production of uniform material such as steel; single element or element ratio monitoring; and determination of carbon in soils (Cremers et al. 2001)

As an example, Cremers (2001), in a comparative investigation analysed soil of 'specific morphology' by conventional techniques and by LIBS, selecting a single emission line from carbon (C(1) emission 247.8 nm, excitation at 1,064 nm, 50 mJ, 10 ns pulses). Detection limits of carbon in soil, based on carbon to silicon ratio calibrations, were determined to be 300 mg C/kg with a precision of 4% to 5% and accuracy of 3% to 14%. A comparison of LIBS values with those from conventional dry combustion agreed, with an r^2 of 0.96 for soils of 'distinct morphology'. The analysis time for LIBS, not including sample preparation (air-drying and sieving to >2 mm), was less than 1 min.

More recently, Gehl and Rice (2007) reported research by Ebinger et al. (2003) to combat the spectral interference of iron at the 247.8 nm line and recommended C analysis at 193 nm. Their comparison of C, measured by LIBS with C determined by dry combustion, for three Colorado soils was highly correlated ($r^2 = 0.99$), and reproducible with an average $r^2 = 0.97$ for 30 analysis (Gehl and Rice 2007).

LIBS has been coupled to a cone penetrometer, by researchers at the Department of Energy (USA) and Los Alamos National Laboratory, to produce a mobile soil-specific LIBS system for depth profiling heavy metal concentrations *in situ*. The initial prototype was mounted in a vehicle (20 to 40 tonne truck) to house the LIBS laser, computer and spectrometric device and the pneumatic drilling system required to sink the LIBS tip to depths of up to 30 feet. Profiles for chromium concentrations from 0 to 1,100 ppm at depths from 6 to 8 feet have been reported (Grisanti and Crocker 1998). This core-penetrating LIBS system may be of use in environmental forensic investigations, or in probing deep burial sites either for contamination by heavy elements, or for other buried disposals where a change in elemental composition is expected.

A scaled-down version, claiming to be the first backpack LIBS system, was also developed (Grisanti and Crocker 1998), and has been followed by a range of portable or hand held devices, which would be most useful for top surface soil investigations. A backpack mounted LIBS system powered by a lithium battery, producing 35 to 50 mJ and weighing less than 10 kg, has been reported by Bogue (2004), and for a hand or foot-force inserted 'Direct Push' FO-LIBS (Fibre optic- LIBS) by Mosier-Boss et al. (2002). At one site, the Naval Air Station North Island, San Diego, CA, LOD for Chromium was 130 ppm and a comparison between FO-LIBS and ICP were claimed to be in '100% agreement' (Mosier-Boss et al. 2002).

Other advantages of the LIBS are its relative low cost to run, with no argon gas (ICPMS) or nitrous oxide/acetylene (AAS) required, and no high voltage power supply needed. This makes for ease of mobility and in-field deployment. Stand-Off LIBS (ST-LIBS) have been developed to enable analysing the composition of a material at significant range (several metres to over 100 m) by transmitting the laser beam using a telescope arrangement (Applied Photonics 2008).

There are, however, some weaknesses of the LIBS system. Organic material produces a characteristic broad hump, analogous to Bremstrahlung in XRF, which may interfere with some spectral peaks and limits accuracy of quantitative analysis in some areas of the spectra especially in soils with high or variable organic matter. In addition, whereas for some elements, mercury for example, a few well-defined and strong lines are produced which make for unequivocal assignment and easier quantitation, other elements, such as iron, give a large number of lines which means that any soil sample containing iron will give very noisy spectra which may limit use of this technique for soil investigations.

Limitations in LIBS applied to soil investigations can be seen in published papers that indicate variability, problematic calibration and quantitation, poor reproducibility, large CV and high detection limits. For example, Lazic et al. (2001) reported an analysis of mercury in soil with a limit of detection of 84 ppm, which is higher than many natural levels of mercury in soil. Capitelli et al. (2002) reported iron in soil at detection limits of 500 ppm with RSD of 3%, whereas Sallé et al. (2005), analysing a range of metals in soils for preparation for a mission to Mars, reported LOD between 3% and 7.5% but with RSD of between 10% and 25%. This was an example of Stand-Off LIBS (ST-LIBS) with analysis carried out at a range of over 5 m.

The increasing number of LIBS suppliers, some dedicated to forensic applications (see, for example, Walker 2005) is indicative of an increasing use of LIBS in forensic investigations.

Elemental Analysis

Further discrimination may be available by consideration of the elemental composition. However, care must be taken as many components of soil do not have a constant elemental composition. Indeed, according to a definition (University of Delaware 2008) "One of the six qualifications to be a mineral is a definite, *but not fixed*, chemical formula". For example, Serpentine (Engrand et al. 2006) has a theoretical formula $Mg_3[Si(OH)_2O_5](OH)_4$ but an empirical formula containing variable amounts of the elements magnesium, aluminium and silicon and oxygen $Mg_{2.6\text{–}2.8}Al_{0.2\text{–}0.5}[Si_{1.7\text{–}2.0}Fe_{0.10}O_{5.8\text{–}4.5}](OH)_{3.2\text{–}4.5}$.

Mass Spectrometric Elemental Techniques

Mass spectrometric techniques not only have the ability to determine the full range of elements present in a sample, but have the added advantage that isotopes of elements can be determined which can add an extra dimension for discrimination.

Stable isotopes of light elements, C, N, O, H and S have been applied to a number of investigations as have isotopes of heavier metals, such as Pb, Zn, St.

Solution ICPMS is a well developed technique but suffers from the need to add acids and water which may contain contaminant elements, and from the fact that it will give averaged results for all of the soil digested. Laser ablation coupled to ICPMS offers the opportunity to select spots or lines of sample for individual analysis.

There are three options for laser ablation which provide different information, *viz.* single spot, depth penetrating spot and line or raster-over-surface.

A single spot gives analysis of a small area and small depth range. Repeated ablation of a single spot enables changes with depth to be determined (for example, in the case of TOF-SIMS of surface profiling, see Coumbaros et al. 2008). Rastered spots give a selection of the surface. Doing a line enables the problems of initial ablation, ridge effects etc. to be smoothed out but, however, produces an 'averaged' result.

The lack of standard procedures and reference material has hindered the acceptance of LA-ICPMS as a legitimate and verifiable analysis. However, various groups of researchers have been working on developing, proving and publishing protocols and certified reference material (Latzchoczy et al. 2005).

In an application for soil analysis (New Wave 2008) of 24 elements (Os, Ru, Rh and W as internal standards), 18 elements were reported to lie within 10% of expected value (Mg, P, K, Sc, Ti, V, Mn, Ni, Cu, Zn, Rb, Y, Nb, La, Ce, Nd, Tl and Th). However, some elements showed deviation in percentage accuracy, specifically: barium 13.5%, Cr 11.6%, Ga 11.6%, As 13.7%, Pb 10.4% and Sr 25.4%. The authors also acknowledge that "accuracy and precision of LA-ICPMS data can be limited by heterogeneity of minerals in some soil types".

Modern laser ablation techniques allowing multiple sampling of small individual minerals has enabled homogeneity within a single mineral to be studied, although at this level the reproducibility of the whole laser ablation and ICPMS are also component factors in perceived variability. In one example, Bouman et al. (2008) investigated a single mineral of 200 μm width with multiple laser pits. Reproducibility for 207Pb/206Pb of six spots of 'Sample 91500' was reported as 0.069801 (1 s.d. = 0.22%, $n = 6$) illustrating that this technique can be used to provide reproducible results from a single crystal/mineral.

Much research has been done on understanding the processes and determining the reproducibility of laser interaction with matter and the resultant 'ejecta' (Chenery et al. 1992). Examples of materials related to soils include papers on gold (Watling et al 1994), diamonds (Watling et al. 1995a) and sulphite minerals of chalcopyrite, galena, stibnite, pyrite, sphalerite and arsenopyrite (Watling et al. 1995b).

Many researchers have used solution ICPMS to obtain elemental fingerprints of materials. An example of modern laser ablation techniques applicable to soil science comes from research undertaken to establish links between soil, water and vegetation. Myors et al. (1998) used elemental composition to profile heroin seizures. Watling, who has pioneered the application of LA-ICPMS and chemometric interpretation for forensic investigations in Perth, Western Australia (Watling et al. 1997), has published work on LA-ICPMS of many matrixes, and investigated the inter-elemental association patterns between cannabis crops by LA-ICPMS (Watling 1997). Walker continued this work using digest solutions of cannabis to enable QAQC and production of a quantitative database of elemental profile fingerprints. Both researchers have extended this approach to investigations of linking wine, grapes, leaves to soil and water in vineyards (Green et al. 2004).

Laser Ablation High-Resolution Multi-Collector Inductively Coupled Plasma Mass Spectrometry

Most research using laser ablation multi-collector ICPMS (LA-HR-ICPMS) and laser ablation high resolution MC-ICPMS (LA-HR-MC-ICPMS) has been concentrated on geological and environmental processes, for example in terrestrial circumstances from detrital zircons from diamond-bearing sediments in Namibia (Klama et al. 2003) and in relation to evidence for early Earth differentiation (Boyet et al. 2003). In relation to other planetary systems, studies have been conducted concerning the origin of the moon (Poitrasson et al. 2004), the age of the moon (Kleine et al. 2003a), cosmogenic tungsten on the Moon (Lee et al. 2002), with respect to the early evolution of the Martian mantle (Kleine et al. 2003b).

The basic principle of high resolution multi collector ICPMS is in providing sufficient resolving power that the interferences can be separated sufficiently such that the specific ions investigated can be entrained onto the collector cups without any signal (or minimum signal) from the interfering ions. For example, ArO (nominally 56 and 58) and Fe 56 and 58 can be separated as the accurate masses are $Ar^{16}O$ (39.9481 + 15.9994) = 55.9475 whereas ^{56}Fe = 55.8452, and $Ar^{18}O$ (39.9481 + 17.9991) = 57.9472 whereas ^{58}Fe = 57.9332.

Two current instruments on the market achieve the necessary separation in two different ways. Nu Instruments (2008) use fixed Faraday cups and use Zoom Focusing to force the ions into the appropriate cups. On the other hand, Thermo-Finnegan (Thermo 2008) have an array of cups that can be moved along rods to the exact position of the ion beams. By having double focusing mass spectrometry and spaced cups, both instruments are able to measure the part of the ion beam (before the shoulder) that corresponds to the ion of interest without the interference of the ‘isobaric’ ions. This opens up the way for in-depth investigations of absolute isotope ratios of many of the ions in the periodic table from lithium 6/7 to uranium 235/238 and ions in between such as zinc, iron, selenium and lead.

HR-MC-ICPMS holds great potential for use in forensic and environmental soil investigations but definitely lies at the extreme end of the spectrum of cost, for both set-up and running, complexity in training, operating and interpretation, complex sample preparation (except for laser ablation) and effort to get acceptance in courts. Some may question if we need the ability to determine isotopes of metals in minerals for forensic investigations and if the extra information gained is worth the considerable extra effort and expense.

Conclusions

As with the nanometre scale Raman spectra of surfaces, the ability to determine isotope ratios of a specific element in a 10 ∝m crater of a single mineral crystal may provide information, not all of which is required. With a non-homogeneous material such as soil, both nano-scale Raman and 10 μm scale isotopic rations of specific elements by LA-HR-MC-ICPMS may be the chemical equivalent of the philosophical/theoretical/debating argument of "how many angels can dance on a pin-head?" Therefore, are we getting into too much fine detail so that there is now doubt on differentiating, not because of genuine differences, but because of non-homogeneity in the sample of variability in the analytical technique? If our analysis can obtain reproducible results from separate spots on a single crystal or mineral, then we have a valid method. However, if our analysis is hampered because we are measuring such a small sample size that we can no longer determine the exact position of sampling, or we are getting varied results because the mineral is not homogeneous at the scale of the sampling, then we have gone too far and should 'back off'. For example with the Confocal Raman surface scanning we can image a 10 × 10 μm surface but cannot be certain where this sample is in relation to the sample as a whole.

To conclude, the example given of Engrand et al. (2006) for the screening of Mg, Si and Fe to allocate mineralogy is also an example where the advanced techniques discussed could be applied. LIBS (Laser Induced Breakdown Spectrometry) could yield a relative peak intensity very quickly (within 1 min) for the individual elements and high-resolution – multi-collector ICPMS would be able to differentiate between ions of $^{56}Fe^{+}$ and $^{28}Si_2^{+}$, which nominally have the same m/e but require resolution of greater than 2000 for separation.

As with other analytical investigations, the conclusions of the comparisons is that not all techniques are needed for all cases, and a hierarchy of techniques can be developed which may change with each case depending on the resources available and the needs of the investigation. The ultimate aim would be to tailor analysis used to optimise results for minimised time, effort and money.

You may agree with Kenneth Williams, in the guise of Arthur Fallowfield in the BBC radio comedy series 'Beyond Our Ken' (1958–1964), who said "Well, the answer lies in the soil". However, we have to accept that not all soil samples will yield forensically discriminating results – as Omar Khayyam said:

> And, strange to tell, among that Earthen Lot
>
> Some could articulate, while others not:
>
> Then said another—Surely not in vain
>
> My Substance from the common Earth was ta'en

Some soil samples will have been taken and analysed in vain. With this in mind these new techniques should see greater levels of uptake. The techniques used now were once new and not routinely accepted. The advancements made with such new techniques in the field of forensic soil analysis is in their testing and if they are sustainable they will prove their worth.

References

Applied Photonics (2008). www.appliedphotonics.com. Accessed 01/02/08.

Benson S, Lennard C, Maynard P and Roux C (2006). Forensic applications of isotope ratio mass spectrometry – a review. Forensic Science International 157(1):1–22.

Bogue RW (2004). Boom time for LIBS technology. Sensor Review 24(4):353.

Bouman C, Schwieters J, Cocherie A, Robert M and Wieser M (2008). *In situ* U-Pb zircon dating using laser ablation-multi ion counting ICP-MS (LA-MIC-ICP-MS)..Thermo, http://www.thermo.com/eThermo/CMA/PDFs/Articles/articlesFile_25756.pdf. Accessed 21/07/08.

Boyet M, Blichert-Toft J, Rosing M, Storey M, Télouk P and Albarède F (2003). ^{142}Nd evidence for early Earth differentiation. Earth and Planetary Science Letters 214:447–442.

Capitelli F, Colao F, Provenzano MR, Fantoni R, Brunetti G and Senesi N (2002). Determination of heavy metals in soils by laser induced breakdown spectroscopy. Geoderma 106:45–62.

Chenery S, Hunt A and Thompson A (1992). Laser ablation of minerals and chemical differentiation of the ejecta. Journal of Analytical Atomic Spectrometry 7(4):647–652.

Coumbaros J, Denman J, Kirkbride KP, Walker GS and Skinner W (2008). An investigation into the spatial elemental distribution within a pane of glass by time of flight secondary ion mass spectrometry. Journal of Forensic Sciences 53(2):312–320.

Creeper NL (2008). Development of a forensic database of the soils in the Adelaide Metropolitan region. Thesis (B.Sc.(Hons.)), School of Chemistry, Physics and Earth Sciences, Flinders University of South Australia. pp46.

Cremers DA, Ebinger MH, Breshears DD, Unkefer PJ, Kammerdiener SA, Ferris MJ, Catlett KM and Brown JR (2001). Measuring total soil carbon with laser-induced breakdown spectroscopy (LIBS). Journal of Environmental Quality 30:2202–2206.

CSIRO (2008). http://www.clw.csiro.au/services/mir/index.html. Accessed 01/02/08.

Dawson L, Morrison A, Ross J, Mayes B and Robertson J (2007). Soil Forensics University Network (SoilFUN). http://www.macaulay.ac.uk/forensics/soilfun/

Ebinger MH, Norfleet ML, Breshears DD, Cremers DA, Ferris MJ, Unkefer PJ, Lamb MS, Goddard KL and Meyer CW (2003). Extending the applicability of laser-induced breakdown spectroscopy for total soil carbon measurement. Soil Science Society of America Journal 67:1616–1619.

Engrand C, Kissel J, Krueger FR, Martin P, Silen J, Thirkell L, Thomas R and Varmuza K (2006). Chemometric evaluation of time-of-flight secondary ion mass spectrometry data of mineral in the frame of future *in situ* analysis of cometary material by COSIMA onboard ROSETTA rapid communications. Mass Spectrometry 20:1361–1368.

Gehl RJ and Rice CW (2007). Emerging technologies for *in situ* measurement of soil carbon. Climatic Change 80:43–54.

Green GP, Bestland EA and Walker GS (2004). Distinguishing sources of base cations in irrigated and natural soils: evidence from strontium isotopes. Journal of Biogeochemistry 68(2):199–225.

Grisanti AA and Crocker CR (1998). Cone Penetrometer for Subsurface Heavy Metal Detection, Report DE-FC-21-94MC31 388-20 EM for Task 13, U.S. Department of Energy, EERC Publication.

Guild GE (2005). Innovative UV-Vis spectrophotometric and chemometric analysis of forensic samples. Thesis (B.Sc.(Hons.)), School of Chemistry, Physics and Earth Sciences, Flinders University of South Australia.

Hartschuh A, Anderson N and Novotny L (2003). Near-field Raman spectroscopy using a sharp metal tip. Journal of Microscopy 210:234–240.

Janik LJ, Merry RH, Forrester ST, Lanyon DM and Rawson A (2007). Rapid prediction of soil water retention using mid infrared spectroscopy. Soil Science Society of America Journal 71:507–514.

Klama K, Lahaye Y, Weyer S and Brey GP (2003). U-Pb geochronology, Hf isotopy and trace element geochemistry of detrital zircons from diamond bearing sediments in Namibia determined by LA-(MC)-ICP-MS. European Journal of Mineralogy 15:100.

Kleine T, Mezger K and Münker C (2003a). Constraints on the age of the Moon from ^{182}W. Meteoritics and Planetary Science Supplement 38:A5212.

Kleine T, Muenker C, Mezger K, Palme H and Bischoff A (2003b). ^{182}Hf-^{182}W constraints on the early evolution of the Martian mantle. Meteoritics and Planetary Science Supplement 38:A111.

Kneipp K, Wang Y, Kneipp H, Perelman LT, Itzkan I, Dasari RR and Feld MS (1997). Single molecule detection using surface-enhanced Raman scattering (SERS), Physical Review Letters 78:1667.

Latzchoczy C, Dücking M, Becker S, Günther D, Hoogewerff J, Almirall J, Buscaglia JA, Dobney A, Koons R, Montero S, van der Peyl G, Stoecklein W, Trejos T, Watling J and Zdanowicz V (2005). Evaluation of a standard method for the quantitative elemental analysis of float glass samples by LA-ICP-MS. Journal of Forensic Sciences 50(6):327–1341.

Lazic V, Barbini R, Colao F, Fantoni R and Palucci A (2001). Self-absorption model in quantitative laser induced breakdown spectroscopy measurements on soils and sediments. Spectrochimica Acta Part B: Atomic Spectroscopy 56:807–820.

Lee D-C, Halliday AN, Leya I, Wieler R and Wiechert U (2002). Cosmogenic tungsten on the Moon. Earth and Planetary Science Letters 198:267–274.

Mosier-Boss PA, Lieberman SH and Theriault GA (2002). Field demonstrations of a direct push FO-LIBS metal sensor. Environmental Science & Technology 36(18):3968 -3976.

Mueller M, Gornushkin IB, Florek S, Mory D and Panne U (2007). Approach to detection in laser-induced breakdown spectroscopy. Analytical Chemistry 79:4419–4426.

Myors R, Wells RJ, Skopec SV, Crisp P, Iavetz R, Skopec Z, Ekangaki A and Robertson J (1998). Preliminary investigation of heroin fingerprinting using trace element concentrations. Analytical Communications 35(12):403–410.

New Wave (2008). http://www.new-wave.com/. Accessed 26/06/08.

Nie S and Emory SR (1997). Probing single molecules and single nanoparticles by surface-enhanced Raman scattering. Science 275:1102.

Nu Instruments (2008). http://www.nu-ins.com. Accessed 02/03/08.

Poitrasson F, Halliday AN, Lee D-C, Levasseur S and Teutsch N (2004). Origin of the moon unveiled by its heavy iron isotope composition. Earth and Planetary Science Letters 223(3–4):253–266.

Sallé B, Cremers DA, Maurice S, Wiens R and Fichet P (2005). Evaluation of a compact spectrograph for *in-situ* and stand-off laser-induced breakdown spectroscopy analyses of geological samples on Mars missions. Spectrochimica Acta Part B: Atomic Spectroscopy 60:805–815.

Smith KA and Cresser MS (Eds.) (2004). Soil and Environmental Analysis. Marcel Dekker, New York, Basel.

Stöckle RM, Suh YD, Deckert V and Zenobi R (2000). Nanoscale chemical analysis by tip-enhanced Raman spectroscopy. Chemical Physics Letters 318(1–3):131–136.

Thermo (2008). http://www.thermo.com/com/cda/product/detail/1,,11753,00.html. Accessed 21/07/2008.

Thomas R (2008). Maths in the dock, Millennium Mathematics Project 2004–2008, University of Cambridge. http://plus.maths.org/issue19/features/forensic/index.html. Accessed 29/06/2008

University of Delaware (2008). University Museums of the University of Delaware, http://www.museums.udel.edu/mineral/mineral_site/education/properties/formula2.html. Accessed 02/03/2008.

Walker GS (2005). Applications of laser induced breakdown spectrometer in forensic investigations,In: Proceedings of International Association of Forensic Science 17, Hong Kong August, IAFS, Hong Kong Police Force.

Watling RJ (1997). Sourcing the provenance of cannabis crops using inter-element association patterns 'fingerprinting' and laser ablation inductively coupled plasma mass spectrometry. Journal of Analytical Atomic Spectrometry 13:917–926.

Watling RJ, Herbert HK, Delev D and Abell ID (1994). Gold fingerprinting by laser ablation inductively coupled plasma mass spectrometry. Spectrochimica Acta Part B: Atomic Spectroscopy 49(2):205–219.
Watling RJ, Herbert HK, Barrow IS and Thomas AG (1995a). Analysis of diamonds and indicator minerals for diamond exploration by laser ablation-inductively coupled plasma mass spectrometry. The Analyst 120(5):1357–1364.
Watling RJ, Herbert HK and Abell ID (1995b). The application of laser ablation-inductively coupled plasma-mass spectrometry (LA-ICP-MS) to the analysis of selected sulphide minerals. Chemical Geology 124(1–2):67–81.
Watling RJ, Lynch BF and Herring D (1997). Use of laser ablation inductively coupled plasma mass spectrometry for fingerprinting scene of crime evidence. Journal of Analytical Atomic Spectrometry 12:195–203.
Wessel J (1985). Surface-enhanced optical microscopy. Journal of the Optical Society of America, B 2:1538–1540.

Chapter 26
Automated SEM-EDS (QEMSCAN®) Mineral Analysis in Forensic Soil Investigations: Testing Instrumental Reproducibility

Duncan Pirrie, Matthew R. Power, Gavyn K. Rollinson, Patricia E.J. Wiltshire, Julia Newberry and Holly E. Campbell

Abstract The complex mix of organic and inorganic components present in urban and rural soils and sediments potentially enable them to provide highly distinctive trace evidence in both criminal and environmental forensic investigations. Organic components might include macroscopic or microscopic plants and animals, pollen, spores, marker molecules, etc. Inorganic components comprise naturally derived minerals, mineralloids and man-made materials which may also have been manufactured from mineral components. Ideally, in any forensic investigation there is a need to gather as much data as possible from a sample but this will be constrained by a range of factors, commonly the most significant of which is sample size. Indeed, there are a very wide range of analytical approaches possible, and a range of parameters that can be measured in the examination of the inorganic components present in a soil or sediment. These may include bulk colour, particle size distribution, pH, bulk chemistry, mineralogy, mineral chemistry, isotope geochemistry, micropalaeontology

---

D. Pirrie(✉) and H.E. Campbell[1]
Helford Geoscience LLP, Menallack Farm
Treverva, Penryn, Cornwall
TR10 9BP, UK
dpirrie@helfordgeoscience.co.uk

D. Pirrie and M.R. Power
Intellection UK Ltd, North Wales Business Park
Abergele LL22 8LJ, UK

G.K. Rollinson
Camborne School of Mines, University of Exeter
Cornwall Campus, Penryn,
Cornwall TR10 9EZ, UK

P.E.J. Wiltshire
Department of Geography and Environment,
University of Aberdeen, Elphinstone Road
Aberdeen AB24 3UF, UK

J. Newberry
Department of Natural and Social Sciences,
University of Gloucestershire, Swindon Road,
Cheltenham GL50 4AZ, UK

K. Ritz et al. (eds.), *Criminal and Environmental Soil Forensics*

and mineral surface texture, amongst a host of others. Whilst it would be prudent to utilise as many parameters as possible, the forensic significance of the resultant data should be carefully considered. In particular, many parameters that could be measured (such as particle size distribution, pH and colour) may be affected by the nature of the sample, and vary temporally, or in response to sample storage conditions and hence, provide misleading results. The inorganic components of soil and sediment are typically relatively inert and therefore, potentially, one of the more valuable parameters to measure. To this end, we have utilised an automated scanning electron microscope with linked energy dispersive X-ray spectrometers (QEMSCAN®) in numerous criminal forensic investigations. The sample measurement is operator independent and enables detailed characterisation of the mineralogy of even small samples. The reproducibility of automated SEM-EDS analysis using QEMSCAN was tested by the repeated analysis of the same samples using the same operating, measurement and processing parameters. The results show that instrumental variability is significantly less than observed natural variability between samples. The observed instrumental variability is interpreted to be a function of the sub-sampling of a different sub-set of the overall assemblage of particles present in the samples analysed. Thus automated mineral analysis can be regarded as a robust, highly repeatable method in forensic soil examination.

Introduction

The analysis of particulate trace evidence which has been passively or actively transferred to an offender, or to a vehicle or an object linked to an offender, has been widely used in the detection of serious crime. For example, trace evidence might include tissue, hair or body fluids from the victim or fibres from a victim's clothing or some other surface. In other cases, trace evidence might include particulate materials present in the environment, either derived from natural biological or geological origins, or man-made materials or mineral products that are also present in the environment. To test an association or otherwise, between a suspect and a place or a victim, consideration of the overall assemblage of particles present can provide a much more distinctive pattern, than an elusive search for so-called rare or distinctive particles. In particular, it can be very compelling in criminal forensics if two populations or assemblages of particles from unrelated classes of trace evidence co-occur in samples from, for example, a crime scene and a possible offender's footwear.

Soil is a valuable type of trace evidence in that it is largely composed of two broadly unrelated groups of particulate trace evidence – the organic and the inorganic components. Soils will contain organic components such as macroscopic and microscopic plant remains, commonly the most useful of which are pollen and spores. Organic components in soil samples might also include micro-organisms such as diatoms, or biological marker molecules. The soil sample will also contain an inorganic component which will be dominantly composed of naturally occurring mineral grains derived from the underlying bedrock geology and its weathering products, along with minerals from overlying superficial deposits. Additionally, even in rural environments, there may be introduced inorganic components, including both naturally occurring minerals and mineral products and also other man-made materials. Although there is an interrelationship between the organic (biological) and the inorganic (geological) particles present, they can largely be viewed as discrete classes of trace evidence. Consequently, if the same populations of both organic and inorganic particulates co-occur in control samples from a crime scene and from samples related to a suspect, then the presence of these two classes of trace evidence together can strengthen any evidential link (e.g. Brown et al. 2002).

There is a wide range of different physical, chemical, mineralogical and biological parameters which can be measured in a soil sample. In both research articles and forensic casework, physical, chemical and mineralogical approaches used to test the similarity, or otherwise, of different soil samples have included: particle size distribution, sand grain surface morphology, bulk soil colour, pH, conductivity, bulk major, minor and trace element geochemistry, individual grain chemistry, bulk mineralogy, individual grain mineralogy and stable isotope geochemistry (e.g. Brown 2006; Brown et al. 2002; Marumo et al. 1986; McCrone 1992; McVicar and Graves 1997; Morgan et al. 2006; Pirrie et al. 2004; Pye 2004; Pye and Croft 2007; Pye et al. 2006a, b; Rawlins et al. 2006; Ruffell and Wiltshire 2004; Sugita and Marumo 1996, 2001). Biological approaches have included the examination of soil palynology, microbiology, such as mycology, diatoms, micro-palaeontology (Mildenhall et al. 2006; Wiltshire 2006), and the examination of organic marker molecules (Dawson et al. 2004).

In the UK there is no agreement on the best protocols or methodologies for the forensic examination of soil samples. The evidential value of many of these approaches can be challenged both scientifically and in court, and the resultant data might lead to erroneous false positive or false negative conclusions as to the similarity or otherwise between two soil samples (c.f. Morgan and Bull 2007). For example, it is clear through forensic casework experience that different fabrics coming into contact with the same soil surface will pick up and retain different particle sizes. The relationships between grain size and mineralogy of the soil, and other indirect parameters which are controlled by the soil mineralogy, such as the bulk soil geochemistry, will affect such phenomena. Consequently, the direct, uncritical comparison of the bulk chemistry of the soil and the particles recovered from clothing would potentially give a false negative correlation.

Of the currently available methodologies for the examination of soil samples, an integrated soil mineralogy and soil palynology approach has been shown to provide the best discriminatory ability and has also proven to be the best approach when examining the provenance of an unknown soil sample (e.g. Brown 2006; Brown et al. 2002; Rawlins et al. 2006). Although soil mineralogy is acknowledged as an appropriate and important discriminant, there are a number of methods which can be used to quantify mineralogical characteristics of soils. In this paper, we test the reproducibility of one such analytical technique based on samples derived from a criminal forensic investigation. The aims of the paper are to evaluate how reproducible the data collected using automated scanning electron microscopy are, and to compare the instrumental variability in measured mineralogy with the observed degree of mineralogical variability seen in control samples collected from a crime scene and from items of footwear and a motor vehicle.

Soil Mineralogy

There are a number of methods used to characterise the mineralogy of a soil sample which have been developed through the geological sciences and are well known and proven techniques. These include: (i) the preparation of thin sections, particularly for particles of the size of sand grains and greater, and their examination using optical polarising light microscopy; (ii) X-ray diffraction; (iii) manual scanning electron microscopy; (iv) electron microprobe analysis; and (v) automated scanning electron microscopy. All of these techniques have both strengths and limitations. The analytical approach adopted must be considered on a case-by-case basis, but will typically be controlled by the size of sample available for analysis. Forensic casework experience has shown that commonly less than 0.1 mg of soil might be available for analysis. Consequently, the analytical approach chosen must be capable of providing quantitative mineralogical data for very small sample volumes.

In this study we have utilised an automated mineral analysis system, the basis of which is grounded in scanning electron microscopy (SEM). This is a widely used technique not only in sedimentary geology, but also in many branches of forensic science. Initially, manual SEM studies focused on the detailed three-dimensional imaging of particles, but that is now usually fully integrated with semi-quantitative or quantitative phase chemistry acquired through the use of energy dispersive or wavelength dispersive X-ray spectrometers (EDS or WDS). By integrating the phase chemistry with the grain morphology, it is possible to identify a very wide range of minerals and man-made particulate grains through SEM-EDS/WDS analysis.

Although SEM-EDS analysis is a standard methodology in some branches of forensic science, such as the analysis and identification of gunshot residue particles (e.g. Brozek-Mucha and Jankowicz 2001), it has not previously been used routinely in soil forensic investigations. This is because manual SEM-EDS analysis is highly time-consuming if a dataset is to be collected which is large enough for statistically robust analysis. The method is also often operator dependent. Consequently, the most common use of manual SEM-EDS analysis has been in the examination of soil samples for rare, 'exotic' particles, for comparative purposes (Pye 2004), and the bulk of the soil sample is generally not analysed (although see Pye and Croft 2007). This can result in the collection of datasets which are dependant on subjective interpretation, the significance of which can be open to question. Manual SEM analysis has also been used in the examination of quartz grain surface morphology (e.g. Bull and Morgan 2006; Morgan et al. in press).

In several areas of both forensic science and other branches of minerals, materials and environmental science, there have been significant advances in the development of automated or computer-controlled SEM-EDS analysis systems (e.g. Camm et al. 2003; Goodall and Scales 2007; Knight et al. 2002; Martin et al. in press; Pirrie et al. 2003, 2004; Xie et al. 2005). In forensic science, the standard analytical approach to the examination of samples for particles which are texturally and chemically indicative of gunshot residue (also referred to as cartridge discharge residues) is the use of automated SEM-EDS analysis (ASTM 2002). Suspect samples are examined so that candidate particles are automatically located based on the chemical composition, size and shape, and all other particles present in a sample are excluded from the analysis. In allied areas of mineral science there has been a significant revolution in both the application and available technology for automated mineral analysis (e.g. Goodall and Scales 2007). In the mining industry there are two widely used technologies, the Mineral Liberation Analyser (MLA) and QEMSCAN® and, although these technologies have been used in industry for many years, only recently have scientific papers detailing these methods and their potential applications been published (e.g. Benvie 2007; Goodall and Scales 2007; Lotter et al. 2003).

The MLA primarily locates and characterises particles based on their average backscattered electron (BSE) coefficient which relates to the mean atomic number of the grain in question (Xiao and Laplante 2004). The MLA system enables the very rapid search of samples for 'bright phases' (Xiao and Laplante 2004). Once candidate particles are located based on their BSE signature they can be characterised more fully by the acquisition of energy dispersive X-ray spectra. However, the limitation to this approach is that phases which have a similar mean atomic number may be difficult to differentiate (e.g. many of the common rock-forming silicate minerals).

QEMSCAN systems map the particles present in a sample in terms of their mineralogy, based on the BSE signal, the rapid acquisition of EDS spectra and the automated comparison of the acquired spectra against a database of known mineral species (Pirrie et al. 2004). Mineral grains are therefore identified on the basis of the chemistry as measured during the automated EDS analysis. The main limitation to QEMSCAN analysis is that the technique, like all e-beam based technologies, cannot distinguish between mineral polymorphs (e.g. calcite versus aragonite) or minerals with very similar chemistry (e.g. topaz, kyanite, andalusite, silimanite).

Despite these limitations there are a number of significant advantages in the use of QEMSCAN technology in the examination of soils and sediments in forensic investigations:

(i) Because the technology can mineralogically map on a particle-by-particle basis, the detection limits are extremely low, i.e. even if there is only a single particle of a specific mineral present it can be analysed.
(ii) Where individual particles are composed of more than one mineral, then the textural relationships of the different phases within the particle are imaged, can be visually interpreted and mineralogical association data evaluated.
(iii) In our forensic casework experience, based on samples from the UK, it is apparent that man-made materials are a ubiquitous component of surface soils.

These phases are commonly amorphous and would be difficult to characterise using other analytical approaches; QEMSCAN characterises both naturally occurring minerals and anthropogenic particles automatically.

(iv) Once a sample has been measured in particle mineral analysis mode, then the resultant data set can be examined either as a whole population of particles, as a sub-set of the overall population of particles, or on a particle-by-particle basis.

(v) Structurally similar minerals (e.g. the plagioclase solid solution series) can be subdivided and characterised based on mineral chemistry.

(vi) Although the reported mineral listing is a simplification of the main dataset, it is case specific and allows a rigorous test of the association, or otherwise, between different samples.

Testing the Reproducibility of Automated Mineral Analysis

In this study we test the reproducibility of automated mineral analysis using QEMSCAN. The data are from the QEMSCAN facility at the Camborne School of Mines, University of Exeter, Microbeam Analytical Facility. The QEMSCAN used is based on a Carl Zeiss EVO50 scanning electron microscope platform with an array of four Gresham energy-dispersive light element X-ray spectrometers combined with other proprietary software and hardware. The latest QEMSCAN systems utilise silicon drift detectors with a much improved light element response.

Samples

The samples used were from a recent criminal forensic investigation. As part of the investigation into the murder of a woman and the disposal of her body, soil samples were recovered from: (a) an item of footwear; (b) a motor vehicle; (c) the body deposition site; and (d) other locations pertinent to the enquiry. Trace amounts of soil were collected by washing both the uppers and the lowers of the left and right item of footwear separately. As part of the forensic strategy in this investigation, these samples were subdivided into separate aliquots for mineralogical and palynological analysis; only the mineralogical data are considered in this chapter.

Soil samples from the front offside and front nearside footwells of a motor vehicle were also examined, as were washings of soil recovered from the accelerator pedal cover of the motor vehicle. A suite of nine surface soils, sediment and vegetation samples were taken from the identified offender approach path at the body deposition site, along with control samples collected in the area around which the offender(s) vehicle was likely to have been in contact.

In this particular case, it transpired that the offenders had not come into contact with soil particles at the body deposition site, so that the soil mineralogy data did not provide an evidential link between the footwear, vehicle and the crime scene. However, the offender's footwear and clothing were in contact with vegetation at the body deposition site, and palynological trace evidence demonstrated a strong evidential link between the clothing, footwear, vehicle and crime scene. One of the offenders admitted guilt; both were found guilty of murder.

Methodology

All of the soil samples for mineralogical analysis were placed within 30 mm diameter moulds and mixed with a laboratory grade epoxy resin. The samples were then placed in a pressure vessel overnight, before being coded and the resin moulds back-filled. The blocks were left to cure in an oven at 50 °C overnight. The surface of the resin blocks were then carefully cut back and polished so that the mineral grains in the sample were exposed on the polished surface of the block. Extreme care has to be taken during the polishing stage: too little polishing and the mineral particles are not exposed; too much and there is the potential for the sample to be destroyed, particularly with small samples which are also usually the most forensically important. Once the samples are polished they are carbon coated prior to automated mineral analysis. The sample preparation method preserves the original soil particles within the resin block so that subsequently they could be re-examined using any electron-beam analytical approach.

There are currently four main modes of QEMSCAN mineral analysis (Pirrie et al. 2004) summarised here. The bulk mineral analysis (BMA) mode enables the complete surface of the polished block to be analysed, with a series of scan lines stepping across the sample at a line spacing slightly greater than the average particle size (so that any one particle is only intercepted once). The system scans along the line until the BSE signal shows that the beam has intercepted a mineral particle rather than the surrounding resin, at which point a series of EDS analyses are collected at an operator-defined beam stepping interval. This analytical approach provides quantitative modal mineralogical data, but does not provide particle by particle data such as shape, texture or particle size.

The most applicable measurement mode for forensic investigations is particle mineral analysis (PMA), whereby the system searches the block for discrete mineral particles. Once a particle has been located, the system rasters the electron beam across the particle at a predefined stepping interval, effectively mapping the surface chemistry of the grain. Once a particle has been mapped, the system moves on to the next grain and repeats the chemical mapping process. The QEMSCAN system has an array of four EDS detectors. The working distance is very short and signal acquisition is very rapid, therefore allowing in excess of 4,000 mineral grains to be analysed in each sample. This estimate is based on typically obtaining in excess of 150,000 discrete EDS analyses in a sample in a 3 to 4 h instrumental run time.

Although data acquisition is operator independent, during a particle mineral analysis the operator is required to define two key parameters: (1) the beam-stepping interval (which can be as small as 0.2 μm); and (2) the size range of particles to be analysed. The smaller the beam-stepping interval, potentially the higher the resolution of the resultant dataset. However, the smaller the stepping interval the longer the analysis takes and therefore the greater the cost of the analysis. There is, therefore, a time/resolution trade-off and, based on unpublished data, a beam-stepping interval of 6 μm provides the optimum balance between data resolution and analytical time.

The spot size for the electron beam is 2 to 3 μm, but utilising the QEMSCAN data processing software, and through careful definition of mineral categories, it is possible to disentangle overlapping analyses such that meaningful data can be measured at a resolution of 0.2 μm. The second key parameter defined by the operator is the size range of the particles to be analysed. Again this is an issue of analytical time versus data resolution, as it is possible to analyse particles down to the micron scale, which has sometimes been required (e.g. in the analysis of atmospheric aerosol particles, Martin et al. in press). In forensic casework we have found that, in most soils, the examination of the 20 to 400 μm particle size range is optimal.

A very significant advantage of QEMSCAN is the ability, if required, to filter the acquired dataset, based on particle size. The QEMSCAN software allows a 'bulk' analysis to be reported by particle size fraction, thus focusing on the modal mineralogy of particles below or above a certain size of particular interest. This enables a like-for-like comparison of the mineralogy of the same particle size range in different types of sample.

The QEMSCAN can also be operated in trace mineral search (TMS) (also referred to as the specific mineral search SMS) mode. In this mode, candidate particles are defined on the basis of their mean atomic number (backscattered electron coefficient). Particles with a mean atomic number outside the defined BSE range are excluded from analysis but those within the range are located and mapped at an operator-defined pixel spacing. This analytical mode is best suited to the search and characterisation of rare 'bright' phases.

Finally, the fieldscan operating mode can be imagined as placing an operator-defined grid over the surface of the sample with EDS analyses being carried out at each intersection point of the grid. This analysis mode is best suited to the characterisation of non-particulate materials such as rock chips and polished sections of ceramics.

In each mode of measurement, each individual EDS analysis is automatically compared with a table of known spectra (the species identification protocol or SIP) and a mineral or chemical phase name is assigned to each analysis point. This raw data file (the primary mineral list) is then processed, so that similar EDS analyses can be grouped together into a manageable mineral listing (the secondary mineral list). For example, the primary mineral list might include a number of spectra that relate to different plagioclase feldspar compositions in the albite to anorthite solid solution series. These can be combined together into a single 'plagioclase' group, or reported separately, depending on the significance or otherwise of this phase in the particular study. The mineral/chemical phase processing categories used in this study are shown in Table 26.1.

Table 26.1 QEMSCAN mineral/chemical phase categories used in the data processing

QEMSCAN processed category	Minerals and phases included within QEMSCAN category
Quartz	Silica group of minerals (e.g. quartz, opal, tridymite, cristobalite, etc.)
Plagioclase	Albite to anorthite solid solution
Orthoclase	Albite to orthoclase solid solution including sanidine and microcline
Muscovite	Muscovite and alteration products such as sericite
Kaolinite	Kaolin group such as halloysite and kaolinite
FeAlK silicates	Fe-bearing K-Al-silicates such as Fe-stained illite. May include anthropogenic materials such as slag and fly ash
FeCaAl silicates	Fe-bearing Ca-Al-silicates. May include Fe-bearing smectites and Fe-stained smectite
MgFeAl silicates	Silicate particles containing Mg, Al and Fe such as Mg-rich chlorite (clinochlore) and hornblende
MgFeK silicates	Silicate particles containing Mg, Fe and K such as biotite, phlogopite and glauconite
MgFe silicates	Mg- and/or Fe-silicates such as olivine and pyroxene. May also include very fine mixtures of Fe oxide and silica
Micrite	Calcium carbonates with admixed silicates
CaAl silicates	Ca- and Al-rich silicates (excluding plagioclase). May include some minerals of the smectite group and also very fine mixtures of calcium carbonate and clay minerals interpreted to be portland cement
Mn silicates	Mn silicates such as spessartine. May include intimate mixtures of Mn oxides and silica
Zircon	Zircon
Mg calcite	High Mg calcite
Calcite	Calcite, aragonite and other Ca carbonates
Apatite	Apatite and Ca phosphates
Ti phases	Ilmenite, rutile (and its polymorph anatase). May also include anthropogenic products that use these minerals as filler such as paints
Fe-sulphides	Pyrite and pyrrhotite
$Fe\text{-}O_x/CO_3$	Fe-oxide group such as magnetite, hematite, goethite and may include anthropogenic Fe-rich phases such as welding fumes
Al oxides	Aluminium oxides. May also include Al metal
Zn mineral	Zn Fe phase (Fe oxides/hydroxides containing Zn – e.g. corrosion products of galvanized steel)
Ca sulphates	Gypsum and anhydrite
Others	Any other phases not included above

With any analytical technique there are two key questions: (1) What is the analytical accuracy? (2) What is the analytical precision (see Pye et al. 2006a)? In this chapter we address the second question; the accuracy of SEM-EDS analysis is already well documented (e.g. Reed 2005). The data presented are based on PMA analyses carried out at a 6 μm beam stepping interval for particles ranging in size between 20 and 400 μm. In the initial forensic investigation, each sample was analysed once using the same operating parameters.

To test the reproducibility of the analyses, three soil samples were selected, one each from: (1) the body deposition site; (2) the sole of an item of footwear; (3) recovered by washing the rubber cover of a vehicle accelerator pedal.

Each of these three samples was analysed ten times using the same operating, measurement and processing parameters. During a PMA analysis, the operator defines the number of particles to be analysed. Commonly, this is a subset of the total population of grains present in the polished block (because the area actually measured is randomised) and during the repeat analysis of the same sample it is unlikely that exactly the same subset of particles will be measured. In most studies the number of particles to be analysed is set at 4,000. However, in some very small samples, fewer particles may be present in the sample volume analysed. In most traditional sedimentological provenance studies, a minimum of 300 grains is usually considered necessary for a dataset is to be considered to be statistically valid. The data for the repeat measurements of the same samples can be compared with the data for the single analyses for the different samples from the different locations.

Results

The analytical reproducibility of the QEMSCAN analysis is tested here by comparing the modal mineralogical data for repeat analyses of the same samples with individual analyses from related samples. All analyses were performed using the same operating, measurement and processing parameters. Data are reported as modal mineralogy (i.e. the relative abundance of each mineral/chemical phase grouping reported as a percentage).

Mineralogy of Soil Samples Recovered from the Body Deposition Site

Previous studies have shown that there is commonly considerable small-scale spatial variation in soil characteristics (McKinley and Ruffell 2007; Pye et al. 2006b). Consequently, the number of samples collected from a crime scene needs to reflect the observed variability at the scene (McKinley and Ruffell 2007), although it should be noted that the variability may not be obvious until the samples are analysed. The mineralogical data for the nine samples are shown graphically in Figure 26.1a. These data are based upon analysis of several thousand individual mineral grains, and up to over one million individual EDS analyses per sample (Figure 26.1a).

The soil samples from the body deposition site are dominated by MgFeAl silicates, quartz, orthoclase and FeCaAl silicates, along with less abundant plagioclase, kaolinite, FeAlK silicates, MgFe silicates, micrite, CaAl silicates, calcite and Fe-O_x/CO_3 (Figure 26.1a). Minor or trace phases present in these samples are: muscovite, MgFeK silicates, Mn silicates, zircon, Mg calcite, apatite, Ti phases, Fe sulphides, Al oxides, Ca sulphates and 'others' (Figure 26.1a). As shown in Figure 26.1a, although there are considerable similarities between some of the samples from the body deposition site (e.g. samples 1, 3, 4, 7 and 8), there is also variation which reflects the degree of spatial variation which typically occurs in even a small area

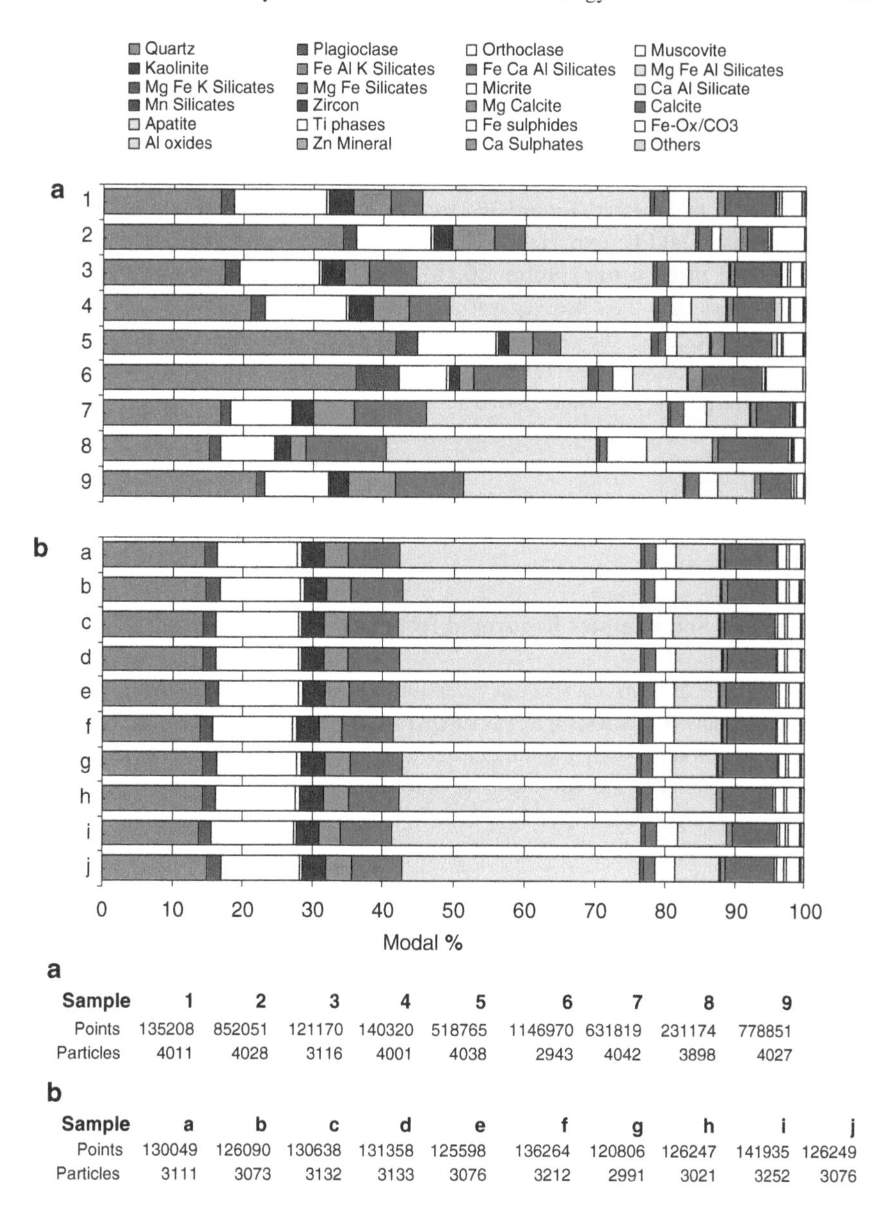

a

Sample	1	2	3	4	5	6	7	8	9
Points	135208	852051	121170	140320	518765	1146970	631819	231174	778851
Particles	4011	4028	3116	4001	4038	2943	4042	3898	4027

b

Sample	a	b	c	d	e	f	g	h	i	j
Points	130049	126090	130638	131358	125598	136264	120806	126247	141935	126249
Particles	3111	3073	3132	3133	3076	3212	2991	3021	3252	3076

Fig. 26.1 Modal mineralogical data based on QEMSCAN analysis. (a) Samples analysed from the crime scene showing the spatial variability in the modal mineralogy. (b) Repeated analyses for the same sample from the crime scene. Legend reads left-to-right with respect to shading of sectors in bars. Indent table shows number of analysis points and number of particles analysed respectively

(in this case the samples were collected from an area of approximately 12 m^2). In terms of the measured mineral abundance there is greater variation for the commonly occurring minerals (e.g. quartz ranges between 15.08% and 41.68%, orthoclase 6.75% to 13.15% and MgFeAl silicates 8.81% to 34.21%), and moderate variation in the less abundant phases (e.g. Ti phases 0.16% to 0.92%, zircon 0.02% to 0.23% and apatite 0.08% to 1.07%). This variation represents the naturally occurring spatial heterogeneity of surface soils and is typical of what we have observed in other criminal forensic investigations.

Sample 3 was re-analysed ten times using the same operating, measurement and data processing parameters (Figure 26.1b). There is some variation in the reported mineralogy, as would be expected from slightly different populations of particles being analysed in each run (Figure 26.1b). However, the degree of variability is significantly less than the observed variability in mineralogy between the different samples analysed from the body deposition site. For example, measured quartz abundance varies between 13.73% and 14.90% with eight of the analyses falling within the range 14.17% and 14.90%. Orthoclase abundance ranges between 11.22% and 11.72%, whilst MgFeAl silicates range between 33.32% and 35.05%. There is also significantly less variance for the minor phases. For example, ranges in measured abundance are 0.92% to 1.14% for Ti phases, 0.04% to 0.06% for zircon and 0.26% to 0.34% for apatite.

Mineralogy of Soil Samples Recovered from Footwear

An item of footwear from a suspect in the inquiry was submitted for palynological and mineralogical analysis. Each sample was subdivided into two aliquots, one for mineralogy and one for palynology. Only the mineralogical data are presented in this chapter.

The mineralogical data for the four samples from this item of footwear are presented in Figure 26.2a. These four samples are all dominated by abundant CaAl silicates, micrite, FeCaAl silicates and quartz (Figure 26.2a). Other relatively common phases are plagioclase, orthoclase, FeAlK silicates, MgFeAl silicates, MgFe silicates, Mg calcite, calcite and Fe-O_x/CO_3. Minor or trace phases present are: muscovite, kaolinite, MgFeK silicates, Mn silicates, zircon, apatite, Ti phases, Fe sulphides, Al oxides, Ca sulphates and 'others'. The four samples are very similar in overall mineralogical profile, although there is a subtle difference between the uppers and the soles (Figure 26.2a). The similarity in the mineralogy of the four samples suggests that both the left and right item of footwear and the uppers and soles have come into contact with particles from the same surface(s).

The mineralogical variability in these four samples is much less than for the samples from the body deposition site with, for example, 9.34% to 11.49% quartz, 22.50% to 28.59% CaAl silicates and 22.25% to 28.92% micrite. With the minor/trace phases, again the observed variability is quite low with, for example, 0.28% to 0.36% Ti phases, 0.02% to 0.03% zircon and 0.10% to 0.15% apatite (Figure 26.2a).

In this study the mineralogical data for each sole are directly comparable. However, to test instrument variability, the sample from the left sole was selected for repeat analysis. This sample was selected because in the original measurement only approximately 700 mineral grains were identified, which is higher than the 300 grains often accepted as a minimum for statistical analysis, but considerably less than the 3,000 to 4,000 grains usually sought.

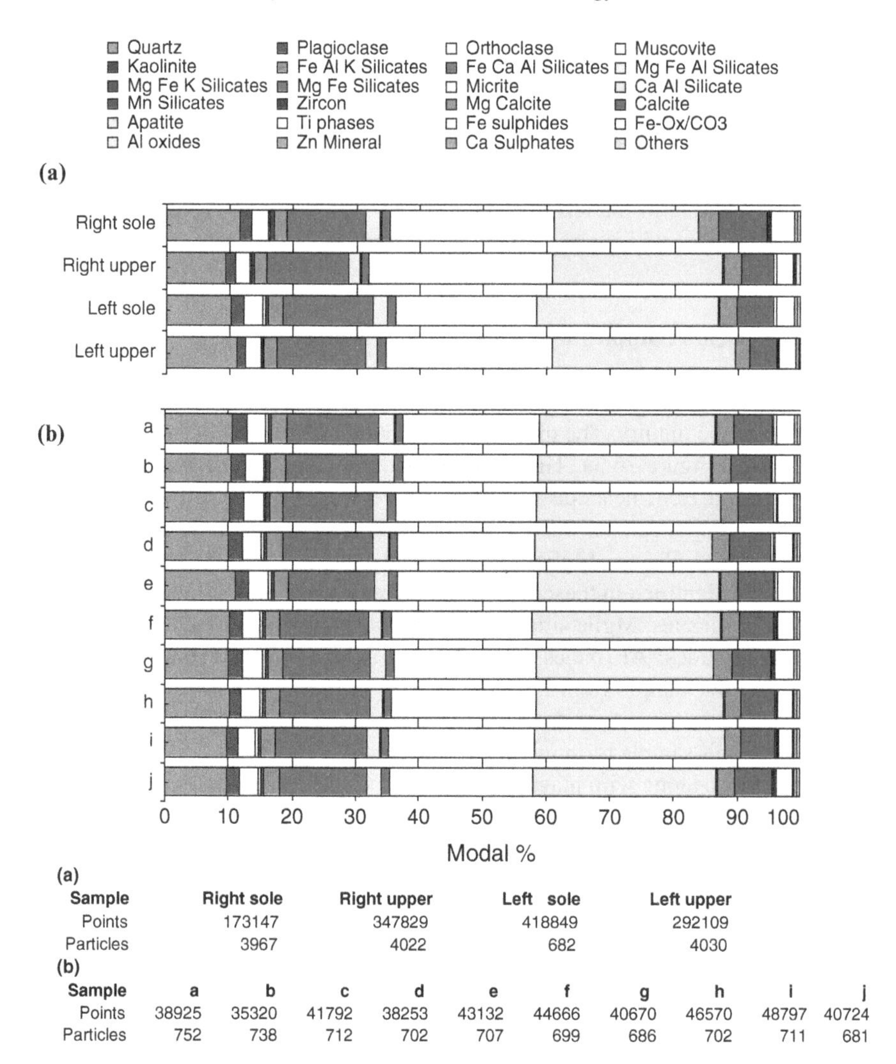

(a)

Sample	Right sole	Right upper	Left sole	Left upper
Points	173147	347829	418849	292109
Particles	3967	4022	682	4030

(b)

Sample	a	b	c	d	e	f	g	h	i	j
Points	38925	35320	41792	38253	43132	44666	40670	46570	48797	40724
Particles	752	738	712	702	707	699	686	702	711	681

Fig. 26.2 Modal mineralogical data based on QEMSCAN analysis. (a) Four samples analysed from the uppers and soles of the left and right item of footwear. (b) Repeated analyses for the sample from the sole of the left item of footwear. Legend reads left-to-right with respect to shading of sectors in bars. Indent table shows number of analysis points and number of particles analysed respectively

The sole of the left item of footwear was quite clean; the sample was removed from the sole only and was subdivided for both the palynological and mineralogical analysis. Consequently, the resultant mass of material for mineralogical analysis was significantly less than 0.01 g. No other available technique could provide quantitative mineralogical data for as small a sample volume.

The ten repeated measurements of the sample using the same operating, measurement and processing parameters resulted in little variation between the ten repeated analyses (Figure 26.2b). Once again, the variation seen in the repeated analyses is less than that observed between the four samples analysed from the items of footwear. Quartz abundance ranges between 9.61% and 11.13%, CaAl silicates 27.16% and 29.76% and micrite 21.19% and 22.92% (Figure 26.2b). Variability for the minor/trace mineral phases is also low (e.g. Ti phases 0.31% to 0.40%, 0% to 0.07% zircon and 0.14% to 0.17% apatite). The dataset for the repeated measurement of the sample from the sole of the left item of footwear shows slightly greater variability than that for the repeated analysis of the soil sample from the body deposition site.

Mineralogy of Soil Samples Recovered from a Motor Vehicle

As described above, a number of samples were collected from a motor vehicle used by a suspect in the inquiry. The mineralogical data for the samples from the vehicle are presented in Figure 26.3a. The modal mineralogy of the two samples from the front offside and front nearside footwells are very similar to each other and are dominated by quartz, calcite, CaAl silicates and Fe-O_x/CO_3 along with plagioclase, orthoclase, FeCaAl silicates, MgFeAl silicates and micrite (Figure 26.3a). Minor/trace mineral phases identified in these two samples are muscovite, kaolinite, FeAlK silicates, MgFeK silicates, MgFe silicates, Mn silicates, zircon, Mg calcite, apatite, Ti phases, Fe sulphides, Al oxides, Ca sulphates and 'others' (Figure 26.3a). The mineralogy of the sample from the rubber cover of the accelerator pedal is markedly different to the sweepings from the footwells of the vehicle (Figure 26.3a). The mineralogy of this sample is dominated by micrite, CaAl silicates, FeCaAl silicates, quartz and calcite along with plagioclase, orthoclase, FeAlK silicates, MgFeAl silicates, MgFe silicates, Mg calcite, Ti phases and Fe-O_x/CO_3 (Figure 26.3a). Trace mineral phases identified in this sample are muscovite, kaolinite, MgFeK silicates, Mn silicates, zircon, apatite, Fe sulphides, Al oxides, Ca sulphates and 'others' (Figure 26.3a). The mineralogical data suggest that a discrete assemblage of particles is present on the vehicle pedal compared to those present in the vehicle footwells.

The sample from the vehicle pedal was selected for repeat analysis; the selection of this sample was because in a 'real' forensic enquiry particles on a vehicle pedal might have greater evidential importance than the overall assemblage of particles in a vehicle footwell. This sample was analysed ten times using the same operating, measurement and processing parameters (Figure 26.3b). The repeated analyses of this sample show the smallest degree of variability of any of the data sets for repeated analyses. Quartz varies between 10.09% and 10.42%, FeCaAl silicates between 14.14% and 14.66%, and CaAl silicates between 24.42% and 25.23%. There is also less variability for the more minor mineral phases (e.g. Ti phases 0.92% to 1.01%, zircon 0.13% to 0.16%, apatite 0.19% to 0.25%).

Comparison of the Mineralogical Results

The overall aim of a forensic mineralogical examination is, as described by Brown (2006), to provide evidence of both the closeness of an association (i.e. how closely

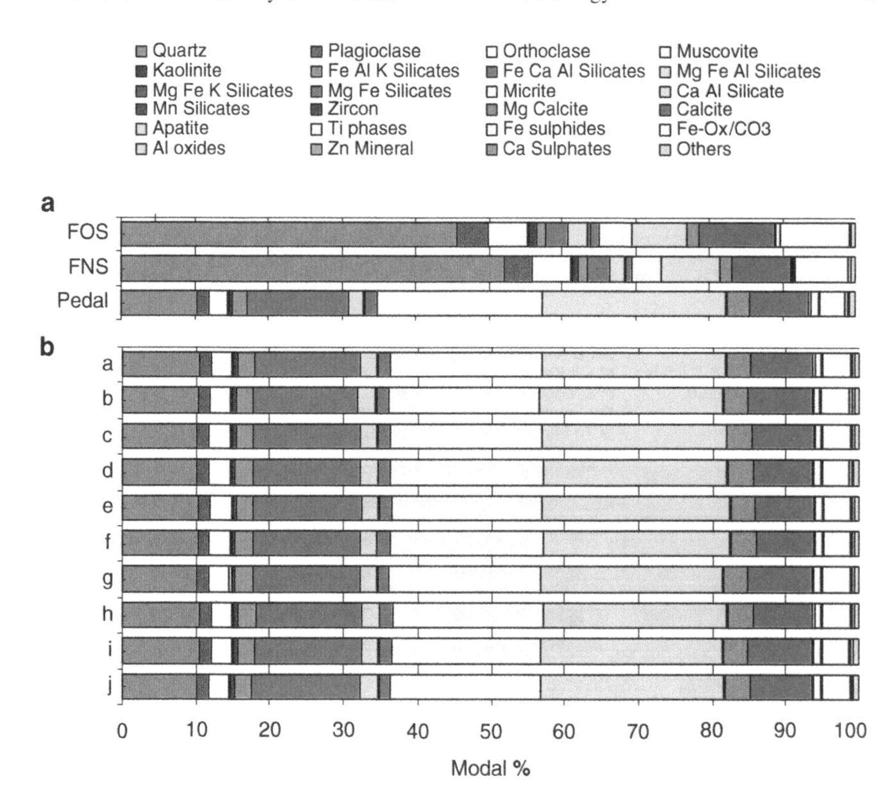

Fig. 26.3 Modal mineralogical data based on QEMSCAN analysis. (a) Samples analysed from the front offside (FOS) vehicle footwell, front nearside (FNS) vehicle footwell and from the rubber accelerator pedal cover (pedal). (b) Repeated analyses for the sample from the rubber accelerator pedal cover from the motor vehicle. Legend reads left-to-right with respect to shading of sectors in bars. Indent table shows number of analysis points and number of particles analysed respectively

two samples 'match') and to test the uniqueness of that association. Others would argue that it is not possible to look for a 'match' or test an association in forensic geoscience, and that geological data are probabilistic and should be used to 'exclude' rather than to 'match' (see Morgan and Bull 2007). The closeness of an association is easy to test, but the absence of a detailed database of soil mineralogy (at a resolution that reflects the observed variability in soil mineralogy of the order of 1 to 10 m) means that testing the uniqueness of an association is more difficult. Commonly in combined forensic mineralogical and palynological studies, the palynological data are of a slightly better spatial resolution (Brown 2006).

The data from the body deposition site, the suspect's footwear and the suspect's motor vehicle are compared in Figure 26.4a and 26.4b. Based on these results the mineralogy of the samples from the footwear is markedly different to the mineralogy of the samples collected from the body deposition site (Figure 26.4a). Based on these data there is no association between these two groups of samples, and the crime scene could be excluded as a likely place for the transfer of this population of soil particles to the footwear. These observations do not necessarily mean that the particular item of footwear had not been at the body deposition site, just that the overall assemblage of mineralogical particles on the footwear was not derived from the site.

The data for the three samples from the motor vehicle when compared with the samples from the body deposition site can also be interpreted to suggest that the sample from the accelerator pedal cover was not derived from the body deposition site (Figure 26.4b). There are, however, some similarities between the assemblage of par-

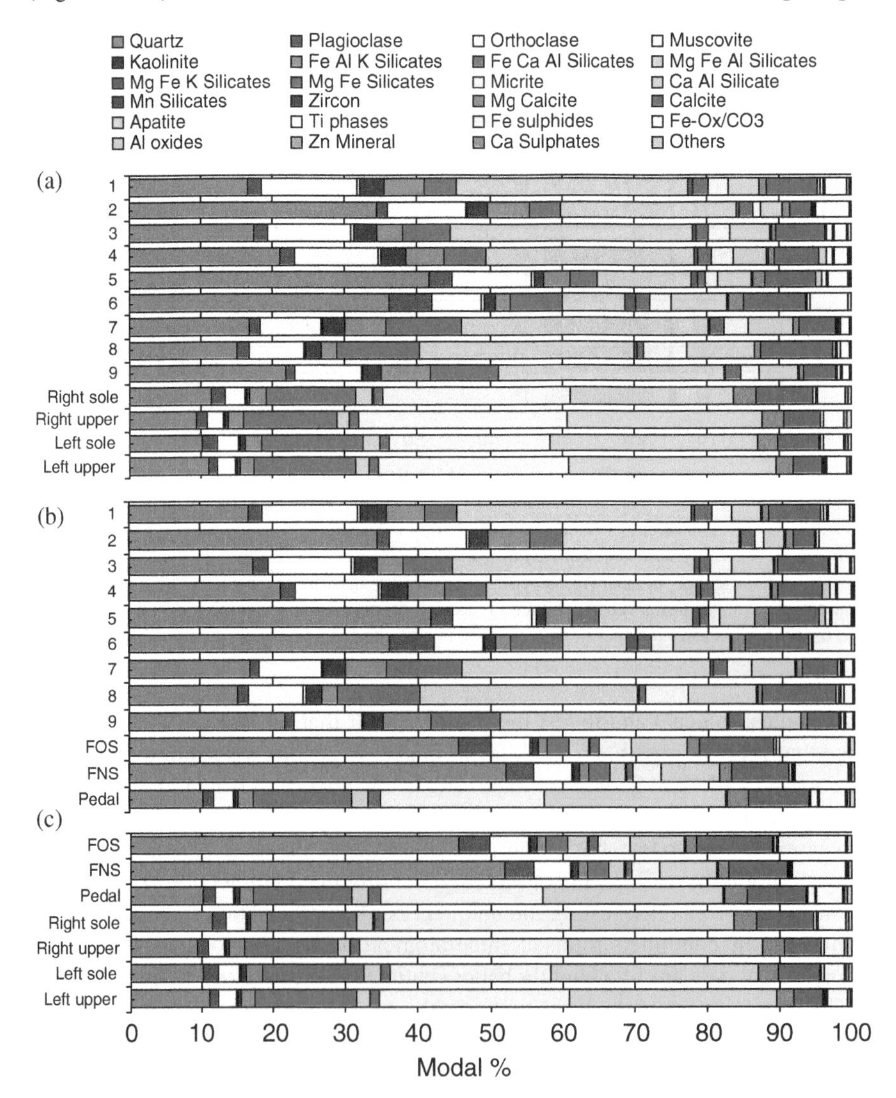

Fig. 26.4 Comparisons of the samples from the crime scene, the motor vehicle and the footwear. (a) Comparison of footwear vs. crime scene. (b) Comparison of vehicle vs. crime scene. (c) Comparison of footwear vs. motor vehicle. Legend reads left-to-right with respect to shading of sectors in bars

ticles in the vehicle footwells and some of the samples from the body deposition site, such that it would not be possible to exclude the possibility that some of the particles in the vehicle footwells could have been derived from the body deposition site.

Figure 26.4c compares the mineralogy of the samples from the item of footwear with the samples from the motor vehicle. There is a distinct difference in the overall mineralogical assemblage of particles present in the front offside and nearside footwells of the motor vehicle and the footwear samples, and a marked similarity between the mineralogy of the samples from the item of footwear and the sample from the rubber cover of the accelerator pedal. These data strongly suggest that the mineral particles on the accelerator pedal and on the footwear could have been derived from the same place. Because the same population of particles are present on both the soles and the uppers of the footwear, then it is probable that they were transferred from the sole of the footwear to the accelerator pedal cover rather than *vice versa.* The source of these particles is unknown.

The data for the repeated analyses of the soil sample from the body deposition site, the sample from the left sole of the footwear and from the accelerator pedal are shown in Figure 26.5. In each graph, all ten analyses are plotted although the correspondence of the data is such that the individual analyses cannot be easily resolved from these graphs. The data clearly demonstrate that, whilst the assemblage of mineralogical particles on the footwear and on the accelerator pedal could not have been derived from the body deposition site, they were probably derived from the same place. Although simplistic, graphs such as this are particularly valuable when presenting complex evidence to a jury in a criminal trial.

Conclusions

The analytical strategy used in any criminal or civil forensic investigation needs to be tailored to the specific details and requirements of that particular investigation. In many cases, the integration of a range of methodologies may provide the best discrimination between different samples, such as the integration of soil mineralogy and palynology (e.g. Brown 2006; Brown et al. 2002; Rawlins et al. 2006). There are both advantages and limitations to all of the techniques of mineral analysis available, but automated SEM-EDS analysis has proven to be a very robust technique in that large datasets can be collected from even very small samples. Data acquisition is rapid and operator independent.

In any analytical technique there are a number of variables which may affect the resultant dataset. This starts with sample collection and the preparation of the sample for analysis; indeed this is often critical to the outcome of the forensic investigation. If instrumental analysis is carried out then we need to consider both instrumental accuracy and instrumental variability. In this study we show that the analytical variability of automated mineral analysis using QEMSCAN is significantly less

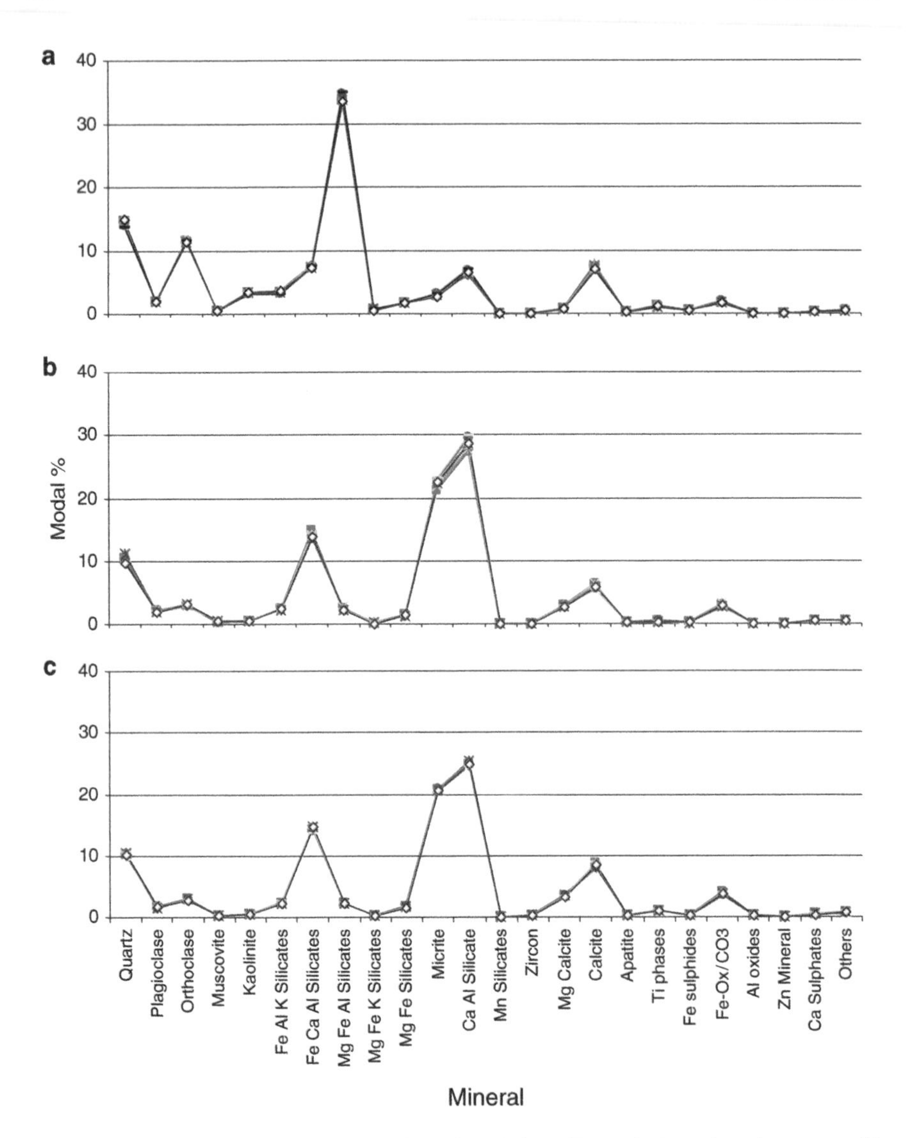

Fig. 26.5 Comparison of the repeated analyses for (a) the soil sample from the crime scene; (b) the suspect's footwear; (c) the vehicle accelerator pedal

than the observed natural variability between different samples, such that a single analysis can be considered representative of that individual sample.

Acknowledgements We gratefully acknowledge discussions regarding aspects of this research with colleagues working in forensic geoscience. Technical support was provided through the Universities of Exeter and Gloucester. DP acknowledges Intellection Pty Ltd. (Australia), and in particular Chief Scientist Dr. Alan Butcher, for their support of this research. Our practical forensic casework experience has been developed in collaboration with many of the UK police forces who are duly acknowledged. We are very grateful to Karl Ritz for significantly improving our figures.

References

ASTM (2002). Standard guide for gunshot residue analysis by scanning electron microscopy/ energy dispersive spectroscopy. Annual Book of ASTM Standards 2002:528–530.

Benvie B (2007). Mineralogical imaging of kimberlites using SEM-based techniques. Minerals Engineering 20:435–443.

Brown AG (2006). The use of forensic botany and geology in war crimes investigations in NE Bosnia. Forensic Science International 163:204–210.

Brown AG, Smith A and Elmhurst O (2002). The combined use of pollen and soil analyses in a search and subsequent murder investigation. Journal of Forensic Sciences 47:614–618.

Brozek-Mucha Z and Jankowicz A (2001). Evaluation of the possibility of differentiation between various types of ammunition by means of GSR examination with SEM-EDX method. Forensic Science International 123:39–47.

Bull PA and Morgan RM (2006). Sediment fingerprints. Science and Justice 46:107–124.

Camm GS, Butcher AR, Pirrie D, Hughes PK and Glass HJ (2003). Secondary mineral phases associated with a historic arsenic calciner identified using automated scanning electron microscopy: a pilot study from Cornwall, UK. Minerals Engineering 16:1269–1277.

Dawson LA, Towers W, Mayes RW, Craig J, Väisänen RK and Waterhouse EC (2004). The use of plant hydrocarbon signatures in characterising soil organic matter. In: Forensic Geoscience: Principles, Techniques and Applications (Eds. K Pye and DJ Croft), pp 269–276. The Geological Society of London Special Publication 232, London.

Goodall WR and Scales PJ (2007). An overview of the advantages and disadvantages of the determination of gold mineralogy by automated mineralogy. Minerals Engineering 20:506–517.

Knight RD, Klassen RA and Hunt P (2002). Mineralogy of fine-grained sediment by energy-dispersive spectrometry (EDS) image analysis – a methodology. Environmental Geology 42:32–40.

Lotter NO, Kowal DL, Tuzun MA, Whittaker PJ and Kormos L (2003). Sampling and flotation testing of Sudbury Basin drill core for process mineralogy modelling. Minerals Engineering 16:857–864.

Martin RS, Mather TA, Pyle DM, Power M, Allen AG, Aiuppa A, Horwell CJ and Ward CPJ (in press). Composition resolved size distributions of volcanic aerosols in the Mt Etna plumes. Journal of Geophysical Research.

Marumo Y, Nagatsuka S and Oba Y (1986). Clay mineralogical analysis using the <0.05 mm fraction for forensic science investigation – its application to volcanic ash soils and yellow-brown forest soils. Journal of Forensic Sciences 31:92–105.

McCrone WC (1992). Forensic soil examination. Microscope 40:109–121.

McKinley J and Ruffell A (2007). Contemporaneous spatial sampling at scenes of crime: advantages and disadvantages. Forensic Science International 172:196–202.

McVicar MJ and Graves WJ (1997). The forensic comparison of soils by automated scanning electron microscopy. Journal of the Canadian Society of Forensic Scientists 30:241–261.

Mildenhall DC, Wiltshire PEJ and Bryant VM (2006). Forensic palynology: why do it and how it works. Forensic Science International 163:163–172.

Morgan RM and Bull PA (2007). Forensic geoscience and crime detection. Minerva Medicolegale 127:73–89.

Morgan RM, Wiltshire PEJ, Parker A and Bull PA (2006). The role of geoscience in wildlife crime detection. Forensic Science International 162:152–162.

Morgan RM, Little M, Gibson A, Hicks L, Dunkerley S and Bull PA (in press). The preservation of quartz grain surface textures following vehicle fire and their use in forensic enquiry. Science and Justice.

Pirrie D, Power MR, Rollinson G, Camm GS, Hughes SH, Butcher AR and Hughes P (2003). The spatial distribution and source of arsenic, copper, tin and zinc within the surface sediments of the Fal Estuary, Cornwall, UK. Sedimentology 50:579–595.

Pirrie D, Butcher AR, Power MR, Gottlieb P and Miller GL (2004). Rapid quantitative mineral and phase analysis using automated scanning electron microscopy (QEMSCAN®); potential applications in forensic geoscience. In: Forensic Geoscience: Principles, Techniques and Applications (Eds. K Pye and DJ Croft), pp 123–136. The Geological Society of London Special Publication 232, London.

Pye K (2004). Forensic examination of sediments, soils, dusts and rocks using scanning electron microscopy and X-ray chemical microanalysis. In: Forensic Geoscience: Principles, Techniques and Applications (Eds. K Pye and DJ Croft), pp 103–122. The Geological Society of London Special Publication 232, London.

Pye K and Croft DJ (2007). Forensic analysis of soil and sediment traces by scanning electron microscopy and energy-dispersive X-ray analysis: an experimental investigation. Forensic Science International 165:52–63.

Pye K, Blott SJ and Wray DS (2006a). Elemental analysis of soil samples for forensic purposes by inductively coupled plasma spectrometry – precision considerations. Forensic Science International 160:178–192.

Pye K, Blott SJ, Croft DJ and Carter JF (2006b). Forensic comparison of soil samples: Assessment of small-scale spatial variability in elemental composition, carbon and nitrogen isotope ratios, colour, and particle size distribution. Forensic Science International 163:59–80.

Rawlins BG, Kemp SJ, Hodgkinson EH, Riding JB, Vane CH, Poulton C and Freeborough K (2006). Potential and pitfalls in establishing the provenance of earth-related samples in forensic investigations. Journal of Forensic Sciences 51:832–845.

Reed SJB (2005). Electron Microprobe Analysis and Scanning Electron Microscopy in Geology. Cambridge University Press, Cambridge.

Ruffell A and Wiltshire PEJ (2004). Conjunctive use of quantitative and qualitative X-ray diffraction analysis of soils and rocks for forensic analysis. Forensic Science International 145:13–23.

Sugita R and Marumo Y (1996). Validity of colour examination for forensic soil identification. Forensic Science International 83:201–210.

Sugita R and Marumo Y (2001). Screening of soil evidence by a combination of simple techniques: validity of particle size distribution. Forensic Science International 122:155–158.

Wiltshire PEJ (2006). Consideration of some taphonomic variables of relevance to forensic palynological investigations in the United Kingdom. Forensic Science International 163:173–182.

Xiao Z and Laplante AR (2004). Characterizing and recovering the platinum group minerals - a review. Minerals Engineering 17:961–979.

Xie RK, Seip HM, Leinum JR, Winje T and Xiao JS (2005). Chemical characterization of individual particles (PM10) from ambient air in Guiyang City, China. Science of the Total Environment 343:261–272.

Chapter 27
Rapid, Reliable and Reviewable Mineral Identification with Infrared Microprobe Analysis

Brooke A. Weinger, John A. Reffner and Peter R. De Forest

Abstract Infrared microprobe analysis (IRMA) is a valuable technology for mineral identification. Prior to the discovery of X-ray diffraction, the polarised light microscope was the principal tool for identifying minerals. While X-ray diffraction and electron beam microanalysis made major contributions to mineralogy, IRMA contributes unique information about the chemistry of complex minerals. Using diamond internal reflection to collect infrared (IR) spectra of minerals is an approach that extends optical mineralogy to new levels. It is simple, direct and rapid. Traditional whole grain mounts or petrographic sections, used for polarised light microscopy, are analysed by IR internal reflection. Attenuated total reflection (ATR) spectra are collected by touching the diamond internal reflection element to areas of interest. Silicates, carbonates, nitrates, phosphates, hydrates and other minerals provide unique spectra for identification. Using the diamond ATR microscope objective allows for the selective isolation of individual minerals for simultaneous collection of microscopic, optical and IR data, thus enabling the indisputable identification of most minerals. Applying IRMA to forensic soil analysis is an example of using this IR spectral method. A library of diamond ATR spectra of common soil minerals was prepared. These reference data let the analyst confirm the identity of mineral grains in soil evidence. Polarised light microscopical examination is the primary step. Each phase is then characterised by its ATR spectrum and searched through the spectral library. IRMA unites light microscopy with IR spectroscopy to create a powerful system capable of identifying mineral grains and correlating microstructure with chemistry. This technique has great utility in the analysis of material in a forensic context, particularly in relation to trace evidence.

B.A. Weinger(✉), J.A. Reffner, and P.R. De Forest
John Jay College of Criminal Justice/CUNY, 445 West 59th Street, New York, USA
email:bweinger@yahoo.com

B.A. Weinger and J.A. Reffner
Smiths Detection, 21 Commerce Drive, Danbury, CT, USA

K. Ritz et al. (eds.), *Criminal and Environmental Soil Forensics*

Introduction

Soil evidence can be, and has been, used in a plethora of ways in both criminal and civil cases. Notwithstanding the value of soil evidence, it is not used to its fullest potential by many forensic science laboratories due in large part to the lack of criminalists trained in forensic geology and the time required for soil analysis by the classical methods. To rectify this problem, there is a need to find supplementary methods to polarised light microscope analysis that decrease the time and increase the precision of mineral identification. This research demonstrates the use of infrared microprobe analysis (IRMA) as an efficient means for mineral identification and a complementary technique to polarised light microscopy.

Mineral identification is one of the most important aspects of soil discrimination because minerals are a major component of many soil varieties. Minerals are defined as "naturally occurring inorganic substance having relatively constant chemical composition and fairly definite physical properties" (Smith 1946). Most soil samples contain between three and five mineral varieties. Although thousands of minerals exist in nature, only about 20 are common in soil specimens (Murray and Tedrow 1992; Petraco and Kubic 2004). However, it is not enough just to know the most prevalent minerals found in soil because it is an unusual mineral variety that often has the potential to be the most useful in a case.

Infrared spectroscopy has been proven to be invaluable in the analysis of organic and inorganic covalent materials because the IR spectrum of a material is a physical constant. Materials that differ in molecular structure give rise to different IR absorption spectra; thus the IR spectrum can be used to identify an unknown by comparison with known standard reference spectra. Minerals are naturally occurring inorganic solids, and thus have characteristic spectra that correlate with their chemical formula and their crystal lattice. Microscopy and IR spectroscopy are united with the application of the IR microprobe to detect, document, evaluate, analyse and identify trace evidence. This technology is well established for the forensic analysis of fibre, illicit drug and paint evidence (e.g. Kirkbride and Tungol 1999; Beveridge et al. 2001). However, it is not widely applied to other types of evidence such as soil samples. Advances in the field of IRMA make it possible to perform fast identification of minerals found in soil evidence.

Instrumentation

The IlluminatIR™ microprobe represents a significant advance in the field of IR microspectroscopy. The original IR microspectrometer was a benchtop IR spectrometer with a microscope attachment, whereas the IlluminatIR is a Fourier Transform (FT) IR attachment for an infinity-corrected light microscope. This creates a unified instrument that combines polarised light microscopy with IR microspectroscopy, which allows for the ability to take quality optical measurements, photomicrographs and FT-IR spectrum simultaneously. The instrument has two objectives capable of IR microprobe analysis: the all-reflecting objective (ARO) which is used for reflection-absorption spectroscopy and the attenuated total internal reflection (ATR) objective, with set magnifications of 15x and 36x, respectively. It is the ATR objective that is used for the analysis of mineral grains.

The ATR objective uses a parabolic reflector to direct the IR radiation onto the diamond-sample interface. The IR beam from the interferometer enters on one side of the diamond and reflects from the diamond-sample interface, returning to a second parabolic reflector and then the detector. The visible light, which is co-axial with the IR light path, travels through the centre of the objective which enables the specimen to be viewed throughout the analysis. Diamond is an ideal material for an internal reflecting element (IRE). It has a high index of refraction (n = 2.4) and is transparent through most of the IR spectral region. It also has superior mechanical properties (such as its extreme hardness) and it is chemically inert, thus it can be used to analyse hard or corrosive samples.

Even though the IR light does not enter the sample, the electric field of the radiation penetrates some distance into the less refractive material. This electric field is called the evanescent wave. The distance the evanescent wave extends into the less refractive material is called the depth of penetration (d_p). Harrick (1987) defines the depth of penetration for incident radiation to be "the distance required for the electric field amplitude to fall to e^{-1} of its value at the surface". The depth of penetration depends on the difference in refractive indices of the two media (n_1 and n_2), the angle of incidence (θ), and the wavelength of the incident radiation (λ), based on Equation 1:

$$d_p = \lambda / [2\pi(n_1^2 \sin^2\theta - n_2^2)]^{1/2} \quad (1)$$

An important factor of this equation is the wavelength dependence. Equation 1 is rewritten as Equation 2 to reflect the relationship between the depth of penetration and the wavenumber (WN), which is the inverse of the wavelength in centimetres.

$$d_p = 1 / \{WN[2\pi(n_1^2 \sin^2\theta - n_2^2)]^{1/2}\} \quad (2)$$

As the wavelength increases (and the wavenumber decreases), so does the depth of penetration. This wavelength dependence on the depth of penetration is an important distinction between ATR and transmission spectra. The peak locations are similar but there is a difference in the peak intensities. ATR spectra, as compared to transmission spectra, will have lesser peak intensities in the high wavenumbers and more intense peaks in the lower wavenumbers. Although an ATR correction can be applied to correct for the path length variation at different wavelengths, it is best to compare an ATR spectrum with other ATR spectra and likewise for a transmission spectrum.

When the diamond IRE is used, the depth of penetration of the evanescent wave is approximately equal to ¼ of the wavelength of incident light. This is calculated from Equation 1, using the refractive index of diamond (n_1 = 2.4), a 45° angle of incidence ($\theta = 45°$), and a reasonable refractive index for the sample (n_2 = 1.5). The depth of penetration has a direct relationship with the wavelength of the infrared light, and increases linearly from approximately 0.5 to 4 μm. The depth of penetration's wavenumber dependence similarly increases from 0.5 to 4 μm (Figure 27.1).

When the evanescent wave penetrates into the sample, the radiation interacts with the sample. If portions of the incident radiation are absorbed by the sample, the reflected radiation will be lessened, or attenuated, at the wavelengths corresponding to the sample's absorption bands. This is the origin of the term 'attenuated total reflection'. In order for a sample to be analysed by ATR, it must have a refractive index lower than that of the IRE. If the refractive index of the sample is greater than that of the IRE, total internal reflection will not occur and the incident radiation will be transmitted across the boundary and into the sample. The refractive index of diamond is relatively high, so this is not usually a problem but, with this study of minerals, it became important.

The IlluminatIR has an internal, motorised and backlit masking aperture that is controlled by the computer software. This aperture defines the sample area that will be analysed using the IR spectroscopy capabilities of the IlluminatIR. The user can control the diameter of the IR beam spot on the sample to be analysed, with a range of apertures from 5 to 42 μm. This enables the user to analyse a mixture composed of varying particle sizes by customising the IR spot size for each particle. The IlluminatIR is a mid-infrared instrument with a spectral range from 4,000 to 650 cm^{-1}. The lower wavenumber is limited by the mercury/cadmium telluride (MCT) photoconductive detector. This precludes the use of the spectral region from 650 to 350 cm^{-1}, which can be highly diagnostic for many mineral varieties. This is a limitation of the instrument.

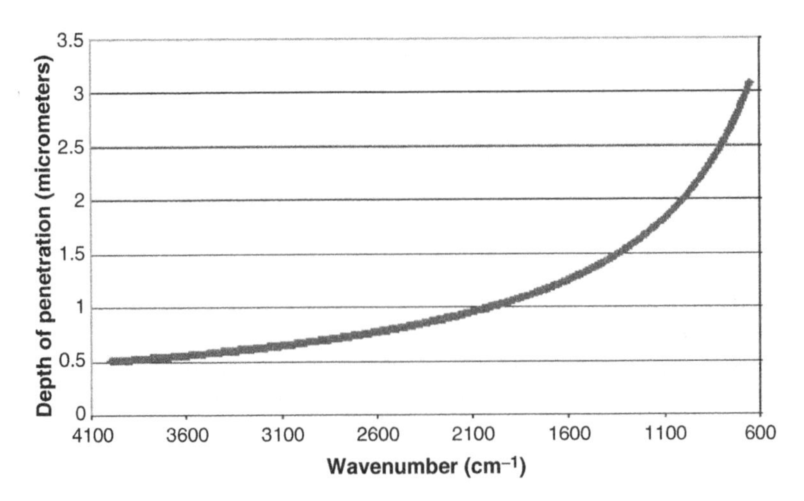

Fig. 27.1 Relationship between the depth of penetration and wavenumber, using the diamond IRE (n = 2.4), assuming that the refractive index of the sample is 1.5 and the angle of incidence is 45°

Sample Preparation Techniques

The preparation of solids for analysis with the diamond ATR objective is largely sample dependent because of the hardness of a sample. High-quality spectra are attained when good contact is made between the ATR objective and the sample, but the amount of force required for good contact can affect solids of varying hardness in different ways. A soft solid is able to yield with contact pressure, thus making its analysis simple in a dry mount preparation. In contrast, brittle solids do not compress with contact pressure, and will shatter when force is applied (Figure 27.2). This makes IRMA quite difficult because of the scattering of the fragments.

A solution to this is to mount brittle solids in immersion oil, as microscopists have done for many years. The mounting of a sample in oil has three major advantages: the highest quality image is obtained because of the reduction in visible light scattering; optical properties, such as refractive index, are able to be measured; and the viscosity of the oil prevents the scattering of the shattered fragments during crushing. However, IR microscopists and spectroscopists are concerned with the effect of the oil on the resulting spectra, since oils are organic liquids with some IR absorption. As contact pressure is being increased, the mineral's peaks increase and the presence of the oil decreases (Figure 27.3). The forcible contact between the diamond and the sample squeezes out the oil from between the two, thus reducing the path length through the oil while increasing the contribution of the mineral to the resulting IR spectrum without significant contribution from the oil.

In forensic soil analysis, the mineral fraction of the soil is separated from the organic fraction using a flotation or washing process, prior to obtaining a mineralogical profile. The IRMA analysis in this research is to be completed on the mineral fraction; thus there would not be any organic soil material in the sample being analysed. In addition, the oil used in the mineral analysis is a simple hydrocarbon with a simple IR spectrum while the organic components of soil have very complex spectra. The characterisation of the organic matter by IRMA could still be

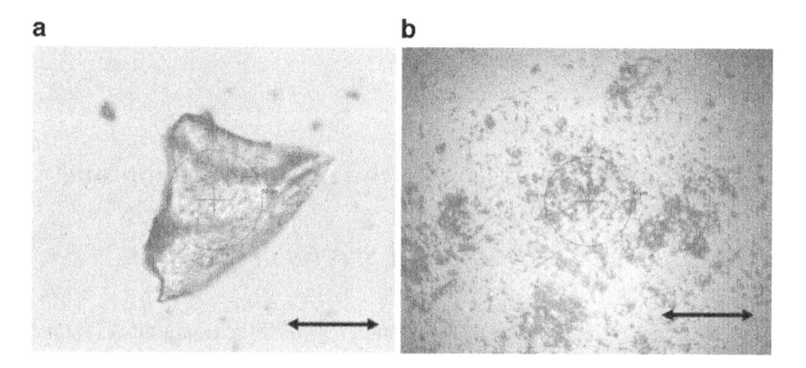

Fig. 27.2 (a) Photomicrograph of a dry-mounted opal sample, prior to analysis. (b) Same dry-mounted opal sample after being shattered and scattered by the diamond ATR objective. Scale arrow = 50μm

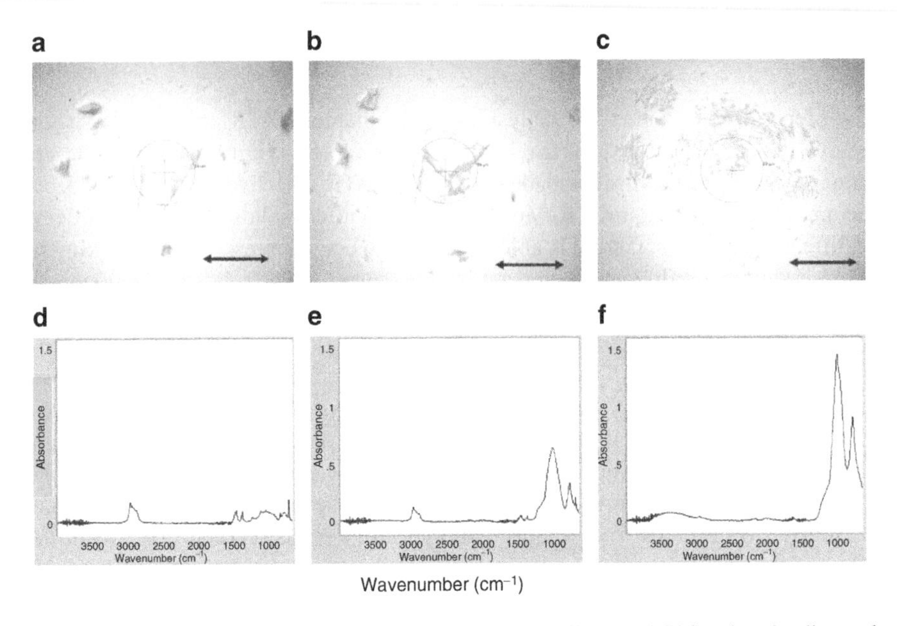

Fig. 27.3 Photomicrographs of an opal sample in mineral oil (nd = 1.516) using the diamond ATR objective, under increasing contact pressure. (a) After initial contact. (b) After significant contact. (c) After being crushed but not scattered. Scale arrow = 50 μm; (d–f) shows corresponding IR spectra, Grams AI software

accomplished by doing a spectral subtraction of the oil's spectrum from that of the organic matter or by removal of the oil.

The forensic analysis of soil seeks to identify the minerals present and, as such, the hardness of a questioned mineral is unknown. To prevent the loss of the sample by scattering during the crushing operation, mineral analysis by diamond ATR analysis should be performed on an oil-mounted sample. Unlike diffuse reflection and transmission IR analysis techniques, such as mulls and pellets, there is no particle size effect in ATR IR spectra. The crushing of minerals, which creates a range of particle sizes, does not alter the nature of the spectrum produced with this IRMA technique.

Results of Mineral Analysis with the Infrared Microprobe

Library of Diamond ATR Infrared Spectra of Minerals

Ninety-six different mineral varieties were analysed with the diamond ATR, and their spectra compiled into a mineral library database. All data were collected with the IlluminatIR, using a 42 μm IR beam diameter, 4 cm^{-1} resolution, 64 co-added scans and 64 co-added background scans. The ATR library was prepared of 85 different mineral varieties from the 96 mineral varieties analysed, with the exclusion of 11 minerals based on the limitations outlined in the next section. For each mineral variety, 10 to 20 spectra were normalised and averaged to make the library spectrum.

The minerals standards were obtained from two Cargille No M-4 Comminuted Minerals Reference Sets (R.P. Cargille Laboratories, Inc., Cedar Grove, NJ), a set of minerals as denoted in the Manual for Forensic Mineralogy (Wehrenberg 1998), clay minerals used by the American Petroleum Institute and Columbia University to compile Analytical Data on Reference Clay Materials (Kerr et al. 1950), and minerals from WARD'S Natural Science (WARD'S Natural Science Establishment, LLC., Rochester, NY). We did not characterise the reference minerals by another technique, such as X-ray diffraction. However, at least two sources of each mineral variety were used to ensure the authenticity of the each of the reference minerals.

After the mineral ATR library was compiled, its validity, reproducibility and robustness were tested with a second source of the same mineral varieties that were used to make the library. Three spectra of each mineral variety were obtained using an IlluminatIR II. All minerals were analysed using a 42 μm IR beam diameter, $4\,cm^{-1}$ resolution, 64 co-added scans and 64 co-added background scans. The data were then evaluated and compared with Grams AI software using two different statistical algorithms: a correlation and first derivative correlation. Based on this comparison, the preferred library-search criteria were determined. It should be noted that the library hit list is not an analytical solution, and the first hit should not be immediately taken to be the 'correct' answer. The analyst should compare the unknown spectrum to those of the top 10 minerals' spectra (at least) on the hit list in order to make a visual confirmation.

The hit quality index (HQI), or Quality, of the spectrum search was the feature used to compare the two library searches. HQI is a numerical rank obtained as a result of the spectrum search, with 0 being the best match and 1.414 or 1.000 being the worst match, depending on the search algorithm (Grams/AI Help Resource).

The correlation algorithm of an unknown to a library entry is "effectively a normalised least squares dot product on the unknown", where "both the unknown and the library data are centred about their respective means before the vector dot products are calculated" (Grams/AI Help Resource). The HQI is then calculated as:

$$\mathrm{HQI} = 1 - \frac{(Lib_m \bullet Unkn_m)^2}{(Lib_m \bullet Lib_m)(Unkn_m \bullet Unkn_m)} \tag{3}$$

In this equation, the library data ($\mathrm{Lib_m}$) and the unknown data ($\mathrm{Unkn_m}$) are defined as:

$$Lib_i = Lib - \left[\left(\sum_{i=1}^{n} Lib_i \right) / n \right] \tag{4}$$

$$Unkn_i = Unkn - \left[\left(\sum_{i=1}^{n} Unkn_i \right) / n \right] \tag{5}$$

The advantage of centring the data is that it enables the HQI to be independent of the normalisation of the spectra. This minimises the effects on the HQI of a noisy baseline and sharp negative dips in the unknown spectrum due to the atmospheric interferences of water vapour and carbon dioxide peaks (Grams/AI Help Resource).

The first derivative correlation algorithm is the same as the correlation algorithm except, prior to calculating the HQI, the first derivative of the library and known spectra are taken. The first derivative calculation is performed after mean centring. The first derivative calculation is done by subtracting previous points (Equations 6 and 7).

$$Lib_i = Lib_i - Lib_{i-1} \tag{6}$$

$$Unkn_i = Unkn_i - Unkn_{i-1} \tag{7}$$

The most common advantage of this is to help correct for bad baselines. Another advantage is with the analysis of broad peaks, as the derivative reveals the slope of the curve (Grams/AI Help Resource).

When the correlation algorithm was used to test the mineral library, 67 of the 85 mineral varieties came up as the first hit on the hit list for at least one of the three trials. For those minerals that were the first hit on the hit list, the average HQI difference between the first and second hits was 0.06714. When the first derivative correlation algorithm was used, 73 of the mineral varieties came up as the first hit on the hit list for at least one of the three trials. For those minerals that were the first hit on the hit list, the average HQI difference between the first and second hits was 0.225039. It was determined that the first derivative correlation is the better search algorithm to use when analysing minerals based on the greater number of minerals identified as the first hit and the larger HQI difference which means a greater discriminating power.

Diamond ATR IRMA alone proved to be successful at identifying most of the mineral varieties analysed, with some limitations detailed in the next section. When IRMA is paired with a preliminary polarised light microscopic analysis of the optical properties, or in some cases a confirmatory PLM analysis, mineral identification is conclusive.

Aragonite and calcite, two polymorphs of calcium carbonate ($CaCO_3$), were readily differentiated by their IR spectra. The IR spectra of these minerals are displayed in Figure 27.4a and 27.4b; all spectra were produced using Grams AI software. Two silicon dioxide minerals, quartz (SiO_2) and opal ($SiO_2{}^*nH_2O$) are easily distinguished with IRMA (Figure 27.4c and 27.4d). Diamond ATR analysis is capable of differentiating between albite (sodic plagioclase, $NaAlSi_3O_8$) and labradorite (intermediate plagioclase, $(Ca,Na)(Si,Al)_4O_8$), two common feldspars (Figure 27.4e and 27.4f).

The clay minerals were readily distinguished with IRMA. The spectra of kaolinite ($Al_2Si_2O_5(OH)_4$) montmorillonite ($(Na,Ca)_{0.3}(Al,Mg)_2Si_4O_{10}(OH)_2{\bullet}n(H_2O)$), saponite ($(Ca/2,Na)_{0,3}(Mg,Fe^{++})_3(Si,Al)_4O_{10}(OH)_2{\bullet}4(H_2O)$), and illite ($(K,H_3O)(Al,Mg,Fe)_2(Si,Al)_4O_{10}[(OH)_2,(H_2O)]$) are shown in Figure 27.5.

Albite (sodic plagioclase, $NaAlSi_3O_8$) and microcline ($KAlSi_3O_8$) had very similar spectra. This explains why microcline is the first hit for the albite tests when using

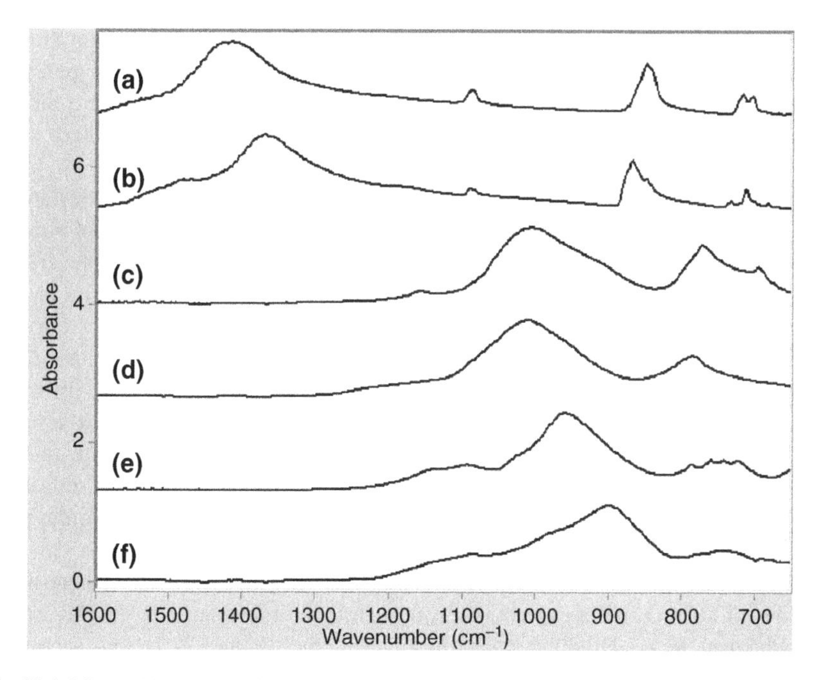

Fig. 27.4 Library ATR spectra of (a) aragonite; (b) calcite; (c) quartz; (d) opal; (e) albite; (f) labradorite.

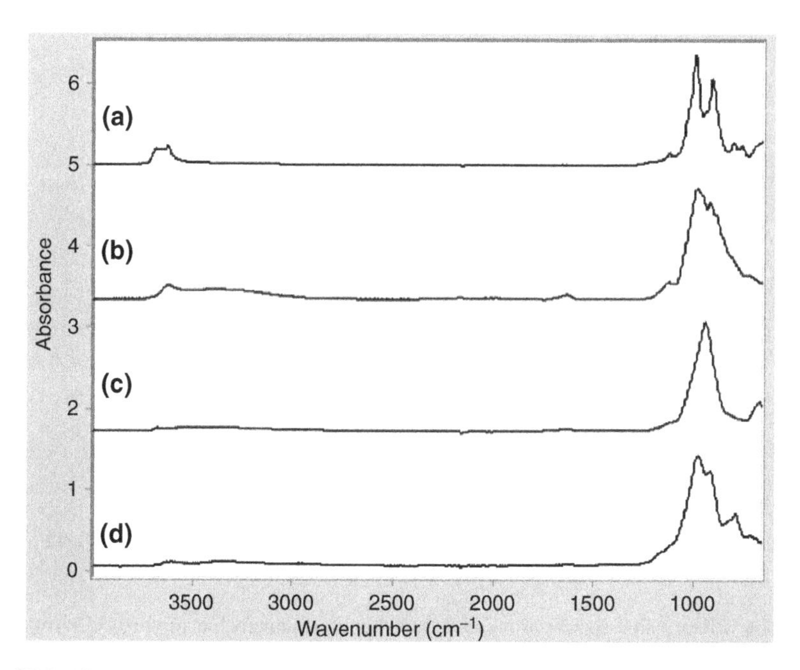

Fig. 27.5 Library ATR spectra of (a) kaolinite; (b) montmorillonite; (c) saponite; (d) illite.

the correlation algorithm and for one of the three albite tests when using the first derivative algorithm. However, when the 700 to 800 cm^{-1} region is compared, there are clearly four peaks for the albite spectrum, whereas both microcline pink and microcline white have two peaks (Figure 27.6). The third microcline studied, which was not distinguished by colour, did not have distinct peak splitting.

This analysis of minerals showed some mineral varieties that have overlapping or indistinguishable spectra. This reveals the fact that the IR spectroscopy of minerals is not an infallible stand-alone method for mineral identification. However, it is very powerful as a complementary technique to polarised light microscopy. The IlluminatIR microprobe enables the polarised light microscope observations to be paired with IR analysis, thus making it an excellent tool for mineral analysis.

Orthoclase and microcline, which are both composed of $KAlSi_3O_8$, proved to have undifferentiated spectra. Both of these minerals have the same hardness and density and have very similar optical properties (refractive index, birefringence, etc). One difference is their crystal structure: microcline is triclinic and orthoclase is monoclinic. Despite the difference in structure, their spectra are too similar to be discriminated. This would be a trivial problem for X-ray diffraction.

Chert is a fine grain, microcrystalline quartz. Chert and quartz are not distinguishable by X-ray diffraction and, as expected, were not able to be differentiated by their IR spectra. Visually, they are readily differentiated from each other (Figure 27.7), thus supporting the need to use the polarised light microscope to observe the sample prior to analysis.

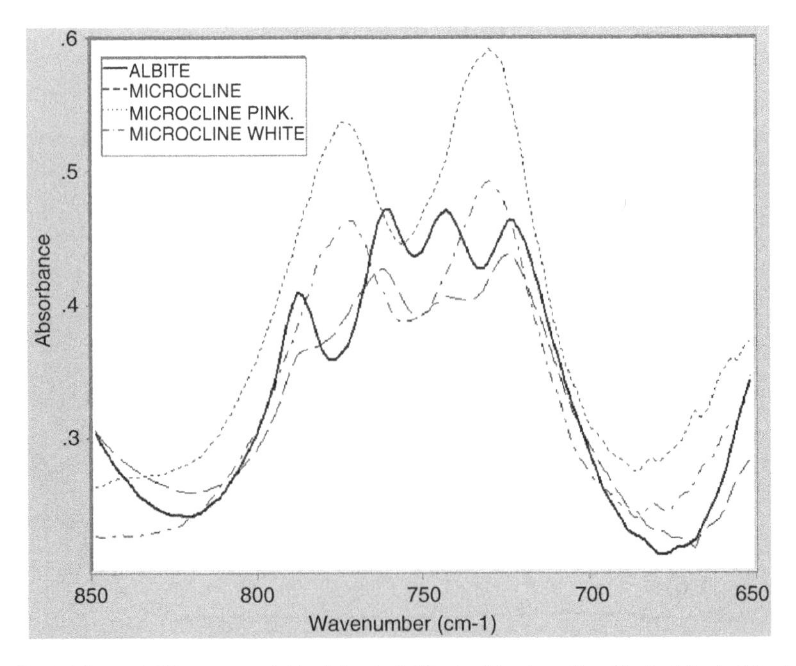

Fig. 27.6 Library ATR spectra of (a) albite (solid line); (b) microcline (dashed line); (c) microcline pink (dotted line); (d) microcline white (dashed and dotted line). Note abscissa scale differs from other figures and ranges 850 to 650 cm^{-1}.

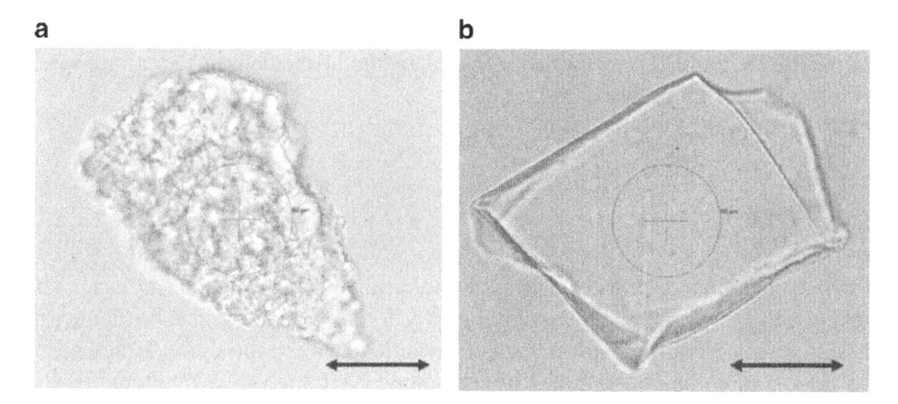

Fig. 27.7 Polarised light photomicrographs of minerals mounted in immersion oil (nd = 1.516 @ 23 °C). (a) chert; (b) quartz. Scale arrow = 50 μm.

Limitations

The diamond ATR analysis proved to be reproducible for each mineral variety analysed, with some limitations. There are three categories of minerals that cannot be identified with this method of analysis, and this includes halides, metal oxides and sulphides, and minerals with refractive indexes greater than that of the diamond IRE. The first category, halides, includes the minerals halite (NaCl), fluorite (CaF_2) and cryolite (Na_3AlF_6), which are transparent in the IR spectral region. The resulting spectra were of atmospheric water vapour only. The second group includes most minerals composed of metal oxides and sulphides, and included magnetite ($FeFe_2O_4$), hematite (Fe_2O_3), pyrite (FeS_2), ilmenite ($Fe^{2+}TiO_3$), rutile (TiO_2), and cassiterite (SnO_2). These produce extraordinarily high absorbance values (greater than 3) or non-repeatable spectra which cannot be used for a positive identification. Two metal oxides that do have unique and reproducible spectra are corundum (Al_2O_3) and zincite (ZnO). Lastly, for total internal reflection of the IR beam to occur inside the diamond ATR objective, the refractive index of the mineral must be less than that of diamond (n = 2.4). As a result, minerals such as goethite ($Fe^{+++}O(OH)$) and sphalerite ((Zn,Fe)S) do not produce reproducible IR spectra. Even though a definitive identification cannot be made for these minerals, the lack of a spectrum or one with exceedingly large absorbance value can provide valuable information on the identity of the mineral in question.

There are two significant explanations for spectral variations within a given mineral variety. The first is that a mineral only has a "relatively constant chemical composition" (Smith 1946). If there are differences in chemical composition, albeit small ones, then this will affect peak heights and can shift peak locations. Secondly, because the analysis is completed on single mineral grains, orientation affects can cause spectral differences (Klima and Pieters 2006). However, because the mineral grains are crushed prior to ATR analysis, the orientation affects are potentially minimised.

A study of orientation effects on the IR spectra of mineral grains collected by diamond ATR could be accomplished by polarising the incident IR radiation, which has potential for future research.

Conclusion

IRMA enables rapid, reliable and reviewable mineral identification by scientists lacking expertise in optical crystallography or forensic geology. The brittleness of an unknown mineral or mixture cannot be anticipated, so the optimal method of preparation of an unknown mineral for diamond ATR analysis is by mounting the sample in immersion oil. The advantage of diamond ATR analysis is that the sample preparation is minimal, and requires no additional preparation beyond that for traditional polarised light microscopy. The time required for analysis is only a few minutes for each mineral in the sample. The work presented here offers the soil examiner an alternate method to confirm the identification of a mineral sample mounted on a slide using IRMA. ATR spectra are not the same as transmission and diffuse reflection spectra, with shifts in band frequency and different band intensities, and thus it is essential to compare the spectrum of a mineral analysed by IRMA to a library of ATR IR spectra. Although there are references for diffuse reflection and transmission IR spectra, such as that prepared by Salisbury (1987) for the U.S. Geological Survey, there is no such reference for the ATR spectra of minerals. The ATR library of the 85 minerals that was created in this research is a start of such a reference for minerals. If a suspect and known soil sample from a crime scene contains a mineral not in the mineral ATR library, for the purposes of the case, the two IR spectra can be compared and shown to be the same despite not being able to identify the mineral variety. Thus IRMA can be used to identify the minerals in a soil sample as well as compare suspect and known samples from a crime scene to show commonalities.

The disadvantage encountered was the inability to analyse halide minerals, metal oxide and sulphide minerals, and minerals with a refractive index greater than that of the diamond IRE ($n = 2.4$). In addition, there were several minerals with IR spectra that were indistinguishable. Although a definite identification of a mineral by its IR spectrum alone is not always achievable, the technique is able to narrow down the mineral possibilities. In addition, when paired with polarised light microscopy to observe optical and morphological properties, a definitive identification would be possible.

In conclusion, IRMA offers an efficient means to identify mineral composition and is a complementary technique to polarised light microscopy. The IlluminatIR enables a sample of interest to be analysed with polarised light microscopy and simultaneously analysed with IR spectroscopy. It has been shown that IRMA enhances and strengthens the analysis of mineral samples, and is a useful tool in the characterisation of soil evidence. The ability to integrate polarised light microscopy with IR microspectroscopy to minerals in soil samples is unprecedented, and offers a powerful tool in the soil forensic toolkit.

Acknowledgements The authors would like to acknowledge the many scientists who contributed to this research. In particular, Pauline Leary from Smiths Detection for her technical assistance with regard to using the IR microprobe, as well as her continual guidance and support. We are also indebted to Dr. Christopher S. Palenik from Microtrace, LLC for supplying us with samples for a single blind study and useful information regarding this research. Dr. Thomas A Kubic provided valuable comments and his knowledge and experience are gratefully acknowledged.

References

Beveridge A, Fung T and MacDougall D (2001). Use of infrared spectroscopy for the characterisation of paint fragments. In: Forensic Examination of Glass and Paint: Analysis and Interpretation (Taylor and Francis Forensic Science Series). (Ed. B Caddy), pp. 183–242. CRC, New York.

Grams/AI Help Resource (N.D.). [Computer Software]. Thermo Scientific. *www.thermo.com/com/cda/landingpage/0,,585,00.html?ca=grams*

Harrick NJ (1987). Internal Reflection Spectroscopy. Harrick Scientific, New York.

Kerr PF, Hamilton PK, Pill RJ, Wheeler GV, Lewis DR, Burkhardt W, Reno WD, Taylor GL, La Habra Laboratory, Mielenz RC, King ME and Schieltz NC (1950). Analytical Data on Reference Clay Materials. American Petroleum Institute, New York.

Kirkbride KP and Tungol MW (1999). Infrared microspectroscopy of fibres. In: Forensic Examination of Fibres (Taylor and Francis Forensic Science Series), 2nd ed. (Eds. JR Robertson and M Grieve), pp. 179–222. CRC, Pennsylvania.

Klima RL and Pieters CM (2006). Near- and mid-infrared microspectroscopy of the Ronda peridotite. Journal of Geophysical Research 111:EO1005.

Murray RC and Tedrow JCF (1992). Forensic Geology, 2nd ed. Prentice-Hall, New Jersey.

Petraco N and Kubic TA (2004). Color Atlas and Manual of Microscopy for Criminalists, Chemists, and Conservators. CRC, New York.

Salisbury JW (1987). Mid-Infrared (2.1–25 μm) Spectra of Minerals, 1st ed. Department of the Interior, U.S. Geological Survey, Virginia.

Smith OC (1946). Identification and Qualitative Chemical Analysis of Minerals. D. Van Nostrand, New York.

Wehrenberg JP (1998). Manual for Forensic Mineralogy. Short Course at Northwest Association of Forensic Scientists; 1998 Spring Meeting. Northwest Association of Forensic Scientists, Montana.

Chapter 28
Preservation and Analysis of Three-Dimensional Footwear Evidence in Soils: The Application of Optical Laser Scanning

Matthew R. Bennett, David Huddart and Silvia Gonzalez

Abstract This chapter explores the application of optical laser scanning to the collection, preservation and analysis of footwear evidence in soils using examples from the archaeological record and from a series of experiments. Optical laser scanning provides a direct, non-invasive method of recording footwear evidence with sub-millimetre accuracy. It allows the original print to be re-visited at any time using a range of view angles and light illuminations at any time within an investigation. Although practical problems remain with the routine deployment of such equipment at a typical crime scene, the potential of such techniques to revolutionise the way in which three-dimensional footwear evidence is recorded is considerable. This point is illustrated using an example from the geo-archaeological record and via three experimental scenarios. The first of these experiments involved the use of series of barefoot impressions and the direct comparison of data obtained via optical laser scanning with that obtained from direct casting methods. The second and third experiments involved artificial crime scenes in which a range of footwear was used to leave a palimpsest of prints. Optical laser scans were used to interpret this evidence and to quantify wear patterns in order to link specific footwear to individual prints. These experiments are used within the chapter as a basis with which to review both the advantages and disadvantages of optical laser scanning. On the basis of this review we argue that the potential of such technology in a criminal context is clear given further technological developments to allow it operational deployment.

M.R. Bennett(✉)
School of Conservation Sciences, Bournemouth University, Talbot Campus, Fern Barrow, Poole BH12 5BB, UK
e-mail:mbennett@bournemouth.ac.uk

D. Huddart and S. Gonzalez
School of Biological and Earth Sciences, Liverpool John Moores University, Byrom Street, Liverpool, L3 3AF, UK

K. Ritz et al. (eds.), *Criminal and Environmental Soil Forensics*

Introduction

Footwear impressions are important but are frequently undervalued as a source of evidence within a range of criminal investigation (Bodziak 2000). Footwear traces come in a variety of different forms from the classic 'footprint in the flower bed' to subtle traces left by muddy shoes or blood-stained feet. Although the materials and techniques used in the collection of footwear evidence have become increasingly sophisticated (e.g. Hueske 1991; Du Pasquier et al. 1996; Bodziak 2000; Buck et al. 2007), the basic premise of photographing, casting, or lifting of footwear evidence remains consistent.

In the case of three-dimensional impressions within soil, which form the focus of this chapter, traditional methods of preservation and analysis involve some form of casting technique. Casting of footwear evidence is not however without some critical challenges. Assuming that the right casting materials and methods are used for the appropriate conditions and substrate, something that is well reviewed in forensic manuals (e.g. Bodziak 2000), then several key issues remain. The first of these involves the very nature of the casting process which is, by definition, invasive with the potential to disturb the evidence (Du Pasquier et al. 1996; Bodziak 2000). The second issue involves the process of lifting a cast which removes trace evidence adhering to the basal surface of the cast and may obscure, at least initially, the footwear evidence preserved by the cast (Bull et al. 2006). This trace evidence has priority in any subsequent investigation and must be removed and analysed before the cast is available for study. The speed with which this is done depends on the resource available and investigation priorities. If this trace evidence is not dealt with as a priority, then the investigator will initially only have photographs of the impression with which to work, and no three-dimensional data. There is also an ever-present risk that until the cast is cleaned it is not known whether the evidence has been successfully lifted and, in some cases, the original evidence may have been lost, especially in transitory environments such as bare earth or soft ground.

The third issue is the way in which the evidence is used. Once a cast is produced, it must then be examined, and information on the shoe impression uploaded by a skilled operator into one of the increasing number of databases which serve to link sole patterns to shoe manufactures (e.g. Ashley 1996; Geradts and Keijzer 1996). Equally, if a 'Schallamach', or similar pattern of sole abrasion, is present, it needs to be examined and recorded (Schallamach 1968; Bodziak 2000). These processes are all based on two-dimensional data, yet many types of foot impressions are three-dimensional and the third dimension is usually discarded in these, increasingly, electronically-driven systems. This reflects the fact that there is no easy way of converting a physical cast into three-dimensional data that can be uploaded into a database. The point is made even more clearly with respect to wear characteristics, which are treated for the most part in a qualitative fashion. The degree of wear from a new reference sole is not easily quantified.

The final issue is one of storage and ease of retrieval. Footprint casts are, by definition, bulky and need to be stored against the need to match them to a suspect's footwear. The evidence is not easily visible via a database, apart from the photographic component.

Comparison of a suspect's footwear with a print is done by a visual and semi-quantitative comparison of a print with the shoe. Again, comparison of three-dimensional objects is done by trained operators rather than by digital comparison. It is often easier to compare the two-dimensional aspects of an impression and shoe rather than treating them as three-dimensional forms.

All of the issues and challenges identified above help explain why footwear evidence is often not as valued as other types of trace evidence, despite the fact that it has powerful advocates and its potential has been frequently documented (Bodziak 2000). In this chapter we argue that many of these issues can be resolved by the application of three-dimensional optical laser scanning to the collection, preservation and analysis of selected types of footwear evidence. By creating high resolution Digital Elevation Models of prints at the crime scene, the investigator has immediate access to three-dimensional data on footwear impressions which can be stored and accessed electronically and treated in a more quantitative fashion.

The collection of three-dimensional data by optical scanning has recently been demonstrated for the collection of footwear impressions in snow (Buck et al. 2007) and the documentation of soft tissue injuries (Thali et al. 2005). However, to date, there has been little work with respect to its application for the preservation of a wider range of footwear impressions in soils. The authors recognise two separate challenges associated with the application of this type of technology to recording and preserving footwear evidence. Firstly, one must demonstrate the potential advantages of such a technique with respect to more traditional approaches and secondly, one must develop the equipment and operational methodology so that it can be routinely and easily deployed at a crime scene. The aim of this chapter is to explore the first of these two issues by showing how three-dimensional scanning has been used to preserve ancient footprints in geo-archaeological contexts, and then to demonstrate its future potential application in modern forensics using a series of simple experiments. It is accepted that the equipment presented here is currently too bulky to be operationally effective, but we wish to draw attention to the potential of the technology for in the future.

Footprints from the Past

The presence of human and animal prints is a rare occurrence in the archaeological and geological record, requiring specific conditions for preservation. Perhaps the most famous example is the Laetoli prints in East Africa with an age of 3.6 Ma BP which provide some of the earliest evidence of bipedalism in early hominids (Leakey and Hay 1979; Leakey and Harris 1987). Human and animal footprints of Mesolithic age occur in a variety of coastal locations around the UK, such as in the Severn Estuary (Aldhouse-Green et al. 1992) and on the Formby coast (Roberts et al. 1996). In Mexico there are several locations where human footprints have been found recently and, in the case of those found in the Valsequillo Basin, they have been used to support a claim for the early human colonisation of the Americas (Gonzalez et al. 2006).

During the work in Valsequillo the authors pioneered the *in situ* scanning of fragile footprint evidence as a means of preserving and analysing human trace fossils preserved in volcanic ash within the geo-archaeological record (Figure 28.1). Traditional casting techniques are not only destructive of the fragile materials in which such footprints are preserved but do not facilitate the accurate study or sharing of digital information with other researchers. Although the technique has been developed while working at the Valsequillo footprint site (Huddart et al. in press), it has now been applied to a number of other footprint sites involving a wide range of substrate sediments (e.g. peats, silts, silty clays, travertine) and environments (e.g. inter-tidal flats, lake shorelines and calcareous pool margins).

The method involves the use of a Konica-Minolta VI 900 three-dimensional optical surface digitiser which uses laser triangulation technology to create an elevation model that can be rendered by a digital photograph. Resolution is of the order of ±0.1 mm in the vertical plane, with x and y accuracy of ±0.22 and ±0.16 mm respectively.

Fig. 28.1 (A) Konica-minolta vi 900 optical laser scanner in the field. Note the small hand held generator. (B) Close-up of part of aluminium scanner framer. The scanner is mounted on a shuttle that runs the length of the central beam. (C) Scanner rig with its light protective shroud for scanning in bright sun light. (D) Neolithic-Bronze age human footprint exposed on the foreshore at formby, UK

In the field, the equipment is powered by a small generator and either mounted on a tripod or on a custom-built aluminium rig (Figure 28.1a, b). In either case, control of ambient light levels is essential since the VI 900 is designed for use in the laboratory rather than in bright sunlight, as it has a red light laser.

The aluminium rig consists of a 3 m long beam suspended from two triangular-shaped adjustable legs at each end. The scanner is mounted on a shuttle which is able to move the length of the central beam. The rig is then enclosed within a canvas outer skin to control light levels (Figure 28.1c). Adjustable flaps, both along the top and at each end of the outer skin, provide control of the light level (Figure 28.1c). This arrangement not only allows continuous scanning irrespective of light levels but also sequential scans to be taken along a swathe 3 m long by 0.75 m wide. Each swathe is also photographed using a high-resolution digital camera under normal light conditions once the rig has been removed. These images can subsequently be used for draping over the digital elevation models obtained from the scanner in preference to the photographic images captured by the scanner itself, which are of lower resolution and may be affected by the light control measures on the rig.

The point cloud captured by the scanner is processed using Konica-Minolta's Polygon Editing Tool (PET). This processing joins all the points, up to 300,000 points per laser scan, producing a network of polygons over which a surface mesh can be draped. More advanced data processing is undertaken in Rapidform 2006, available from INUS Technology, or by exporting the point cloud as xyz data and subsequently analysing it in ArcGIS-9 by ESRI. Within these software packages it is possible to make measurements and optimise viewing angles for interpretation of the surface. It is important to note that in processing the data the point cloud, and therefore the surface topography, of the scan is not edited or modified in any fashion and that a log of all processing steps is kept along with the raw data output from the scanner. The scanner is routinely calibrated by the manufacturer and test objects – small metal cubes – are placed within all scans to provide an accurate cross-check on the calibration and measurements made.

It should also be noted that it is possible to convert the digital information into a solid object for public display using rapid prototyping technology (Huddart et al. in press). The design and engineering technology used is normally associated with the development of new consumer products, like mobile phone cases, where design drawings are turned into solid objects for consumer trials or visualisation.

To illustrate the applications of the technique within geo-archaeology, data are presented from the Cuatrociénegas Basin in Coahuila State, Mexico (Rodríguez-de la Rosa et al. 2004; Gonzalez-Gonzalez et al. in press). This basin contains a number of freshwater pools fed by hydrothermal ground-water which passes through a basement of limestone. As a consequence the pools are associated with fresh-water limestone and travertine. Human footprints of uncertain age have been found in travertine around one of the pools within the Cuatrociénegas Basin (Rodríguez-de la Rosa et al. 2004). Figure 28.2 shows one footprint trail exposed on the floor of an old quarry and appears to represent a single individual walking along a sub-horizontal surface. Photographs and isopleth maps of specific prints are shown and demonstrate how the scanning process can be used to retrieve foot impressions.

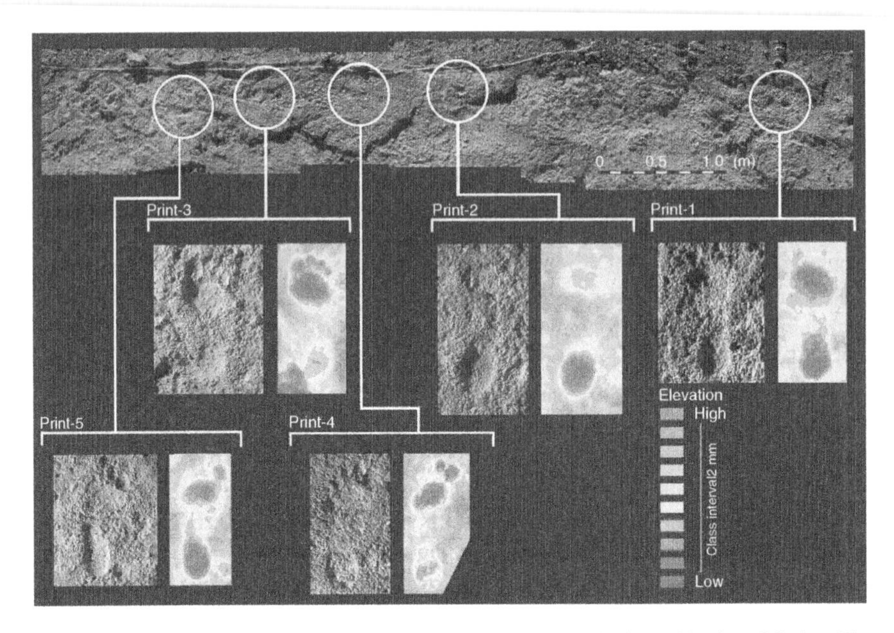

Fig. 28.2 Footprint trail from the Cuatrociénegas Basin in Coahuila State, Northern Mexico. Five footprints are shown along a single discontinuous trail, illustrated via photomontage. This trail is exposed in the floor of a travertine quarry. Close-up photographs and isopleth maps created from scanned images are shown for each of the five prints (*see colour plate section for colour version*)

Comparison of the scans also reveals the degree of morphological variability within any given trail.

Forensic Experimentation

In order to explore the potential of three-dimensional optical laser scanning in the collection, preservation and analysis of footwear evidence, a series of three experiments were conducted. The overall rationale of these experiments was firstly, to illustrate the quality of the data obtained via laser scanning, secondly, to demonstrate the measurement accuracy possible when using digital scans, and finally, to provide an evidence base with which to compare directly laser scanning with more traditional methods. All the experiments involved the use of a shallow sand tray (2.41 m long, 0.71 m wide and 0.09 m deep) filled with builder's soft sand with a moisture content of 2.88% and a mean grain size of 0.6 μm. Moisture content was determined by repeated sampling for the duration of the experiment and determined by a water loss method. When necessary, adjustments were made by the addition of distilled water using a hand-held crop spray. The surface of the sand tray was raked between experiments, and levelled using planar paddle. The choice of sand in these experiments was made for ease, but any soil-based material could have been used.

Experiment 1

Experiment 1 involved the use of unshod feet which, although uncommon at crime scenes in developed countries, provided a vehicle with which to illustrate the quality of the data obtained by laser scanning and to compare it with that obtained from more conventional casting methods. Initially, one male (Male 1; 83 kg in weight, 181 cm tall) walked the length of the sand tray bare-foot, on four separate occasions, using a similar pace and stride length. Between each traverse the footprints were photographed and scanned using the Konica-Minolta VI 900 mounted 0.6 m above the sand tray on the aluminium rig illustrated in Figure 28.1a, after which the sand tray was raked and levelled. The exercise was repeated with a second male (Male 2; 98 kg in weight, 177 cm tall), and the footprints photographed, scanned and cast. The casts were made using Crownstone, following fixing of the impression with Unifix-16.

Figure 28.3 shows a composite scan of the full length of the sand tray showing three sequential foot prints made during the first traverse (Male 1). Close-up scans of each of the footprints are also shown with 1 mm colour isopleths affixed to the surface to enhance visualisation. The scans resolve the surface-texture of the sand, and can resolve features to an accuracy of ±0.1 mm. The variability in the foot impressions made by this individual can be interpreted from the data, illustrating the level of detail which can be obtained.

To contrast the data quality obtained by conventional casting with that obtained by laser scanning, the prints made by Male 2 were both scanned and cast.

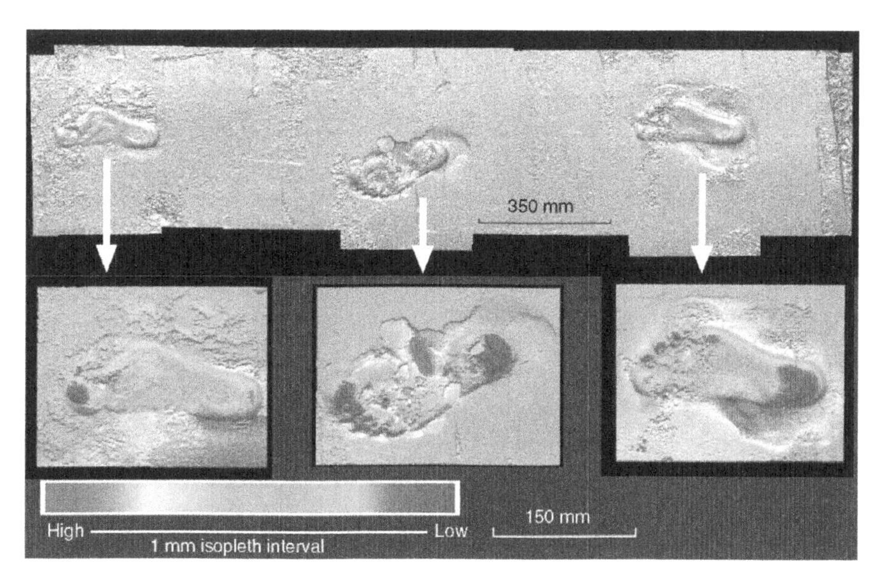

Fig. 28.3 Sequence of footprints created by Male 1. Upper panel is a composite scan of the sand tray while the lower panels are close-up scans of each footprint rendered with 1 mm isopleth to enhance visualisation (*see colour plate section for colour version*)

These different methods are compared visually in Figure 28.4 for one footprint and illustrate the presence of casting artefacts within the plaster cast. To provide a quantitative comparison, the cast was scanned, the output from which was then registered in Rapidform using a series of anatomical landmarks on both the footprint scan and the cast scan. This procedure creates an interferometric image showing the degree of relative, rather than absolute, difference between the two scans across the footprint (Figure 28.5).

Areas with little or no difference are not displayed in the interferometric image, whilst those showing surface deviation between the two scans can be identified. This analysis was repeated several times using different anatomical landmarks to ensure consistency and reproducibility of the interferometric image. The image in Figure 28.5 shows that, although for significant areas of the print there is no visible difference between the two scans and therefore techniques, there is in other locations a marked difference, especially around the toes and lateral margins of the print. This approach does not directly determine which of the two techniques is the more accurate, but simply demonstrates that a difference exists. However, it should be noted that some of the areas of greatest difference correspond to areas of pouring artefacts within the plaster cast.

Further comparison of the two approaches is possible by looking at biometric measurements obtained from the scan and from the cast. In the case of the scans, a series of standard footprint biometrics (Robbins 1985; Laskowski and Kyle 1988; Kennedy et al. 2005) were measured in Rapidform with a precision of two decimal places; all measurements were made in millimetres. The same footprint biometrics

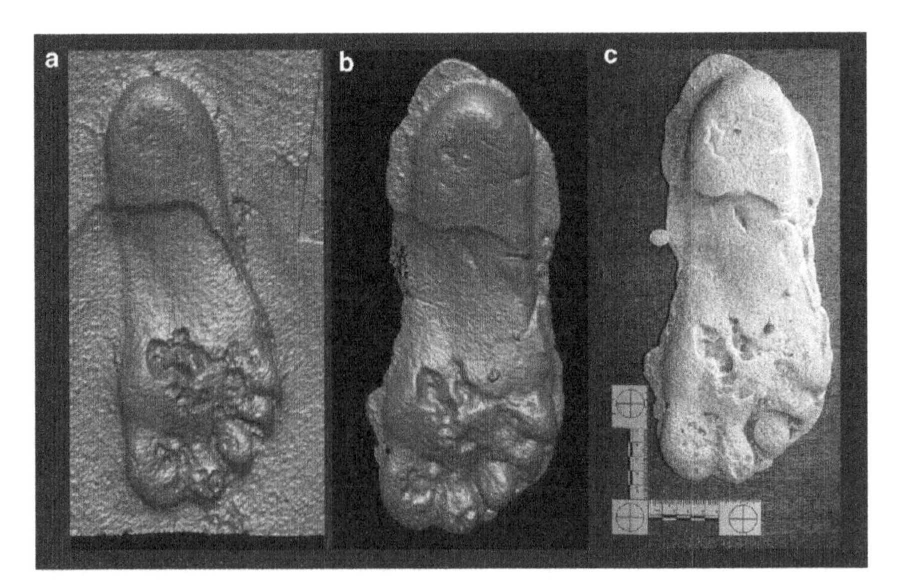

Fig. 28.4 Comparative photographs of the same print. (a) Image taken from a scan of the original footprint, inverted here to allow comparison. Note that the scanner is sensitive enough to pick out the writing on the paper label. (b) Scan of the plaster cast. (c) Photograph of the plaster cast

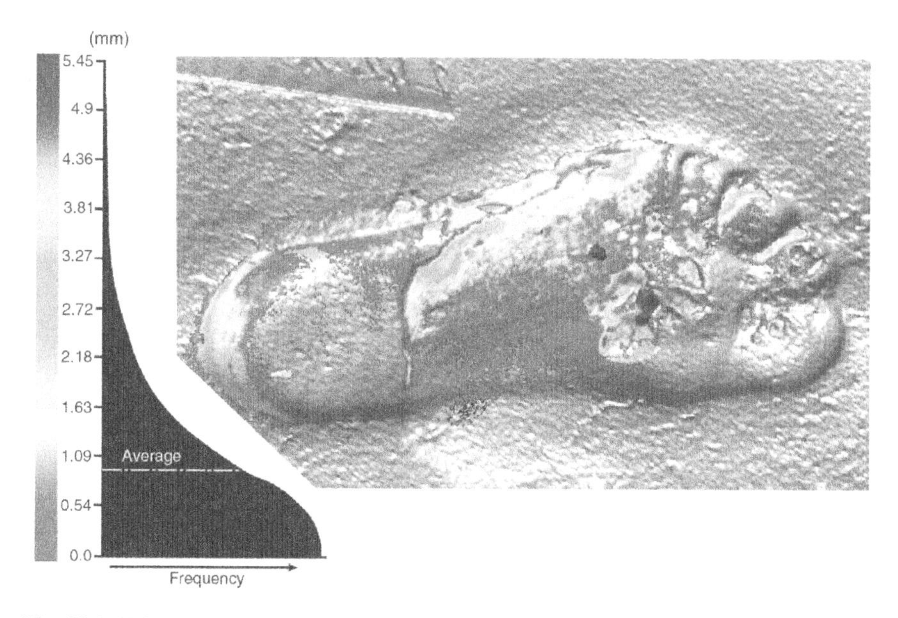

Fig. 28.5 This interference image of two scans one taken from the original print and one taken from the cast of that print. Coloured areas are those with a measurable difference between the two scans following image registration in Rapidform. The interference image clearly demonstrates that the two scans are not identical. The difference can be accounted for by the casting rather than the scanning process (*see colour plate section for colour version*)

were measured from each of the casts using a set of digital Vernier callipers with a precision of two decimal places (Table 28.1). There is insufficient data to allow statistical comparison but there is a marked degree of difference between the metrics obtained, reflecting the greater precision and accuracy that is possible with digital measurements.

Experiment 2

In Experiment 2 one male (83 kg in weight, 181 cm tall) walked the length of the sand tray wearing four different pairs of shoes creating a palimpsest of footwear impressions (US Size 9, UK Size 8.5, European Size 43, or 267 mm; Figure 28.6). Two of these pairs of footwear were identical makes of boot, but with different degrees of wear. The sand tray was scanned and photographed as described above but, in addition, the footwear used was then mounted on an improvised cobbler's last and the soles scanned. The aim was to demonstrate that each of the four footwear types could be distinguished on the basis of the scanned images and to demonstrate how wear characteristics can be documented in a quantitative fashion using optical scanning.

Table 28.1 Comparison of footprint biometrics obtained manually from footprint casts and digitally from optical laser scans of the original evidence. Footprints were created by Male 2 and all measurements are in millimetres and footprint indices are those of Kennedy et al. (2005)

	Male 2 right		Male 2 left		Male 2 right		Male 2 left	
Metrics (mm)	Cast	Scan	Cast	Scan	Cast	Scan	Cast	Scan
Length	272	275	268	271	284	300	277	279
Width ball	110	117	110	114	118	115	116.3	109
Heel width	64.7	64.7	66.7	68.6	71.3	72.9	73.2	76.5
Minimum width	61.5	53.7	58.1	51.5	66.9	58.2	71.9	48.2
Toe-1 length	46.8	50.7	44.3	51.2	57.1	59.1	50.4	47.4
Toe-1 width	35	35.9	34.3	35.8	31.5	41.2	36.7	37.8
Toe index	0.74	0.7	0.77		0.55	0.69	0.72	0.79
Instep-foot index	0.22	0.19	0.21	0.19	0.23	0.19	0.25	0.17
Ball-foot index	0.4	0.42	0.41	0.42	0.41	0.38	0.41	0.39
Heel-ball index	0.58	0.55	0.6	0.59	0.6	0.63	0.62	0.7
Heel-foot index	0.58	0.55	0.6	0.59	0.6	0.63	0.62	0.7
Toe-ball index	0.31	0.3	0.31	0.31	0.26	0.35	0.31	0.34

Figure 28.6 shows the palimpsest of footprints created in the tray during the second experiment and the soles of the associated footwear. All four shoes used in the experiment show varying degrees of wear which are characteristic of a flat-footed individual. Two identical shoes were used in this experiment, with different degrees of wear. These impressions can be distinguished easily on the basis of heel wear and the level to which marginal cleats and circular domes in the sole have been removed by abrasion.

Such matching could be achieved by conventional forensic photography and casting. However, the key advantage of the use of the scanned data is the speed at which matching can be accomplished using a single three-dimensional digital image, employing viewing from a variety of angles and different levels of illumination. Multiple photographs and casts would be required to achieve the same results by more conventional methods. In terms of time, it took approximately 1 h to process the digital data and prepare the scans of both the footwear and sand tray for visual analysis; having done so, the matching of footwear to prints was achieved in a matter of minutes.

The use of the scans offers another significant advantage in that they allow the amount of wear to be quantified, which is often an important factor in linking a suspect with a common shoe type to a crime scene (Bodziak 2000). Figure 28.7 shows the right heel of the two identical shoes used in the experiment and the different levels of wear are revealed by coloured isopleths aided by the addition of a 1 mm contour map of the more worn heel. Using Rapidform it is possible to sample data along a given cross-section on each of the scans taken from the footwear (Figure 28.8) and to compare this with a similar cross-section taken from one of the foot impressions. This provides a quantitative basis by which to link a foot impression to a worn shoe. The two cross-sections were co-referenced using a series of control points from areas of the sole not subject to wear in the boot instep and by measuring down from the upper surface of the rubber sole exposed in the lateral sides of the boot.

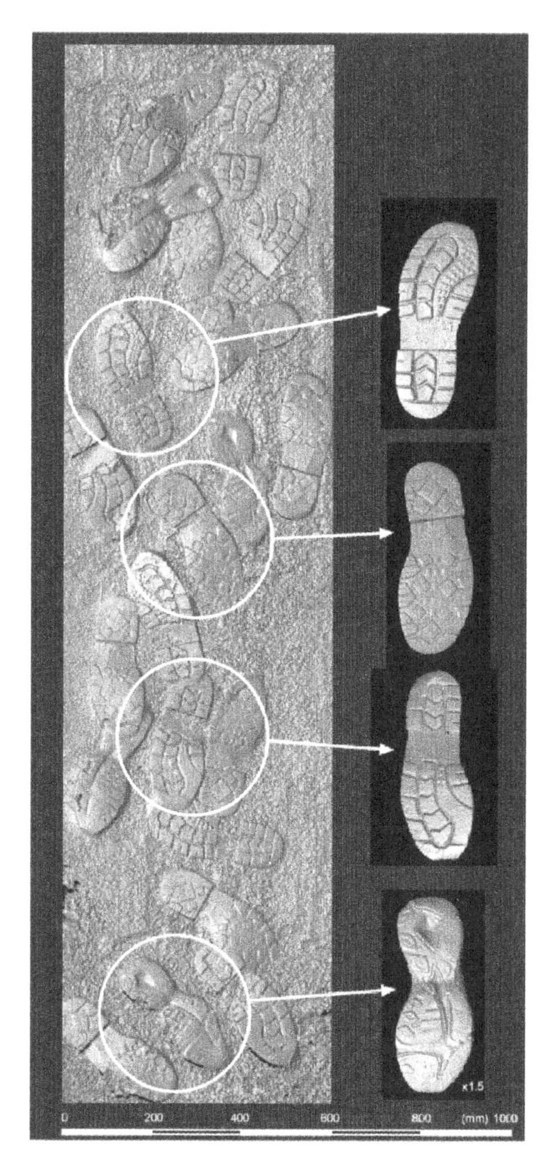

Fig. 28.6 Footprint palimpsest within the sand tray used for Experiment 2. All images are derived from three-dimensional optical laser scans. Note that the two items of similar footwear can be separated on the basis of their wear characteristics

Experiment 3

In Experiment 3, one female (95 kg in weight, 153 cm tall) walked the length of the sand tray wearing four different pairs of shoes creating a palimpsest of footwear impressions (US Size 7, UK Size 6, European Size 39, or 248 mm; Figure 28.9).

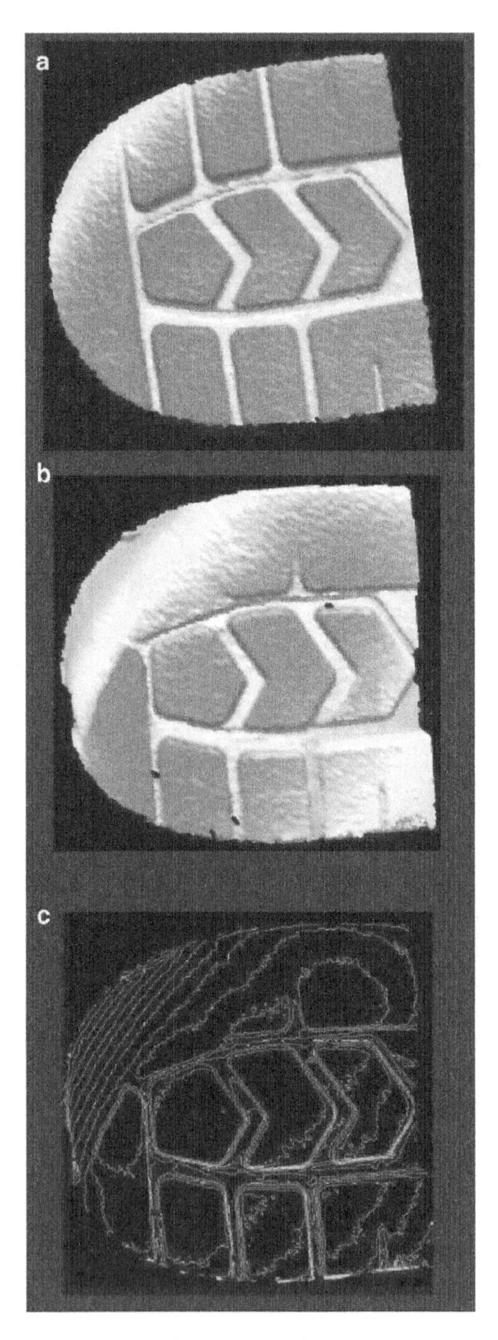

Fig. 28.7 Three-dimensional images of two similar right heels showing different levels of wear depicted by the 1 mm colour isopleths. (a) Right heel showing limited wear. (b) Right heel showing pronounced wear on the outer heel edge. (c) Contour map, 1 mm interval, of Heel B allowing direct quantification of the level of wear (*see colour plate section for colour version*)

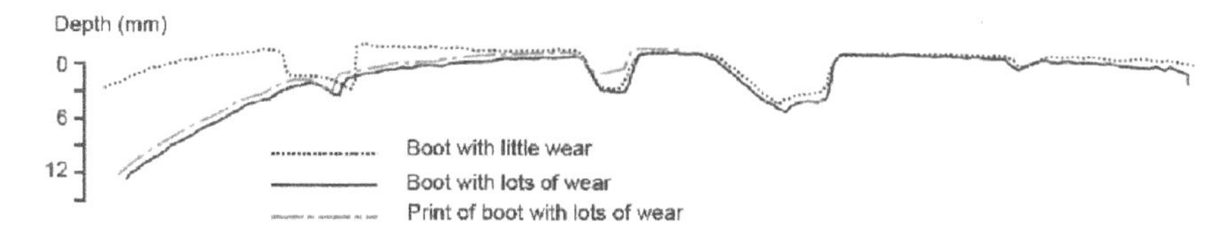

Fig. 28.8 Cross-sections taken from scans of the right heels shown in Figure 28.7 and from a print made by one heel. The cross-section taken from the print can clearly be assigned to the boot exhibiting considerable wear

The footwear selected was more uniform in terms of its tread-pattern, but the density of foot impressions was increased compared with experiments one and two to make recognition harder and more consistent with a heavily trampled crime scene. The order in which the footprints were made and the number of traverses of the sand tray were not disclosed to those undertaking the analysis until after the data had been collected and the sequence established. The aim was to demonstrate how a series of footprint trails can be separated and sequenced in time.

Figure 28.9 shows the palimpsest of footprints created in the tray by the operative and the scans of the soles of the associated footwear. The different footwear is clearly distinguishable and all prints can be assigned easily to one of the four pairs of shoes using the tread patterns and sole outlines.

Having assigned each foot impression to a particular shoe, the temporal sequencing was established by looking for the cross-cutting relationships between individual foot impressions (Figure 28.10). With one exception, the sequence shown in Figure 28.10 was subsequently shown to be correct. In making the second sequence of foot impressions, the operator walked in both directions along the length of the tray, but there is evidence for only one direction of foot impressions. The evidence for this second traverse has been lost under later impressions.

Such interpretation could have been undertaken from vertical photographs, especially ones taken from varying angles and light levels. However, the advantage of the tools used here is that image analysis and presentation techniques, such as rotation, enlargement, rendering with different colours, changing the levels and direction of illumination, and measuring sole width and depth or pattern, broadens the scope for interpretation and recognition of features.

There are also advantages in having a single digital model of the sand tray rather than successive photographs. These advantages become particularly important in dealing with the numerous partial prints present within this palimpsest.

Discussion and Conclusion

The data presented from the geo-archaeological record and the three experiments described above illustrate the potential of optical laser scanning in the collection, preservation and analysis of footwear evidence. If this approach is to be embraced by the forensic community for practical applications, then one must be clear about

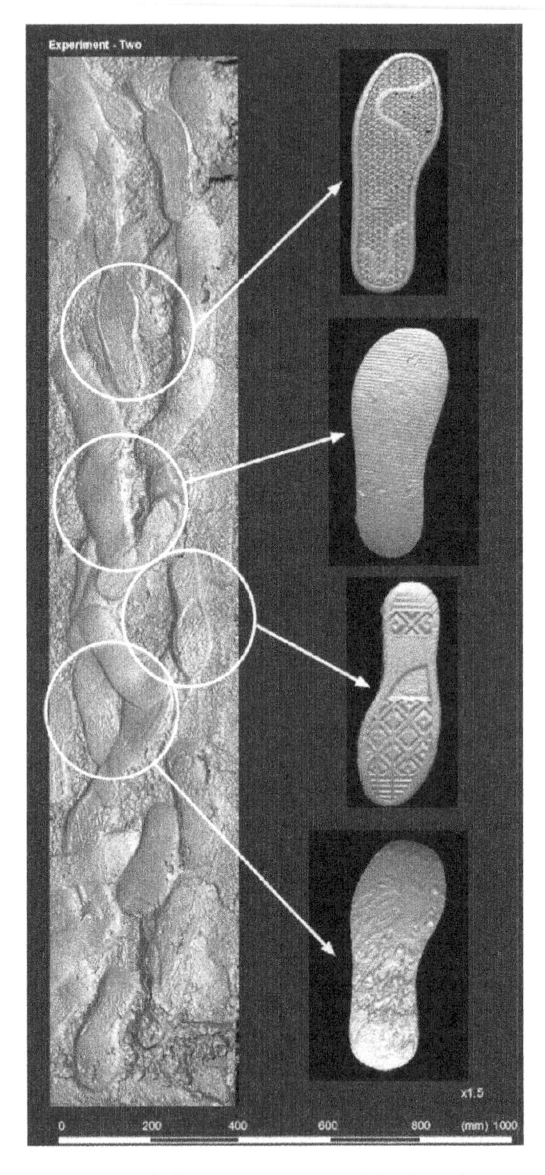

Fig. 28.9 Footprint palimpsest within the sand tray used for Experiment 3. Note the extent to which the prints have been superimposed. All images are three-dimensional scans

how it either complements, or replaces, existing tools, in this case the casting and photography of three-dimensional wear evidence. In light of this, the technology has been evaluated using the evidence presented above in relation to three typical phases within a criminal investigation, namely the collection of evidence at the crime scene, the analysis of that evidence and finally the presentation of that evidence as part of a criminal case.

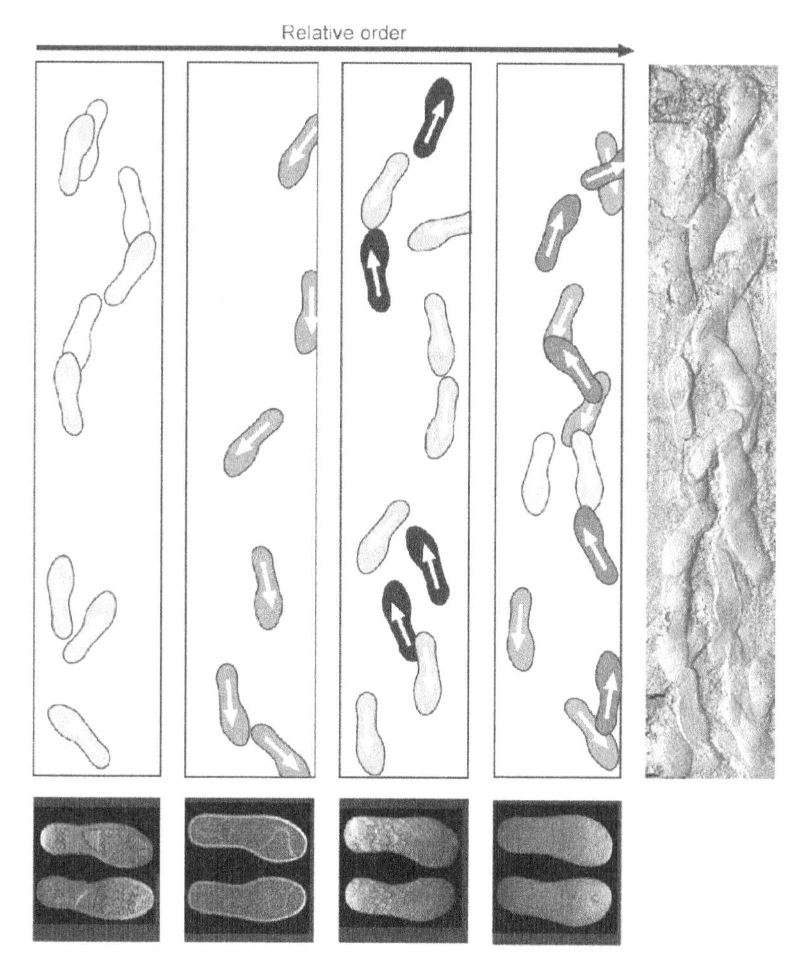

Fig. 28.10 Footprint sequence in which the palimpsest on right-hand panel (also in Figure 28.9) was created (*see colour plate section for colour version*)

During evidence collection, the scanning of footwear traces provides an accurate non-invasive, fast and effective method of capturing evidence as illustrated in this chapter and elsewhere (Buck et al. 2007). The digital data obtained from scanning can be evaluated immediately and, if necessary, fed into an ongoing forensic collection process. In most cases, only one scan per print is needed, which can be viewed from alternative angles once captured. Digital photography also provides immediate reassurance to an investigator that the data has been captured, but conventional photography and casting does not offer such reassurance and has the inherent risk that, if unsuccessful, evidence may be lost.

Casting is, by its nature, an invasive process and has the potential to introduce artefacts into evidence that are the result of the casting process (Figure 28.5). In addition, casts pick up trace evidence which must be dealt with before they can be cleaned for analysis (Bull et al. 2006). Scanning also provides a means of quickly processing a suspect's footwear in three dimensions and capturing that evidence at one moment in time. This is particularly pertinent where evidence cannot be retained.

The current equipment is bulky and currently impractical for some operational settings (Figure 28.1). It also has a high capital cost. While these are critical operational issues, there is potential for the use of such equipment and the new opportunities for analysis it offers. However, such issues can be resolved over time with developments by manufacturers provided that there is a clear demand from the forensic community.

Once an investigation leaves the crime scene, footwear evidence collected by laser scanning can be used to create quantitative measurements of foot dimensions or wear characteristics. This is particularly important since it is possible to make a quantitative case with which to link a suspect's footwear to the evidence left at a crime scene. The data is also more easily stored and shared between investigators than a physical cast, or mould, which is bulky and cannot be transferred digitally. The key advantage of a digital scan is the ability to rotate the image so that it can be viewed from any direction and to vary the angle and degree of illumination, as shown in Experiments Two and Three. The analytical tools involved in analysing a scan involve training on the relevant software, and dexterity to control viewing of the imagery. All the software used in this chapter is commercially available and easily used.

The use of digital scans, and the associated photography, allows the investigator to re-analyse records of the crime scene and footwear evidence with greater flexibility and with a lowered risk of incomplete data capture at the site. Finally, output from a scan can be prepared as a series of screen images, or a three-dimensional model can be created from the digital data using a variety of commercial three-dimensional printing solutions (Huddart et al. in press).

There exist two separate challenges associated with the application of this type of technology to recording and preserving footwear evidence in soils. Firstly, there is a need to demonstrate the potential advantages of such techniques with respect to more traditional approaches, and secondly to develop the equipment and operational methodology to make its application a routine occurrence at a scene of crime. The latter challenge is one which requires further consideration by manufacturers, in conjunction with the forensic community to enable the use of this technology to be an operational reality.

Acknowledgements This work is a product of the NERC grant (NA/C519446/1) into the investigation of the Early Human Colonization of the Americas. The authors would like to acknowledge the assistance of Sophie Fisher and Gareth Roberts in the preparation of this chapter.

References

Aldhouse-Green SHR, Whittle AWR, Allen JRL, Caseldine AE, Culver SJ, Day MH, Lundquist J and Upton D (1992). Prehistoric human footprints from the Severn Estuary at Uskmouth and Magor Pill, Gwent, Wales. Archaeology Cambrensis 141:14–55.

Ashley W (1996). What shoe was that? The use of computerised image database to assist in identification. Forensic Science International 82:7–20.
Bodziak WJ (2000). Footwear impression evidence: detection, recovery and examination. CRC, Boca Raton, FL.
Buck U, Albertini N, Naether S and Thali M (2007). 3D documentation of footwear impressions and tyre tracks in snow with high resolution optical surface scanning. Forensic Science International 171(2–3):157–164.
Bull PA, Parker A and Morgan RM (2006). The forensic analysis of soils and sediment taken from the cast of a footprint. Forensic Science International 162:6–12.
Du Pasquier E, Hebrard J, Margot P and Ineichen M (1996). Evaluation and comparison of casting materials in forensic sciences. Applications to tool marks and foot/shoe impressions. Forensic Science International 82:33–43.
Geradts Z and Keijzer J (1996). The image-database REBEZO for shoeprints with developments on automatic classification of shoe outsole designs. Forensic Science International 82:21–31.
Gonzalez-Gonzalez A, Lockle MG, Rojas CS, López-Espinoza J and Gonzalez (in press). Human Tracks from Quaternary tufa deposits, Cuatrocienegas, Mexico. Ichnos.
Gonzalez S, Huddart D, Bennett MR and Gonzalez-Huesca A (2006). Human footprints in Central Mexico older than 40,000 years. Quaternary Science Reviews 25:201–222.
Huddart D, Bennett MR, Gonzalez S and Velay X (in press). Documentation and preservation of Pleistocene human and animal footprints: an example from Toluquilla, Valsequillo Basin (Central Mexico). Ichnos.
Hueske E (1991). Photographing and casting footwear/tiretrack impressions in snow. Journal of Forensic Identification 41:92–95.
Kennedy RB, Chen S, Pressman IS, Yamashita AB and Pressman AE (2005). A large-scale statistical analysis of barefoot impressions. Journal of Forensic Sciences 50:1–10.
Laskowski GE and Kyle VL (1988). Barefoot impressions – a preliminary study of identification characteristics and population frequency of their morphological features. Journal of Forensic Sciences 33:378–388.
Leakey MD and Harris JM (1987). Laetoli: a Pliocene site in Northern Tanzania. Clarendon Press, Oxford.
Leakey MD and Hay RL (1979). Pliocene footprints in the Laetoli Beds at Laetoli, Northern Tanzania. Nature 278:317.
Robbins LM (1985). Footprints: collection, analysis and interpretation. Charles C. Thomas, Springfield, IL.
Roberts G, Gonzalez S and Huddart D (1996). Intertidal Holocene footprints and their archaeological significance. Antiquity 70:647–51.
Rodríguez-de la Rosa RA, Aguillón-Martínez MC, López-Espinoza J and Eberth DA (2004). The fossil record of vertebrate tracks in Mexico. Ichnos 11:27–38.
Schallamach A (1968). Abrasion, fatigue and smearing of rubber. Journal of Polymer Science 12:281.
Thali MJ, Braun M, Buck U, Aghayey E, Jackowski C, Vock P, Sonnenschein M and Dirnhofer R (2005). VIRTOPSY-scientific documentation, reconstruction and animation in forensics; individual and real 3D data based geo-metric approach including optical body/object surface and radiological CT/MRI scanning. Journal of Forensic Sciences 50:424–428.

Chapter 29
Discrimination of Domestic Garden Soils Using Plant Wax Compounds as Markers

Robert W. Mayes, Lynne M. Macdonald, Jasmine M. Ross and Lorna A. Dawson

Abstract Much of the forensic comparison of soils has focused on mineralogy at the expense of soil organic components. This chapter discusses how the nature of soil organic matter could be a useful tool in forensic comparison of soil. The chemical diversity and resistant nature of most plant-derived wax markers in soil result in the potential to provide robust profiles that vary with the characteristics of the vegetation that contributed to its formation. The epicuticular wax of most plants contains mixtures of long-chain aliphatic lipids, including hydrocarbons (mainly *n*-alkanes), long-chain fatty alcohols and fatty acids. The composition of these wax compounds differ between plant species, and these patterns persist in the soil. Plant wax markers may have value in characterising soils for forensic investigation where they could be used to indicate vegetation characteristics of an unknown sample, or for evaluative comparison of a 'questioned' evidence sample to that of a known scene of crime/reference site. We conducted a pilot study in North East Scotland to determine whether individual patches of garden planting bed soil could be discriminated from one another based on their plant wax marker profiles. Results indicated that both *n*-alkane and long-chain fatty alcohol profiles in garden soils were generally specific to the individual planting bed. The soils from the planting beds could be discriminated from one another, with the exception of two beds from within the same garden which differed little in broad-scale vegetation type. These results demonstrate the potential of plant wax markers in discriminating patches of soil at a scale relevant to forensic evaluative comparison. Further studies are required on soils and vegetation from a wider population of gardens and vegetation types, to extend the application of these plant wax markers to facilitate investigative approaches in potentially indicating plant community characteristics associated with an unknown source.

R.W. Mayes(✉), L.M. Macdonald, J.M. Ross and L.A. Dawson
The Macaulay Institute, Craigiebuckler, Aberdeen AB15 8QH, UK
e-mail: r.mayes@macaulay.ac.uk

K. Ritz et al. (eds.), *Criminal and Environmental Soil Forensics*

Introduction

Soil is a complex mixture of weathered mineral rock and decomposing organic materials originating from plants, animals and micro-organisms. Natural variation in the soil environment results from differing geophysical and environmental factors such as underlying geology, climate, landscape topography, and land-use/vegetation (past and present). Since traces of soil can easily be transferred to clothing, vehicles, tools and other objects, soil has long been considered to be a potential source of forensic information (Dawson et al. 2008; Locard 1930). Most cases using soil in forensic work have used mineralogical characteristics (Ruffell and McKinley 2005) and palynological information (Mildenhall 2006) to support association of questioned samples and scene of crime. With the exception of such palynological investigations, there have been very few forensic cases where detailed information about the organic components of soil has been used. Generally, the major mineralogical differences of soil vary at a regional scale, over geological time frames, while the organic component of soil is influenced by associated vegetation and soil management decisions that are made over more local scales and timeframes. Owing to these differences and the independent information that they can provide, soil organic matter profiles and mineralogical techniques can potentially complement each other in the investigation of forensic problems.

Organic Matter in Soil

Soil organic matter (SOM) can simply be described as all organic material found in soil, irrespective of origin or state of decomposition (Baldock and Skjemstad 2000; Krull et al. 2003). The composition of SOM is related primarily to present and past vegetation inputs, and processes of decomposition. In most soils the main organic inputs are derived from local vegetation litter (e.g. leaf litter, root litter and detritus), although managed soils (agricultural, municipal, or residential) may also receive organic inputs arising from the application of manures, slurry, composts or mulches. These decomposing inputs provide a wide diversity of biochemical signatures which offer strong potential for developing novel investigative and/or evaluative tools for forensic application. Due to the high degree of variability in plant cover through land-use management options, plant derived chemical profiles within the soil have potential in providing forensic intelligence at a scale suitable to distinguish specific locations that differ in plant community composition.

The SOM is a dynamic component where chemical changes occur in parallel with decomposition processes. The composition of SOM is difficult to measure directly (Baldock and Skjemstad 2000) and is commonly divided chemically or physically into fractions that have different chemical properties and turnover rates (Krull et al. 2003). Dissolved and particulate SOM fractions include carbohydrates and recognisable plant fractions, and are considered to have a relatively rapid turnover rate (<10 years). The humus fraction comprises a variety of organic materials (complex polysaccharides, lipid and wax compounds, humic acids, suberin, cutin, etc.) and usually constitutes the largest pool of the SOM. These compounds are considered relatively resistant to decomposition. Aromatic carbon compounds and charcoal are considered largely inert and recalcitrant, having residency times in the region of thousands of years. The soil organic component also encompasses the microbial biomass, the community structure of which responding to soil type, land-management and environmental conditions (Girvan et al. 2003; Wakelin et al. 2008). Several studies have examined the potential use of soil microbial community profiles for forensic application (e.g. Horswell et al. 2002; Lerner et al. 2006; Macdonald et al. 2008).

Selecting suitable markers from these diverse SOM components requires careful consideration. For example, compounds with high turnover rates, such as most carbohydrates, proteins and nucleic acids, are likely to be less useful for forensic comparison than those compounds with slower turnover rates that are likely to provide a more reliable profile. Alternatively, while the complex nature of polymeric materials such as humic compounds and tannins may render them appropriate for profiling purposes, they prove difficult to characterise and analyse quantitatively in soil. Analytical methods which involve fragmenting the polymeric substances into complex mixtures of lower molecular weight compounds are available and have the potential to provide highly specific profiles of the organic component of soil. These procedures include pyrolysis (Ceccanti et al. 2007) and thermally-assisted hydrolysis and methylation with tetramethyl ammonium hydroxide (see Challinor 2001 for TMAH). However, such profiles can be very complex, difficult to interpret, and are dependent on the specific conditions used for analysis bringing into question analytical reproducibility which may limit forensic applications.

Plant Wax Compounds as Biomarkers

Within the humus fraction, plant wax compounds are complex mixtures of lipids consisting mainly of aliphatic compounds with relatively long carbon chains (Ca. C_{20}–C_{60}) and comprise a number of classes (Table 29.1). The most common and widely studied are hydrocarbons, including *n*-alkanes, and free and esterified long-chain fatty alcohols and fatty acids. Most plant wax components can be quantitatively analysed as individual compounds by gas chromatography (GC) or gas chromatography-mass spectrometry (GCMS), following solvent extraction and separation into different compound types by liquid chromatography, most conveniently by using solid phase extraction columns. Such analytical methods are both reliable and relatively straightforward and the facilities are commonly available in most forensic laboratories. Furthermore, they have the advantage that a series of individual compounds can be analysed in a single chromatographic run. Improved methods for automated extraction and separation have also been reported (Wiesenberg et al. 2004). Depending on the sample matrix, extracts prepared for GC or GCMS analysis may contain compounds (e.g. various hydrocarbons), not of plant wax origin, but with similar properties, which can also be separated and quantitatively analysed.

Table 29.1 Main classes of plant wax compounds which may have value as biomarkers in soil

Compound class	Carbon number	Properties	Comments
Hydrocarbons			
n-Alkanes	C_{21}–C_{37}	Odd-numbered carbon chains predominate	Very common
Branched-chain alkanes	C_{28}–C_{32}	Iso-alkane, methyl branch at C_2 Anteiso-, methyl branch at C_3	Rare
n-Alkenes	C_{27}–C_{33}	Odd-chain monenes with double bond at C_{7-8}	In inflorescences and also for C_{9-10} alkenes, in leaves of some woody plants
		Odd-chain monenes with double bond at C_{9-10}	
	C_{24}–C_{30}	Even-chain monoenes with double bond at C_{1-2}	In leaves of some woody plants
Free long-chain alcohols			
Primary alcohols	C_{20}–C_{34}	Saturated, unbranched, mainly even-chain	Common, at high concentrations
Secondary alcohols		Odd-chain, mainly 10-nonacosanol (C_{29})	High concentrations in gymnosperms
Free long-chain fatty acids	C_{16}–C_{34}	Saturated, unbranched, mainly even-chain	Common, C_{20}–C_{34} useful as markers
Wax esters	C_{35}–C_{64}	Esters of long-chain fatty acids and long-chain fatty alcohols	Common
Long-chain fatty aldehydes	C_{16}–C_{34}	Saturated, unbranched, mainly even-chain	High concentrations in some species
Long-chain ketones		Odd-chain, mainly 10-nonacosanone	In some plants
β-Diketones	C_{29}–C_{33}	Odd-chain, with C = O groups on even-chain positions	High concentrations in some species
Other aliphatic compounds and triterpenoids		Polyesters of hydroxyacids, secondary ketones, hydroxy-β-diketones, etc.	Variable across the plant kingdom

Plant wax compounds are found on all external surfaces of plants, including leaves, roots, and flowers, but primarily on the surfaces of leaves. Plant *n*-alkanes have, over the last 20 years, been widely used as faecal markers in grazing farm livestock and a range of mammalian herbivores, to measure digestibility and intake of grazed vegetation (Mayes and Dove 2006). Furthermore, differences in *n*-alkane profiles in the faeces of herbivores can be used to determine the plant species composition of a herbivore's diet from the pattern found in the faeces, together with knowledge of the patterns in the individual dietary plant species. The reliability of this approach has been enhanced by using additional plant wax compounds, such as long-chain fatty alcohols, together with *n*-alkanes (Ali et al. 2005a). The discrimination between plant-species can be expected to increase as the number of compounds used as markers is increased (Bugalho et al. 2004). However, the suitability of *n*-alkane and long-chain fatty alcohol profiles over other plant wax compounds (long-chain fatty acids, aldehydes and wax esters) for discriminating between numerous plant species has been demonstrated (Jansen et al. 2006).

Similar methodology and principals have been applied to soils, to provide evidence of the vegetative inputs to the soil system. The patterns of plant wax compounds in soil are relatively long-lived, and tend to reflect the patterns of the same compounds found in plants associated with that soil (Dawson et al. 2004). Wiesenberg et al. (2006) demonstrate decreasing lipid yield with soil depth, reflecting the decreasing organic carbon in lower soil horizons. Evidence for the long-term survival of plant wax marker patterns in soil has been provided by samples of buried soil layers (^{14}C-dated to 5000–6000 years BP) from the Trotternish Ridge in Skye, Scotland, where *n*-alkane and alcohol analysis data indicated the presence of heather in a buried horizon, matching independent evidence from pollen identification (Dawson et al. 2004).

Because they are considered part of the relatively stable carbon fraction of soil, plant wax compounds should be suitable as biomarkers and may have several advantages as a robust profiling method as they: (i) reflect the vegetative inputs, and therefore land-use characteristics of soil; (ii) are relatively long-lived in soil and stable over long temporal periods and (iii) are relatively easily analysed as discrete identifiable compounds. Profiling soil plant wax compounds could be useful for the two forensic contexts of investigative intelligence, aiming to use landscape level indicators within plant wax compound profiles to indicate land use or vegetation characteristics, and evaluative comparison of a 'questioned' soil sample to reference samples taken from a scene of crime.

Study of the Plant Wax Composition of Surface Soils from Planting Beds in Residential Gardens

Plant wax compound profiles could have a very useful role in providing a supportive link in associating a suspect/possession to a specific soil source, for example, a residential garden may be of interest in a scenario such as burglary. The wax profiles of surface soils in residential gardens will likely be a consequence of both present and past vegetation, and reflect localised soil management practices (cultivation, etc.). Therefore, it can be expected that the profiles will vary greatly from garden to garden, and likely within different planting beds within a garden.

We conducted a pilot study to explore the hypothesis that plant wax marker profiles of residential garden soils would be strongly dependent on the patch-specific vegetation of individual planting beds within gardens.

Materials and Methods

Soil was sampled from seven planting beds in four gardens situated in and around the city of Aberdeen, Scotland. The flowerbeds were not selected according to management, plantings or location (Table 29.2). Five replicate soil samples (0 to 2 cm depth) were sampled from each flowerbed. The sampling grid was a 25 × 25 cm square, and soil samples were taken from each corner, with the fifth sample taken from the centre of the square. Such a sampling arrangement could be expected to provide information at a scale relevant to that of a suspect's footprint. Each sample was freeze-dried and milled (Retsch Mixer Mill MM301) prior to analysis for *n*-alkanes and long-chain fatty alcohols using modifications to the method described by Dove and Mayes (2006). The amount of soil used in the analysis was adjusted according to the SOM, 1.0 and 0.5 g for mineral soils (SOM <12% dry soil) and organic soils (SOM >12%), respectively. The concentrations of internal standard solutions used for *n*-alkane (tetratriacontane, C_{34}) and fatty alcohol (1-heptacosanol, C_{27}) analyses were reduced (0.1295 and 0.1242 mg/g, respectively). Before use, solvents (*n*-heptane, ethanol and ethyl acetate) were redistilled and glass columns containing silica-gel (Sigma Aldrich Ltd.) were used to avoid potential contamination sources. Without further purification, the alcohol fraction collected from these columns was treated with BSTFA to form the trimethylsilyl derivatives, and directly analysed using GCMS in the selected ion mode (ThermoFinnegan Trace DSQ).

Table 29.2 Outline details of planting beds in selected Scottish domestic gardens providing soil samples for comparative study on plant wax composition

Address	Location[a]	Flowerbed	Brief description
Hopetoun Crescent	Bucksburn Aberdeen	Front bed	Annual plants for last 15 years; garden compost dug in annually; overhanging apple tree
Hopetoun Crescent	Bucksburn Aberdeen	Rear bed	Annual plants for last 15 years; garden compost dug in annually; overhanging rowan and cherry trees
Hopetoun Avenue	Bucksburn Aberdeen	Front bed	Uncultivated border with annual weeds; hoed once or twice each year. Large mock orange bush (*Philadelphus*) at side of border
N. Deeside Road	Aberdeen City	Front bed	Border around lawn; spring bulbs and different annual plants each year; close to main road
N. Deeside Road	Aberdeen City	Rear bed	Border around lawn; small shrubs, bulbs and clematis; damp, shaded site
Kinneff	Roadside, Kinneff	Shrub bed	Shrubs for 14 years (*Vibernum, Hebe, Philadelphus, Hydrangea*); understore of lesser celandine (in spring); overhanging wych elm and sycamore trees
Kinneff	Roadside, Kinneff	Vegetable bed	Plot 14 years old, brassicas, broad beans, leeks, peas and potatoes in recent years; garden compost dug in annually; chickweed problem; overhanging wych elm and sycamore trees

[a]All locations in the city of Aberdeen, except for Kinneff, which is in Aberdeenshire.

Individual *n*-alkanes and fatty alcohols were identified, and calculated as concentrations in dry soil (μg/g). Multivariate data approaches were used in order to evaluate general relationships between samples based on overall profile differences. Bray Curtis similarity matrices were generated on log transformed *n*-alkane and fattyalcohol datasets, before non-metric multidimensional scaling (MDS) was carried out in order to display the relative relationships between samples (Clarke 1993). The analysis of similarity (ANOSIM) procedure was carried out to test the null hypothesis that there were no differences between planting beds (Clarke 1993). ANOSIM reports an R statistic which indicates the level of segregation between groups, with a value close to unity indicating complete group separation, and a value close to zero implying little or no segregation. The associated significance level (P) is analogous to the univariate P-value. SIMPER (similarity percentage) analysis was used to determine the percentage contribution of individual *n*-alkanes/fatty alcohols towards describing planting bed dissimilarity (Clarke 1993). All data analyses were carried out using PRIMER6 (Clarke and Gorley 2006; Primer-E Ltd., Plymouth, UK).

Results

The *n*-alkane and fatty alcohol concentrations for the individual soil samples are shown in Figures 29.1 and 29.2, respectively. It can be seen that for both *n*-alkanes and fatty alcohols, there were noticeable differences between the planting beds in the patterns of these compound types and there were similarities in patterns of these compound types for soil samples within beds.

The MDS ordination plots of soil *n*-alkane profiles and long-chain fatty alcohol profiles are shown in Figures 29.3 and 29.4, respectively. These demonstrate the dominant influence of individual flowerbeds over plant wax compound profiles, showing clear discrimination between most flowerbeds for both the *n*-alkane profiles ($R = 0.86$, $P < 0.001$) and the fatty alcohol profiles ($R = 0.85$, $P < 0.001$). Pair-wise comparisons indicated that the *n*-alkane profile of the Kinneff shrub-bed showed the greatest degree of dissimilarity from all of the other planting beds (Table 29.3), which was mainly due to the contributions of C_{29} and C_{27} *n*-alkanes (SIMPER). The fatty alcohol profiles of the two Kinneff planting beds showed strong discrimination from the other garden bed soils, with C_{24} and C_{30} fatty-alcohols contributing strongly. The rear N. Deeside Rd. bed also separated clearly from all other flower-beds, with a high level of segregation (Table 29.3). Although some pair-wise comparisons of fatty-alcohol profiles showed lower levels of dissimilarity than comparisons of *n*-alkane profiles (e.g. the two Kinneff beds), greater dissimilarity was observed for fatty alcohol profiles between the front gardens of Hopetoun Crescent and N. Deeside Road., than for the comparable *n*-alkane profiles. The two Hopetoun Crescent planting beds could not be discriminated from one another. Although the Kinneff vegetable bed differed from the Kinneff shrub bed, it could not be discriminated from the rear bed at Hopetoun Crescent based on *n*-alkane profiles alone.

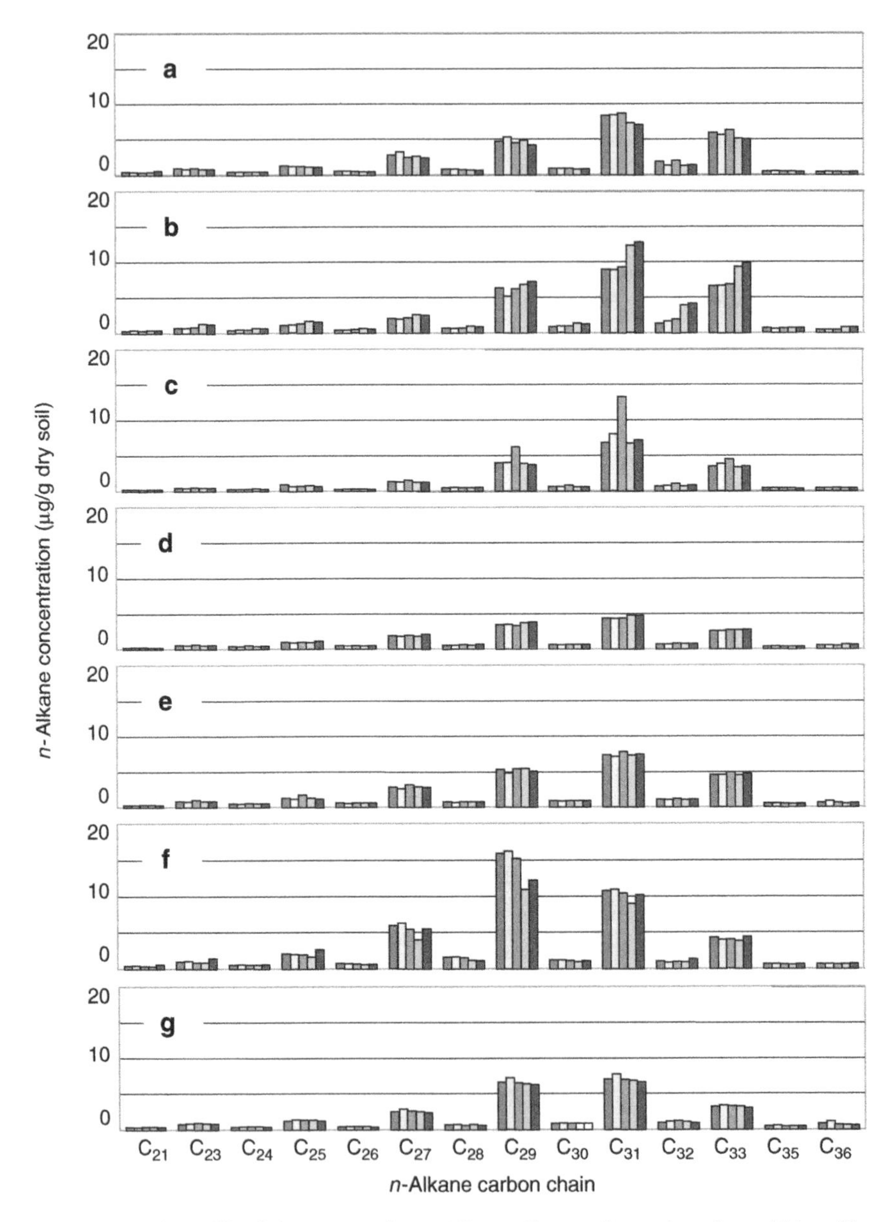

Fig. 29.1 *n*-Alkane (C_{21}–C_{36}) concentrations of five replicate soil samples taken within a 25 × 25 cm area from planting beds in four domestic garden sites. (a) Hopetoun Crescent front. (b) Hopetoun Crescent rear. (c) Hopetoun Avenue front. (d) N. Deeside Rd. Front. (e) N. Deeside Rd. Rear. (f) Kinneff shrub bed. (g) Kinneff vegetable bed. Different grey shades in bars consistently denote the different replicates within each site

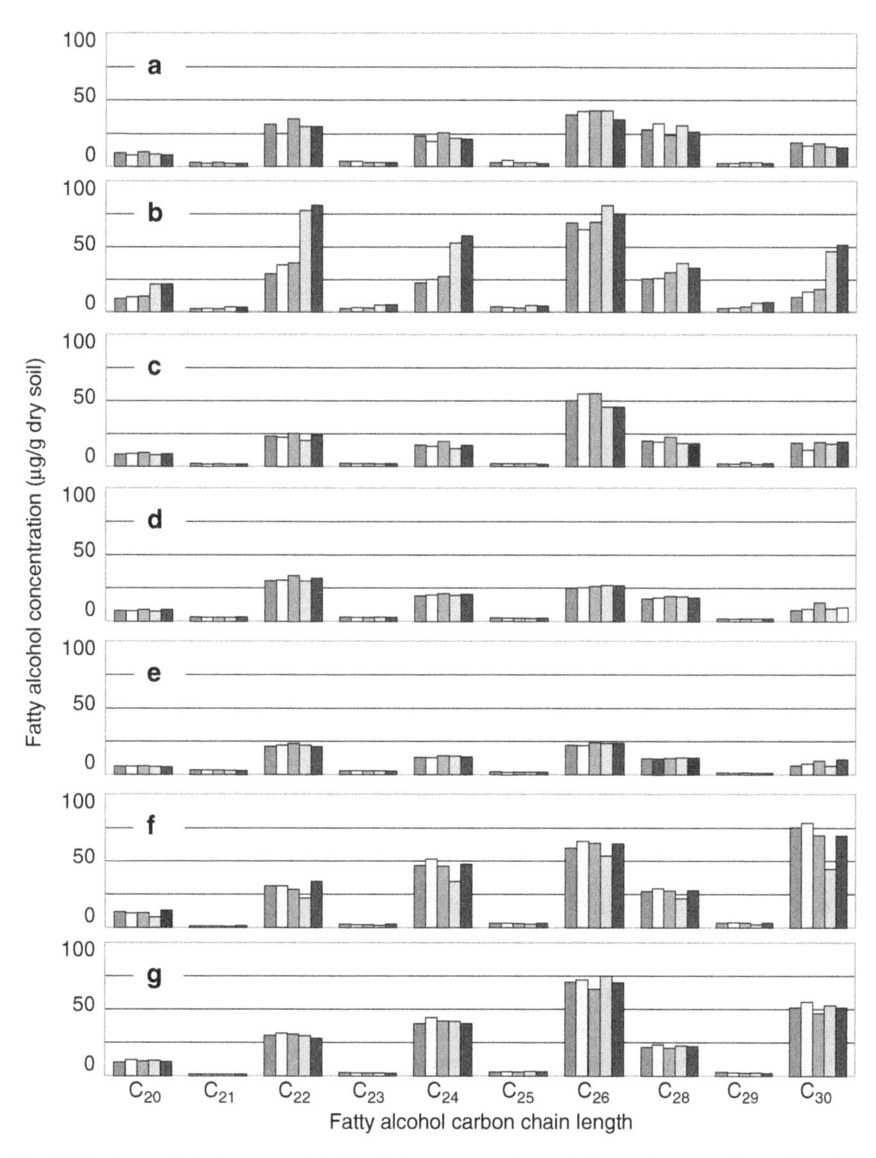

Fig. 29.2 Long-chain fatty alcohol (C_{20}–C_{30}) concentrations of five replicate soil samples taken within a 25 × 25 cm area from planting beds in four domestic garden sites. (a) Hopetoun Crescent front. (b) Hopetoun Crescent rear. (c) Hopetoun Avenue front. (d) N. Deeside Rd. Front. (e) N. Deeside Rd. Rear. (f) Kinneff shrub bed. (g) Kinneff vegetable bed. Different grey shades in bars consistently denote the different replicates within each site

Table 29.3 ANOSIM pair-wise comparison of dissimilarity in soil *n*-alkane and long-chain fatty alcohol profiles between planting bed sites. A 'global R' statistic indicates overall group segregation and associated significance, while the pair-wise comparisons (body of table) indicate the level of segregation between individual planting beds. An R statistic close to unity indicates complete separation of groups, and close to zero implies little or no segregation

Planting bed	Hopetoun Cres. front	Hopetoun Cres. rear	Hopetoun Ave. front	N. Deeside Rd. front	N. Deeside Rd. rear	Kinneff Shrub bed
(a) *n*-Alkanes	Global R = 0.855 (P < 0.001)					
Hopetoun Cres. rear	0.47					
Hopetoun Ave. front	0.81	0.85				
N. Deeside Rd. front	0.55	0.64	0.72			
N. Deeside Rd. rear	1.00	1.00	0.71	1.00		
Kinneff shrub bed	1.00	0.98	1.00	1.00	1.00	
Kinneff vegetable bed	1.00	0.69	0.72	0.99	1.00	1.00
(b) Fatty alcohols	Global R = 0.845 (P < 0.001)					
Hopetoun Cres. rear	0.49					
Hopetoun Ave. front	0.96	0.64				
N. Deeside Rd. front	0.97	0.64	1.00			
N. Deeside Rd. rear	1.00	0.94	1.00	1.00		
Kinneff shrub bed	1.00	0.44	0.99	1.00	1.00	
Kinneff vegetable bed	1.00	0.49	1.00	1.00	1.00	0.60

Discussion

These results demonstrate the potential of markers within the soil organic matter to discriminate soils that have originated from local-scale soil patches having correspondingly different vegetation characteristics. Variation between planting beds was greater than variation within beds, allowing clear differentiation of soil patches based on the soil *n*-alkane and long-chain fatty alcohol profiles. These site locations were selected without prior knowledge of vegetation or soil type, and therefore represent a small but unprejudiced population of gardens. Ultimately the outcome of any forensic evaluation of a questioned sample will likely depend on the characteristics of a reference soil population. However, our results suggest that plant wax compound profiles could provide a distinctive soil 'profile' at a scale relevant to forensic comparison in an evaluative context. Where plant wax profiles prove distinctive, they could provide a tool to definitively exclude potential source sites from further consideration. Further studies with a larger soil population of urban gardens would help to address how likely it would be to find similar plant wax profiles in a local area, helping to assess the weight that can be placed on any evaluative comparison of soil.

Whilst the garden soil study described above gave promising results based on *n*-alkane and long-chain fatty alcohol profiles, it may be expected that discrimination would be further improved by using additional plant wax compounds in the soil. The very long-chain fatty acids (C_{20}–C_{34}), found in plant cuticular waxes, have been shown, like the *n*-alkanes and fatty alcohols, to differ in composition between plant

Fig. 29.3 Non-metric multidimensional scaling (MDS) ordination plots of soil *n*-alkane profiles of domestic garden sites. Five replicate samples collected within a 25 × 25 cm area from each planting bed at the sites: (▲) Hopetoun Crescent front; (△) Hopetoun Crescent rear; (■) Hopetoun Avenue front; (●) N. Deeside Rd. Front; (○) N. Deeside Rd. Rear; (◆) Kinneff shrub bed; (◇) Kinneff vegetable bed

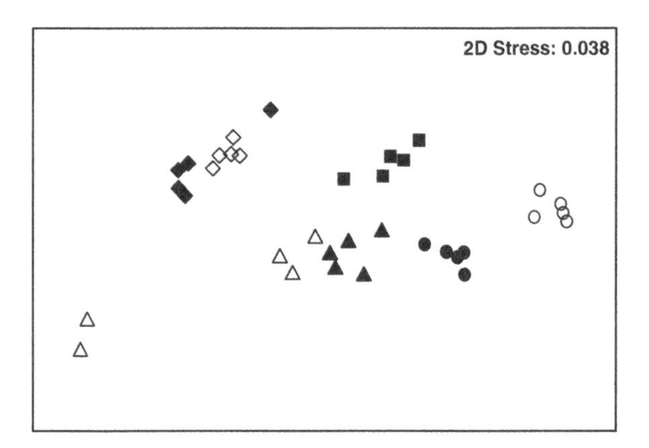

Fig. 29.4 Non-metric multidimensional scaling (MDS) ordination plots of soil long-chain fatty alcohol profiles of domestic garden sites. Five replicate samples collected within a 25 × 25 cm area from each planting bed at the sites: (▲) Hopetoun Crescent front; (△) Hopetoun Crescent rear; (■) Hopetoun Avenue front; (●) N. Deeside Rd. Front; (○) N. Deeside Rd. Rear; (◆) Kinneff shrub bed; (◇) Kinneff vegetable bed

species (Ali et al. 2005b) and their presence in soil has been well documented (e.g. Bull et al. 2000). Also found in soil, and originating from plant wax, are fatty aldehydes (Stephanou 1989) and triterpenoids (e.g. Naafs et al. 2004). Although Jansen et al. (2006) report that *n*-aldehydes and wax ester compounds in plant leaves and roots are less suited as biomarkers due to their absence in a significant number of plant species, the absence of such compounds could indeed provide diagnostic information. Additional hydrocarbons, such as branched chain alkanes and *n*-alkenes, may also be of value as contributors to a soil organic profile, although relatively little is known about their survival in soil.

Our work has specifically targeted wax markers of plant origin. However, it is also possible that non-plant derived hydrocarbons can be identified during the analysis, depending on the sample matrix, extraction method and preparation. Non-plant derived hydrocarbons may be useful in determining if an unknown soil sample has originated from a pristine soil environment or from one which is under anthropogenic influences, for example as indicated by the presence of petrochemical hydrocarbon compounds.

Forensic soil samples are likely to be of limited size, and potentially originate from surface or near sub-surface soils. Although the quantities of soil likely to be transferred to footwear or clothing are likely to be smaller than used in the conventional analysis of plant wax compounds, our recent work has shown that reproducible analysis can be achieved using as little as 40 mg of a low organic-matter soil (<20% carbon) and 13 mg of a soil of high OM content (>20% carbon). Low concentrations of plant wax compounds may indicate that a 'questioned' soil sample may have originated from a lower soil horizon, or from a soil under little vegetative influence. The concentrations recovered from these residential garden soils were consistent with that previously found in top-soils of agricultural nature (Dawson et al. 2004).

The plant wax compound profiles in soil are undoubtedly related to the patterns in present and past vegetation growing on the soil, together with compounds originating from any soil adjuncts. There may be forensic value, especially for intelligence gathering, in obtaining information on the plant wax composition of cultivated garden plants, garden weed species and the analysis of composts, manures and other soil additives. With such information, it may be possible, for example, to predict the type of garden bed (in terms of plantings and soil management) from a soil sample gathered from a suspect's shoe. At present, little is known about the plant wax composition of garden plants, and considerable further work and development of comprehensive database approaches are required to gather a suitable base of information on vegetation and soil plant wax profiles, in order to facilitate investigative approaches.

Although largely neglected in the forensic comparison of soils, methodological advances and improved understanding of the properties of SOM bring new opportunities for the development of innovative techniques in this field. Plant wax markers, and other organic substances, have the potential to complement other more widely used techniques in forensic geosciences, such as mineralogical or palynological based techniques. The use of these techniques in combination could provide stronger associative evidence, and be more readily applied than currently practised.

Acknowledgements The authors acknowledge the receipt of funding from the Rural and Environment Research and Analysis Directorate (RERAD) within the Scottish Government. Lynne Macdonald was funded by the Engineering and Physical Sciences Research Council (EPSRC) through the SoilFit project.

References

Ali HAM, Mayes RW, Lamb CS, Hector BL, Verma AK and Ørskov ER (2005a). The potential of long-chain fatty alcohols and long-chain fatty acids as diet composition markers. Journal of Agricultural Science 143:85–95.

Ali HAM, Mayes RW, Hector BL and Ørskov ER (2005b). Assessment of *n*-alkanes, long-chain fatty alcohols and long-chain fatty acids as diet composition markers: The concentrations of these compounds in rangeland species from Sudan. Animal Feed Science and Technology 121:257–271.

Baldock JA and Skjemstad JO (2000). Role of the soil matrix and minerals in protecting natural organic materials against biological attack. Organic Geochemistry 31:697–710.

Bugalho MN, Dove H, Kelman W, Wood JT and Mayes RW (2004). Plant wax alkanes and alcohols as herbivore diet composition markers. Journal of Range Management 57:259–268.

Bull ID, Nott CJ, van Bergen PF, Poulton PR and Evershed RP (2000). Organic geochemical studies of soils from the Rothamsted classical experiments – VI. The occurrence and source of organic acids in an experimental grassland soil. Soil Biology and Biochemistry 32:1367–1376.

Ceccanti B, Masciandaro G and Macci C (2007). Pyrolysis-gas chromatography to evaluate the organic matter quality of a mulched soil. Soil and Tillage Research 97:71–78.

Challinor JM (2001). Review: The development and applications of thermally assisted hydrolysis and methylation reactions. Journal of Analytical and Applied Pyrolysis 61:3–34.

Clarke KR (1993). Non-parametric multivariate analysis of changes in community structure. Australian Journal of Ecology 18:117–143.

Clarke KR and Gorley RN (2006). PRIMER v6: User Manual/Tutorial. PRIMER-E, Plymouth Marine Laboratory, UK.

Dawson LA, Towers W, Mayes RW, Craig J, Vaisanen K and Waterhouse EC (2004). The use of plant wax signatures in characterising soil organic matter. In: Forensic Geoscience: Principles, Techniques and Applications (Eds. K Pye and DJ Croft), pp. 269–276. Geological Society Special Publication 232, London.

Dawson LA, Campbell CD, Hillier S and Brewer MJ (2008). Methods of characterising and fingerprinting soils for forensic application. In: Soil Analysis in Forensic Taphonomy: Chemical and Biological Effects of Buried Human Remains (Eds. M Tibbett and D Carter), pp. 271–315. CRC, Boca Raton, FL.

Dove H and Mayes RW (2006). Protocol for the analysis of *n*-alkanes and other plant-wax compounds and for their use as makers for quanitifying the nutrient supply of large mammalian herbivores. Nature protocols 1:1680-1696. [online Journal].

Girvan MS, Bullimore J, Pretty JN, Osborn AM and Ball AS (2003). Soil type is the primary determinant of the composition of the total and active bacterial communities in arable soils. Applied and Environmental Microbiology 69:1800–1809.

Horswell J, Cordiner SJ, Maas EW, Martin TM, Sutherland BW, Speir TW, Nogales B and Osborn AM (2002). Forensic comparison of soils by bacterial community DNA profiling. Journal of Forensic Sciences 47:350–353.

Jansen B, Nierop KGJ, Hageman JA, Cleef AM and Verstraten JM (2006). The straight-chain lipid biomarker composition of plant species responsible for the dominant biomass production along two altitudinal transects in the Ecuadorian Andes. Organic Geochemistry 37:1514–1536.

Krull ES, Baldock JA and Skjemstad JO (2003). Importance of mechanisms and processes of the stabilisation of soil organic matter for modelling carbon turnover. Functional Plant Biology 30:207–222.

Lerner A, Shor Y, Vinokurov A, Okon Y and Jurkevitch E (2006). Can denaturing gradient gel electrophoresis (DGGE) analysis of amplified 16s rDNA of soil bacterial populations be used in forensic investigations? Soil Biology and Biochemistry 38:1188–1192.

Locard E (1930). Analyses of dust trace parts I, II and III. American Journal of Police Science, 1:276–298; 401–418; 496–514.

Macdonald LM, Singh BK, Thomas N, Brewer MJ, Campbell CD and. Dawson LA (2008). Microbial DNA profiling by multiplex terminal restriction fragment length polymorphism for forensic comparison of soil and the influence of sample condition. Journal of Applied Microbiology. 105:813–821.

Mayes RW and Dove H (2006). The use of *n*-alkanes and other plant-wax compounds as markers for studying the feeding and nutrition of large mammalian herbivores. In: Herbivores: The Assessment of Intake, Digestibility and the Roles of Secondary Compounds (Eds. CA Sandoval-Castro, FD de B Hovell, F Torres Acosta and A Ayala-Burgos), pp153–182. BSAS Publication 34, Nottingham University Press, Nottingham, UK.

Mildenhall DC (2006). An unusual appearance of a common pollen type indicates the scene of the crime. Forensic Science International 163:236–240.

Naafs, DFW, van Bergen PF, de Jong MA, Oonincx A and de Leeuw JW (2004). Total lipid extracts from characteristic soil horizons in a podzol profile. European Journal of Soil Science 55:657–669.

Ruffell A and McKinley J (2005). Forensic geoscience: applications of geology, geomorphology and geophysics to criminal investigations. Earth Science Reviews 69:235–247.

Stephanou E (1989). Long-chain aldehydes: An overlooked but ubiquitous compound class of possible geochemical and environmental significance. Naturwissenschaften 76:464–467.

Wakelin SA, Macdonald LM, Rogers SL, Gregg AL, Bolger TP and Baldock JA (2008). Habitat selective factors influencing the structural composition and functional capacity of microbial communities in agricultural soils. Soil Biology and Biochemistry 40:803–813.

Wiesenberg GLB, Schwark L and Schmidt MWI (2004). Improved automated extraction and separation procedure for soil lipid analyses. European Journal of Soil Science 55:349–356.

Wiesenberg GLB, Schwark L and Schmidt MWI (2006). Extractable lipid contents and colour in particle-size separates and bulk arable soils. European Journal of Soil Science 57:634–643.

Chapter 30
Environmental Forensic Investigations: The Potential Use of a Novel Heavy Metal Sensor and Novel Taggants

Pat Pollard, Morgan Adams, Peter K. J. Robertson, Konstantinos Christidis, Simon Officer, Gopala R. Prabhu, Kenneth Gow and Andrew R. Morrisson

Abstract This chapter presents a novel hand-held instrument capable of real-time *in situ* detection and identification of heavy metals, along with the potential use of novel taggants in environmental forensic investigations. The proposed system provides the facilities found in a traditional laboratory-based instrument but in a hand held design, without the need for an associated computer. The electrochemical instrument uses anodic stripping voltammetry, which is a precise and sensitive analytical method with excellent limits of detection. The sensors comprise a small disposable plastic strip of screen-printed electrodes rather than the more common glassy carbon disc and gold electrodes. The system is designed for use by a surveyor on site, allowing them to locate hotspots, thus avoiding the expense and time delay of prior laboratory analysis. This is particularly important in environmental forensic analysis when a site may have been released back to the owner and samples could be compromised on return visits. The system can be used in a variety of situations in environmental assessments, the data acquired from which provide a metals fingerprint suitable for input to a database. The proposed novel taggant tracers, based on narrow-band atomic fluorescence, are under development for potential deployment as forensic environmental tracers. The use of discrete fluorescent species in an environmentally stable host has been investigated to replace existing toxic, broadband molecular dye tracers. The narrow band emission signals offer the potential for tracing a large number of signals in the same environment. This will give increased data accuracy and allow multiple source environmental monitoring of environmental parameters.

P. Pollard(✉), M. Adams, P.K.J. Robertson, K. Christidis, S. Officer, G.R. Prabhu and K. Gow
Centre for Research in Energy and Environment, The Robert Gordon University, Schoolhill, Aberdeen AB10 1FR, UK
e-mail: p.pollard@rgu.ac.uk

A.R. Morrisson
School of Life Sciences, The Robert Gordon University, Aberdeen AB25 1HG, UK

K. Ritz et al. (eds.), *Criminal and Environmental Soil Forensics*

Introduction

Heavy Metals Sensor and Contaminated Land

These metals are most commonly measured by inductively coupled plasma atomic emission spectroscopy (ICPAES) following microwave digestion, as the detection limits required (Table 30.1) are not demanding, at 40 to 2000 mg/kg (air-dried soil), for even domestic use. ICPAES detection limits range from 0.008 mg/l arsenic to 0.0004 mg/l nickel in aristar nitric acid digest. Inductively coupled plasma mass spectroscopy (ICPMS) is also sometimes used, but detection limits in the μg/kg range are not necessary. Hydride generation atomic absorption spectroscopy (HGAAS) has been used for the very selective measurement of metals which forms volatile hydrides, and cold vapour atomic absorption spectroscopy (CVAAS) can measure mercury at as low as 5 ng/l. However, all of these methods use laboratory instruments which require highly trained staff and are not suitable for on-site monitoring. Two methods which could be used for the measurement and analysis of a range of heavy metals in the field are electrochemistry and X-ray fluorescence (XRFt) (Table 30.2).

The use of small, safe, X-ray tubes to replace dangerous radioactive sources has allowed the development of field portable X-ray fluorescence (XRF) systems. When an X-ray passes through a sample, absorption by an inner shell electron of an X-ray quantum produces an excited ion which spontaneously emits a characteristic x-ray fluorescent spectrum. Current field portable XRF systems are capable of monitoring 25 elements simultaneously. Detection limits for a field portable XRF system are typically: 10 to 100 mg/kg for Cu, Ni, Pb, Cr, Hg, Se, Zn, As, and 50 to 150 mg/kg for Cd (Innovxsys 2007). All systems have their own advantages and disadvantages. For example, XRF does not have the potential to differentiate between the oxidation states of an element which is a very significant factor in its level of toxicity. Cr(VI) has an oral toxicity of 0.003 mg/kg/d compared to Cr(III) which has an oral toxicity of 1.5 mg/kg/d. One of the

Table 30.1 Excerpt from ICRCL 59/83 Trigger Concentrations

Class	Contaminant	Planned use	Concentration range mg/kg (air-dried soil) threshold
Metals	Cu, Ni, Pb, Cd, Hg, Zn, As	Domestic gardens, allotments, parks and playing fields	40–2000
Anions	S, SO_4^{2-}, S^{2-}, CN^-, B	Domestic gardens, any uses where plants grow	3–5000
Organics	TEM, TPHs, PAHs, PCBs, BTEX, Phenols, Volatile Organics	Domestic gardens, landscaped areas	5–1000
Others	pH, Asbestos	Domestic gardens, landscaped areas, hardcover	<5 pH

Table 30.2 Comparison of analytical techniques used for heavy metal detection

Analytical technique	Sensitivity	Cost	Real time detection	Ease of use	Instrument size
GFAA	Good	Medium	No	Expert required	Large, complex laboratory instrument
ICPAES	Good	High	No	Expert required	Large, complex laboratory instrument
ICPMS	Good	High	No	Expert required	Large, complex laboratory instrument
LCMS	Good	High	No	Expert required	Large, complex laboratory instrument
XRF (laboratory instrument)	Good	Medium	No	Expert required	Large, complex laboratory instrument
XRF (field portable instrument)	Good	High	Yes	Non-expert	Small, simple field portable instrument
Electrochemistry (laboratory instrument)	Good	Medium	No	Expert required	Large, complex laboratory instrument
Electrochemistry (field portable)	Good	Medium	Yes	Non-expert	Small, simple field portable instrument

major problems of quantitative XRF is matrix effects, the extent of which depends on the mass absorption coefficients of all the elements present in the matrix.

Stripping analysis is an electroanalytical method which involves a pre-electrolysis step to pre-concentrate a substance onto the surface of an electrode. After the electro-deposition the material is stripped from the electrode, in this case, by differential pulse voltammetry. The potential at which a metal is stripped (oxidised) from the electrode is characteristic of not only the metal but also of its oxidation state. The peak is proportional to the concentration of the metal species and their separation is important for quantification. Research on the detection of heavy metals using electrochemical measurements has focused on computer-based systems (Bersier et al. 1994; Rossmeisl and Nørskov 2008). The advantages of these systems are summarised thus: the ability to manage experiments involving sophisticated measurement techniques; storage of large amounts of data; data can be manipulated in complex ways; digital filtering can be performed; and results can be presented in a convenient format.

Several general purpose electro-analytical instruments based on computer systems are available commercially. However, these instruments are expensive, complicated to operate and bulky. Some 'portable' instruments are impractical for use in the field, requiring procedures normally carried out in a laboratory, such as pipetting. It follows that a fast, reliable, relatively inexpensive, portable (hand-held) and independent instrument, capable of the direct monitoring of heavy metal contaminants *in situ* (Christidis et al. 2006, 2007), without the need for an analytical chemist, is very desirable.

This chapter describes the development of such an electrochemical instrument which is capable of gathering real-time quantitative data on a range of heavy metal contaminants, and recent research on the potential application of novel fluorescent taggant tracers (Pollard et al. 2007) as an additional tool in environmental forensic investigations.

Environmental Taggants

One of the largest problems associated with current fluorescent dye tracers is their eco-toxicological effects within the environment and their effect on human health. Commonly used tracers such as Fluorescein, Lissamine Flavine FF, Rhodamine WT, Rhodamine B, Sulpho Rhodamine G, Sulpho Rhodamine B, and eosin, amongst others, have been studied for the toxicological effects (Field et al. 1995). Many of these may have breakdown products and synergistic effects with compounds within soil and aqueous systems which could produce increased dangers to human health and the environment. So, the need for a new type of tracer which can provide environmental stability and minimal toxicological effects is great. The application of such tracers in soil and sediment transport monitoring in industrial environments, where there is already a number of possible background contaminant sources, makes monitoring molecular fluorescent tracers challenging.

In environments where there might be multiple sources of contamination, the use of multiple tracers, exhibiting discrete fluorescent atomic emissions with wavelength and lifetime discrimination, would be a distinct advantage. This would allow the monitoring of the origin of contaminants from more than one source and for real time fluorescence measurements to be made with reduced background interference.

Examples of glass-based environmental taggants already in use are glass microspheres, which have been applied to the identification of the source of explosives. In this, the chemical composition of the glass microsphere allows tracking of the place and date of manufacture of the explosives, allowing quicker analysis of residue.

The current research glass matrix, developed for security applications, is a borosilicate matrix which provides an environmentally stable host into which lanthanide ions can be placed. Borosilicate glass and rare earth ions can resist temperature and pressure variations under which traditional molecular dye tracers would degrade. When compared to phosphate or fluoride based glasses (Wada and Kojima 2007) borosilicate is a far more stable host and much less susceptible to water attack. This is important when a taggant is applied in soil forensics, where there may be a range of environmental conditions which may significantly compromise fluorescence from dye tracers.

Heavy Metal Sensor

Heavy Metal Solutions and Samples

The heavy metal solutions used in the initial part of the research were prepared from standard 1000 ppm in 1 M Aristar nitric acid (HNO_3) solutions, with the supporting electrolyte of sodium chloride (NaCl), also from a 1.0M stock solution. Test solutions of lead^{II}, cadmium^{II}, zinc^{II}, nickel^{II}, mercury^{II} and copper^{II} were prepared at different concentration levels in the range of 1 to 50mg/kg and used to dope sub-samples of a clean sediment sample. The ions selected for examination were chosen due to their availability, and are all in the ICRCL (1997) suite of measurements required for contaminated land. Two modes of operation were used in the initial experiments. Firstly, to maintain controlled conditions, the doped sediment samples were placed in a sample holder with a 0.1M NaCl supporting electrolyte at a controlled pH prior to measurement. Secondly, when the calibration had been established, containing the data at a wide range of pH values, and the electrode system had been adapted to incorporate a silver/silver chloride (Ag/AgCl) reference electrode, pH was measured. Then, the samples were measured directly.

The heavy metals were measured by pushing the screen-printed electrode cartridge directly into the doped sediment sample (or directly in to the soil in the field), without the need for extraction of metals from the soils, as required by conventional laboratory methods such as ICPAE measurements. The pre-concentration (deposition) time was 60s at −1400mV potential. The scanning voltage range used was from −1400 to +1000mV, with a step potential of 2mV, pulse height of 25mV, a scan rate of 120ms and pulse duration of 50ms. Thirty two independent measurements of the electrical potential and peak amplitude of each of the metals species were recorded. The instrument displays the concentration on the liquid crystal display (LCD) display but the data is stored for download onto a personal computer.

The detection and identification of metals in multi-metal mixtures was also investigated. Further subsamples of the clean sediment were doped with 5ppm of each of the four different metal species. 30mg/kg doped samples were also made and measure as above.

Instrumentation and Methods

The schematic diagram for the electrochemical instrument as developed is shown in Figure 30.1 (Gow et al. 2008). It consists of five main units: the waveform generator, the potentiostat, the cell (sensor), the data acquisition system (comprising the programmable gain amplifier, ADC and data interfaces) and the microcontroller. From the different electrochemical techniques suitable for heavy metal detection, differential-pulse anodic stripping voltammetry (DPASV) was chosen in preference over the others. It is a precise analytical method which has been widely used for the trace determinations of several heavy metals (Wang 1985; Bersier et al. 1994) with excellent limits of detection.

The excitation signal for DPASV consists of two stages. The first stage is a pre-concentration step of a fixed duration (60s) for the application of a negative potential. For example, −1.4V causes the deposition stage where the analyte species are reduced and plated onto the working electrode surface. The second analysis stage is where each metal is oxidised back into its original state by means of a

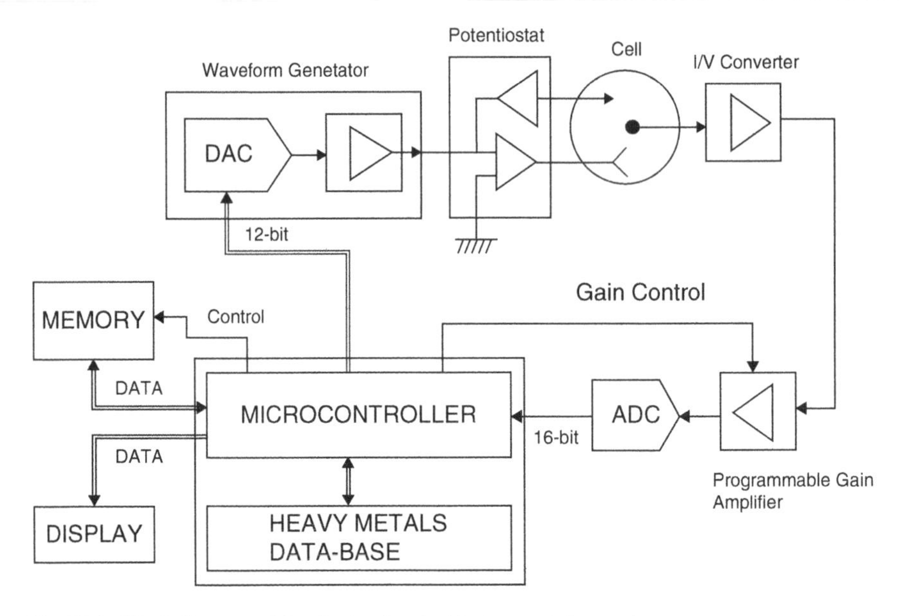

Fig. 30.1 Block diagram of the electrochemical instrumentation system

time-controlled excitation waveform with the applied potential scanned from the negative starting voltage towards a positive end voltage of +1.0 V. The excitation waveform consists of small pulses (120 ms period and 50 ms duration) of constant amplitude (25 mV) superimposed on a staircase waveform of 2 mV step potential. This waveform is provided by a specially designed waveform generator. The signal generator function is provided by the 16-bit microcontroller which synthesises the excitation signal from its digital data equivalent. The approach allows signal generation to be very flexible, as all of the parameters can be re-programmed and a complex waveform can be generated very easily.

The signal in a digital form undergoes a digital to analogue conversion with a 12-bit resolution. The excitation signal is applied to the sensor via a potentiostatic circuitry, which controls the voltage potential applied to the cell. It also acquires and amplifies the current flowing through the cell. The current is sampled twice in each pulse period (once before the potential step, and then again just at the end of the pulse). The difference between these two current samples is recorded. This process is repeated for all pulses of the signal.

The sensor (cell) uses solid (screen-printed) working electrodes rather than the more common hanging mercury drop electrode (HMDE) used by most traditional laboratory instruments (Figure 30.2). Apart from the size and cost difference, an important advantage of solid electrodes is that they do not require the use of mercury which is toxic.

In the data acquisition process, the electrochemical response current is firstly converted to a voltage, and then amplified before being converted, at fixed time

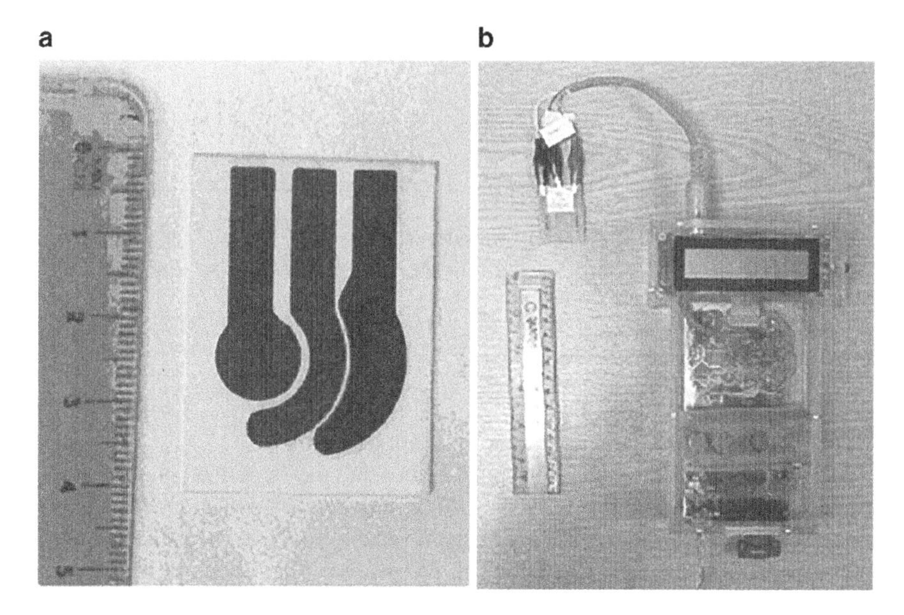

Fig. 30.2 First prototype of heavy metal sensor. (a) Screen printed carbon electrode. (b) Electrochemical instrumentation

intervals, to digital data using a 16-bit analogue to digital converter. The resultant 16-bit binary words which convey the current information are stored in the microcontroller memory. The full scale range (FSR) and resolution of the data acquisition unit is set by a programmable amplifier under the control of the microcontroller.

For very small signal amplitudes, the system FSR is set to 400 μA corresponding to a 6.1 nA resolution; for higher amplitudes the FSR is set to 30 mA giving a resolution of 458 nA. An important characteristic of the system is its portability, so it is designed for battery operation. DC to DC converters supply internal logic (5 V) and linear (±12 V) supplies from a single battery. The battery capacity of 1.4 Ah delivers approximately eight hours of continuous operation at a power consumption rate of approximately 2 W.

Detection and Identification of Heavy Metals

An identification technique based on the probability density functions (PDF) of oxidation potential measurements has been developed. PDF curves have been used in the development of decision algorithms for feature selection in various applications (Gunarathne and Christidis 2002; Bahlman 2006). Data are first arranged into a numerical order from which various statistical features are obtained, such as minimum, maximum, mean, median and standard deviation. These are then used as a basis for classification.

The probability (p) for a feature which assumes a value between f1 and f2 is given by Stanley (1973):

$$p(f_1 \leq f \leq f_2) = \int_{f_1}^{f_2} p(f)df \quad (1)$$

where *p(f)* is the probability density function of *f*.

In the case of a Gaussian distribution, the probability density function *p(f)* is given below:

$$p(f) = \left[1/(\sigma\sqrt{2\cdot\pi})\right]\cdot e^{-(f-\mu)^2/2\sigma^2} \quad (2)$$

where μ is the mean value and σ is the standard deviation.

As the potential of the excitation signal approaches the oxidation potential of one of the metals plated onto the electrode surface, it undergoes oxidation, and the current increases rapidly and reaches a maximum value (peak current) when the applied potential approximates to the metal's oxidation potential (E_p). When an actual test is carried out, the oxidation potential E_p is assessed and examined against the PDF to determine the probability of membership of that analyte with all the analytes stored in a database of PDF measurements. The analyte representing the highest probability of likeliness is thus identified. Analytes identified in this way are automatically given a probability of likeliness, indicating the prediction accuracy.

From the dataset obtained from the 32 measurements described above, the points of the peak current amplitude, and associated oxidation potential, can be extracted.

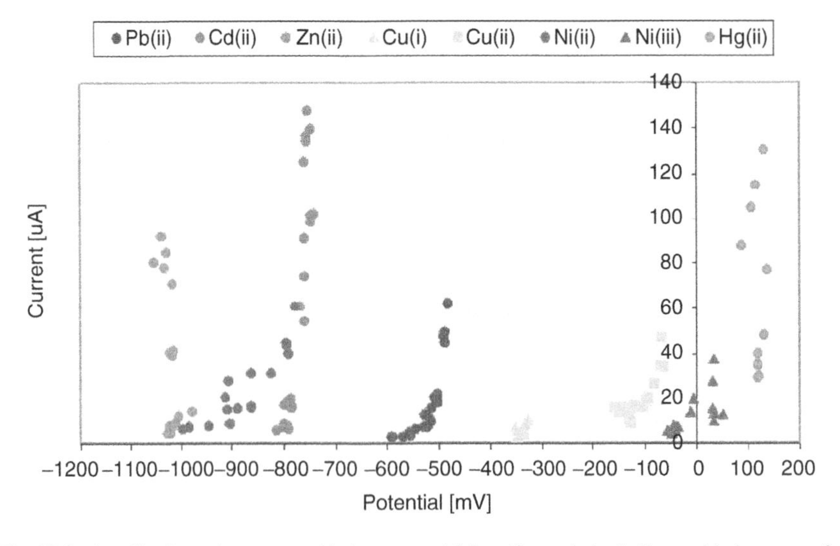

Fig. 30.3 Amplitude peak versus oxidation potential for all metals including oxidative states for Cu and Ni

The plot shown in Figure 30.3 is a space scatter diagram, as used in the field of pattern recognition (Prutton et al. 1990), allowing interpretation of any clustering of the data into groups that have strong likeness or similarity The separation of clusters can then be undertaken using tools such as artificial neural networks (De Carvalho et al. 2000; Cukrowska et al. 2001).

Measurements and Results

Initial calibration curves were obtained using the single element test solutions, example graphs of which are shown in Figure 30.4. The coefficients of the first order curve-fitting equations (calibration curves) were stored in system memory and used to obtain the concentration level of the metal, if identified. These equations have been updated as the instrument has been refined and the concentration range extended, down to 250 μg/kg, and up to 100 mg/kg.

Peak position and peak height can vary with pH. Measurements of a 30 mg/kg lead standard doped clean sediment were carried out for 28 runs on 28 days to check the reproducibility of the peak position and peak height. The peak position was -0.564 ± 0.005 mV above pH -0.7. More variation was seen with peak height but this was overcome when the electrode system was adapted to incorporate an Ag/AgCl central electrode to replace the central carbon reference electrode. This allowed the pH to be measured prior to the metal analysis cycle. A set of experiments to measure the peak height for each metal as the pH was varied was carried out.

Figure 30.5 shows the differential pulse voltammogram taken by the instrument. The four peaks representing the oxidation potential of the metals can be interpreted. Using the identification algorithm, the system reported the following results: cadmium^{II} at 4 mg/kg concentration, lead^{II} at 7 mg/kg concentration, copper^{II} at 6 mg/kg concentration and zinc^{II} at 8 mg/kg concentration with a maximum error of 3 mg/kg. The instrument was also used to examine possible contamination of field samples. Four contaminated samples (supplied by SureClean Ltd., Aberdeen) were examined for heavy metal contamination as described earlier. The samples were tested using the first prototype instrument, and were analysed using a Perkin Elmer Optima 3300 DV ICPAES which gave the following presented results (Table 30.3).

Novel Taggants

A typical borosilicate glass (Zhu et al. 2007) consists of silicon dioxide, boric oxide, sodium oxide and alumina. The ratio of these components can affect the glass network formed and therefore the emission wavelengths from dopant ions. When compared with phosphate or fluoride, glass using borosilicate as a host matrix is far more stable to water attack. The use of borosilicate formulations as glass hosts for lanthanide doping has been studied with eight rare earths which

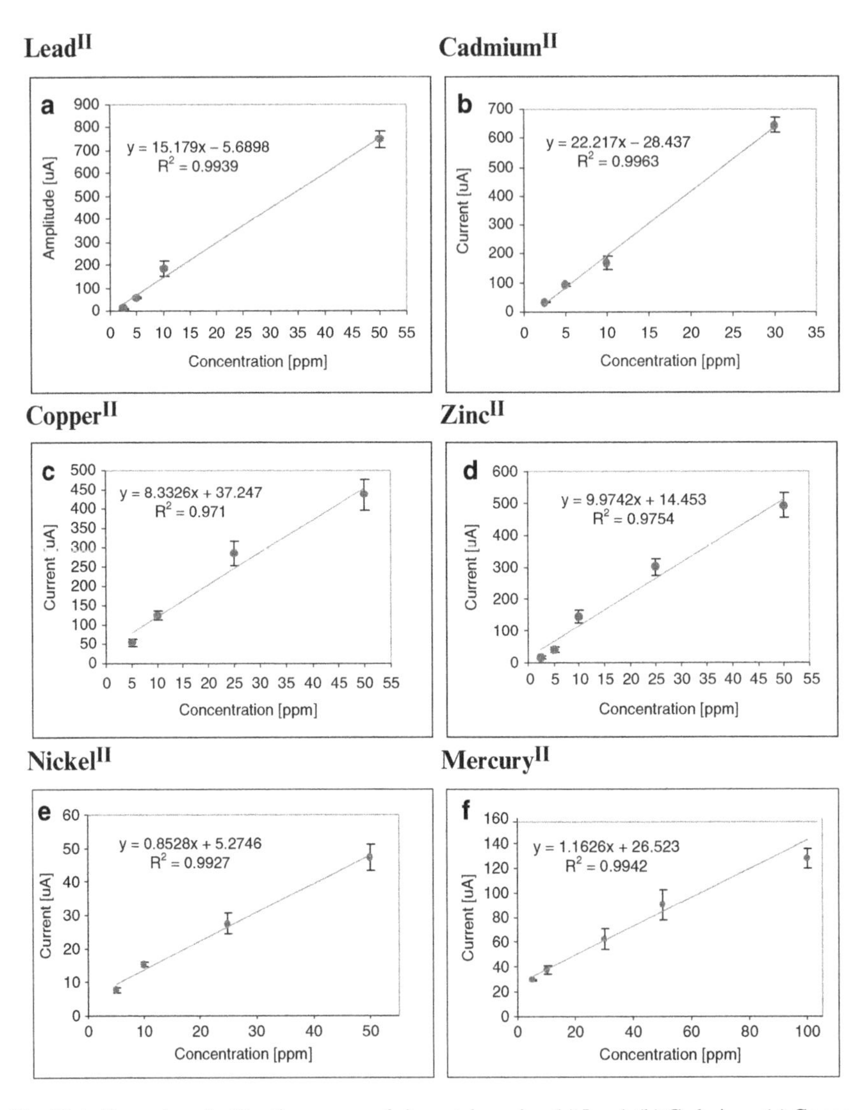

Fig. 30.4 Examples of calibration curves of six metal species. (a) Lead. (b) Cadmium. (c) Copper. (d) Zinc. (e) Nickel. (f) Mercury

exhibit a visible fluorescent emission e.g. europium, terbium, dysprosium, cerium, samarium, praseodymium, erbium and thulium (Zhoutang Li et al. 1988).

In light of these above advantages, production of the glass taggants used a borosilicate glass composition doped with rare earth salt, and the samples were produced using the traditional ‘melt quench’ method (Ramkumar et al. 2008). These samples were then ball milled to produce taggant particles in the region of 5 to 75 μm.

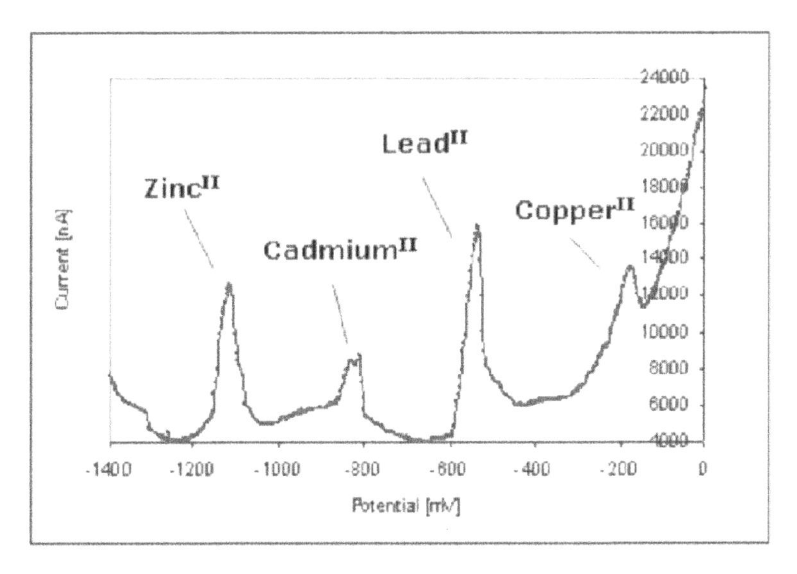

Fig. 30.5 Differential pulse voltammogram for a solution with a mutli-metal mixture containing Cd, Pb, Cu at 5 mg/kg and Zn at 10 mg/kg

Table 30.3 Comparison of zinc II contaminated sample using the heavy metal sensor and ICPAES

Sample no.	Metal present	Concentration mg/kg heavy metal sensor	Concentration mg/kg ICPAES
1	$Zinc^{II}$	137	140
2	$Zinc^{II}$	117	119
3	$Zinc^{II}$	98	101
4	$Zinc^{II}$	83	88

Novel Taggant Detection

The proposed use of novel taggants in environmental forensic investigations is made possible by the discrete, highly intense, fluorescent emissions. These make it possible to differentiate between background molecular fluorescence signals and the target taggant fluorescence. These taggants have the potential to be used to trace the source and/or final destination of contaminants in environmental contamination litigation to ‘make the polluter pay’.

Taggant Detection System

The taggant detection system, developed for the trial investigation, utilised a pulsed 532 nm laser as the excitation source with an optical fibre collector coupled to an avalanche photodiode, fitted with a 10 nm band pass filter at 610 nm for Europium fluorescence. To replicate the process of scanning across contaminated land, a contaminated soil sample with integrated taggants was scanned past the fixed excitation and detection system.

It is possible to determine that from a single doped Europium taggant sample there are a large number of potential excitation wavelengths, with several emission wavelengths, which could be monitored to track the movement of contamination (Figure 30.6). The strongest emission wavelength observed for Europium is 615 nm, a range of excitation wavelengths produce emission; 362, 381, 393, 412, 465, 531 and 579 nm.

The added advantage of using any of the previously mentioned lanthanide series as a doped taggant is the potential to improve the sensitivity of the tracer measurements by using fluorescent lifetime discrimination. In the environment, background fluorescent signals predominately originate from the molecular fluorescence signal of organics. These molecular fluorescent signals are very short lived, in the region of nanoseconds, compared to atomic fluorescence where the lifetime of the signal is in the region of milliseconds. This means that lifetime discrimination in the instrument can effectively remove the background signals improving the sensitivity of the measurement system. Figure 30.7 shows an example of a Europium lifetime profile.

Conclusions

A novel hand-held, easy to use electrochemical instrument capable of real-time *in situ* detection, identification and measurement of concentrations of heavy metals directly in soil samples has been developed. The sensitivity of the system has been assessed and showed that metals can be detected at low concentrations currently down to 1 mg/kg for Pb^{II}, Cd^{II}, Cu^{II} and 3 mg/kg Zn^{II}, which fits the ICRCL requirements.

The system was field tested, where it was able to assay the contamination of four samples in a fast, easy and accurate procedure, and the results verified by a commercially available laboratory instrument. The surveyor was able to locate hotspots on site, thus avoiding the expense and time delay in having to make a return trip to site after laboratory analysis has highlighted hotspots. This is particularly important in environmental forensics when a site may have been released back to the owner and samples on a return visit could be compromised. The system can be used in a variety of situations in environmental assessment and the data acquired provides a metals fingerprint which can be input to databases for subsequent forensic investigations.

The proposed novel taggants, based on narrow band atomic fluorescence, are under development for potential deployment as forensic environmental tracers. The use of discrete fluorescent species in an environmentally stable host has been investigated to replace existing toxic, broad band molecular dye tracers. The narrow band emission signals offer the potential for the tracing of a large numbers of signals in the same environment. This will give increased data accuracy and also allow multiple source environmental monitoring of environmental parameters.

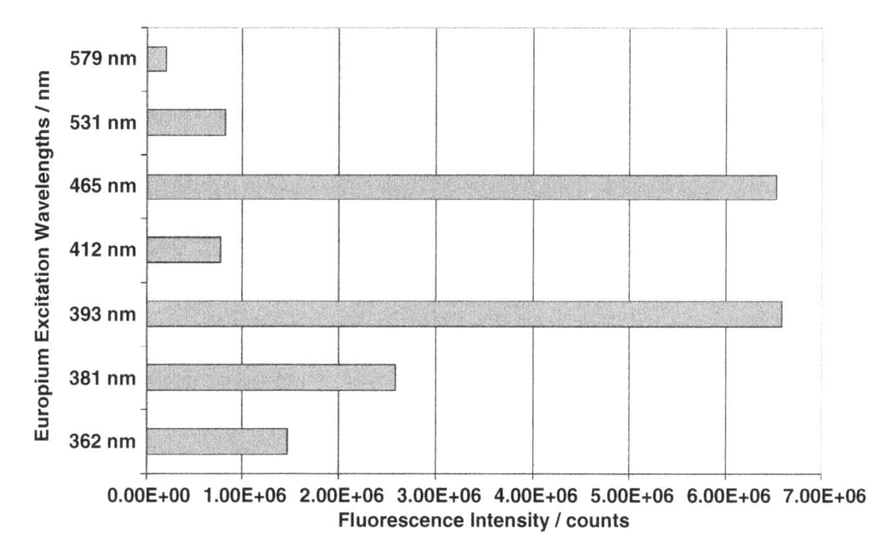

Fig. 30.6 Europium 615 nm intensity variations with a range of excitations wavelengths

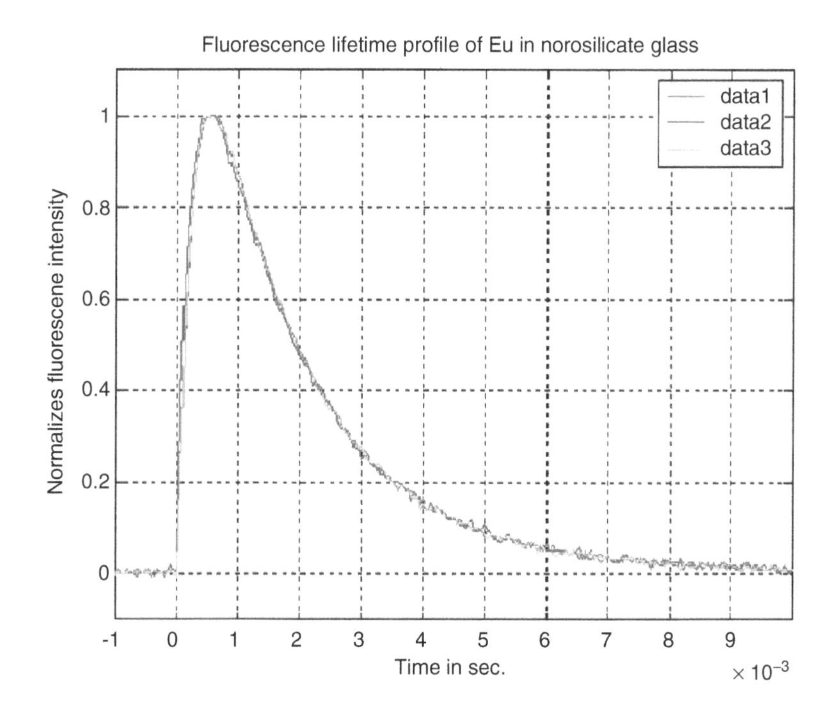

Fig. 30.7 Fluorescent lifetime profile of europium ions in borosilicate glass measured in triplicate; Data 1, Data 2 and Data 3

References

Bahlman C (2006). Directional features in online handwriting recognition. Pattern Recognition 39:115–125.

Bersier PM, Howell J and Brunlett C (1994). Tutorial review. Advanced electroanalytical techniques versus atomic absorption spectrometry, inductively coupled plasma atomic emission spectrometry and inductively coupled plasma mass spectrometry in environmental analysis. Analyst 119:219–232.

Christidis K, Pollard P, Gow K and Robertson PKJ (2006). On-site monitoring and cartographical mapping of heavy metals. Instrumental Science and Technology 34:489–499.

Christidis K, Robertson P, Gow K and Pollard P (2007). Voltammetric *in situ* measurements of heavy metals in soil using a portable electrochemical instrument. Measurement 40:960–967.

Cukrowska E, Trnková L, Kizek R and Havel J (2001). Use of artificial neural networks for the evaluation of electrochemical signals of adenine and cytosine in mixtures interfered with hydrogen evolution. Journal of Electroanalytical Chemistry 503:117–124.

De Carvalho RM, Mello C and Kubota LT (2000). Simultaneous determination of phenol isomers in binary mixtures by differential pulse voltammetry using carbon fibre electrode and neural network with pruning as a multivariate calibration tool. Analytica Chimica Acta 420:109–121.

Field MS, Wilhelm RG, Quinlan JF and Aley TJ (1995). An assessment of the potential adverse properties of fluorescent tracer dyes used for groundwater tracing. Environmental Monitoring and Assessment 38:75–96.

Gow K, Robertson PKJ, Pollard P and Christidis K (2008). Measurement method and apparatus for analysing the metal or metal ion content of a sample. European Patent: EP1904838.

Gunarathne GPP and Christidis K (2002). Material characterisation *in situ* using ultrasound measurements. IEEE Transactions on Instrumentation and Measurement 50:368–373.

ICRCL (1987). Inter-departmental Committee on the Redevelopment of Contaminated Land (ICRCL), 59/83 Guidance on the Assessment and Redevelopment of Contaminated Land. 2nd Edition. Available from: http://www.ContaminatedLAND.co.uk/std-guid/icrcl-l.htm. Accessed 21-07-2008.

Innovxsys (2007). Detection Limits for Innov-x Classic Field Portable XRF Instrument in Low Density Soil Types. http://www.innovxsys.com/products/detect. Accessed 21-07-08.

Pollard P, Adams M, Prabhu GR, Officer S and Hunter C (2007). Sensitive Novel Fluorescent Tracers for Environmental Monitoring. OCEANS 2007 – Europe, 18–21 June 2007:1–6.

Prutton M, Gomati MME and Kenny PG (1990). Scatter diagrams and hotelling transforms: application to surface analytical microscopy. Journal of Electron Spectroscopy and Related Phenomena 52:197–219.

Ramkumar J, Sudarsan V, Chandramouleeswaran S, Shrikhande VK, Kothiyal GP, Ravindran PV, Kulshreshtha SK and Mukherjee T (2008). Structural studies on boroaluminosilicate glasses. Journal of Non-Crystalline Solids 354:1591–1597.

Rossmeisl J and Nørskov JK (2008). Electrochemistry on the computer: Understanding how to tailor the metal overlayers for the oxygen reduction reaction. Surface Science, 602(14):2337–2338.

Stanley LT (1973). Practical Statistics for Petroleum Engineers. The Petroleum Publishing, USA:15–20.

Wada N and Kojima K (2007). Glass composition dependence of Eu3 + polarization in oxide glasses. Journal of Luminescence 126:53–62.

Wang J (1985). Stripping Analysis: Principles, Instrumentation, and Applications. VCH Publishers Deerfield Beach, FA: 119.

Zhoutang Li PW, Xueyin J, Zhilin Z and Shaodong X (1988). The synthesis of rare earth borosilicate glasses and their luminescence properties. Journal of Luminescence 40&41:135–136.

Zhu D, Zhou W, Day DE and Ray CS (2007). Preparation of fluorescent glasses with variable compositions. Ceramics International 33:563–568.

Chapter 31
Separation and Concentration of Trace Evidence from Soils Using a Hydropneumatic Elutriation Trace Evidence Concentrator (TEC)

Alvin J.M. Smucker and Jay A. Siegel

Abstract Trace evidence is a forensic science term that describes any evidence that occurs in very small quantities and which usually requires a microscope for visualisation and analysis. Common types include hairs, textile fibres, paint fragments, glass, bomb parts, plastics, metal fragments and other similar materials. Methods of analysis for trace evidence are well accepted and validated. The major challenge has been, and continues to be, the discovery and isolation of fragments of trace evidence from a matrix such as soil or similar debris. Evidence in outdoor crimes such as bombings and fires is especially susceptible to difficulties in evidence collection. Traditional methods for trace evidence collection from soils include hand sorting under a low power microscope and the use of nested sieves of decreasing mesh size. Both of these methods suffer from being extremely time consuming and recovery rates are generally very low owing to the difficulty in picking out soil coated fragments. The Trace Evidence Concentrator (TEC) system utilises the concept of density gradient separation by combining centrifugal forces which separate more dense materials from lower density materials. The lighter density materials of trace evidence are carried by surface tension forces at the air-water interface of small bubbles to a stack of multiple sized screens which retain the trace evidence. Additionally, the TEC cleans glass fragments, vehicle headlamp filaments, paint chips, and unexploded ordnance lost at the scenes of violent accidents and crimes. The result is that recovery rates can be dramatically increased and search times significantly reduced.

A.J.M. Smucker (✉)
Soil Biophysics, Department of Crop and Soil Sciences, Michigan State University, East Lansing, Michigan MI, USA
e-mail: smucker@msu.edu

J.A. Siegel
Forensic and Investigative Sciences Program, Indiana University, Purdue University Indianapolis, IN, USA.

K. Ritz et al. (eds.), *Criminal and Environmental Soil Forensics*

Background

Many crimes occur in outdoor urban and rural areas. Through direct or indirect contact, people and objects deposit various types of trace evidence in soil in and around the crime scene (Locard 1930). Such materials may be of human origin such as hairs, skin cells, clothing fibres, etc. Other trace evidence arises from objects such as automobiles (glass, plastic, paint), explosive devices (explosive residue, parts of the device), and guns (bullet fragments or particles of shot). Much of this evidence is microscopic and cannot easily be seen with the unaided eye, especially when it is in or on a matrix such as soil. Particles of trace evidence are not only mixed with soil but can become coated with clay or other soil particles, making them practically invisible. This potentially valuable evidence is often never discovered or recovered because it is too difficult to spot. If it is recovered, it is too difficult to separate by conventional evidence recovery methods such as common sticky tape (Flinn 1992) and electrostatic lifting methods most commonly used at crime scenes (SWGMAT 1999).

The integrity and significance of trace materials as associative evidence rely on proper collection of the sample, sample size, and sample preservation (SWGMAT 1999). Transfers of trace evidence, e.g. apparel fibres, body hair, skin, paint chips, glass, rubber and other polymers, unexploded ordnance and gunshot particles, between individuals often occur in cases where there is little person-to-person contact, causing much of the evidentiary sources to fall onto soils. Many of the more than 500 unidentified victims of the World Trade Center should have been identified by a more comprehensive collection of human tissue large enough to complete DNA identifications. No infrastructure exists for coordinating this rapid and comprehensive recovery, sorting and separation of human remains from massive disasters resulting in delays of months or years before effective victim identification can be achieved in large-scale disasters (Biesecker et al. 2005). Large quantities of data gathering, sample washing, analyses and archiving of victims' remains are simply unavailable and must be developed before the next major disaster (Budowle et al. 2005). Holland et al. (2003) suggested several major needs for improving sampling and processing technology and infrastructure. Their concluding comments were "…an enhanced mass fatality forensic infrastructure would lead to more rapid and efficient identifications in the event of future mass disasters or terrorist attacks". The Trace Element Concentrator (TEC) system of extraction would greatly contribute to this national mass fatality forensic infrastructure.

Trace Evidence Separation and Concentration

Most of the present day methods for collecting trace evidence go back to the 1970s. They were not designed to recover evidence from difficult matrices such as soil. The methods used for separating trace evidence from soils are hand searching and nested sieves. No known apparatus has been developed that concentrates small quantities of trace evidence from soils at scenes of crimes without the requirement a human being physically searching the soil for evidence.

A newly patented hydropneumatic elutriation technology has been designed to separate the vast majority of trace evidence from soils at rates up to 25% faster than conventional police laboratory methods of dry sieving (Johnston 1997). During development of the TEC, graduate students at Michigan State University were able to recover 95% to 100% of trace evidence from 'doped' soil samples in controlled experiments. This TEC system is a highly modified hydropneumatic elutriation device whose design is based upon the 1982 polyvinyl chloride model used for removing fine plant roots from soils (Smucker et al. 1982). It is believed the new TEC hydropneumatic technology will contribute more to criminal justice than the 1982 root washer's contributions have to the rapidly expanding rhizosphere and environmental sciences. The root washer technology greatly improved the cost efficiency and recovery efficiency of root removal from soils (Gillison Variety Fabrication, Inc. http://www.gillisons.com/root_washer.htm). Trace evidence is concentrated by accumulating textile fibres and fragments on a nest of specifically designed low energy separation filters. Additionally, the TEC thoroughly cleans and removes many coarser fragments, e.g. glass and bone fragments, paint chips, asphalt, unexploded ordnance etc. from mineral soils.

Configuration, Operation and Benefits of the TEC System

The patented TEC system consists of three components: (i) a serpentine diffusion chamber (Figure 31.1), which generates both a high kinetic energy vortex of water and a uniform mixture of fine air bubbles. The continuously-running hydropneumatic vortex quantitatively separates heterogeneous lower density materials from the more dense mineral soils via a combination of centrifugal energy and water-air elutriation upward through; (ii) a columnar elutriation chamber (80 × 20 cm), according to the differential density principle. Particles

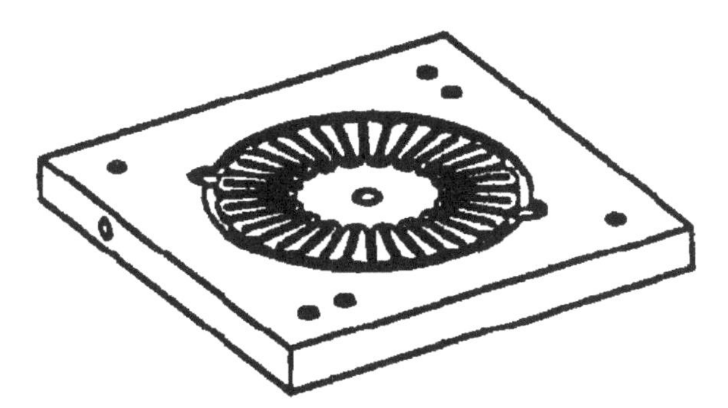

Fig. 31.1 Diagrammatic representation of the serpentine vortex-dispersion chamber that generates a high kinetic energy water vortex and disperses fine gas bubbles into the rotating vortex for the most rapid and quantitative elutriation of trace evidence materials through the elutriation chamber and into a nest of filters. The size of this base ranges from 20 by 20 cm in laboratories up to current maximum of 80 by 80 cm square

such as hair, textile fibres and many biological materials are less dense than water and are carried up and over to the filters. Materials such as glass, metal and plastic are washed free of soil particles and settle to the bottom of the chamber. The combined upward movement of a rotating vortex and gas-water interface of small bubbles within the vortex develop centrifugal forces within the hydropneumatic elutriation vertical tube cylinder. The diameter and length of this chamber and the running time for complete separation of trace evidence depends upon the specific soil textures containing the trace evidence. This vertical chamber moves the washed soil particles, having particle densities of 2.65 g/cm^3, to the cylinder walls where the drag forces of the chamber wall reduce the vortex forces, causing the mineral materials to drop to the base of the hydropneumatic elutriation chamber by gravity. Particles lighter than soil minerals are separated from the mineral soil particles by combinations of vortex and elutriation forces against gravity and washed free of mineral residues as they are moved upward by buoyant forces of the water/air mixture within the vortex. The lower density trace evidence particles, eg. $<2.5\,g/cm^3$ are retained in the upward flowing vortex by the surface tension forces at the gas-liquid interface of the gas bubbles and elutriated upward against gravity across; (iii) a connection tube equipped with an air vent and onto the low-kinetic energy manifold of nested sieves with the largest above the smallest mesh screen sizes of sieves submersed in water. This filtration arrangement produces a very low kinetic energy that retains the trace evidence for future storage and examination. The more dense fine clay materials, $<2\,\mu m$, pass through the filters by the forces of gravity. Pressure-regulated air and water pressure lines, connected to the serpentine vortex-dispersion chamber, produce the desired flow rates required to produce the high-energy vortex filled with an appropriate quantity of small air bubbles.

Mode of Operation of the TEC

Immediately following the sampling and transport of crime scene soil samples, the samples are mechanically dispersed and multiple samples, approximately 500cc each, are run through the hydropneumatic elutriation system and elutriated onto sieves for 10min. Water flow is stopped and the elutriation chamber is emptied while the nest of sieves is replaced. The procedure is repeated until the entire soil sample has been processed for trace evidence.

Contributions of TEC to Forensic Soil Science

The new TEC system requires less labour to concentrate more trace evidence 50% faster by a single technician (Johnston 1997), reducing the human error accrued by several technicians using conventional sampling, storage and dry sieving methods for extracting trace evidence from soils. The extraction of personal trace evidence, at a faster rate, from more crime scenes, will eliminate suspects from enquiries and strengthen the prosecution to bring more criminals to justice more rapidly. The TEC extraction system results in a faster, more robust, and less labour-intensive collection of DNA based evidence from crime scene soils. At the end of each sample extraction cycle, relatively clean trace evidence materials are preserved for identification by microscopic, chemical, or DNA analyses (Figure 31.2).

This hydropneumatic elutriation-based technology involves a novel concept that enables the rapid and quantitative collection of trace evidence from soil samples as small as a few grammes up to kilogrammes. Upon expanded utilisation, the TEC has some potential to replace conventional manual screening methods for extracting trace evidence from crime scene soils. Training protocols and thorough testing of the TEC will demonstrate the proof-of-concept that fully complies with the Fry-Davis legal criteria and Daubert Standard requirements for collecting and presenting credible and legally reliable trace evidence admissible to the courts (SWGMAT1999). The TEC system promotes integrated multiple analyses of current soil forensic soil evidence, e.g. clay minerals, pesticides, ion chemistry, and delta ^{13}C and ^{18}O signatures typical for soil samples from specific crime sites. Additionally, TRFLP and other DNA and mRNA analyses of plant species, pollen, mesofauna, fungal, and bacterial communities associated with certain personal samples, e.g. used cigarette filters in soils sampled from specific crime sites can be coupled with conventional trace evidence listed above. Furthermore, the TEC sepa-

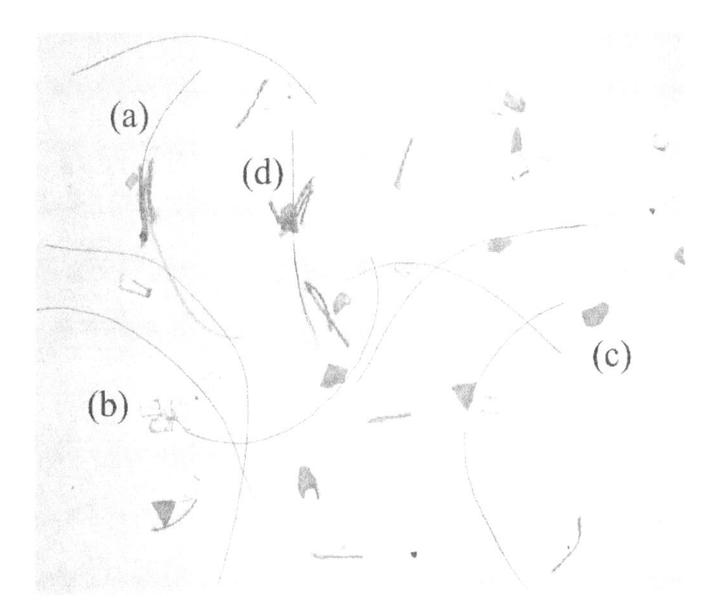

Fig. 31.2 Separation of very clean trace evidence from a loam soil using trace evidence concentrator (TEC) hydropneumatic elutriation process of (a) hair; (b) glass fragments; (c) paint chips; (d) textile fabrics (from Johnston, 1997)

ration of trace evidence can be combined with current analytical techniques for identifying specific soil components as outlined in Figure 31.3.

Applications and Conclusions

Although every person loses personal trace evidence, e.g. body hair, skin fragments, chewing gum, cigarette butts and clothing fibres, during struggles or stressful activities, most onsite crime scene investigators are unable to extract adequate quantities of these critical objects from the scenes of crimes, especially those crimes completed over soils. Additionally, this lack of evidence often delays the separation of suspects from actual criminals keeping innocent victims for prolonged periods of time.

Preliminary trials using the TEC renaissance show great potential for collecting evidence previously missed by the screening of dry soil samples, increasing the recovery time of trace evidence by 220% to 50% (Johnston 1997).

One of the goals for expanding the TEC processing is to complete the testing required for establishing a highly efficient TEC prototype that will operate both separately and in tandem with a series of TECs which could extract trace evidence of explosives and human remains from very large volumes (tonnes) of soil. Such TEC systems and their manifolds can be commercially produced for delivery

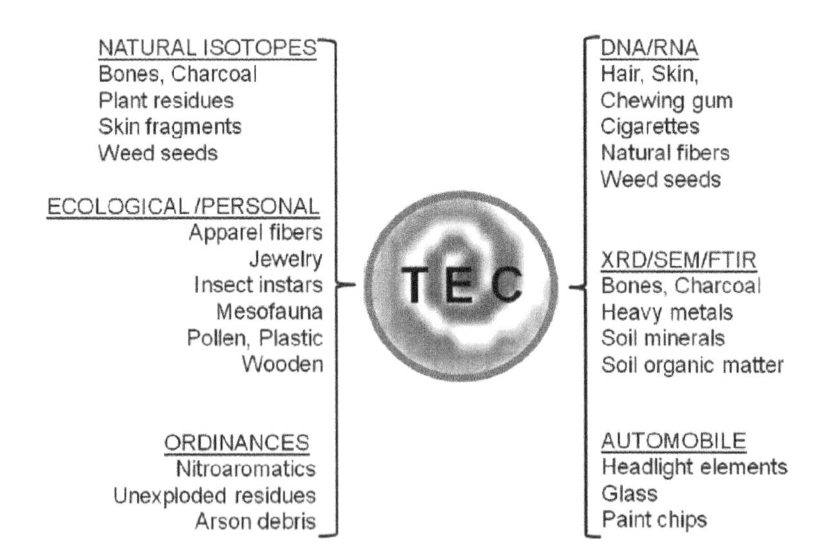

Fig. 31.3 Diagrammatic representation of an integrated analytical and morphological identification of trace evidence that includes individual evidence profiles combined with site specific profiles of evidence that can be aggregated into a comprehensive evidence profile useful for specific crimes. Furthermore, combining sampling protocols with these evidence profiles across specific soil textures can be incorporated into crime-solving models involving mineral soils

worldwide. Homeland security could be improved by the system. For example, immediate extractions of unexploded ordnance could lead to earlier identification of explosive sources leading to earlier arrests and convictions of true perpetrators before they can cause additional major disasters. Data collection that includes the majority of small human remains at scenes of major disasters such as the World Trade Center, following tsunamis and earthquakes that result in mass fatalities and the commingling of victim remains and building materials cause unprecedented delays in the identification of victims. The TEC system requires substantial testing for developing a computer-controlled and history-recording system that will bring the TEC to a useful and legally credible criminal justice apparatus that enable procurement of trace evidence from crime scene soils more quickly than current approaches.

References

Biesecker LG, Bailey-Wilson JE, Ballantyne J, Baum H, Bieber FR, Brenner C, Budowle B, Butler JM, Carmody G, Conneally PM, Duceman B, Eisenberg A, Forman L, Kidd KK, Leclair B, Niezgoda S, Parsons TJ, Pugh E, Shaler R, Sherry ST, Sozer A and Walsh A (2005). DNA identifications after the 9/11 World Trade Center attack. Science 310:1122–1123.

Budowle B, Baechtel CT, Comey AM, Giusti AM and Klevan L (1995). Simple protocols for typing forensic biological evidence: Chemiluminescent detection for human DNA quantification and RFLP analyses and manual typing of PCR amplified polymorphisms. Electrophoresis 16:1559–1567.

Budowle B, Bieber FR and Eisenberg AJ (2005). Forensic aspects of mass disasters: Strategic considerations for DNA-based human identification. Legal Medicine (Tokyo) 7(4):230.

Flinn L (1992). Collection of fiber evidence using a roller device and adhesive lifts. Journal of Forensic Sciences 37:106.

Holland MM, Cave CA, Holland CA and Bille TW (2003). Development of a quality, high throughput DNA analysis procedure for skeletal samples to assist with the identification of victims from the World Trade Center attacks. Croatian Medical Journal 44:264.

Johnston JG (1997). The trace evidence concentrator: A method for isolating trace evidence from soil samples. M.S. Thesis. School of Criminal Justice, Michigan State University.

Locard E (1930). The analysis of dust traces. Part II and III. American Journal of Police Science 1:401–418 and 496–514.

Smucker AJM, McBurney SL and Srivastava AK (1982). Quantitative separation of roots from compacted soil profiles by the hydropneumatic elutriation system. Agronomy Journal 74:500–503.

SWGMAT (1999). Trace Evidence Recovery Guidelines. Developed by the Scientific Working Group on Materials Analyses Evidence Committee. Forensic Science Committee Oct. 1999. 1(3). www.fbi.gov/hq/lab/fsc/backissu/jan2000/swgmat.htm Accessed 12/02/2002.

Part V
Postscript

Chapter 32
Soils in Forensic Science: Underground Meets Underworld

A. David Barclay, Lorna A. Dawson, Laurance J. Donnelly, David R. Miller and Karl Ritz

Abstract Security and environmental health issues are high on international political and social agendas, with public policy leading to legislation on environmental liability, and support for improved technologies and capabilities for use by relevant authorities. The role of soils and geological material, and their associated ecological properties, as significant sources of intelligence and physical evidence is increasingly recognised by those involved in criminal investigations. The transference of soil, or objects held within the soil matrix, has enabled associations between combinations of victim or artefact, perpetrator, and locations; demonstrable to a standard of proof and transparency required to substantiate evidence in courts of law. Formalisation of approaches to areas of search, site identification, measurement and excavation has provided a framework for structuring strategic and tactical planning of investigations. Greater attention is now being paid to new means of data capture and its interpretation, including the production of robust databases of the properties and spatial distribution of soils and underlying geological influences. Public investment is being made in research, development and operational use by investigating authorities of a toolbox of soil forensic techniques. These include isotopic analysis, non-invasive means of measurement and mapping, and the identification of properties using biological, pedological, sedimentological and geographical approaches. The advantages of integration and appropriate coupling of complementary and independent measures is now being recognised. Some tools are at a conceptual stage, such as contingency planning for mass graves, and others experimental, such as novel taggants.

A.D. Barclay
Caorainn, Laide IV22 2NP, UK

L.A. Dawson and D.R. Miller
The Macaulay Institute, Craigiebuckler, Aberdeen AB15 8QH, UK

L.J. Donnelly
Halcrow Group Ltd., Deanway Technology Centre, Wilmslow Road, Handforth, Cheshire SK9 3FB, UK

K. Ritz(✉)
National Soil Resources Institute, Natural Resources Department, School of Applied Sciences, Cranfield University, Bedfordshire MK43 0AL, UK
e-mail: kritz@cranfield.ac.uk

K. Ritz et al. (eds.), *Criminal and Environmental Soil Forensics*

Others are subject to testing in different environmental conditions to assess their capacity to discriminate between contrasting anthropogenic influences. Yet, in countries such as Australia, the UK, and the USA the analysis of soils is being used successfully in legal submissions on issues from site verification to body decomposition. Although some statistical principles are becoming incorporated into forensic soil science there is a need for further development in this area, and greater dialogue between statisticians, environmental scientists, forensic scientists and practitioners. However, the co-authorship of the chapters in this book, across a wide range of disciplines and levels of practice, demonstrates the significance of developing close working relations and understanding for greatest societal benefit to be gained from the work on soil forensics, both in a criminal and an environmental context. The challenges are to identify and address interdisciplinary research questions, and to advance and prove technological developments such that they appropriately support intelligence gathering and the provision of reliable evidence. The maturing disciplines of criminal and environmental soil forensics will then deliver on the expectations and demands of public policy around the world.

Introduction

Internationally, the issues of security and environmental health are high on political and public agendas, leading to needs for effective mechanisms for responses to threats to our quality of life and safety. The European Security Research Advisory Board (2006), in their report on the European Security Research Agenda, called for improved forensic technologies and analysis capabilities for both digital and physical crime scene investigation. The European Commission's Directive on Environmental Liability (European Commission 2004) is due for implementation, under which perpetrators of activities causing damage to habitats, waters or contamination which puts at risk human health, are financially liable.

Until recently, at least in the United Kingdom, the analysis of soils for forensic purposes almost exemplified the term 'cottage industry', with scientists applying individual disciplines, techniques and instruments to individual cases, normally at the behest of a senior investigating officer (SIO). Often such a request would arise because of some perceived deficiency in the evidential strength against the main suspect in a serious crime. Soil analysis was rarely applied systematically as a routine source of intelligence or physical evidence. As soil is a variable matrix, it was not possible to maximise the information available to SIOs unless a suite of techniques could be purposefully employed in a given case, raising concomitant issues of identifying the factors of greatest use, the reliability, availability and cost of the tools. These issues all contributed to the difficulty in promoting the analysis of soil amongst police services nationally, as opposed to that of individual techniques, such as palynology, to local forces.

Soil, like footwear, is a useful form of evidence simply because the offender cannot commit the crime without standing and moving at the scene, inevitably leading to some item being contaminated, often with layering of deposits from other locations (Locard 1930). At a specific scene soil may be transferred to the suspect's footwear, clothing or vehicles, or to victims, caches of drugs or weapons. Or, it may be transferred from a victim to a scene, and occasionally between victim or suspect.

Interpretation of the observations of soils, and the associated environment, can provide a clarification of events, and inform the prioritisation of police resources during an investigative phase of enquiry, which is the first and major part of any investigation. Crucially, knowledge of the characteristics of soil can provide inceptive intelligence, to help indicate where a spade was used, a car driven, or where other evidence may be located. This process of providing physical intelligence is an open one, just one aspect of many lines of enquiry which assist the investigation before any suspect has been identified (Auchie, Chapter 2).

However, it is intelligence, not evidence, that is the lifeblood of any police investigation, whether it be a stranger murder, drug dealing, or a burglary series. The majority of the involvement of physical analysis in development of intelligence will occur during the investigative phase. The police recognise a clear distinction between the standard of proof and transparency necessary to substantiate evidence for the Court (Auchie, Chapter 2; Aitken, Chapter 3), and the 'best guess' assessment which is of prime importance during the investigative phase. Inevitably individual practitioners who, because of the nature of their involvement tend to concentrate on evidence, are largely unaware of their potential to assist the investigative phase. That distinction between physical intelligence and forensic evidence was not fully appreciated until recently by the wider forensic science community, who had little direct knowledge of investigations or indeed of the misconception that soils can ever be 'matched' (Robertson, Chapter 1).

Soil intelligence, like other analysis in forensic science, can usually be re-assessed to provide direct evidence during the later evidential phase of an investigation, where it is being assembled against a specific suspect. Such soil evidence is often comparative, but clearly it is necessary for the soil expert to assist the court as to the likelihood of the results being derived from some other, unrelated, location (Fitzpatrick et al., Chapter 8; Wiltshire, Chapter 9).

However, the investigative environment is changing, enhancing the potential for soil analysis, particularly in parallel with other forms of physical intelligence. For example, in the UK, the inception of national investigative support for police through the National Crime Faculty in 1996 (now the National Policing Improvement Agency, NPIA), quickly led to forensic scientists becoming an integral part of investigations. This, in turn, produced a paradigm shift in forensic science in the UK, which is paralleled in Australia (Robertson, Chapter 1). By 2005, the UK Parliamentary Select Committee (House of Commons Science and Technology Committee 2005) was able to confirm this key finding, stating that "The main contribution that forensic science makes to the criminal justice system is the generation of intelligence to assist investigations; the provision of actual evidence is a small part of its role." The role and potential importance of soil as intelligence and evidence is also becoming recognised in other countries (e.g. Gradusova and Nesterina, Chapter 5).

Around the globe, interest is developing for the application of geologically related material in forensic science, with a number of books published recently. For example, principles, techniques and applications of forensic geoscience have been illustrated by Pye and Croft (2004), from macro-scale field application to micro-scale laboratory studies. The use of geoscience in forensic analysis, with many case examples, is presented by Pye (2007), who instructs on sample handling, preparation and the characterisation of physical, chemical and mineralogical traits and their subsequent forensic application. Following on from a comprehensive review of forensic geoscience by Ruffell and McKinley (2005), a wide coverage of the many applications and case studies of interrelated geo-disciplines has been presented (Ruffell and McKinley 2008). Murphy and Morrison (2007) also provide a comprehensive treatise on environmental forensics. Murray (2004) wrote that "The challenge for forensic geology, as with all scientific evidence, lies with education. Investigators and evidence collectors must be educated on how earth materials can make a major contribution to justice when properly collected, properly studied, and properly presented in a judicial setting".

This volume is part of the inevitable and necessary maturing of the science of forensic soil analysis. The SIOs' needs are already clear, which are primarily rigorous standards of analysis and assessment of significance, utilising a suite of techniques appropriate to the context of each case, delivered in a cost and time effective way. The real benefit for soil scientists is that the 'hidden' intelligence market is by far the larger, and SIOs already see the need for both evidential materials and intelligence (Harrison and Donnelly, Chapter 13).

Evidence

Expert evidence of a scientific nature should be based on science which is testable, has a known error rate, has been published in peer-reviewed literature and is generally accepted within the relevant community (Aitken, Chapter 3). In line with this, the reporting of statistical analysis plays an increasingly important role in the administration of justice generally, and in the evaluation of evidence. In reporting on forensic science, the UK House of Commons Select Committee on Science and Technology (2005) sought greater awareness of statistics, also stressing the need for good quality, reliable data.

Aitken (Chapter 3) provides a comprehensive explanation of the concepts behind weighing evidence from the perspectives of competing hypotheses. A candidate hypothesis may be related to the likelihood of contact between the suspect and the scene of a crime. Under Bayes' Theorem, probabilities are considered as a measure of belief. The likelihood ratio (LR) produced can then be translated into a common English phrase which equates to the strength of evidence (Rose 2002; Robertson, Chapter 1). However, this approach has not yet found wide acceptance in UK Courts where the judiciary seem to take a conservative approach to its adoption.

When handling evidence, in addition to sample continuity and integrity which are paramount, extreme care is essential at the initial examination stage. It is at this point that other potential sources of physical evidence, such as fibres, hairs, and flakes of paint can be present in a soil sample (e.g. Gradusova and Nesterina, Chapter 5). Smucker and Siegel (Chapter 31) describe an automated system to allow removal of such items, which is currently then carried out by microscopic examination. It is also at this stage that any layering (organic or mineral), or differential staining (such as between uppers and soles of shoes) can be ascertained (Fitzpatrick et al., Chapter 8; Wiltshire, Chapter 9) and individual samples collected if relevant.

In relation to environmental forensic investigations, similar issues arise in the provisions of evidence for prosecution (Mudge, Chapter 10). The construction of environmental profiles enables the reconstruction of environments by interpreting the presence of proxy indicators, which are fragments of the original environment, and may consist of whole (or parts of) animals, plants, microbes, soils and sediments (see for example, Sensabaugh, Chapter 4; Mayes et al., Chapter 29). Such profiling has been applied in the application of a suite of biological and microbiological, pedological, sedimentological and geographical techniques, for application in examinations for either criminal or environmental investigations. For example, Fernandes et al. (Chapter 11) describe applications in which analysis of the composition, and structure, of a substance, specifically concrete, can be used in the eventual demonstration of liability, and subsequent prosecution of environmental crime, and Martens and Walraevens (Chapter 12) describe the use of geophysical techniques to trace and delimit soil and groundwater pollution. Conversely, Williams et al. (Chapter 7) consider the environmental profile, including the spatial distribution of the characteristics of an area for the containment of pollutants from a site, with specific reference to the identification of burial sites.

In Chapters 11 (Femandes et al.) and 12 (Martens and Walraevens), the use of tools for *in situ* measurements is a necessity, reflecting the importance of developing establishing standard operating procedures for data capture and handling, an issue highlighted by Morrisson et al. (Chapter 6). As, almost by definition, the scenes of crime, be it criminal or environmental, offer some form of risk to the investigator, the requirement for responsible consideration of their health and safety whilst in the process of observation cannot be minimised, as recognised by Williams et al. (Chapter 7), and Martens and Walraevens (Chapter 12).

Geoforensics

Geoforensics provides scientific and application knowledge in the fields of geosciences to assist with a wide range of investigations both at the intelligence or the evidential phase (e.g. Bull et al. 2008). With new applications of science, individual tools may be adopted for use, possibly without full evaluation of their robustness, and outwith an overall framework which would guide thinking on the users' aims, purpose, and structuring of approaches to problem solving. This may be particularly so in fields where research and development may produce sophisticated and technologically advanced solutions to detection, measurement, interpretation or analysis.

Harrison and Donnelly (Chapter 13) provide, for the first time, an overview of the history, evolution, philosophy and development of geoforensics, outlining conceptual issues to structure approaches to laboratory investigation, and field search, for the purpose of locating concealed objected or graves, and the provision of evidence for court, or to link people or items to a crime scene or victim. The role of geographic data in the form of maps, databases and archives contributes to a conceptual geological model of an area, formalising background intelligence and knowledge in such a way as to bound areas of search and assist in the choice of strategy adopted, the appropriate choice of instrumentation and determination of the search methodology.

In developing the conceptual geological model of an area, Geographic Information Systems (GIS) are used in the sieving and analysis of spatial data, informing the investigator of the characteristics of an area with which they can then assess the likelihood of success in conducting searches. Such data includes digital maps of soils, vegetation, terrain height, and ownership, often augmented by additional ground measurements. Examples of the potential or actual use of such tools in casework are evident at national and regional levels, such as in the contingency planning for pandemic events (Williams et al., Chapter 7), or at more local levels such as the search for murder victims in Ireland (McKinley et al., Chapter 18), or mass graves from the Nazi era in Germany (Fielder et al., Chapter 19).

In all of the examples using GIS tools, the principles of sieve mapping are employed and the formal, or informal, application of rules which have refined candidate areas of search to target areas in which specific field measurements can then be deployed. In each case the approach taken has been based upon 'thinking spatially', taking into account what would be possible where (e.g. accessibility or the ability to dig and choose candidate sites for body disposal), fitting into the high-level conceptual model of areas of interest (Harrison and Donnelly, Chapter 13).

As discussed by Williams et al. (Chapter 7), so the issue of the availability of data of appropriate attributes, scale and resolution is rehearsed by McKinley et al. (Chapter 18) and Fielder et al. (Chapter 19). This is largely resolved by data capture using relatively new techniques for measurement, such as that of terrain height using Lidar (McKinley et al., Chapter 18) or increasingly common tools such as ground penetrating radar (GPR), and high accuracy field surveys using global positioning systems (GPS).

Although the investigators' armoury of tools continues to grow, issues of cost- and time- effectiveness are also of importance, weighed against their potential to provide timely information, in environments which are subject to potentially rapid change. Keaney et al. (Chapter 14) and Hanson et al. (Chapter 15) provide examples of where conditions at sites of interest may be subject to change, or present limitations to the interpretation of data gathered using field observations. These include the use of non-destructive analysis of soil, rock and other crystalline materials adhering, as a means of rapid screening prior to further forensic-based scientific analysis, which may have significant resource overheads (Fitzpatrick, Chapter 8).

Currently, the choice of relevant analytical techniques is determined by the size of the available sample. However, this leads to problems in comparing trace material to bulk comparator samples as this needs to be 'like-with-like'. The analysis of 'bulk soil' can mask the individuality of sample characteristics, thus stressing the importance of sample examination using several levels of magnification prior to any bulk analysis. Jarvis et al. (2004) and Rawlins and Cave (2004) also identified problems of analysing small sub-samples by chemical analysis and have debated the representative nature of the results with bulk samples and therefore their relevance in forensic enquiry. Morgan et al. (Chapter 16) discuss similar issues in relation to sampling at a scene for the provision of accurate and reliable data, for example in relation to soils from footwear.

Having chosen tools suited to tasks, such as the observation and recording *in situ*, account needs to be taken of their ultimate use, such as the provision of robust evidence in court (Auchie, Chapter 2). Testing is required across a range of environmental conditions to produce a body of evidence about the robustness of the tools themselves. One example in relation to the use of GPR is that of Jervis et al. (Chapter 17), with the targeted application of measurements of clandestine graves.

Where greater sophistication is employed by the criminal or terrorist, so must the investigating team be suitably equipped. These experimental studies contribute to the body of empirical evidence which underpins the technological developments and aspirations of the criminal investigator. As the situations faced may usually be unique in themselves, the tools and skills upon which they may draw should be credible, reliable and cost-effective in fitting the level of importance of the crime.

Taphonomy

A common subject of serious crime is that of investigation, recovery and interpretation of evidence from human bodies. The impact of soil and its components on the post-mortem fate of remains was presented in a seminal volume by Tibbett and Carter (2008). The chapters in this volume that relate to taphonomy demonstrate the rapidity with which this discipline is now advancing (Tibbett and Carter, Chapter 20). The early perception that soil type has little impact on cadaver decomposition (Mant 1987) has been replaced with the increasing application of soil science (e.g. Carter et al., Chapter 21; Janaway et al., Chapter 22), and notably soil biology (Stokes et al., Chapter 23; Parkinson et al., Chapter 24), to understand stages and timings of body decomposition in a forensically useful manner. As many chapters show, the differences between taphonomic processes where bodies are located *upon* or *within* soil are profound.

Soil systems process cadavers as 'just' another form of organic matter; it is the human perspective that makes them a particularly special form of such material. Tibbett and Carter's call for a more contrived approach to taphonomic research is certainly apposite, and there would be much to be gained from interactions between taphonomists and ecologists. Indeed, carcass decomposition of any soil animal is also relatively poorly understood from an ecological perspective. Therefore, advances in understanding the role of soil systems in taphonomy will inform and improve search-and-locate systems, aging of bodies, timing of death predictions, and in a reciprocal manner to potentially inform disposal procedures following pandemic (Williams et al., Chapter 7).

Technology

The history of science and technology is marked by a continued development in the ability to measure entities and their properties with increased sophistication, accuracy, precision and robustness at ever wider scales. This is of importance to forensic science from two main perspectives. Firstly, more comprehensive analyses and characterisation of material provides a potentially richer frame to establish the degree of similarity or dissimilarity between samples. Secondly, the ability to analyse ever smaller pieces of trace evidence increases the potential power of forensic deduction.

Soil, or indeed sediment analysis is multi-faceted and may include mineralogy, chemistry, colour, organic matter and particle size. However, due to the (often limited) size of the evidential sample, the choice of appropriate methods may be concomitantly limited. One of the most suitable approaches is through use of spectroscopy, as such a technique can be applied directly on to soil which has adhered to a suspect or victim such as in fabrics (e.g. Keaney et al., Chapter 14). Additional approaches are in the morphological study of the individual grains, which can be carried out using electron microscopy, or by characterising the individual quartz particles (e.g. Morgan et al., Chapter 16; Pirrie et al., Chapter 26; Weinger et al., Chapter 27).

Geochemical techniques using isotope ratios and such geochemical signatures have been utilized in forensic work (Trueman et al. 2003), with the determination of the isotopic composition of organic substances occurring at trace level in very complex mixtures, such as sediments and soils. This has been made possible during the last twenty years due to the rapid development of molecular-level isotopic techniques (Lichtfouse 2000). Used in combination with other analytical techniques and, indeed, coupled with other elemental analyses, isotopic analysis may prove to be a useful tool in a forensic context (Croft and Pye 2003; Walker, Chapter 25). These methods alone, however, do not always discriminate samples effectively (Sugita and Marumo 1996), and the integration of such isotopic approaches with traditional methods can help (Pye et al. 2006). Transfer of new developments from the discipline of soil science, such as coupling the isotopic analysis of specific compounds (e.g. Glaser 2005), also show promise.

Advances in multi-variate analysis of soil-derived materials of both inorganic and organic nature are finding application in forensic casework. For example, techniques such as laser-ablation inductively-coupled mass spectrometry (Walker, Chapter 25), QEMSCAN (Pirrie et al., Chapter 26), and IRMA (Weinger et al., Chapter 27) offer hitherto unattainable spatial resolution in the elemental mapping of trace evidence. Emergent technology, such as nano-scale secondary ionisation mass spectrometry (nanoSIMS; Guerquin-Kern et al. 2005), takes this orders-of-magnitude lower still.

The analysis of organic molecules from soils is also increasing in sophistication, such as plant-derived alkanes (Mayes et al., Chapter 29), and DNA (Sensabaugh, Chapter 4), thus potentially providing an independent component value to complement inorganic characterisation. Increasingly detailed analysis of the 'soil genome', arguably one of the most complex biological constructs on the planet, can be expected with the advent of ultra-high throughput sequencing systems. Given such complexity, it remains to be seen what the forensic utility of soil DNA, at least at a community scale, will be, and this is an arena in which soil microbial ecologists and forensic scientists should be interacting.

The sophistication of technological and data analysis is only part of the requirement of the forensic practitioner and end-user. There is also a need for instrumentation that provides ease-of-use, particularly in the field, and that enables novices to determine multivariate parameters in soil and environmental samples (e.g. Pollard et al., Chapter 30). Refinement of technology for sampling is also occurring, and Smucker and Siegel (Chapter 31), provide an example of technology transfer from a device originally designed by soil scientists studying plant roots, to a trace-evidence concentrator.

Reliance is placed on the scientific method to prove capacity to provide accurate, reliable and repeatable observations. In support of the selection of suitable approaches for searching sites, experimental approaches can be used to prove the robustness of tools such as 3D profiling of imprints, and provide guidance on the level of statistical confidence which may be achieved under different conditions. Such experimental work is presented by Bennett et al. (Chapter 28), providing supporting observations on the potential of 3D laser scanning tools, showing their capability to differentiate between people of different weights and the rate of movement across an area. Clearly, it is not possible to test observations under all possible ground conditions, but the laboratory studies give credence to the choice of tool, and an awareness of the limitations which also might be expected. When there is a restricted extent of experience of their use, this also creates a demand for experiments into the variables which may limit the interpretation or role of new applications of tools, which in relation to environments strongly influenced by weather and groundwater conditions can be significant.

However powerful and beguiling the technology, there remains the crucial role of the expert (e.g. Wiltshire, Chapter 9). It must be recognised that technology is but a tool and is never a single means to an end. Expert knowledge and experience encapsulates both skills and cognisance of environmental conditions at scenes of crime. one of the key requirements. A requirement of taking materials through to evidence in court will be the involvement of an expert of the appropriate discipline.

Databases

One theme which arises through several chapters is the suitability of existing data on soils for application to forensic purposes (e.g. Morrisson et al., Chapter 6; Fitpatrick et al., Chapter 8; Wiltshire, Chapter 9). Soil surveys were designed to understand the spatial distribution of soil types, particularly from an agricultural perspective, and were originally expressed as maps, with associated physico-chemical data also being recorded stored and structured (e.g. European Commission 2005), but at small geographic scales (Williams, Chapter 7; Heineke et al. 1998). However, such surveys are becoming much more sophisticated, with the increasing establishment of national soil inventories, which usually involve, or are intended to involve, some degree of re-sampling, to add a temporal dimension to the data. However, gaps exist in the statistical analysis of such data, and mechanisms for making such analysis accessible to different end-users. The spatial resolution of soil data is also increasing, and urban soils, of clear pertinence in a forensic context, but historically sparse in soil surveys, are now being incorporated (e.g. British Geological Survey 2008; USDA Natural Resources Conservation Service 2008). Geological surveys, which generally have a historical precedent over soil surveys have also been carried out in most countries, which adds an important depth-related component to such environmental data, and provides the geological foundation for much of the mineralogy of superficial soil layers. As for soils, initiatives are also underway to unify geological data (OneGeology 2008).

Most of the data relating to soil properties in soil surveys or national inventories is physico-chemical in nature. However, there is now an increasing trend to measure biological properties in such schemes, and more sophisticated inorganic and organic analysis, such as DNA, XRDS and FTIR spectroscopy. Sophistication is also increasing as both extant and new data are digitised. All such data have utility in a forensic context, particularly in terms of provenancing samples. Furthermore, by combining such information with other databases such as above-ground environmental audits and applying them in GIS systems, there is considerable potential to develop a quantitative provenancing framework for soils. However, this still requires substantial improvements in the spatial resolution and breadth of properties being measured.

Extant soil surveys and data systems were not constructed from a forensic perspective. There is a key requirement for integration of databases at national/international levels. There is also an issue relating to the nature of the samples taken for soil surveys, which are usually stratified of the order of several-cm increments, typically 15 cm, or taken at the mid point of a soil horizon. Surface soils, attached to shoes or clothing by pressure contact, are arguably likely to constitute a large proportion of forensic samples. Jeffery et al. (2007) show how, in arable soils, the microbial signal at least is very different in the top 1 mm, and unifies thereafter. Such signals would be lost in samples bulked across almost any greater depth.

Considering the chapters dealing with geoforensics, and the capabilities for analyses of soils summarised above, some key questions for the future of soils databases in a forensics context can be identified, including: what properties are most useful? Are they being measured in national soil inventories? At what resolution should they be measured? Can such databases be integrated internationally? Are the main soils well represented? Can the surface soil characteristics relate to archived data? Some of these issues remain to be addressed, with investigating agencies increasingly interested in understanding the potential such databases may offer in combating crime and terrorism (e.g. European Security Research Advisory Board 2006).

Statistics

Whilst statistical principles are becoming incorporated into forensic soil science, there is much room for further development in this area.[1] Aitken (Chapter 3) provides a thorough generic consideration of probabilistic reasoning, as well as pertinent considerations specifically in relation to soil. However, there is a huge body of statistical knowledge and research into the sampling, comparison, inference and quantification of probability that could be apposite, drawn upon and used to stimulate advances in both disciplines. In experimental research terms, the need to replicate appropriately in taphonomic studies to provide robust statistical bases to the data is only starting to become appreciated (Tibbett and Carter, Chapter 20). As Lark (2008) states, "in a proper statistical investigation, the way in which we propose to make inferences determines how we sample", and goes on to argue that an appropriate inferential framework for soil forensic statistics is lacking. Statistically robust sampling, at least from the statistician's perspective, whilst always desirable, is not necessarily always tenable, particularly in relation to trace evidence in an investigation. 'Optimised' sampling would be where the compromises that have to be made between the number and dimensions of samples and their potential or actual availably are reconciled as far as practicable. There needs to be greater dialogue between statisticians, soil and geoscientists, forensic scientists and practitioners. We see this is as an area that offers particular potential and for which there is much need in the development of the maturing discipline of soil forensic science.

Conclusions

The development of a 'forensic toolbox' to aid in investigations which can be called upon by authorities charged with addressing threats to public security demands development and application of a wide range of knowledge. Core to the creation of such a toolbox is consideration of the question 'what needs to be known?', and its corollary, 'how wrong can we afford to be?' Whether it is in the provision of evidence, the exploitation of existing or new tools technologies, or the preparation of materials for use in court, an understanding is required of the requirements of the investigative process, and the thresholds of acceptability of evidence.

The increased move towards the formalisation of approaches to the reduction in areas of search, site identification, measurement and excavation has provided a framework for structuring the strategic and tactical planning of criminal investigation teams. The transdisciplinary nature of such teams draws strength from both the

[1] A Science Citation Index (www.isiwebofknowledge.com) search based on the keywords (soil and (forensic* or geoforensic*) and statistic*) yielded 13 returns as of July 2008. The majority of these cover application of statistics rather than review. Removing 'soil' from this cluster returned 945 references.

component scientific disciplines and the practitioner or end-user of the tools. Amongst the key gaps identified is the relatively limited exploitation of statistical tools in the acquisition and provision of evidence, and the resolution of databases detailing soil properties.

A number of different disciplines working together necessitate the exchange of knowledge between scientists, and between the practitioners and end-users, in order for the outcomes to be innovative and effective. The nature of the co-authorships of chapters in this book, across disciplines and levels of practice, demonstrates the significance of developing close working relations and understanding for greatest societal benefit to be gained from the work on soil forensics both in a criminal and an environmental context. Soil forensics continues on its path from a predominantly local and individual enterprise to a global- and community-based discipline, thus considerably enhancing its future application. As we state in the Preface, soils support a wide range of goods and services; there can be few more important than the recognition of their support of public security and health.

References

British Geological Survey (2008). Geochemical Baseline Survey of the Environment (G-BASE). www.bgs.ac.uk/gbase/home.html. Accessed 30 July 2008.

Bull PA, Morgan RM and Freudiger-Bonzon J (2008). A critique of the present use of some geochemical techniques in geoforensic analysis. Forensic Science International 178:35–40.

Croft DJ and Pye K (2003). The potential use of continuous flow isotope ratio mass spectrometry as a tool in forensic soil analysis: a preliminary report. Rapid Communications in Mass Spectrometry 17:2581–2584.

European Commission (2004). Environmental Liability – Directive 2004/35/EC. Office for Official Publications of the European Communities, L-2995 Luxembourg. 56 pp.

European Commission (2005). Soil Atlas of Europe. Office for Official Publications of the European Communities, L-2995 Luxembourg. 128 pp.

European Security Research Advisory Board (2006). Meeting the Challenge: The European Security Research Agenda. Office for Official Publications of the European Communities, L-2995 Luxembourg. 84 pp.

Glaser B (2005). Compound-specific stable-isotope (delta C-13) analysis in soil science. Journal of Plant Nutrition and Soil Science 168:633–648.

Guerquin-Kern JL, Wu TD, Qintana C and Croisy A (2005). Progress in analytical imaging of the cell by dynamic secondary ion mass spectrometry (SIMS microscopy). Biochimica et Biophysica Acta 1724:28–238.

Heineke HJ, Eckelmann W, Thomasson AJ, Jones RJA, Montanarella L and Buckley B (Eds.) (1998). Land Information Systems: Developments for Planning the Sustainable Use of Land Resources. European Soil Bureau Research Report No.4, EUR 17729 EN. Office for Official Publications of the European Communities, L-2995 Luxembourg.

House of Commons Science and Technology Committee (2005). Forensic Science on Trial. Seventh Report of Session 2004-05. HC 96-1. The Stationery Office, London.

Jarvis KE, Wilson HE and James SL (2004). Assessing element variability in small soil samples taken during forensic investigation. In: (Eds. K Pye and DJ Croft), Forensic Geoscience: Principles, Techniques and Application. Geological Society, London, Special Publications, 232:171–182.

Jeffery S, Harris JA, Rickson RJ and Ritz K (2007). Microbial community phenotypic profiles change markedly with depth within the first centimetre of the arable soil surface. Soil Biology and Biochemistry 39:1226–1229.
Lark RM (2008). Book review of Geological and Soil Evidence: Forensic Applications by K. Pye (CRC Press). European Journal of Soil Science 59:419.
Lichtfouse E (2000). Compound-specific isotope analysis. Application to archaeology, biomedical sciences, biosynthesis, environment, extraterrestrial chemistry, food science, forensic science, humic substances, microbiology, organic geochemistry, soil science and sport. Rapid Communication in Mass Spectrometry 14:1337–1344.
Locard E (1930). Analyses of dust traces parts I II and III. American Journal of Police Science 1:276–298 401–418 and 496–514.
Mant AK (1987). Knowledge acquired from post-war exhumations. In: (Eds. A Boddington, AN Garland and RC Janaway), Death, Decay and Reconstruction: Approaches to Archaeology and Forensic Science. Manchester University Press, Manchester, pp. 65–78.
Murphy BL and Morrison RD (2007). Introduction to Environmental Forensics: A Forensic Approach. Academic, New York.
Murray RC (2004). Evidence from the Earth. Forensic Geology and Criminal Investigation. Mountain Press, PO Box 2399, Missoula, USA. 226 pp.
OneGeology. http://www.onegeology.org/. Accessed 12 July 2008.
Pye K (2007). Geological and Soil Evidence. Forensic Applications. CRC, Taylor & Francis Group, London. 335 pp.
Pye K, Blott SJ, Croft DJ and Carter JF (2006). Forensic comparison of soil samples: Assessment of small-scale spatial variability in elemental composition, carbon and nitrogen isotope ratios, colour, and particle size distribution. Forensic Science International, 163:59–80.
Pye K and Croft DJ (Eds.) (2004). Forensic Geoscience: Principles, Techniques and Applications. (Eds.) Geological Society, London, Special Publications, 232.
Rawlins BG and Cave M (2004). Investigating multi- element soil geochemical signatures and their potential use in forensic studies. In: (Eds. K Pye and DJ Croft), Forensic Geoscience: Principles, Techniques and Application. Geological Society, London, Special Publications, 232:197–206.
Ruffell A and McKinley J (2005). Forensic Geology & Geoscience. Earth Science Reviews 69:235–247.
Ruffell A and McKinley J (2008). Geoforensics. Wiley, Chichester. 332 pp.
Sugita R and Marumo Y (1996). Validity of color examination for forensic soil identification. Forensic Science International 83:201–210.
Tibbett M and Carter DO (2008). Soil Analysis in Forensic Taphonomy: Chemical and Biological Effects of Buried Human Remains. CRC, Taylor & Francis Group, Boca Raton, FL, USA. 340 pp.
Trueman C, Chenery C, Eberth DA and Spiro B (2003). Diagenetic effects on the oxygen isotope composition of bones of dinosaurs and other vertebrates recovered form terrestrial and marine sediments. Journal of The Geological Society, London, 160:895–901.
USDA Natural Resources Conservation Service (2008). Urban Soil Mapping and Inventory. http://soils.usda.gov/use/urban/mapping.html. Accessed 30 July 2008.

Index

T

U

V

W

X

Lightning Source UK Ltd.
Milton Keynes UK
UKOW06n0557171215

264770UK00015B/28/P